# Ensuring safety and quality in the production of beef

## Volume 2: Quality

It is widely recognised that agriculture is a significant contributor to global warming and climate change. Agriculture needs to reduce its environmental impact and adapt to current climate change whilst still feeding a growing population, i.e. become more 'climate-smart'. Burleigh Dodds Science Publishing is playing its part in achieving this by bringing together key research on making the production of the world's most important crops and livestock products more sustainable.

Based on extensive research, our publications specifically target the challenge of climate-smart agriculture. In this way we are using 'smart publishing' to help achieve climate-smart agriculture.

Burleigh Dodds Science Publishing is an independent and innovative publisher delivering high quality customer-focused agricultural science content in both print and online formats for the academic and research communities. Our aim is to build a foundation of knowledge on which researchers can build to meet the challenge of climate-smart agriculture.

For more information about Burleigh Dodds Science Publishing simply call us on +44 (0) 1223 839365, email info@bdspublishing.com or alternatively please visit our website at www.bdspublishing.com.

**Related titles:**
Ensuring safety and quality in the production of beef Volume 1: Safety
Print (ISBN 978-1-78676-056-2); Online (ISBN 978-1-78676-058-6, 978-1-78676-059-3)

Chapters are available individually from our online bookshop: https://shop.bdspublishing.com

climate•SMART•publishing
IN AGRICULTURAL SCIENCE

BURLEIGH DODDS SERIES IN AGRICULTURAL SCIENCE

NUMBER 12

# Ensuring safety and quality in the production of beef

## Volume 2: Quality

Edited by Professor Michael E. Dikeman,
Kansas State University, USA

burleigh dodds
SCIENCE PUBLISHING

Published by Burleigh Dodds Science Publishing Limited
82 High Street, Sawston, Cambridge CB22 3HJ, UK
www.bdspublishing.com

Burleigh Dodds Science Publishing, 1518 Walnut Street, Suite 900, Philadelphia, PA 19102-3406, USA

First published 2017 by Burleigh Dodds Science Publishing Limited


Library of Congress Control Number: 2016960975

British Library Cataloguing in Publication Data
A catalogue record for this book is available from the British Library

ISBN 978-1-78676-060-9 (print)
ISBN 978-1-78676-062-3 (online)
ISBN 978-1-78676-063-0 (online)
ISSN 2059-6936 (print)
ISSN 2059-6944 (online)

Typeset by Deanta Global Publishing Services, Chennai, India
Printed by Lightning Source

# Contents

# Series list

| Title | Series number |
|---|---|
| **Achieving sustainable cultivation of maize - Vol 1**<br>From improved varieties to local applications<br>Edited by: Dr Dave Watson, CGIAR Maize Research Program Manager, CIMMYT, Mexico | 001 |
| **Achieving sustainable cultivation of maize - Vol 2**<br>Cultivation techniques, pest and disease control<br>Edited by: Dr Dave Watson, CGIAR Maize Research Program Manager, CIMMYT, Mexico | 002 |
| **Achieving sustainable cultivation of rice - Vol 1**<br>Breeding for higher yield and quality<br>Edited by: Prof. Takuji Sasaki, Tokyo University of Agriculture, Japan | 003 |
| **Achieving sustainable cultivation of rice - Vol 2**<br>Cultivation, pest and disease management<br>Edited by: Prof. Takuji Sasaki, Tokyo University of Agriculture, Japan | 004 |
| **Achieving sustainable cultivation of wheat - Vol 1**<br>Breeding, quality traits, pests and diseases<br>Edited by: Prof. Peter Langridge, The University of Adelaide, Australia | 005 |
| **Achieving sustainable cultivation of wheat - Vol 2**<br>Cultivation techniques<br>Edited by: Prof. Peter Langridge, The University of Adelaide, Australia | 006 |
| **Achieving sustainable cultivation of tomatoes**<br>Edited by: Dr. Autar Mattoo, USDA-ARS, USA & Prof. Avtar Handa, Purdue University, USA | 007 |
| **Achieving sustainable production of milk - Vol 1**<br>Milk composition, genetics and breeding<br>Edited by: Dr Nico van Belzen, International Dairy Federation (IDF), Belgium | 008 |
| **Achieving sustainable production of milk - Vol 2**<br>Safety, quality and sustainability<br>Edited by: Dr Nico van Belzen, International Dairy Federation (IDF), Belgium | 009 |
| **Achieving sustainable production of milk - Vol 3**<br>Dairy herd management and welfare<br>Edited by: Prof. John Webster, University of Bristol, UK | 010 |
| **Ensuring safety and quality in the production of beef - Vol 1**<br>Safety<br>Edited by: Prof. Gary Acuff, Texas A&M University, USA & Prof.James Dickson, Iowa State University, USA | 011 |
| **Ensuring safety and quality in the production of beef - Vol 2**<br>Quality<br>Edited by: Prof. Michael Dikeman, Kansas State University, USA | 012 |
| **Achieving sustainable production of poultry meat - Vol 1**<br>Safety, quality and sustainability<br>Edited by: Prof. Steven C. Ricke, University of Arkansas, USA | 013 |
| **Achieving sustainable production of poultry meat - Vol 2**<br>Breeding and nutrition<br>Edited by: Prof. Todd Applegate, University of Georgia, USA | 014 |
| **Achieving sustainable production of poultry meat - Vol 3**<br>Health and welfare<br>Edited by: Prof. Todd Applegate, University of Georgia, USA | 015 |

# Introduction

Beef production faces a range of challenges. There is an ongoing need to ensure **safety** in the face of threats from zoonoses and other contaminants. Consumers also have ever-higher expectations of sensory **quality**. There is a sharper focus on the role of meat in nutrition and health, as well as growing interest by consumers in animal welfare and the environmental impact of beef farming as components of quality. These topics and challenges are addressed in two volumes. Volume 1 looks at key research trends in production and processing to ensure **safety** of beef. This volume, *Ensuring safety and quality in the production of beef Volume 2: Quality*, reviews developments related to beef **quality**.

The chapters of this volume suggest a number of common themes. They show the progress and continuing challenges in understanding and exploiting the genetic basis of factors affecting the growth and composition of muscle in cattle, and of meat quality traits in particular. They also show how an understanding of muscle development and fat deposition, and the process of post-slaughter conversion of muscle to meat, is critical to the understanding and improvement of individual quality traits.

## Part 1 Breeding and growth

Chapter 1 looks at the relative importance of genetic compared to other factors such as sex, age and husbandry in determining carcass and meat quality. The evolution of each breed represents a particular combination of traits, which can be used to tailor the combination of carcass size, conformation and finish demanded in any specific market. From a carcass quality point of view, the important breed traits are muscularity and propensity to fatten. Indicators of carcass quality can include dressing percentage (the weight of the carcass expressed as a percentage of the live weight), cutability or meat yield (the proportion of the carcass that is saleable as meat), and finish (subcutaneous and intermuscular fatness). It has also been shown that meat quality characteristics such as appearance, texture, fat colour, marbling, flavour and, in particular, tenderness are heritable.

While these traits are genetically influenced, until recently it has only been possible to select for them on the basis of their phenotypic expression. The mapping of the bovine genome in 2009 has led to an ongoing effort to find quantitative trait loci (QTL), markers, genes and single-nucleotide polymorphisms (SNPs) associated with aspects of carcass and meat quality. A number of QTL for traits of importance to carcass and meat quality have been identified and are now being incorporated into commercial breeding programmes. As the science and practice of genomics advances, so too will opportunities to select for these traits genotypically, allowing for earlier selection and a more precise tailoring of carcass size and composition to match market preferences. Chapter 1 concludes by suggesting that the role of the traditional breeds as repositories of traits of interest to beef breeders will continue, but increasingly, commercial breeders will be searching for individual animals carrying the genotype for desirable carcass and meat quality traits, regardless of the breed type itself. This topic is discussed in more detail in Chapter 17.

Chapter 1 provides a foundation for Chapter 2, which explores traditional animal breeding of cattle to improve carcass composition and meat quality. It focuses, in particular, on both marbling and tenderness that are moderately to highly heritable traits. Genetic selection for carcass attributes such as marbling and tenderness is relatively new compared to selection for growth traits. They have traditionally been difficult and expensive to measure phenotypically. However, structured progeny testing schemes have aided in the development of estimated breeding values (EBV) and expected progeny differences (EPDs), which allow for the comparison of animals within a breed for their genetic potential for a given trait. These progeny tests have now given way to the use of techniques such as ultrasound that have allowed EBV for marbling to become pervasive in breeding programmes. The chapter includes case studies of successful breeding for such traits. It concludes by suggesting the advantages of augmenting genomic information with traditional EPDs in the future, a topic which is further reviewed in Chapter 17.

Chapters 2 and 3 discuss factors affecting muscle development and fat deposition in cattle and how they impact on carcass and meat quality. An understanding of these biological processes is critical to understanding the effects of cattle nutrition, for example. As Chapter 3 shows, bovine skeletal muscle is mainly composed of three types of oxidative and glycolytic muscle fibres that are found in varying proportions in different muscles. The proportion of these different fibre types varies with biological and environmental factors (breed, age, sex, rearing conditions), resulting in variability of carcass and meat qualities. The chapter reviews the various factors that determine muscle composition and the relationships between muscle fibre type composition, carcass composition and meat quality. It shows ways in which the proportion of different fibre types can be altered by genetic selection, rearing conditions and slaughter age.

Beef cattle are quite efficient at depositing fat: first abdominal, then seam (intermuscular), then subcutaneous and finally marbling (intramuscular and interfascicular) fat. The development and composition of these fat deposits are influenced primarily by the diet of the cattle, although marbling is under significant genetic control. Chapter 4 addresses the origin of adipocytes, the ontogeny of fat development, factors influencing fat development (such as the relative effects of grain and pasture-based diets) and finally factors influencing fatty acid composition of fat deposits. As it shows, research has helped to establish the appropriate balance between forage feeding and finishing on a high-concentrate diet in optimising the quantity and quality of fat deposition in beef cattle.

## Part 2 Management of cattle

Chapter 5 builds on Chapter 4 by looking in more detail at cattle nutrition. This chapter focuses on nutrition of growing cattle post-weaning through the intervening backgrounding or stocker phase and, finally, the finishing phase. It looks at what research suggests about the relative merits of pasture-based and feed-based diets (including the effects of different forage types, grain-based diets and feed supplements), as well as combinations of pasture and feed, together with issues such as composition and timing of high-concentrate feeding on quality traits. The chapter focuses primarily on how the nutrition in each phase impacts marbling or quality grade of the final products, but also discusses other quality traits such as tenderness, flavour, and lipid or nutrient profiles, and shelf life. As the chapter shows, there is evidence that nutrition at all stages of growth, rather than just the finishing phase, has the potential to impact final beef quality. It is clear, for example, that marbling can be developed before the finishing phase by providing sufficient energy intake for muscle growth and marbling development. This suggests that a better understanding of lifetime management on quality could lead to both greater efficiency and better and more consistent quality.

Metabolic modifiers are compounds that are either fed to animals or implanted to improve rate of gain, feed efficiency, dressing per cent, carcass meat yield, visual meat quality and shelf life, with varying effects on meat palatability. Chapter 6 presents a general review of current research on the effects of metabolic modifiers on carcass composition, fresh meat quality and cooked meat palatability. It focuses on the effects of single use and re-implant of anabolic steroids; beta-adrenergic agonists (ractopamine hydrochloride and zilpaterol hydrochloride (ZH)); vitamins $D_3$, E and A; and somatotropin on both carcass composition and meat quality. None of the implants or neither of the beta agonists improve marbling or tenderness, but when used properly, have minimal negative effects. They are used because they are very cost effective.

Handling, transport, lairage and slaughter are stressful for cattle, and can have consequences on the quality of the beef produced. Chapter 7 considers what we know about potential stresses and cattle responses directly before and during transport as well in the lead-up to slaughter. The chapter addresses the complexity of each pre-harvest management stage, arising from the interactions between different types of cattle, different physical environments, the manner in which cattle are handled and the methods used for slaughter. The chapter reviews the effects of pre-slaughter management on beef quality, including the risk of dark, firm, dry meat and bruising. It then discusses the main factors during transportation which have been identified as affecting welfare and meat quality, such as loaded density and thermal environment. Finally, the chapter provides an overview of the extensive research on handling and movement of cattle through lairage, to the stunning pen and onto slaughter and how these stages can be optimised for both animal welfare and meat quality.

The final chapter in Part 2, Chapter 8, looks at post-slaughter processes. The conversion of muscle to meat during the rigor mortis process and the conditions under which this process occurs are important for quality attributes of beef including colour, water-holding capacity and tenderness. Electrical stimulation (ES) is used to prevent cold-induced toughening of beef, especially hot-boned meat, but is also applied to enhance the ageing process and, therefore, improve tenderness of aged beef. ES can also impact on other quality parameters such as colour and moisture characteristics as a result of its effect on glycolysis and the rigor mortis process. However, poor use of ES combined with improper chilling may have detrimental effects on the tenderness, colour and water-holding capacity of meat. This chapter discusses current research on the factors influencing the effectivity of ES in optimising meat quality. It also reviews current research on the effects of chilling, including very fast chilling and blast chilling on meat quality, including ways to adjust chilling parameters and other inputs during and after slaughter to improve the economics of carcass chilling while maintaining or improving the quality of the final product.

# Part 3 Quality traits

The chapters in Part 3 consider both individual quality traits and ways they can be measured. One group of chapters discusses individual traits such as colour, tenderness and flavour (Chapters 9, 12 and 13). They are accompanied by Chapter 14, which looks at the role of packaging in optimising visual appearance and palatability, and Chapter 16, which reviews the role of beef in human nutrition.

Meat colour is a critical visual sensory attribute influencing purchasing decisions. Post-mortem muscle is biochemically active and any deviation in ultimate post-mortem muscle pH can influence processes affecting meat colour, such as metmyoglobin reducing activity (MRA), oxygen consumption and lipid oxidation. Recently, much research focus has been given to the role of mitochondria in meat colour. As the chapter shows, mitochondria remain active in post-mortem muscle and can influence myoglobin redox state via oxygen consumption and MRA. Hence, characterising factors that influence mitochondrial function is helping to develop pre- and post-harvest technologies that minimise the economic losses due to discolouration. Research has shown the ways in which minimising ante-mortem stress of cattle, utilising optimum chilling and proper packaging can optimise beef colour development and stability.

In addition to protecting the product, packaging is largely responsible for maintaining myoglobin in the bright red, oxygenated state consumers expect to see. Chapter 14 reviews the strengths and weaknesses in achieving these objectives of traditional aerobic packaging, vacuum packaging, modified atmosphere packages and active packaging (incorporating oxygen scavenging or carbon dioxide emitting technology, films with antioxidant or antimicrobial properties, or colour-enhancing nitrite embedded in the film). In addition, 'intelligent packaging' (e.g. with sensors or labelling to monitor or measure product freshness) is discussed. It also discusses the role of centralised packaging systems and reducing packaging waste using biodegradable materials.

Tenderisation through ageing of meat results from proteolysis, the degradation of proteins (myofibrillar, cytoskeletal and costameric) into smaller subunits. Chapter 12 reviews factors affecting this process, such as the development of rigor mortis, chilling and muscle variation. The biochemical understanding of tenderisation has helped in the design of more targeted approaches to improve beef tenderness. The chapter reviews current research on carcass methods such as ES, the 'tenderstretch' and 'tendercut' techniques as well as foreshank manipulation. It also surveys the effectiveness of other methods such as SmartStretch™ and the Pi-Vac Elasto-Pack System®, as well as pulsed electric fields and newer techniques such as ultrasound and high hydrostatic pressure. Finally, it discusses chemical methods such as the use of ionic solutions, acids and enzymes. Picking up on Chapter 10, it suggests that grading systems able to grade individual cuts or even individual muscles are required so that the inherent variation within and between carcasses can be accounted for and value maximised. Several countries have 'carcass classification' systems that have little relationship to visual appearance of beef and cooked palatability. In other words, they do not provide information consumers desire when purchasing and cooking beef. 'Grading' systems attempt to identify quality traits and to assign value that are important to consumers.

As Chapter 13 indicates, the complex structure and chemistry of raw meat provides the foundation for the distinctive flavour of cooked beef, down to individual amino acid side chains and fatty acids. The generation of the aromas associated with beef flavour during cooking is complex and involves hundreds of chemical reactions taking place simultaneously to form the aroma we recognise. The chapter reviews what we know about the chemistry of beef flavour; the thermal degradation of components such as lipids, which have a particularly important effect on flavour; and the role of Maillard reactions. The chapter also discusses the way we perceive flavour and ways of measuring it, whether through sensory testing or instrumental measurement of flavour compounds. It also surveys the range of factors that help to determine flavour, from genetic variations in fatty acid composition, through the role of cattle nutrition, particularly finishing on grass and forage, as well as the role of marbling and ageing.

As discussed in Chapter 16, beef is an important source of both macronutrients and micronutrients. Macronutrients include protein, and both saturated and unsaturated fatty acids. Micronutrients in beef include haeme iron; zinc; selenium; and vitamins D, B1 (thiamin), B2 (riboflavin), B3 (niacin), B5 (pantothenic acid), B6 (pyridoxin) and B12 (cobalamin). In each case, the chapter discusses factors affecting amounts and bioavailability. The chapter also reviews the evidence linking beef with reported increased risk of cancer, from intrinsic factors that include fat and iron, to extrinsic factors such as heterocyclic amines, polycyclic aromatic hydrocarbons and *N*-nitroso compounds (nitrosamines). The role of beef in nutrition and health is also discussed in Chapter 19.

Building on the discussion of individual quality traits, a second group of chapters in Part 3 looks at ways quality is measured and communicated through grading systems and branded programmes. Chapter 10 reviews the three main beef carcass grading or classification systems used across three continents. These are used for a large proportion of beef production and processing in the world. One is the 'EUROP' system for EU countries, another is the USDA Beef Carcass Grading system in the United States and the third is the Meat Standards Australia system in Australia. The chapter also discusses grading in Japan. The EUROP 'classification' system does little to relate to the visual appearance or palatability of beef. Instrument grading using video image analysis is predominant in the United States for assessment of marbling. The Meat Standards Australia system is the only one in the world that 'grades' individual cuts or muscles and is based on extensive consumer evaluations. The chapter reviews the way the different schemes work and how effective they are in ensuring beef meets consumer expectations of quality.

The emergence of branded beef programmes has had a tremendous impact on the beef industry as a tool for differentiation. Chapter 11 discusses what we know about the factors that define successful brands and the effects of brands on consumer perceptions of meat quality. It reviews different kinds of brands, from those based on breed, and those focused on particular producers, retailers or countries, to those focused on methods of production such as organic brands. Picking up on Chapter 2, it discusses the use of EPDs and instrumental techniques to validate quality aspects involved in branding.

As discussed in Chapter 15, verifying qualities prior to consumption that contribute to a pleasurable eating experience remains a challenge for the beef industry. There are several ways to assess eating quality, such as trained sensory evaluation, consumer taste tests, shear force measurement and relying on quality grading schemes. Most recently, research has focused on flavour evaluation and muscle profiling. As well as sensory evaluation and quality grading, the chapter reviews instrumental assessment of marbling, shear force measurements for tenderness, and assessment of juiciness and flavour (e.g. using measurement of moisture content). The chapter highlights the continuing challenges in successful, non-destructive assessment of beef sensory properties. It also discusses what consumer research suggests about attitudes related to meat quality. As an example, research has indicated that the beef industry has done a good job of increasing the proportion of carcasses that are considered tender. Better husbandry to reduce stress, components of finishing diets and judicious use of finishing implants have resulted in a more consistent product. Research also shows that marbling plays a significant role in flavour.

# Part 4 Emerging trends

Genetic improvement of carcass and meat quality traits by conventional selection has been limited by the difficulties and cost of measuring and recording these traits, as well as the complexity of implementing progeny tests that require individual traceability from birth to harvest. As discussed in Chapter 17, significant benefits in the genetic improvement of meat quality are expected through the use of genomic selection (GS), due to the possibility of providing more accurate estimations of genetic values at early ages, as well as including quality-related traits in breeding programmes that have received less attention due to the difficulties of collecting the phenotypic information needed for the estimation of genetic merit.

Significant developments in DNA technology and genome sequencing have led to the detection of millions of SNPs, making it possible to obtain a very dense coverage of the genome. Genome sequencing and genotyping have also become faster and cheaper. Dense, high-throughput genotyping assays have been developed for many livestock species, allowing the screening of tens to hundreds of thousands of SNPs simultaneously. The availability of genomic information for large numbers of animals allows the implementation of genome-wide association studies (GWAS), which provides a deeper and more comprehensive understanding of the genetic basis for traits of interest.

The implementation of GS depends on building training or reference populations that comprise a large number of genotyped and phenotyped animals, providing the information to estimate SNP effects accurately. However, the improvement of accuracy for many traits included in beef cattle breeding programmes by GS has so far been slow compared, for example, to dairy cattle. This chapter reviews results from current investigations of GS and GWAS for marbling, fatty acid profiles and tenderness. Current results indicate that accuracies of EBVs for these traits are moderate with the current sizes of training populations, but that these could be increased further by increasing the size of such populations. There is also a need to develop faster, non-invasive methods to map and predict meat quality traits in training populations. Genome sequencing data will also greatly improve the understanding of the genetic basis of these traits, and, together with transcriptomic and other 'omics' sciences, identify genes and gene networks. Future research could focus on traits such as flavour, palatability or micronutrient content.

In the face of growing concern about climate change, as well as government and consumer pressure for more sustainable methods of production, the beef industry faces a significant challenge in both producing sufficient food to feed a growing population and reducing environmental impacts per unit of beef produced. Chapter 18 addresses these issues by first discussing definitions of sustainability. It then looks at ways of assessing environmental impact, such as measuring the carbon footprint of beef production or tracking greenhouse gas emissions. A final section reviews practical steps in achieving 'sustainable intensification' to increase production whilst reducing environmental impact, ranging from improving growth rates and finishing weights (e.g. through the use of metabolic modifiers) to improving feed efficiency, reproductive efficiency and animal health.

The debate continues about whether 'red' or 'processed meat' intake is related to adverse health outcomes in the US population and other developed countries. 'Processed meat' can be minimally processed with no added ingredients, or further processed in which chemical additives and heat processing can be involved. Beef is often a focus of such discussions, because of its contribution to saturated fat intake in the diet, although beef has almost as much monounsaturated fat as it does saturated fat. Chapter 19 provides a detailed examination of the evidence for a possible connection between beef intake and cardiovascular disease, and beef intake and the development of diabetes. It also reviews the fatty acid composition of beef, and outlines the dietary recommendations considered to be best supported by the evidence.

It is important to note that epidemiologic studies of nutrition and chronic disease do not establish whether the relationships are real or simply random associations. Concerns about whether red or processed meat causes adverse outcomes cannot simply be justified because epidemiologic studies might not adequately account for fat content, meat processing treatments, other dietary components and/or cooking methods. These factors can modify the health effects of lean beef in the diet. For example, it is important to distinguish the health outcomes that result from further processed meats versus unprocessed red meat, and for the purposes of this chapter, lean beef. As the chapter indicates, an important message to convey is that lean beef in recommended amounts can be included in a healthy dietary pattern.

## Summary

These chapters suggest a number of common themes. They show how an understanding of muscle development and fat deposition is critical to the understanding and improvement of individual quality traits. They also highlight the role of different aspects of production in improving quality traits, for example, the effect of animal management practices to minimise stress on colour, or the role of post-slaughter processes in tenderisation. Grading systems and/or branded programmes enhance consumer confidence and demand for beef. Many of the chapters emphasise the central role of diets of cattle, including the relative merits of pasture-based and feed-based diets in enhancing flavour as well as the significance of the 'finishing' phase compared to earlier phases of nutrition. A major theme is the continuing challenges in measuring and grading quality, which impacts, for example, on further improvements in breeding as well as the enhancement of individual aspects of quality. Finally, beef production can be environmentally friendly and sustainable. Looking at the book as a whole, it is clear that achieving significant and sustainable improvements in beef quality requires an understanding of the importance of each step in the value chain in beef production, and appropriate interventions at each of these stages. This requires understanding and collaboration between different specialists and disciplines. It is hoped that this volume will help to support this kind of interdisciplinary work.

# Part 1

# Breeding and growth

Chapter 1

# Biological types of cattle: carcass and meat quality

*M. A. Price, University of Alberta, Canada*



## 1 Introduction

Cattle have always held a very important place in human society; the word 'cattle' itself comes from the same Latin stem as words like 'capital' and 'chattel', referring to wealth and headship. Cattle have been used for many purposes around the world and throughout recorded history, the most important probably being as a source of draught power, with leather, beef and offal meats as valuable by-products. However, during their lives, in addition to draught power, their milk and blood have been harvested for food and their dung used for manure, as a building material and as a fuel source. The list of by-products from cattle would fill many pages (e.g. Kinsman, 1994; Ockerman and Basu, 2014); they contribute to practically every aspect of the lives of humans from glues, oils, lubricants and soaps to ornaments, fine china, pharmaceuticals, explosives and chew toys for dogs. It is little wonder that peoples around the world treasured cattle and developed their own particular types, with specific identifying traits, naturally adapted and suited to their local environments.

Cattle are thought to have been first domesticated in several locations, including Anatolia in modern-day Turkey, around 8000–10000 years ago (Konarzewski, 2014). Cattle are now considered to be a single species (*Bos taurus*) with several subspecies, including the progenitor of modern cattle, the now extinct aurochs (*B. taurus primigenius*), European cattle (*B. taurus taurus*) and the Zebu or humped cattle of India (*B. taurus indicus*). The native cattle of sub-Saharan Africa, known as Sanga, may be a further subspecies (*B. taurus africanus*). The last known aurochs died in Poland in 1627 (Konarzewski, 2014).

Over time, cattle have evolved through both natural and artificial selection into a wide variety of shapes, sizes and body compositions. In the case of artificial selection (selective

http://dx.doi.org/10.19103/AS.2016.0009.01

breeding), the traits favoured were sometimes more cosmetic than useful. Indeed, some traits (white faces on Herefords, black hides on Angus, large horns on Longhorns and Ankole-Watusis) became a matter of commercial importance in their own right. Most breeders also felt that their breed had the best shape (conformation) and produced the best meat, but in truth, these things have not really been an identified part of any technically sound breeding programme until quite recently.

# 2 The origins of cattle breeds

In most of the world, the local types of cattle became fully adapted through natural selection to the roles they had to play and the management and local environments to which they would be subjected. However, in Europe, by the middle of the eighteenth century (e.g. Robert Bakewell, 1726–95), efforts began to be made to 'improve' cattle (and other species of livestock) by 'breeding the best to the best' in each region and thus developing named breeds (the name often indicating the region of origin) and trying to fix their 'type' usually by inbreeding.

Eventually, the informality of local breeders naming and developing local breeds was replaced by a formal designation of certain animals as being foundation members of the breed and use of the chosen breed name being restricted to the progeny of those animals. The acceptable cattle were listed in a breed registry or herdbook, listing the pedigree of every animal qualified to carry the breed name (i.e. purebreds), and the breeder was issued with a pedigree certificate, indicating its ancestry. A closed herdbook meant that no animals other than those whose ancestry could be traced back to the foundation stock could claim to be members of the breed. Clearly, this would keep the breed members true to type, but also led to problems associated with inbreeding. Consequently, many breed societies opened their herdbooks at one time or another to allow the entry of non-purebreds (heifers usually), provided that they conformed to a very strict set of criteria developed and enforced by the breed societies to ensure that all new registrants conformed to the breed type. These open herdbooks, nevertheless, typically separated the upgraded cattle from the purebred members by placing them in an appendix to the herdbook, at least until they had been bred back to purebred bulls for a certain number of generations.

Because each breed was different, each breed represented a certain identifiable basket of traits that commercial breeders could use to improve their own cattle. Thus, if a breeder felt that his cattle were too small, he could use a bull of a larger breed, or if his cattle were too fat, he could use a bull of a leaner breed. Thus, the breeds were, and remain, a repository for a particular set of traits.

By the middle of the twentieth century, there were hundreds of named breeds of cattle: Porter and Mason (2002) list almost 300 'important and recognisable' breeds of cattle, and Konarzewski (2014) claims that there are 800 extant breeds of cattle worldwide, of which approximately 480 are recognized in Europe. Clearly, as cross-breeding has become more acceptable, particularly in the commercial beef industry, the number of breed types has become very large with a similarly large range of sizes and shapes, but we can cluster many of these breeds together into biological or breed types that can then be used to discuss their carcass and meat quality traits.

From a carcass quality point of view, the important traits are mature body size, muscularity and propensity to fatten (Price, 2014). Although many other traits such as calving ease, mothering ability (particularly milk production), feed conversion efficiency and temperament are important for profitable production, it is the combination of mature body size, muscularity and propensity to fatten that contribute most directly to the commercial value (quality) of the carcass produced. Although these three factors are all at least moderately heritable (Utrera and Van Vleck, 2004; Lambe and Simm, 2014), and therefore, subject to selection pressure, the actual expression of the characteristic will depend on the environment, and in particular, the type, quality and availability of feed. Gender too will affect the expression of these traits, with bulls being larger, more muscular and having a lower propensity to fatten than heifers and cows. Steers (castrated bulls) are generally intermediate, though if allowed to grow out, they will eventually become longer and taller than bulls.

Although these three traits are genetically influenced, until recently it has only been possible to select them on the basis of their phenotypic expression. As the science and practice of genomics advances, so too will opportunities to select for these traits genotypically, allowing for earlier selection and a more precise tailoring of carcass size and composition to match market preferences (Allan and Smith, 2008). However, the ability to use genomics to select for meat quality traits is where its main advantage will lie, because these traits have hitherto been difficult or impossible to identify phenotypically in the live animal. A number of quantitative trait loci (QTLs) for traits of importance to carcass and meat quality have been identified (Allan and Smith, 2008) and are being incorporated into commercial breeding programmes, though difficulties have often arisen in mapping the QTLs to specific positions in the genome (Allan and Smith, 2008). Piedmontese breeders have recognized the association between a deletion in the myostatin gene (responsible for 'double muscling') and tenderness, and have identified cattle within their breed bearing zero, one or two copies of this gene. They use DNA testing to identify homozygous double muscled (DM) (mh/mh or 2-copy) cattle, which then qualify for registration in the *Naturalean Piedmontese Herdbook*.

# 3 Breeds and biological types of cattle

Although the aurochs (*B. taurus primigenius*) were of very large mature body size, domesticated cattle derived from them and kept for draught purposes in most of the world are smaller bodied, presumably in order to limit their maintenance cost (particularly the feed bill) and also to make them more tractable. Indeed, temperament was probably the main or even sole basis of selection by the early breeders (Allan and Smith, 2008). Neither carcass nor meat quality would have been a priority in breed development in those cases.

## 3.1 Dairy breeds

In some parts of the world, dairy cattle have been developed specifically for milk production with no attention to beef production traits. These cattle have a variety of mature body sizes, from the large Holstein Friesian types (Fig. 1) to the smaller Channel

**Figure 1** Holstein cow. Photograph: Holstein Canada.

**Figure 2** Jersey cow. Photograph: Jersey Canada.

Island cattle (Jersey (Fig. 2) and Guernsey), but they all tend to be lightly muscled, which is exacerbated by the heavy nutrient (particularly energy) strain of milk production resulting in little opportunity to accumulate fat, giving dairy cattle an overall angular appearance. However, given the opportunity they readily deposit fat, particularly internally (around the kidneys and other internal organs) and within the muscle (intramuscular or marbling fat) (Berg and Butterfield, 1976; Koch et al., 1976).

## 3.2 Dual-purpose breeds

A dedicated dairy herd is regarded as a luxury in many parts of the world, and dual-purpose dairy/beef animals are more common in those places. Cattle such as the Simmental (Fig. 3), Braunvieh (Fig. 4), Gelbvieh and Red Poll were developed with both beef and milk production in mind. Typically, cows were kept for milk production, but their sons and spare daughters were used for commercial beef production. These cattle tended to be medium to large sized and have moderate muscling and propensity to fatten.

**Figure 3** Simmental cattle. Photograph: Shutterstock.

**Figure 4** Braunvieh bull. Photograph: Ian and Verena Peden.

## 3.3 British breeds

The British breeds of cattle, as exemplified by Hereford (Fig. 5), Angus (Fig. 6) and Shorthorn (Fig. 7), dominated the world of commercial beef production and, by the middle of the twentieth century, were found in all areas of the world with extensive grasslands and established beef industries. These cattle were of medium body size and muscling, with a moderate to high propensity to fatten. This made them particularly useful for finishing on grass, though their British origin meant that they were not well adapted to the local challenges in many parts of the world and needed careful management to avoid falling victims to extremes of heat, drought and local parasites and diseases.

**Figure 5** Hereford bulls. Photograph: Canadian Hereford Association.

**Figure 6** Black Angus bull. Photograph: Canadian Angus Association.

## 3.4 Continental breeds

During the course of the twentieth century, it became increasingly common to feed cattle on cereal grains, particularly maize and barley, in feedlots, but the moderate to high propensity of the British cattle to fatten meant that their carcasses were either too fat at the appropriate weight or too light at the appropriate fat level. To correct this, beef producers around the world began to explore the use of the larger bodied, more heavily muscled and lower propensity to fatten Continental breeds like Charolais (Fig. 8), Maine-Anjou (Fig. 9) and Limousin (Fig. 10). They were used either pure or, increasingly commonly, cross-bred with the British breeds to achieve the desired blend of feed efficiency and carcass quality. The largest of the existing cattle breeds are the Italian white breeds, particularly

**Figure 7** Shorthorn bull. Photograph: Canadian Shorthorn Association.

**Figure 8** Charolais bull. Photograph: Canadian Charolais Association.

**Figure 9** Maine-Anjou cow and calf. Photograph: Heidi Voegeli-Bleiker, Canadian Maine-Anjou Association.

**Figure 10** Limousin bull. Photograph: Canadian Limousin Association.

the Chianina (Fig. 11) (Price, 2014), and the most muscular are the Piedmontese (Fig. 12) and Belgian Blue cattle.

## 3.5 Zebu breeds

During the twentieth century, beef producers in hot climates began to look for ways to improve heat, drought and parasite (particularly tick) tolerance in their hitherto straight-bred British cattle. Some switched to the American Brahman (Fig. 13), a breed developed in the United States from Zebu cattle imported from India. Others cross-bred their British cattle, with the Brahman taking advantage of both hybrid vigour and the complementarity of traits from the parents to combine local adaptation with beefing characteristics. New breeds like Santa Gertrudis, Brangus and Braford were developed from these crossings and

**Figure 11** Chianina bull. Photograph: Walt Browarny.

**Figure 12** Piedmontese bull. Photograph: North American Piedmontese Association.

**Figure 13** Brahman bull. Photograph: Shutterstock.

**Figure 14** Afrikaner bull. Photograph: Afrikaner Breed Society.

**Figure 15** Tuli bull. Photograph: Wayne Southwood.

**Figure 16** Bonsmara bull. Photograph: Afrikaner Breed Society.

became popular in the southern United States and a number of other hot countries, notably Australia, Argentina and Brazil. In Africa, the Sanga-type Afrikaner (Fig. 14) and Tuli (Fig. 15) cattle have been crossed with European cattle to produce the Bonsmara (Fig. 16) (Strydom, 2008) and Tulim breeds, respectively. In general, these new breeds are similar in mature body size and muscling to the British breeds, with a lower propensity to fatten.

## 3.6 Composite breeds

Although cross-breeding, and particularly hybrid breeding, has been slow to develop in Western beef industries, the obvious advantages of hybrid vigour (heterosis) and combining complementary traits from several breeds have meant that recently more multibreed composites are being developed and used in the beef industry. In some cases, the use of multibreeds is simply a part of a breeding system (Beefbooster (Tang et al., 2011), the University of Alberta Beef and Dairy Synthetic (Berg et al., 1990)) rather than the development of a new breed. In other cases, new breeds (Murray Grey, Leachman Stabilizer, Gelbvieh Balancers) have been developed by combining two or more *B. taurus taurus* breeds with complimentary traits together, resulting in cattle suited to a particular environment and market.

# 4 Carcass and meat quality

Although great attention was paid by British cattle breeders to the idea of carcass and meat quality, they were not always successful in identifying the traits for which they were looking. A common error was to confuse fat for muscle, particularly in the ultimately fruitless search for a greater proportion of meat in the high-priced cuts. The early breeders were able to move fat around to 'fill out the corners' on their cattle and produce a blocky shape that was assumed to mean more 'meat' had been deposited. However, the search for more meat in the high-priced cuts resulted in no actual change in muscle distribution with the exception of cattle exhibiting the 'double-muscled' (DM) syndrome (Fig. 17), a mutation in

**Figure 17** DM bull. Photograph: Shutterstock.

the myostatin gene on chromosome 2 (McPherron and Lee, 1997; Fiems, 2012). Although these cattle have a measureable increase in the proportion of meat in the expensive cuts (Berg and Butterfield, 1976; Shahin and Berg, 1985) and produce more tender beef (Wheeler et al., 2001) this comes at the cost of a lack of fitness (e.g. low fertility, calving difficulty, stress susceptibility and walking problems) caused by the enlarged musculature and requiring considerably greater management inputs (Fiems, 2012).

## 4.1 Carcass quality

The definition of 'carcass quality' is variable but can include dressing percentage (the weight of the carcass expressed as a percentage of the live weight), cutability (the proportion of the carcass that is saleable as meat) and finish (the subcutaneous fat thickness). These will be strongly influenced by biological type (mature body size, muscularity and propensity to fatten), and because these traits are under genetic control (Utrera and Van Vleck, 2004; Lambe and Simm, 2014), it is possible to relate potential carcass quality to breed or breed type. Each breed represents a particular combination of traits, and thus, the breed type can be used to tailor the combination of carcass size, conformation and finish demanded in any specific market. For example, when it was found that lot feeding British breeds of cattle on high-energy feeds like barley or maize was leading to carcasses that were either too fat at the preferred weight or too light at the preferred fat level, one solution was to add larger, leaner, more muscular Continental breeds of cattle such as Charolais and Limousin to the mix. In markets that reward muscling, there is also an effort to maximize the muscle-to-bone ratio sometimes even by using the extremely heavily muscled breeds like Piedmontese and Belgian Blue. However, the producer has to also consider traits such as feed conversion efficiency, hardiness and temperament that will influence the overall cost of production. Cow/calf or suckler herd managers must also pay attention to mothering characteristics (fertility, milk production, ease of calving) and select breed combinations for cross-breeding to optimize these traits with beef production.

### 4.1.1 Dressing percentage

Dressing percentage is sometimes considered a measure of carcass quality. Although moderately heritable (Utrera and Van Vleck, 2004), it can be influenced by the environment: lower energy feeds such as grass, hay and silage, if fed to appetite, result in greater gut fill reducing the proportion of carcass (Price et al., 1978); fatter animals will generally have higher dressing percentages because more fat is found in the carcass than in the non-carcass (offal) (Berg and Butterfield, 1976). Heavy muscling also increases the dressing percentage. The weight and condition (wet, dry, mud or manure caked) of the hide can be a major contributor to the dressing percentage, though, in general, this will not be much affected by the breed type. However, Zebu types have been shown to have relatively heavier hides, whereas Holstein and DM cattle have lighter hides than average (Terry et al., 1990; Kambadur et al., 1997; De Smet, 2014). If market-ready cattle are traded on a carcass weight basis, the dressing percentage is removed from the equation, making it less important.

The internal fats (kidney, pelvic and heart fat) are sometimes considered part of the offal and sometimes part of the carcass; regardless of how they are defined, they are of lower value. The partitioning of fats to depots appears to have a genetic component, and some breeds, notably the dairy breeds such as Holstein Friesian and Jersey, have greater quantities of these internal fats at a constant subcutaneous fat thickness than the traditional beef breeds (Berg and Butterfield, 1976; Koch et al., 1976).

## 4.1.2 Cutability or meat yield

The proportions of muscle, fat and bone in the carcass will determine the amount of trimmed meat available for sale (cutability). Therefore, another measure of carcass quality is cutability. Although the propensity to fatten is genetically determined, the actual fat level is determined environmentally, mainly by the amount and quality of feed, so fatness can be relatively easily controlled by the manager (Price et al., 1978). However, the muscle-to-bone ratio is genetically determined and is not much influenced by the environment (Berg and Butterfield, 1976), so some breeds of cattle, for example, the Continental breeds compared to the British breeds, are naturally more muscular (meaty) (Koch et al., 1976), and bulls are more muscular than steers (Price, 1971), which, in turn, are more muscular than heifers and cows (Berg and Butterfield, 1976; Purchas et al., 2002). Some breeds, such as the Piedmontese and the Belgian Blue, are extremely heavily muscled, and selection for this trait has favoured selection for the DM genotype, which carries with it a number of deleterious traits, particularly low fertility and calving difficulty, making them expensive to manage as breeding animals (De Smet, 2014). However, DM cattle also have more tender meat, giving them an important advantage in the beef industry.

Although traditional cattle breeders felt that 'you need considerable bone to grow considerable muscle', there is little evidence to support the idea (Berg and Butterfield, 1976). Indeed, the most muscular cattle of all, those exhibiting the DM syndrome, are known to be fine-boned and have very high muscle-to-bone ratios (Berg and Butterfield, 1976), whereas Holstein Friesian types, despite having relatively light muscling, have large bones, and hence a low muscle-to-bone ratio. Jerseys, on the other hand, while having very light muscling, and fine bones, have moderate muscle-to-bone ratios (Purchas et al., 2002). As noted by Dikeman et al. (1976), bone size, typically estimated from the external diameter of the cannon bones in the live animal, may not be an accurate predictor of bone weight because bone densities vary depending mainly on the diameter of the marrow cavity. They found that Jerseys, despite their fine-boned appearance, had the densest bones, whereas the more coarsely boned Hereford, Simmental and Charolais had the least dense bones. At constant fat levels (easily achieved by management), the muscle-to-bone ratio will influence cutability, so in markets that reward cutability, more heavily muscled (e.g. Continental) breed types may be favoured.

Early beef breeders placed a great deal of emphasis on the proportion of high-priced cuts in the carcass, arguing that selection for cattle with more meat in the expensive parts of the carcass would make them more valuable commercially. The work of Berg and Butterfield (1976) in the middle of the twentieth century largely dispelled that notion, when they showed that muscle distribution is influenced slightly by gender (males having more weight in the forequarter muscles than females, with castrates intermediate) but hardly at all by breed types, with the exception of DM cattle as noted earlier. If the early breeders were successful at all in selecting for shape, it was probably only by moving fat around, to give the appearance of meatiness in the high-priced cuts (Berg and Butterfield, 1976).

## 4.2 Meat quality

Meat quality is a combination of appearance and eating quality (palatability), but may also include factors of production ('clean and green' or welfare friendly (Warner et al., 2010)) that are not directly related to the genetics of the animal itself. Although visual appearance is known to be poorly related to palatability (Price, 1995, 2014), it is nevertheless a prime determinant of saleability. Appearance factors include colour and texture of the muscle, colour and texture of fat and the amount and distribution of marbling within the muscle. Different markets value each of these factors differently. For example, some consumers, generally those accustomed to older, grass-fed beef, prefer an amber or yellow tinge to the fat, whereas others prefer the fat to be white. Some consumers prefer to see more marbling in the meat, whereas some prefer to see less. Thus, although these factors need to be evaluated or scored in a grading or classification system, the scores indicate differences, not necessarily superiority or inferiority (Price, 1995; Warriss, 2000).

### 4.2.1 Appearance

The colour of the muscle is mainly a function of its myoglobin and residual haemoglobin content (Lawrie, 1998), and can be measured by using photographic scales or colorimetry (Warriss, 2000). It is more affected by the animal's age and gender (older and more masculine animals having darker and redder muscle) than by genotype (Lawrie, 1998), and although there are reports in the literature of breed differences in muscle colour, these are inconsistent and may be more associated with management, age and fatness (particularly marbling content) at slaughter than with breed *per se* (Koch et al., 1976; Wheeler et al., 2004; Warner et al., 2010). Colour at the cut muscle surface is also affected by the degree of oxygenation of myoglobin (purple) to oxymyoglobin (red) from initial exposure to oxygen and by the degree of oxidation of oxymyoglobin to metmyoglobin (brown) from prolonged exposure to oxygen (Lawrie, 1998). These are environmental not genetic effects.

The surface texture of the cut muscle is more of a function of muscle bundle size than muscle fibre size (the former being visible to the naked eye, the latter not) and the dryness or stickiness of the cut surface (Lawrie, 1998). The number of fibres per bundle and the number of bundles in the muscle are fixed soon after birth (Joubert, 1956; Lefaucheur, 2010) so, as the animal grows, the fibres and bundles become larger (coarser in appearance), but the number of fibres does not increase. The number of fibres in a primary bundle has been found to be greater in more heavily muscled genotypes, suggesting that they would exhibit a coarser texture; however, it is also clear that this does not necessarily result in a difference in eating quality of the meat (Swatland, 1984). Larger muscle fibres will 'dilute' the proportion of connective tissue in the muscle, potentially reducing background toughness (Koohmaraie et al., 2002). However, larger muscle fibres have also been related to less tender meat, whether via muscle shortening, through greater calpastatin content or by genetic differences (Astruc, 2014).

Texture is also affected by immediate pre-slaughter factors, particularly stress, which may be severe enough to result in reduced muscle glycogen, and hence a high ultimate pH and/or dark cutting and stickiness in the beef. Consequently, temperament, which has a genetic component (Burrow, 2001), can be associated with differences in muscle colour and texture, particularly when stress-susceptible animals are allowed to become stressed in the immediate pre-slaughter period (Lawrie, 1998; Price and Tennessen, 1981).

Fat colour is affected mainly by its content of β-carotene that, although orange in colour, appears yellow in low concentrations. The animal ingests β-carotene in green leaves, and it is then largely converted to colourless vitamin A in the intestinal mucosa before absorption. Both β-carotene and vitamin A are fat-soluble and are stored in body fat (Yeates et al., 1975; Warriss, 2000). However, some breeds of cattle, notably the so-called Channel Island breeds (Guernsey and Jersey) and Zebus, can absorb β-carotene without converting it to vitamin A, giving their fat a distinctive creamy-yellow tinge, which is highly prized in milk and butter, but less so in carcass fat (Yeates et al., 1975). Older cattle, particularly those that have experienced periodic weight loss and recovery, also tend to have yellower fat (Lawrie, 1998; Warriss, 2000). Fat colour can also be influenced by residual blood in the fine blood vessels of the adipose tissue, giving the fat a red or brown colouration. Cattle that are excited or hot (or both) prior to slaughter will increase the flow of blood to the superficial vessels through vasodilation, and this can lead to a condition known as fiery fat (Yeates et al., 1975). Clearly, this is more a consequence of pre-slaughter environment than genotype, but cattle with an excitable temperament are more likely to exhibit it. It should be pointed out that when beef is closely trimmed, fat colour, unless extreme, is of negligible importance.

Marbling is an appearance factor, but is commonly treated as a measure, albeit a weak one, of eating quality (Wheeler et al., 1994). It is the visual appearance of intramuscular fat, especially in the longissimus muscle; marbling scores take into account both amount and distribution. The intramuscular fat depot is one of several in the body, and the partitioning of fat to these depots is under genetic control. Thus, at a constant total fat weight, different breed types will have different proportions of subcutaneous, intermuscular, body cavity and marbling fat. Some breeds, mostly the dairy breeds and notably the Wagyu, have higher proportions of marbling relative to subcutaneous fat and will, therefore, receive higher marbling scores at a constant subcutaneous fat level than the traditional British beef breeds, which, in turn, will have higher marbling scores than the more muscular Continental breeds (Koch et al., 1976). There appears to be an inverse relationship between muscularity and marbling. Zebu cattle have lower marbling scores than European cattle.

All other things being equal, heifers and cows will receive higher marbling scores than steers, which will have higher marbling scores than bulls. Marbling is a moderately heritable trait (Utrera and Van Vleck, 2004), but until recently has been difficult to assess in the live animal, so with some notable exceptions such as the Wagyu breed, little effort has been made until recently to select for it. The development of sophisticated ultrasound equipment to assess marbling in live animals, and the increasing availability of genomic methods to identify the genes and markers associated with marbling, has made it an important part of current selection programmes.

## 4.2.2 Eating quality

The ultimate measure of meat quality is taste or palatability, the actual eating experience. Eating quality resolves into texture or tenderness, juiciness and flavour, though other factors such as mouthfeel and overall liking are sometimes included (Warner et al., 2010). Of these, tenderness, or its opposite, toughness, is consistently shown to be the most important in deciding consumer satisfaction in beef, though not always in other species (Warner et al., 2010). Toughness comes from two sources: myofibrillar toughness from the state of the contractile tissue of the muscle itself and background toughness, which results from the amount and nature of the connective tissue framework supporting the muscle fibres. Myofibrillar toughness is strongly influenced by post-mortem treatment of the muscle and particularly the degree of cold shortening prior to rigor mortis and by the tenderizing effects of ageing after full rigor has been reached (Koohmaraie, 1994; Warriss, 2000). Background toughness varies widely, but consistently from muscle to muscle (and even within some of the larger muscles) within the animal, giving rise to the well-known differences in toughness of different cuts of meat (Swatland, 1984). Background toughness increases with the age of the animal, mainly because of increased number and stability of the cross-links that form within and between the collagen fibres (Lawrie, 1998).

Tenderness has been shown to be highly heritable (Dikeman et al., 2005). The American Simmental Association publishes Expected Progeny Differences (EPDs) for tenderness, so genetic progress can be made through selection. Generally, any genetic differences in tenderness can be negated by the effects of post-mortem factors, such as too rapid a chilling rate, short ageing and/or improper cooking of the meat (Warner et al., 2010). Indeed, there is an old saying in reference to beef that 'the quality goes in after the hide comes off'. Nevertheless, as the industry increasingly adopts best practices to ensure optimum post-mortem treatment of the carcass and meat, genetic differences in tenderness begin to be identifiable (Dikeman et al., 2005; Plastow and Bruce, 2014). The most obvious and consistent of these is that *B. taurus indicus* cattle yield slightly tougher meat than *B. taurus taurus* cattle (Warner et al., 2010). This is related to differences in calpastatin, the enzyme that inhibits calpains, the main agent of myofibrillar degradation during ageing (Koohmaraie, 1994).

Fast twitch, glycolytic muscle fibres have been shown to have a higher relative proportion of calpastatin (Astruc, 2014), and because these fibres have a larger diameter than slow twitch, oxidative fibres, they are generally favoured in more muscular types of cattle (Plastow and Bruce, 2014), but this may lead to tougher meat. British cattle have been shown under carefully controlled conditions to give more tender beef than Continental cattle (Wheeler et al., 2005), but the differences are small and unlikely to be meaningful in any commercial sense.

However, in addition to Wagyu, the most tender beef, measured instrumentally as shear force and confirmed by taste panels, comes from Jersey cattle, a breed that is also known to marble easily, though not nearly as readily as Wagyu (Wheeler et al., 2004). In between these extremes, there are no clear, consistent breed effects on tenderness, whether measured by instrument or by human tasters. However, there are individual animal differences, and as the science of genomics advances, the genetic basis of tenderness will be clarified and sires with a tenderness advantage will be identified regardless of breed (Plastow and Bruce, 2014).

Juiciness is a combination of the water-holding capacity and lipid content of the meat and the response of the taster to the meat in the form of salivation (Lawrie, 1998). For this reason, correlations between instrument measurements of juiciness and perceived juiciness by the taster are poor. Fat in the meat can stimulate salivation (Lawrie, 1998), giving the sensation of juiciness, which, in turn, can lubricate the meat and improve perceived tenderness. Juiciness is also influenced strongly by preparation and cooking methods. With this level of complexity, it is not surprising that juiciness is not consistently affected by breed *per se*, but it is commonly reported that tasters find more marbled meat to be juicier, presumably because the fat increases salivation in the taster.

Flavour is a combination of taste and smell or aroma. The former is perceived on the tongue and the latter in the nasal passages. Although there are only five basic tastes (sweetness, sourness, bitterness, saltiness and umami), there are thousands of flavours, most of which are only perceived when meat is heated (Lawrie, 1998). In general, the flavour of meat is water-soluble and resides in the muscle itself, whereas the species flavour (beef in this case) is fat-soluble (Warriss, 2000). Although several reports have shown breed effects on individual flavour notes (Sato et al., 1995; Insausti et al., 2005) detected by instrument analysis, individual tasters show varying abilities to detect individual flavours (Lawrie, 1998), but in any case, the total sensation is a combination of all the tastes and all the flavours, and it is very difficult to show consistent differences attributable to breed. Generally speaking, flavour is affected by the age and condition (fatness) of the animal (Lawrie, 1998), and the breed differences in flavour that have been reported are likely to be complicated by both pre- and post-slaughter factors associated with the breed, rather than the breed itself.

# 5 Future trends

Mapping of the bovine genome was completed in 2009, and this has led to an ongoing effort to find QTLs, markers, genes and Single-Nucleotide Polymorphisms (SNPs) associated with aspects of carcass and meat quality (Allan and Smith, 2008; Plastow and Bruce, 2014). As technology and the science of genomics move forward together to identify and measure meat quality traits and map them to the bovine genome, it will become possible to identify more accurately males of superior genotype to use as sires and females of inferior genotype to cull from the breeding population. Traditional breeding methods have been limited to traits that could be measured in the live animal, and selection for these traits, through either performance or progeny testing, was slow and required large numbers of animals. However, considerable improvements have been made in growth, carcass quality and more recently feed efficiency traits using traditional techniques. Improvements in meat quality (tenderness, juiciness and flavour) can now become the focus for future cattle breeders by applying genomic information to the breeding programme.

The role of the traditional breeds as repositories of traits of interest to beef breeders will continue, but increasingly commercial breeders will be searching for individual animals carrying the genotype for desirable carcass and meat quality traits regardless of the breed type itself.

# 6 Where to look for further information

For a comprehensive listing of the breeds and types of cattle in the world the reader is referred to 'Mason's World Dictionary of Livestock Breeds, Types and Varieties' 5th edition (see Porter and Mason (2002) in the following References section for full details). For material on individual breeds of cattle an internet or library search of the breed name will usually yield more than enough information. All of the major breeds have herdbook associations or societies at regional, national and world-wide levels.

Perhaps the most comprehensive study of cattle cross-breeding was the Germplasm Evaluation Program undertaken by the United States Department of Agriculture at what became the Roman L. Hruska Meat Animal Research Center at Clay Center, Nebraska. The results, involving thousands of cattle studied over a period of more than 30 years, were reported in a series of publications under the general title 'Characterization of biological types of cattle' (see for example Koch et al. (1976) and Wheeler et al. (2004) in the References section).

An excellent general reference for meat science is: *Encyclopedia of Meat Science*, C. Devine and M. Dikeman, M. (eds), 2nd Edition. Elsevier, Amsterdam, and there are a number of introductory textbooks on the subject such as Aberle (2012), *Principles of Meat Science*, 5th Edition, published by Kendal Hunt; Dubuque, Iowa and Warriss (2000) (see the References section). For a more advanced treatment of the subject Lawrie's Meat Science is a respected and reliable source.

Many countries and international regions have scientific societies of animal science and of meat science that hold annual or biennial meetings or conferences; probably the most important international meeting for meat science is the annual International Congress of Meat Science and Technology (ICoMST). The proceedings of this conference are published annually as a special issue of 'Meat Science', the leading peer-reviewed scientific journal in the discipline.

# 7 References


Allan, M. F. and Smith, T. P. L. (2008), 'Present and future applications of DNA technologies to improve beef production', *Meat Science*, 80, 79–85.

Astruc, T. (2014), *Muscle fibre types and meat quality*, in C. Devine and M. Dikeman (eds), *Encyclopedia of Meat Science*, 2nd Edition. Elsevier, Amsterdam, 2, 442–8.

Berg, R. T. and Butterfield, R. M. (1976), *New Concepts of Cattle Growth*, University of Sydney Press.

Berg, R. T., Makarechian, M. and Arthur, P. F. (1990), 'The University of Alberta beef breeding project after 30 years – A review', *University of Alberta Annual Feeders' Day Report*, 69, 65–9.

Burrow, H. M. (2001), 'Variances and covariances between productive and adaptive traits and temperament in a composite breed of tropical beef cattle', *L. Prod. Sci.*, 70, 213–33.

De Smet, S. (2014), 'Double muscled animals', in C. Devine and M. Dikeman (eds), *Encyclopedia of Meat Science*, 2nd Edition. Elsevier, Amsterdam, 1, 465–70.

Dikeman, M. E., Albrecht, M. D., Crouse, J. D. and Dayton, A. D. (1976), 'Visual appraisal of bovine cannon bone size related to performance, carcass traits and actual metacarpus measurements', *J. Anim. Sci.*, 42, 1077–88.

Dikeman, M. E., Pollak, E. J., Zhang, Z., Moser, D. W., Gill, C. A. and Dressler, E. A. (2005), 'Phenotypic ranges and relationships among carcass and meat palatability traits for fourteen cattle breeds, and heritabilities and expected progeny differences for Warner-Bratzler shear force in three cattle breeds', *J. Anim. Sci.*, 83, 2461–7.

Fiems, L. O. (2012), 'Double muscling in cattle: genes, husbandry, carcasses and meat', *Animals*, 2, 472–506.

Insausti, K., Goni, V., Petri, E., Gorraiz, C. and Beriain, M. J. (2005), 'Effect of weight at slaughter on the volatile compounds of cooked beef from Spanish cattle breeds', *Meat Sci.*, 70, 83–90.

Joubert, D. M. (1956), 'An analysis of factors influencing post-natal growth and development of the muscle fibre', *J. Agric. Sci.*, 47, 59–102.

Kambadur, R., Sharma, M., Smith, T. P. L. and Bass, J. J. (1997), 'Mutations in myostatin (GDF8) in double-muscled Belgian Blue and Piedmontese cattle', *Genome Res.*, 7, 910–15.

Kinsman, D. M. (1994), *Animal By-products from Slaughter.* In Encyclopedia of Agricultural Science Volume 1, Academic Press, 65–74.

Koch, R. M., Dikeman, M. E., Allen, D. M., May, M., Crouse, J. D. and Campion, D. R. (1976), 'Characterization of biological types of cattle III. Carcass composition, quality and palatability', *J. Anim. Sci.*, 43, 48–62.

Konarzewski, M. (2014), 'Species of meat animals: meat animals, origin and domestication', in C. Devine and M. Dikeman (eds), *Encyclopedia of Meat Science*, 2nd Edition. Elsevier, Amsterdam, 3, 357–62.

Koohmaraie, M. (1994), 'Muscle proteinases and meat aging', *Meat Science*, 36, 93–104.
Koohmaraie, M., Kent, M. P., Shackelford, S. D., Veiseth, E. and Wheeler, T. L. (2002), 'Meat tenderness and muscle growth: is there any relationship?', *Meat Science*, 62, 345–52.
Lambe, N. and Simm, G. (2014), 'Traditional animal breeding', in C. Devine and M. Dikeman (eds), *Encyclopedia of Meat Science*, 2nd Edition. Elsevier, Amsterdam, 1, 19–26.
Lawrie, R. A. (1998), *Lawrie's Meat Science*, 6th Edition, Woodhead Publishing.
Lefaucheur, L. (2010), 'A second look into fibre typing – Relation to meat quality', *Meat Science*, 84, 257–70.
McPherron, A. C. and Lee, S. -J. (1997), 'Double muscling in cattle due to mutations in the myostatin gene', *Proc. Natl. Acad. Sci. U.S.A.*, 94(23), 12457–61.
Ockerman, H. W. and Basu, L. (2014), 'By-products: inedible', in C. Devine and M. Dikeman (eds), *Encyclopedia of Meat Science*, 2nd Edition, Elsevier, Amsterdam, 1, 125–36.
Plastow, G. and Bruce, H. (2014), 'Genome projects: modern genetics and genomic technologies and their application in the meat industry – red meat animals, poultry', in C. Devine and M. Dikeman (eds), *Encyclopedia of Meat Science*, 2nd Edition, Elsevier, Amsterdam, 2, 37–42.
Porter, V. and Mason, I. L. (2002), *A World Dictionary of Livestock Breeds, Types and Varieties*, 5th Edition, CAB Bucks.
Price, M. A. (1971), 'Infertile bulls versus steers. III. Carcase characteristics', *J. agric. Sci. Camb.*, 77, 317–23.
Price, M. A. (1995), 'Development of carcass grading and classification schemes', in S. D. Morgan Jones (ed.), *Quality and Grading of Carcasses of Meat Animals*, CRC Press, Baton Rouge, 173–99.
Price, M. A. (2014), 'Species of meat animals', in C. Devine and M. Dikeman (eds), *Encyclopedia of Meat Science*, 2nd Edition, Elsevier, Amsterdam, 3, 328–35.
Price, M. A. and Tennessen, T. (1981), 'Preslaughter management and dark-cutting in the carcasses of young bulls', *Can. J. Anim. Sci.*, 61, 205–8.
Price, M. A., Mathison, G. W. and Berg, R. T. (1978), 'The effects of dietary roughage level on the feedlot performance and carcass characteristics of bulls and steers', *Can. J. Anim. Sci.*, 58, 303–11.
Purchas, R. W., Fisher, A. V., Price, M. A. and Berg, R. T. (2002), 'Relationships between beef carcass shape and muscle to bone ratio', *Meat Science*, 61, 329–37.
Sato, M., Nakamura, T., Numara, M., Kuwahara, K., Homma, S., Sato, A. and Fujimaki, M. (1995), 'Study on factors related to beef quality on the flavor and umami taste of Japanese Black cattle branded beef', *Anim. Sci. Tech.*, 66, 274–82.
Shahin, K. A. and Berg, R. T. (1985), 'Growth and distribution of muscle in Double Muscled and normal cattle', *Can. J. Anim. Sci.*, 66, 625–36.
Strydom, P. E. (2008), 'Do indigenous Southern African cattle breeds have the right genetics for commercial production of quality meat?', *Meat Science*, 80, 86–93.
Swatland, H. J. (1984), *Structure and Development of Meat Animals*, Prentice Hall.
Tang, G., Stewart-Smith, J., Plastow G., Moore S., Basarab, J., MacNeil, M. D. and Wang, Z. (2011), 'Optimizing a beef production system using specialized sire and dam lines', *Can. J. Anim. Sci.*, 91, 353–61.
Terry, C. A., Knapp, R. H., Edwards, J. W., Mies, W. L., Savell, J. W. and Cross, H. R. (1990), 'Yields of by products from different cattle types', *J. Anim. Sci.*, 68, 4200–5.
Utrera, A. R. and Van Vleck, L. D. (2004), 'Heritability estimates for carcass traits of cattle: a review', *Genetics and Molecular Research*, 3(3), 380–94.
Warner, R. D., Greenwood, P. L., Pethick, D. W. and Ferguson, D. M. (2010), 'Genetic and environmental effects on meat quality', *Meat Science*, 86, 171–83.
Warriss, P. D. (2000), *Meat Science an Introductory Text*, CABI Publishers.
Wheeler, T. L., Cundiff, L. V. and Koch, R. M. (1994), 'Effect of marbling degree on beef palatability in *Bos taurus* and *Bos indicus* cattle', *J. Anim. Sci.*, 72, 3145–51.
Wheeler, T. L., Cundiff, L. V., Shackelford, S. D. and Koohmaraie, M. (2004), 'Characterization of biological types of cattle (Cycle VI): carcass, yield, and longissimus palatability traits', *J. Anim. Sci.*, 82, 1177–89.

Wheeler, T. L., Cundiff, L. V., Shackelford, S. D. and Koohmaraie, M. (2005), 'Characterization of biological types of cattle (Cycle VII): carcass, yield, and longissimus palatability traits', *J. Anim. Sci.*, 83, 196–207.

Wheeler, T. L., Shackelford, S. D., Casas, E., Cundiff, L. V. and Koohmaraie, M. (2001), 'The effects of Piedmontese inheritance and myostatin genotype on the palatability of longissimus thoracis, gluteus medius, semimembranosus and biceps femoris', *J. Anim. Sci.*, 79, 3069–74.

Yeates, N. T. M., Edey, T. N. and Hill, M. K. (1975), *Animal Science: Reproduction, Climate, Meat, Wool,* Pergamon Press.

Chapter 2

# Traditional animal breeding of cattle to improve carcass composition and meat quality

*Matt Spangler, University of Nebraska, USA*



## 1 Introduction

Genetic selection in livestock species can arguably be traced to Robert Bakewell in the nineteenth century who developed a system whereby his stock were leased to other farmers in return for phenotypic measurements on progeny. These data were then used to inform his own breeding decisions. In the latter half of the twentieth century, structured progeny testing arose and eventually advancements in statistical methodology allowed for the estimation of an animal's genetic merit. This, coupled with the use of artificial insemination, allowed for connectedness across herds and the formation of genetic evaluations. These evaluations initially centred on growth traits (birth, weaning and yearling) simply because of their ease of measurement.

Genetic selection for carcass attributes such as marbling and tenderness is relatively new as compared to selection for growth traits. Despite their economic importance to either the individual enterprise or the beef industry complex as a whole, marbling, and especially tenderness, represent hard- and expensive-to-measure phenotypes and, as a consequence, the development of genetic selection tools for them lagged behind that of

http://dx.doi.org/10.19103/AS.2016.0009.02

easier-to-measure traits. However, structured progeny testing schemes aided in the initial development of estimated breeding values (EBVs) for these traits. Due to the extreme time lag, these expensive progeny tests gave way to advancements in technology including both ultrasound and genomics that has allowed for EBV for marbling to become pervasive across beef cattle breeds.

# 2 EBVs and expected progeny differences: fundamentals

EBVs allow for the comparison of animals within a breed for their genetic potential as parents for a given trait. EBVs have existed in the beef industry for decades and their use has produced intended genetic change in many traits. EBVs incorporate multiple sources of information, including full pedigree, an animal's own record, genomic data and progeny information. As additional sources of information become available, the accuracy of the EBV increases.

As early as conception, an EBV for an animal exists. With only information from the parents, a pedigree EBV can be calculated as simply the average EBV of the two parents. In an example, if a sire has an EBV of 0.20 and a dam has an EBV of 0.10, the pedigree EBV of the corresponding calf would be 0.15. In the United States, beef breed organizations publish expected progeny differences (EPDs), which are simply half of an EBV. The benefit of an EPD is that it quantifies the genetic value of an animal as a parent, given that only half of their genetic merit will be passed to the next generation.

Once an animal has a record for the trait itself, the EPD is updated as follows:

$$EPD_I = (0.5*EPD_S) + (0.5*EPD_D) + (0.5*\text{Mendelian Sampling Effect})$$

where $EPD_I$ is the EPD for some I individual, $EPD_S$ is the EPD for the sire of animal I and $EPD_D$ is the EPD for the dam of animal I. The phenomena of Mendelian sampling arise due to the fact that each parent passes a random sample of half of its alleles to its offspring and every allele has an equal likelihood of being passed on. This effect can be quantified using contemporary group deviations and is a measure of how much better or worse an animal is compared to the average of his parents. One could envision a scenario in which an animal could receive only the most desirable alleles from both parents, resulting in a favourably large Mendelian sampling effect or the exact opposite, which could result in an unfavourably large sampling effect. Perhaps the best example is a set of flush mates. Although all of them have the same pedigree estimate, they could differ considerably in terms of performance and, consequently, once they have a record, their EPDs differ due to Mendelian sampling.

When using EPD, it is important to understand that the role of EPD is to provide a measure of comparison between individuals within the same population, often the same breed. To compare animals across breeds, estimates from the U.S. Meat Animal Research Center (USMARC) can aid in determining differences between EPDs of different breeds (Table 1). These across-breed adjustment factors, adjusted to an Angus base, are updated annually. It is important to note that these additive EPD adjustment factors are not breed differences, but rather reflect differences in EPDs that can arise due to the combination of inherent breed differences, genetic trends and the subjective base that breed associations elect to consider as their average. Missing values in Table 1 arise due to the particular

**Table 1** Adjustment factors to add to EPDs of 18 different beef breeds to estimate across-breed EPDs for carcass traits

| Breed | Marbling score | Rib-eye area | Fat thickness | Carcass wt. |
|---|---|---|---|---|
| Angus | 0.0 | 0.0 | 0.0 | 0.0 |
| Hereford | −0.32 | −0.10 | −0.053 | NA |
| Red Angus | −0.32 | 0.03 | −0.023 | −6.2 |
| Shorthorn | −0.24 | 0.31 | −0.107 | −11.6 |
| South Devon | −0.09 | 0.21 | −0.129 | −22.3 |
| Beefmaster | NA | NA | NA | NA |
| Brahman | −0.83 | −0.11 | −0.146 | −28.5 |
| Brangus | NA | NA | NA | −12.5 |
| Santa Gertrudis | −0.62 | −0.06 | −0.097 | −5.4 |
| Braunvieh | NA | NA | NA | −44.9 |
| Charolais | −0.39 | 0.98 | −0.207 | 5.4 |
| Chiangus | −0.40 | 0.34 | −0.114 | −20.9 |
| Gelbvieh | −0.33 | 0.65 | −0.117 | −22.6 |
| Limousin | −0.60 | 0.98 | NA | −13.4 |
| Maine-Anjou | −0.60 | 0.78 | −0.192 | −23.6 |
| Salers | −0.14 | 0.85 | −0.203 | −29.7 |
| Simmental | −0.38 | 0.43 | −0.137 | 3.8 |
| Tarentaise | NA | NA | NA | NA |

Source: Adapted from Kuehn and Thallman (2015).

breed not having an EPD for the given trait or a breed has an EPD but is reported on a different scale than the majority of breeds. An example would be a breed that reports an intramuscular fat (IMF) percentage EPD. If a breed does not report EPD on a carcass marbling scale, they would not have an adjustment factor estimated.

Across-breed adjustments for EPDs are needed for producers to compare and make selection decisions among sires from different breeds. Different breeds have emphasized different selection priorities over time, thus each breed differs in its genetic trend. In addition to differences in genetic trend, each breed association sets an arbitrary base for each EPD, creating another source of differences among breeds. Notter and Cundiff (1991) developed methods to compare birth weight, weaning weight and yearling weight EPDs from different breed associations. The Germplasm Evaluation (GPE) Program was utilized to estimate breed differences that are utilized in estimating across-breed adjustment factors. The development of a common base to adjust growth traits, combined with breed association EPDs, allows producers to compare sires from different breeds and make informed selection decisions. Over several years, these original methods have undergone slight statistical changes to add random sire and dam effects, the use of mixed models to estimate regression coefficients and to include estimates of heterosis (Van Vleck et al., 2007). Currently, the USMARC estimates across-breed adjustment factors for 18 breeds for four or more traits including birth weight, weaning weight, yearling weight, maternal milk, marbling score, rib-eye area and fat thickness. The additive factors adjust the EPDs to an Angus base. Across-breed adjustments are updated annually because breed differences change over time due to differing selection pressure within a breed, changes in statistical models used by different breeds calculating EPDs and as GPE continues to collect data (Kuehn and Thallman, 2012).

The calculation of across-breed adjustments relies on breed solutions from the analysis of GPE records at the USMARC and on the average of within-breed EPDs from the respective breed associations (Kuehn and Thallman, 2012). The basic calculations are as follows:

$$M_i = USMARC(i)/b + [EPD(i)_{YY} - EPD(i)_{USMARC}]$$

and the breed table factor $A_i$ to add to the EPD of breed i is equal to:

$$A_i = (M_i - M_x) - (EPD(i)_{YY} - EPD(x)_{YY})$$

where USMARC(i) is the USMARC breed of sire solution (half of the breed solution) of breed i that is converted to an industry scale and adjusted for genetic trend. The pooled regression coefficient (b) of progeny performance at USMARC is regressed on EPD of sire. $EPD(i)_{YY} - EPD(i)_{USMARC}$ is the difference between the average within-breed EPD for breed i to a base year (YY, which is two years before the update) and the weighted average EPD for sires of breed i that have descendants with records at the USMARC. The base breed (x) in this case is Angus. $EPD(i)_{YY} - EPD(x)_{YY}$ is the difference between the average within-breed EPD for breed i and the average within-breed EPD for Angus.

Although multi-breed genetic evaluations currently exist (e.g. American Simmental and International Genetic Solutions), accurate estimation of breed and heterosis effects require that contemporary groups contain multiple breeds and crosses. Without properly accounting for heterosis and breed effects, the resulting EPD are not truly comparable across breeds. Consequently, multi-breed evaluations that do not have adequate data structure to estimate these effects will need to pre-adjust phenotypic records entering into the genetic evaluation using the heterosis and breed effects estimated from USMARC. Otherwise, breeders would be advised to continually use the across-breed adjustment factors even though the EPD were derived from a multi-breed evaluation.

# 3 Advantages of using EPDs over using raw data and ratios

Many producers mistakenly place more emphasis on raw measurements than on EPDs. Raw measurements include the confounded effects of genetics and environment, and consequently, the genetic ability of the animal is unknown. Below is a very simplistic equation describing the phenotype of an animal.

$$P = G + E$$

Where P is the phenotype, G is the genetic effect and E is the environmental effect.

The phenotype is what is seen, or measured, such as the actual scan data for IMF. Both genetics and the environment influence these values, and because we are interested in identifying animals on the basis of their potential as parents, the environment should not be included in the tool used to select animals. Furthermore, actual ultrasound measurements are not comparable from animal to animal because they have neither been adjusted nor do they provide any clue as to how much better or worse an animal is compared to others. A contemporary group ratio does allow for comparison of animals and provides an idea of how much better or worse a particular animal's adjusted record is compared to others within the same contemporary group. The problem is that a ratio is not useful in comparing animals across herds or outside of the defined contemporary group.

The genetic and environmental components of phenotype can be further divided into additive (A), dominance (D) and epistatic (I) genetic effects and both permanent (P) and temporary (T) environmental effects.

$$P = G_A + G_D + G_I + E_P + E_T$$

Generally speaking, we only become concerned with permanent environmental effects when we think about the environmental influence a dam has on her offspring (e.g. a young dam develops mastitis and loses function in one quarter, resulting in reduced weaning weights of subsequent offspring). Contemporary groups account for some of the temporary environmental effects. In genetic evaluations, we are able to predict the additive genetic component, which is presented as an EPD. This is used in determining the heritability ($h^2$), which is simply the fraction of the variance in phenotype ($\sigma^2_P$) that is explained or caused by variation in additive values ($\sigma^2_A$).

$$h^2 = \sigma^2_A/\sigma^2_P$$

The objective of buying a bull is to purchase an animal that will enhance the genetics of the offspring in a herd. Selection based on raw phenotypes, such as actual weights or ultrasound scan values, places selection pressure not only on the genetic potential of an animal but also on environmental influences (herd, year, season, management, etc.). If you look at two drastically different management scenarios: (1) forage tested bulls and

**Table 2** Examples of how to interpret EPDs for several end-product traits

| Traits | Bull A | Bull B |
|---|---|---|
| Carcass weight | 2.0 | 20.0 |
| Percent retail cuts | 0 | 0.2 |
| Marbling | 0 | −0.3 |
| IMF | 3.0 | 1.0 |
| Rib-eye area | 0.06 | 1.6 |
| Fat thickness | −0.01 | −0.09 |
| Rump fat thickness | −0.03 | −0.10 |
| Tenderness | −0.1 | 0.1 |

(2) high-concentrate-fed bulls, it would be expected that the high-concentrate bulls would have greater IMF measurements. The question remains: Are the more desirable IMF scan measurements due to genetics or the fact that they received more feed energy? We know that the environmental benefits will not be passed from parent to offspring, only the genetics. Consequently, selection on the basis of EPDs will help sort the 'wheat from the chaff' because EPDs eliminate environmental differences and quantify genetic differences.

*Carcass weight* – Bull B should produce calves that have 18 pounds more adjusted carcass weight.

*Percent retail product* – The calves from Bull B should yield 0.2% more closely trimmed, boneless retail cuts from the round, loin, rib and chuck. Some breeds may report a yield grade (YG) EPD. The same factors (back fat, rib-eye area and carcass weight) would be included, but a lower YG is more desirable, as opposed to percent retail product, where a higher value is more desirable. In either percent retail product or YG, fat thickness contributes the most to these two calculations. Consequently, selecting for decreased YG or increased percent retail product will lead to leaner animals and so caution should be used to avoid extremely lean replacement females.

*Marbling* – Calves from Bull A should have a 0.3 higher marbling score. Marbling scores range from 1.0, which is devoid of marbling and a utility quality grade to 10.9, which is abundant marbling and a Prime$^{+}$ quality grade. For example, if calves sired by Bull B had a marbling score of 5.0, then we would expect calves sired by Bull A to have a marbling score of 5.3 if bred to similar cows and the calves are managed in a similar environment. Ultrasound EPDs were calculated for a number of breeds for traits of rib-eye area, fat and IMF, which is correlated to marbling, but now the majority of breeds use these ultrasound measurements in the calculation of carcass EPDs. So, instead of seeing both an IMF EPD and a marbling EPD, you just see the marbling EPD, but it has ultrasound measurements included as a correlated trait in the EPD estimation.

*IMF* – Calves from Bull A should produce calves with 2% more IMF than calves sired by Bull B. IMF percentage is measured by ultrasound and is a proxy for carcass marbling. Most breeds incorporate this measurement into their respective carcass marbling EPD as an indicator trait.

*Rib-eye area* – At a given end point, calves from Bull B should have rib-eye areas that are 1.54 square inches larger than Bull A's calves.

*Fat thickness* – At a given end point, calves from Bull A should be 0.08 inches fatter when measured at the 12th rib. This would be less desirable on a carcass animal, but extremely lean females going back into a cowherd may also be undesirable.

*Rump fat thickness* – At a given end point, calves from Bull A should be 0.07 inches fatter when measured between the hooks and pins. This measurement is taken solely via ultrasound.

*Tenderness* – Calves sired by Bull A should produce meat that is more tender than that of calves sired by Bull B by 0.2 pounds of shear force. Tenderness is measured by Warner–Bratzler shear force (WBSF) that is reported in pounds of force required to cut through a 1-inch thick piece of cooked meat (*longissimus* muscle). A lower value is more desirable. Currently, marbling is the driver of the USDA quality grading system. Although tenderness is not considered in assigning a USDA quality grade, it is the most important factor affecting palatability. The two U.S. beef breeds that currently publish an EPD for tenderness are Simmental and Brahman.

# 4 Selection response

The amount of change that can be made in genetic value per unit of time ($\Delta G$) is related to the accuracy of the EPD, the intensity of selection, the amount of genetic variation in the population and the generation interval.

Bourdon (2000) explained the factors that impact the rate of genetic change ($\Delta G$) using the breeder's equation:

$$\Delta G / yr = \frac{r_{EBV,BV} i \sigma_a}{L}$$

Genetics changes in any trait, marbling and tenderness included, are determined by the four factors included in the equation above. Consequently, the rate of genetic change for these traits is the cumulative effect of decisions that breeders and beef breed organizations make. Each of these four factors is discussed in more detail with particular regard to their impact on improving carcass marbling and tenderness.

## 4.1 Accuracy of selection and generation interval

These two components of the breeders' equation are often antagonistic to each other, but both have a substantial impact on the rate of genetic change. High accuracy of selection (the correlation between the true and EBVs; $r_{EBV,BV}$) can be achieved for highly heritable traits such as carcass attributes (Table 3). However, accumulating the data necessary to reach high levels of accuracy can take several years and thus lengthens the generation interval (*L*). Younger sires and dams can be used, but will have lower accuracy. The greatest opportunity for beef cattle breeders to increase the rate of genetic change for carcass traits is to use advances in technology that break the antagonism between selection accuracy and generation interval. This requires the ability to garner substantially more information on an animal's genetic merit much earlier in life.

**Table 3** Approximate number of progenies needed to reach accuracy levels (*r*) for three heritabilities ($h^2$)

| Accuracy | Heritability levels | | |
|---|---|---|---|
| *r* | $h^2$ (0.1) | $h^2$ (0.3) | $h^2$ (0.5) |
| 0.1 | 1 | 1 | 1 |
| 0.2 | 2 | 1 | 1 |
| 0.3 | 4 | 2 | 1 |
| 0.4 | 8 | 3 | 2 |
| 0.5 | 13 | 5 | 3 |
| 0.6 | 22 | 7 | 4 |
| 0.7 | 38 | 12 | 7 |
| 0.8 | 70 | 22 | 13 |
| 0.9 | 167 | 53 | 30 |
| 0.999 | 3800 | 1225 | 700 |

Traditional EPD for marbling required progeny carcass data, making it highly reliable but not timely. In this scenario, a sire must reach a minimum age of four years before carcass data from his offspring could be included in his EPD. Ultrasound technology greatly aided in reducing the generation interval by allowing a parent to have an indicator of carcass marbling (IMF) record themselves. As with any technology, if used correctly, it is helpful, but if used incorrectly, it can be counterproductive. If those practising selection rely on an EPD that includes ultrasound phenotypes, then there is a reduction in generation interval and genetic gain can be made. However, if those practising selection use the actual or adjusted ultrasound data and not the corresponding EPD, then selection accuracy erodes substantially. Traditionally, some U.S. beef breed associations published both an IMF and a marbling EPD. This was not only confusing, but often led to inflated expectations relative to genetic gain. Although the accuracy of the IMF EPD may be high, it is not the economically relevant trait, and selection accuracy for improved carcass marbling is reduced, given that the genetic correlation between the two is not unity.

## 4.2 Genetic variation

There exists a pervasive thought that variation in marbling and tenderness genotypes is undesirable. However, animal breeders view this diversity as an asset that can be exploited to make more rapid genetic change. Increased genetic variation allows for more rapid genetic change because superior animals have a greater advantage over the rest of the population. Over time, with strong directional selection for only one trait, the expectation is that the genetic variation will be reduced and as a consequence the rate of genetic change in the short term may be considerably steeper than that in the longer term.

## 4.3 Selection intensity

Selection intensity (*i*) is the standardized proportion of individuals selected to be the parents for the next generation. As selection intensity increases, the average merit of the selected individuals becomes more dissimilar to the average merit of the population from which they were chosen. Selection intensity can be maximized if only one trait is considered to be under selection and a very few animals are selected to be parents. Selection intensity varies substantially in the beef industry, given the plethora of traits under selection and the fact that a large number of selection candidates are required to maintain herd size. As the number of traits under selection increases, the progress made in any one trait decreases. Given that marbling is not the only trait that impacts profitability, it stands to reason that other traits are also under selection simultaneously, thus reducing the amount of annual genetic progress that can be made in marbling.

# 5 Genetic parameter estimates

Carcass attributes are generally considered to be moderately to highly heritable, meaning that between 30% and 50% of the phenotypic differences observed are due to underlying genetic differences. Bertrand et al. (2001) summarized 17 studies published between 1981 and 1998 and reported the mean heritability estimate for carcass marbling to be 0.46 with a range of 0.15–0.93. Similarly, the average heritability across the two studies reviewed by Bertrand et al. (2001) for IMF was 0.41. The mean reported by Bertrand et al. (2011) was greater than the more recent estimate from Angus cattle reported by MacNeil et al. (2010) of 0.31. Of equal interest is the genetic relationship between carcass marbling and IMF, given that the majority of phenotypes entering into genetic evaluations are proxies via ultrasound and the targeted trait for improvement is carcass marbling. The range in genetic correlations between these two traits is generally between 0.53 and 0.80 (MacNeil et al., 2010; Devitt and Wilton, 2001; Crews et al., 2003; Meyer, 2007), although much larger estimates have been reported (e.g. 0.90; Kemp et al., 2002).

Bertrand et al. (2001) also summarized genetic parameter estimates from literature published since 1980 relative to measures of beef tenderness. In total, 10 studies were summarized, which included the trait of WBSF with a range of heritability values of 0.04–0.53 with a mean of 0.22. Six studies that included taste panel score were also summarized with a range in heritability estimates of 0.10–0.50 and a mean of 0.28. The authors also reported the average genetic correlation between carcass marbling and WBSF and taste panel tenderness scores to be −0.55 and 0.40, respectively. These genetic correlations suggest that as carcass marbling increases, WBSF and taste panel tenderness scores decrease and increase, respectively. Furthermore, the average genetic correlations reported by Bertrand et al. (2001) suggest that genetic variation in carcass marbling explains 30.25% and 16% of the genetic differences in WBSF and taste panel tenderness, respectively. Dikeman et al. (2005) reported the heritability of WBSF to be 0.40, within the range of estimates reported by Bertrand et al. (2001). The phenotypic correlation between WBSF and carcass marbling reported by Dikeman et al. (2005) was favourable but low ($r_p = -0.21$), suggesting that at the phenotypic level improving marbling would have little effect on tenderness. Although the directionality of these correlations is favourable, it is clear that much of the phenotypic and genetic variation that exists for tenderness-related phenotypes is not explained by differences in marbling.

# 6 Examples of successful selection

Vieselmeyer et al. (1996) conducted a breeding experiment whereby six low and six high marbling EPD Angus bulls were randomly mated to MARC II (¼ Hereford, ¼ Angus, ¼ Simmental and ¼ Gelbvieh) cows to produce terminal steers ($n = 131$) and heifers ($n = 125$) over a two-year period. Marbling EPD classification of sire had no effect on fat thickness, USDA YG, carcass weight, finishing daily gain or finishing dry matter intake. The authors did report that the calves sired by the high marbling EPD sires had more ($P < 0.05$) carcass grade choices as compared to calves sired by low marbling EPD bulls (74% and 47%, respectively). Using the same population, Gwartney et al. (1996) reported no differences in shear force values or taste panel ratings between the two sire groups among steer carcasses. However, they did find that among the heifer carcasses, progeny of the high marbling EPD sires had better ($P < 0.05$) overall tenderness. The authors reported that these differences among sexes could simply be a sex effect or related to the fact that the heifers were older on average at harvest than their steer counterparts. The results from this single study provide empirical evidence that selection for increased marbling EPD will lead to improved carcass quality. Yet another example of the utility of EPD in generating response to selection comes from USMARC. As a part of calculating across-breed EPD adjustment factors on an annual basis, the regression coefficient of progeny performance at USMARC on the breed of sire EPD is estimated. According to Kuehn and Thallman (2015), the pooled regression coefficients were 0.55, 0.80, 0.96 and 0.90 for carcass marbling, rib-eye area, fat and weight, respectively. These estimates are pooled across all 18 breeds in the evaluation, and there exists substantial variation among breeds in information content to estimate this parameter. Among the seven largest beef breeds in the United States (Angus, Hereford, Red Angus, Charolais, Gelbvieh, Limousin and Simmental), the regression coefficients (+/− SE) for marbling range between 0.68 (0.14) and 1.23 (0.19).

# 7 Index-based selection

Hazel (1943) summarized the need to formalize a method of multiple trait selection in the opening paragraph of his landmark paper on the topic of selection indexes:

> *The idea of a yardstick or selection index for measuring the net merit of breeding animals is probably almost as old as the art of animal breeding itself. In practice several or many traits influence an animal's practical value, although they do so in varying degrees. The information regarding different traits may vary widely, some coming from an animal's relatives and some from the animal's own performance for traits which are expressed once or repeatedly during its lifetime. ... These factors make wise selection a complicated and uncertain procedure; in addition fluctuating, vague, and sometimes erroneous ideals often cause the improvement resulting from selection to be much less than could be achieved if these obstacles were overcome.*

Despite this meaningful advancement of genetic selection tools in the early 1940s, the U.S. beef industry was slow to adopt a tool that had the potential to greatly simplify sire selection and place emphasis on that which is economically important. Economic indices are the preferred tool for multiple trait selection. A bio-economic index ($H$) is simply a collection of EPDs that are relevant to a particular breeding objective, whereby each EPD is multiplied by an associated economic weight ($a$). For example, the economic index value $H$ can be written as

$$H = \text{EPD}_1 a_1 + \text{EPD}_2 a_2 + \text{EPD}_3 a_3 + \ldots + \text{EPD}_n a_n$$

where EPDs 1, 2 and 3 are multiplied by their corresponding economic weight and summed. Consequently, a high index value does not necessarily mean that an animal excels in all EPD categories, given that superiority in traits can compensate for inferiority in other traits depending on how the EPDs are weighted in the index. A high index value should be thought of as excelling in the ability to meet a breeding objective. Such bio-economic indices can allow for beef cattle producers to select towards profitability. However, to correctly define the weighting factors associated with the genetic values in the index, there must be a quantifiable pricing structure associated with the trait. This can be done for carcass measures such as marbling, weight and traits related to yield, but tenderness represents a potential problem. Until beef producers are confronted with the opportunity to garner premiums (or suffer discounts) based on tenderness, there is not a clear economic signal to place selection pressure on this trait and the value of including it in a selection index is less clear.

Below are examples of how to use and interpret bio-economic selection indices. The first example uses the American Angus Association's Dollar Beef Index ($B). This index assumes that the user will not retain any replacement heifers and will retain ownership of all offspring through the finishing phase and market them on a grid-based system targeting Certified Angus Beef (CAB). The second example uses the American Hereford Association's Baldie Maternal Index (BMI$). This index assumes that producers will use Hereford bulls to breed British-cross cows and heifers and that some replacement females will be retained with the remainder of offspring being sold targeting a Certified Hereford Beef grid. Both indices include EPDs associated with carcass merit, but an all-terminal index (e.g. $B) will place more emphasis on carcass traits as compared to a multipurpose index such as BMI$.

*Example 1:*

| | $B |
|---|---|
| Bull A | 105 |
| Bull B | 90 |

From this example, we would expect that on average, calves sired by Bull A would generate $15/hd. more profit than calves sired by Bull B if the calves are fed through the finishing phase and marketed towards a CAB grid.

*Example 2:*

| | BMI$ |
|---|---|
| Bull A | 42 |
| Bull B | 38 |

From this example, we would expect that on average, calves sired by Bull A would generate $4/hd. more profit than calves sired by Bull B if the producer's breeding goals aligned with the assumed breeding goal of BMI$ detailed above.

# 8 Summary and future trends

## 8.1 Summary

Carcass traits, including marbling and tenderness, are moderately to highly heritable, suggesting that they would respond to direct selection. Marbling and tenderness are moderately and favourably correlated, suggesting that selection for improved marbling could lead to a correlated response in the improvement of tenderness. EPDs exist for marbling and tenderness, although the number of breeds that publish an EPD for tenderness is limited. Both of these trait complexes represent expensive phenotypes to collect, and this can be directly measured only post-mortem. As a consequence, the availability of data has historically required progeny tests that have substantially lengthened the generation interval, thus reducing the rate of genetic change. For marbling, the advent and pervasive use of ultrasound has enabled the indicator trait of IMF to be collected on parents themselves as well as their offspring. IMF is at least moderately genetically correlated to marbling and thus serves as a valuable indicator trait. For tenderness, outside of genomic data, a reliable proxy that can be measured on live animals has not yet been identified.

In the context of genetic selection at the individual enterprise level, increased profit must be a focus. Given that more than on trait impacts profitability of any enterprise, selection for increased profit must include selection for multiple traits simultaneously. The most efficient way to do this is through the use of a bio-economic selection index. Many such indices exist, but it is critical that the user understand the intended use of the index and the traits included in it. Using an index designed for a breeding objective that differs from that which impacts profit of the user could prove counterproductive.

## 8.2 Future trends

Currently, genomics is playing a large role in increasing the accuracy of EPD for numerous traits, including carcass marbling and tenderness. The advantages of augmenting genomic information with traditional EPD will become greater in the future. The increased benefits will be due to improved genomic predictions from knowledge of causal variants that are identified through whole genome sequencing efforts currently underway and advances in statistical methodology to better make use of existing genomic data.

The availability of phenotypic data for both carcass marbling and tenderness is a concern. There must be concerted efforts to increase the amount of data for these two traits that enter genetic evaluations not only for the production of traditional EPD but also for the development of genomically enhanced EPD. For traditional EPD, having carcass marbling phenotypes can improve the accuracy of current EPD. For genomically enhanced EPD, carcass data is required to adequately develop genomic predictors.

A clear market signal is needed at the cow/calf producer level to improve tenderness before this trait will reasonably enter breeding schemes. The segmented nature of the beef industry, coupled with the expense of collecting tenderness phenotypes, has hindered the widespread selection for improvement of this trait. It is clear that this trait will respond to selection, but until there is economic incentive to select for it, it will remain largely ignored.

# 9 Where to look for further information

In addition to the studies referenced in this chapter, there are several review articles that summarize genetic parameters reported in the literature. The list below represents comprehensive resources that summarize a large fraction of the available literature.

Koots, K. P., J. P. Gibson, C. Smith and J. W. Wilton. (1994). Analyses of published genetic parameter estimates for beef production traits. 1. Heritability. *Animal Breeding Abstracts* 62: 309.

Koots, K. P., J. P. Gibson, C. Smith and J. W. Wilton. (1994). Analyses of published genetic parameter estimates for beef production traits. 2. Phenotypic and genetic correlations. *Animal Breeding Abstracts* 62: 825.

Rios Utrera, A. and L. D. Van Vleck. (2004). Heritability of carcass traits in beef cattle: A review. *Genet. Mol. Res.* 3: 380–94.

In addition to the resources above, the annual meeting of the Beef Improvement Federation serves as a valuable resource. Past proceedings from this meeting, as well as guidelines for phenotypic data collection and genetic evaluations, can be found at www.beefimprovement.org.

# 10 References

Bertrand, J. K., R. D. Green, W. O. Herring and D. W. Moser. (2001). Genetic evaluation for beef carcass traits. *J. Anim. Sci.* 79(E. Suppl.): E190–E200.

Bourdon, R. M. (2000). *Understanding Anima Breeding*. 2nd edition. Prentice Hall, Inc., Saddle River, NJ, p. 200.

Crews, D. H. Jr., E. J. Pollak, R. L. Weaber, R. L. Quaas and R. J. Lipsey. (2003). Genetic parameters for carcass traits and their live animal indicators in Simmental cattle. *J. Anim. Sci.* 81: 1427–33.

Devitt, C. J. B. and J. W. Wilton. (2001). Genetic correlation estimates between ultrasound measurements on yearling bulls and carcass measurements on finished steers. *J. Anim. Sci.* 79: 2790–7.

Dikeman, M. E., E. J. Pollak, Z. Zhang, D. W. Moser, C. A. Gill and E. A. Dressler. (2005). Phenotypic ranges and relationships among carcass and meat palatability traits for fourteen cattle breeds, and heritabilities and expected progeny differences for Warner-Bratzler shear force in three beef cattle breeds. *J. Anim. Sci.* 83: 2461–7.

Gwartney, B. L., C. R. Calkins, R. J. Rasby, R. A. Stock, B. A. Vieselmeyer and J. A. Gosey. (1996). Use of expected progeny differences for marbling in beef: II. Carcass and palatability traits. *J. Anim. Sci.* 74: 1014–22.

Hazel, L. N. (1943). The genetic basic for constructing selection indexes. *Genetics* 28: 476–89.

Kemp, D. J., W. O. Herring and C. J. Kaiser. (2002). Genetic and environmental parameters for steer ultrasound and carcass traits. *J. Anim. Sci.* 80: 1489–96.

Kuehn, L. A. and R. M. Thallman. (2012). Across-breed EPD Tables for the year 2012 adjusted to breed differences for birth year 2010. Pages 152–77 in *Beef Improv. Fed. Res. Anu. Meet.*, Houston, TX.

Kuehn, L. A. and R. M. Thallman. (2015). Across-breed EPD Tables for the year 2015 adjusted to breed differences for birth year 2013. Pages 97–124 in *Beef Improv. Fed. Res. Anu. Meet.*, Bioloxi, MS.

MacNeil, M. D., J. D. Nkrumah, B. W. Woodward and S. L. Northcutt. (2008). Genetic evaluation of Angus cattle for carcass marbling using ultrasound and genomic predictors. *J. Anim. Sci.* 88: 517–22.

Meyer, K. (2007). Multivariate analyses of carcass traits for Angus cattle fitting reduced rank and factor analytic models. *J. Anim. Breed. Genet.* 124: 50–64.

Notter, D. R. and L. V. Cundiff. (1991). Across-breed expected progeny differences: Use of within-breed expected progeny differences to adjust breed evaluation for sire sampling and genetic trend. *J. Anim. Sci.* 69: 4673–776.

Van Vleck, L. D., L. V. Cundiff, T. L. Wheeler, S. D. Shackelford and M. Koohmaraie. (2007). Across-breed adjustment factors of expected progeny differences for carcass traits. *J. Anim. Sci.* 85: 1369–76.

Vieselmeyer, B. A., R. J. Rasby, B. L. Gwartney, C. R. Calkins, R. A. Stock and J. A. Gosey. (1996). Use of expected progeny differences for marbling in beef: I. Production traits. *J. Anim. Sci.* 74: 1009–13.

Chapter 3

# Muscle fibre types and beef quality

*Thierry Astruc and Annie Vénien, INRA, France*



## 1 Introduction

Meat comes from the transformation of the skeletal muscle of meat animals after slaughter. Cattle are generally reared to produce either milk or meat, and breeds are purpose-adapted to each type of production. However, dairy-herd animals such as culled cows, bullocks or heifers represent a large share of meat consumed (more than 50% in the European Union (EU)). In the United States and Australia, most beef from males is from steers. The age of the slaughtered animals is, therefore, highly variable, as it extends from calves aged a few weeks to cull cows that may count up to 15 years. The upshot is that there is high variability in the cattle providing meat and, consequently, high variability in meat quality. Pre-slaughter stress, slaughtering technology and the differences in muscle composition are known to be partly responsible for the heterogeneity of meat quality.

'Carcass quality', which depends on meat yield and fat content, is one of the most important criteria for farmers. The commercial grading system in the European Union is heavily weighted towards carcasses with the highest muscle content. In contrast, sensory properties are characterized by all that appeals to consumers' senses, such as colour, tenderness, juiciness and flavour of the product. Nutritional qualities are also increasingly taken into account, and are usually linked to the proportion of amino acids, essential fatty acids and small yet nutritionally valuable molecules such as vitamins. Technological qualities reflect processing ability, often associated with technological yields such as

http://dx.doi.org/10.19103/AS.2016.0009.03

percent water loss during cooling, storage or cooking meat. Health quality reflects the fitness of the product for safe consumption and is often associated with the presence of bacterial flora and also the presence of microcontaminant residues coming from the environment (pesticides, air, ground pollution) or from medical treatments (antibiotics, growth promoters) given to the animals during rearing.

Excluding health quality, most meat qualities depend on muscle characteristics, chiefly their physiological and biochemical specificities, composition and structure.

# 2 Body and muscle composition and structure

## 2.1 Body composition

An animal is composed of muscle, fat and bone, plus offal (heart, liver, lungs, digestive tract), skin, feet and horns. Proportional animal body composition varies not only with age and weight, but also with biological factors such as sex and genotype. This compositional variability is particularly high in cattle, because the cattle population is increasingly made up of various breeds worldwide gathering specialized beef breeds, specialized dairy breeds and dual-purpose crosses. Body composition can be partially modulated through rearing factors such as production system, feed modes, including energy value of feed, and in some cases growth promotants. These compositional changes affect commercial carcass value, as illustrated by the results of Laville et al. (1995), who observed carcass weight variation ranged from 248 to 498 kg for cows, 253 to 385 kg for heifers, 316 to 483 kg for steers and 251 to 414 kg for young cattle slaughtered in a commercial slaughterhouse.

Dressing percentage is expressed as carcass weight/live weight ratio or sometimes as carcass weight*100/empty body weight to eliminate the influence of gut contents (Geay and Robelin, 1979). There is broad variation in dressing percentage due to genotype (Cole et al., 1964; Damon et al., 1960; Geay and Robelin, 1979). A study by Geay and Robelin (1979) put dressing percentage at 58% for Holstein bulls, 63% for Charolais bulls and 67% for Limousin bulls, whereas other studies report dressing percentages from 60% for Hereford to 69.5% for Limousin (Fiems et al., 2003). Geay and Robelin (1979) concluded that there is a hierarchy between dressing percentages of dairy breeds and dressing percentages of large-sized or late-maturing breeds. From their study, meat yield (expressed by the ratio muscle weight*100/carcass weight) was 65% for Holstein bulls and 73% for Limousin and Charolais bulls. Holstein carcasses contained more lipids and less protein than Limousin and Charolais bulls. Double-muscled Belgian Blue gives dressing percentages of 66% but with a meat yield close to 75% (Clinquart et al., 1998; Fiems et al., 2005). These data highlight the scale of potential heterogeneity in body composition due to genetic type.

## 2.2 Muscle composition and structure

While carcass yield is a point of interest from a trade and economics point of view, meat comes from muscles and its quality is partly related to muscle characteristics. The skeletal muscle of farm animals is roughly 75% water, 19% protein and 2.5% intramuscular fat, the remainder being essentially composed of carbohydrates and other soluble non-protein substances.

Skeletal muscle is essentially composed of muscle cells, called 'muscle fibres', due to their elongated shape (from 20 to 100 μm in diameter and up to several centimetres long). Under a microscope, muscle fibres show alternating light isotropic bands (I-bands) and dark anisotropic bands (A-bands) that give skeletal muscle its striated appearance. Muscle fibres account for up to 90% of muscle volume, the remainder being lipids and connective tissue.

Each muscle fibre is delimited by the sarcolemma, which is the plasma membrane of the fibre. The intracellular part contains cellular organelles, nuclei and sarcoplasmic reticulum that are bathed in the sarcoplasm (the cytoplasm of the muscle cell). Sarcoplasm contains glycogen granules and a medley of soluble proteins, including glycolytic enzymes and myoglobin. However, the intracellular volume is almost entirely occupied by the myofibrils, which have an elongated shape and measure 1–2 μm in diameter.

Myofibrils are composed of a succession of sarcomeres, which are delimited by the Z-disks that cross the I-band. The sarcomere is the contractile unit of the muscle. It measures about 2.5 μm from one Z-disk to the next when relaxed. The sarcomere contains at least 28 different proteins (Craig and Padron, 2003; Schiaffino and Reggiani, 2011), but consists mainly of the thin myofilaments, essentially composed of actin, troponin and tropomyosin proteins, and the thick myofilaments, mainly composed of myosin molecules built from four myosin light chains (MyLC; 20000 g $mol^{-1}$) and two myosin heavy chains (MyHC; 230000 g $mol^{-1}$). Myosin molecules account for about 50% of the myofibrillar proteins. Other fibrillar proteins called titin, myomesin, M protein and nebulin lend myofibrils internal structure and cohesion.

Four different isoforms of MyHC – I, IIa, IIb and IIx – have been identified in adult mammals (Schiaffino and Reggiani, 2011). Pure muscle fibre types I, IIA, IIX and IIB contain the isoforms I, IIa, IIb and IIx, respectively, but hybrid fibre types containing different myosin isoforms are also found. These hybrid fibres, such as type I–IIA, type IIA–IIX or type IIX–IIB fibres, are transition fibres between two pure types driven by physiological or environmental changes such as ageing or physical activity levels. The fibre type change always follows the order I→I–IIA→IIA→IIA–IIX→IIX→IIX–IIB→IIB and back in reverse (Pette and Staron, 2000; Schiaffino and Reggiani, 2011).

# 3 Muscle fibre types

## 3.1 Muscle fibre characteristics

The different types of muscle fibres are characterized by their contraction speed (slow-twitch or fast-twitch) and by their preferred metabolic pathway for glycogen degradation (glycolytic or oxidative). They consequently differ in ATPase activity (that reflects contraction speed) and in the organelle, substrate and enzyme activities involved in the characteristic metabolic pathway of different fibre types. Several features of the different muscle fibre types are given in Table 1.

Muscles are composed of these different fibre types in variable proportions according to their function. Muscles likely to provide violent efforts but in short bursts usually contain a high proportion of type IIX and/or IIB fibres, whereas muscles that need to resist fatigue, such as the respiratory muscles or postural muscles, have a high proportion of type I and IIA fibres.

## 3.2 Identifying muscle fibres

The different types of muscle fibres are usually determined by histology (Fig. 1). The most common histochemical method exploits the difference in sensitivity of ATPase activity after exposure to either high or low pH (Brooke and Kaiser, 1970). Oxidative Succinate dehydrogenase (SDH) and Nicotinamide adenine dinucleotide (NADH) tetrazolium reductase staining can be combined with ATPase staining to improve fibre type identification (Ashmore and Doerr, 1971; Peter et al., 1972). Efforts led in the 1970s to combine the histochemical stains for the oxidative enzyme and ATPase staining identified three major fibre types that were given different names, depending on histological technique developed and research team involved. For example, slow-twitch oxidative fibres, fast-twitch oxidative/glycolytic fibres and fast-twitch glycolytic fibres got named slow oxidative (SO), fast oxidative glycolytic (FOG) and fast glycolytic (FG) fibres (Peter et al., 1972), β-R, α-R and α-W fibres (Ashmore and Doerr, 1971) or I, IIA and IIB fibres (Brooke and Kaiser, 1970), respectively.

However, while SO fibres are similarly identified regardless of the technique used, fast-twitch fibres are sometimes differentially identified (Klont et al., 1998). The development of monoclonal antibodies raised against the heavy-chain isoform of myosin allowed the detection of an additional fast-twitch fibre type that was called IIX (Pette and Staron, 1990, 2000; Schiaffino et al., 1989). Type IIX fibres share similar acidic resistance to glycolytic fibres, but their aerobic metabolism is midway between glycolytic and IIA fibres.

In recent years, chemical imaging methods based on localized microspectroscopy have been developed, which makes it possible to acquire molecular maps from histological sections. Chief among these methods are infrared, Raman and fluorescence

**Table 1** Physiological and biochemical characteristics of muscle fibre types

| Peter et al. (1972) | Slow oxidative (SO) | Fast oxidative glycolytic (FOG) | Fast glycolytic (FG) | |
|---|---|---|---|---|
| Ashmore and Doerr (1971) | β-R | α-R | α-W | |
| Brooke and Kaiser (1970) | I | IIA | IIX | IIB |
| Contraction speed | Slow | Fast | Fast | Fast |
| ATPase activity | + | ++ | ++++ | +++++ |
| Fatigue resistance | +++++ | ++++ | ++ | + |
| Glycogen | + | +++ | +++ | ++++ |
| Glycogen metabolism | Oxidative | Oxidative/glycolytic | Glycolytic | Glycolytic |
| Myoglobin | +++++ | +++ | ++ | + |
| Red colour | +++++ | +++ | ++ | + |
| Mitochondria | ++++ | +++ | ++ | + |
| SDH activity | +++++ | ++++ | ++ | + |
| Lipid droplets | ++++ | ++ | + | + |
| Cross-sectional area | + | ++ | ++++ | +++++ |

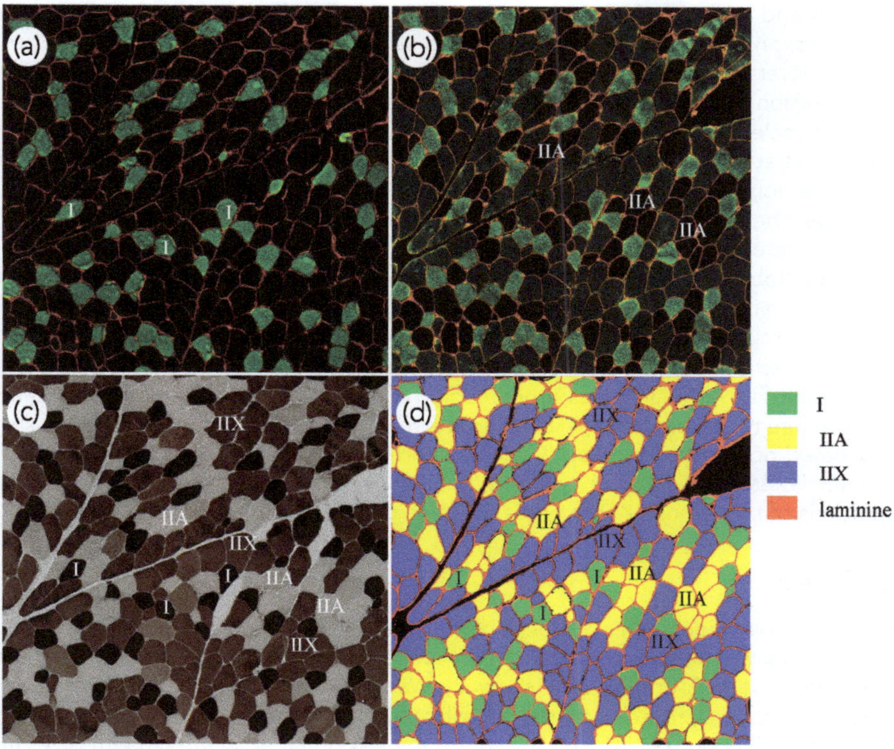


**Figure 1** Fibre type identification by histology (bovine muscle). (a) and (b) Myosin heavy chains isoforms I and I + IIx identified by immunohistofluorescence with BAD5 (a) and S58H2. (b) monoclonal antibodies, respectively. Type IIA fibres are identified by difference. (c) ATPase staining after acidic pre-incubation of the section. Type I, IIA and IIX fibres are stained in black, white and grey, respectively. (d) Virtual image highlighting the fibre type distribution on the section.

microspectroscopies and mass spectrometry imaging. Development work is underway to try to identify the types of fibres on label-free histological sections (Chagnot et al., 2015).

# 4 Variation in muscle fibre type proportions: intrinsic factors

## 4.1 Species

Whatever the species, muscle is composed of muscle fibres of different types. The average fibre diameter ranges from 20 to 85 microns depending on species; in cattle, the average fibre diameter is between 55 and 67 μm (Rehfeldt et al., 1984).

The proportion of each fibre type can vary substantially depending on species (Pette and Staron, 1990) and it can be reflected by muscle and meat colour. White meat species (chicken, turkey, to a lesser extent, pork) are composed of muscles that are poor in myoglobin and thus have low type I and IIA fibre content. These species generally have high proportions of fast-glycolytic type IIX and IIB fibres and are typically not resistant to fatigue. Conversely, species whose musculature is red (cattle, sheep, horse, duck) have a high proportion of type I and IIA oxidative fibres that are rich in myoglobin and, therefore, stain the muscle in red. These species are characteristically resistant to fatigue; cattle and horses were traditionally used as draft animals as their muscle characteristics allowed them to work over long periods.

It is widely held that the number of muscle fibres in various species, including cattle, is definitively determined at birth (Ashmore and Doerr, 1971; Ashmore et al., 1972; Rehfeldt et al., 1984; Robelin et al., 1991; Wegner et al., 2000). In cattle, it is assumed that muscles are overwhelmingly composed of type I, IIA and IIX fibres (Francisco et al., 2011; Tanabe et al., 1998; Young, 1984; Young and Foote, 1984), although a study found evidence of glycolytic fibres present in some individuals from the Blonde d'Aquitaine breed (Picard and Cassar-Malek, 2009). To facilitate further reading of this chapter, the different classifications of muscle fibre types found in the scientific literature will be mostly converted into a single consolidated nomenclature as per Schiaffino and Reggiani (2011), that is, type I, IIA and IIX (as IIB fibres are a rarity in cattle). Within a given species (cattle in this book), the proportions of these muscle fibre types vary with certain biological factors.

## 4.2 Breed

Fibre type proportions can vary depending on breed, but this is not systematic. In general, for a given muscle, domestic cattle breeds that are phenotypically close will share similar fibre type proportions. In a study by Waritthitham et al. (2010), Braham bovine *longissimus dorsi* (LD) muscles shared a similar muscle fibre composition to Charolais LD. No significant difference in fibre type composition was detected in the same muscles from Aubrac, Charolais, Limousin and Salers breeds (Jurie et al., 2005) and between muscles from Holstein and Salers breeds (Jurie et al., 2007). An earlier study found no difference in terms of fibre type frequency between Angus, Holstein and Galloway breeds (Wegner et al., 2000). Likewise, Johnston et al. (1981) found no large differences in muscle fibre type proportions in LD and *semimembranosus* (SM) from Angus, Angus crossbreeds and the Simmental breed, whereas *semitendinosus* (ST) and *biceps femoris* (BF) muscles showed several differences in I and IIA fibre type proportions across breeds. Clancy et al. (1986) observed that Holstein-breed LD muscles showed a greater amount of type IIA fibres and a lesser amount of type IIX fibres than Charolais crosses.

Meta-analysis by Chriki et al. (2012) and Schreurs et al. (2008) on a large population of cattle, including six different breeds (Aubrac, Charolais, Limousin, Montbéliard, Salers and Holstein), also highlighted several differences in muscle fibre type percentage. These studies showed that, as a rule, beef breeds that have been selected for muscle mass (Aubrac, Charolais, Limousin, Salers) had higher glycolytic metabolism associated with a higher proportion of fast glycolytic fibres compared to dairy breeds (Holstein and Montbéliard). Belgian Blue, a breed selected for meat production, also has a high proportion of glycolytic fibres compared to Holstein (Stavaux et al., 1994; Wegner et al., 2000) and to Angus and Galloway breeds (Wegner et al., 2000). However, these metabolic differences do not affect all muscles to the same extent, as proportions of type IIA and IIX fibres were not significantly different between ST and *longissimus thoracis* (LT) muscles in dairy breeds, unlike in beef breeds (Chriki et al., 2012). Animals with larger differences in genotypes show significant differences in terms of muscle fibre type composition. The

Afrikaner breed from South Africa contains a higher proportion of red (types I and IIA) fibres and a lower proportion of white (type IIX) fibres than Holsteins (Dreyer et al., 1977). Interestingly, the LD muscle of water buffalo contains no white fibre, unlike the LD muscle of domestic Angus and Charolais breeds (Solomon et al., 1985).

Muscle fibre size can also vary with cattle breed. Spindler et al. (1980) reported that Holstein steers had smaller fibres than Hereford or Angus steers and Johnston et al. (1975) reported that Charolais steers had larger fibre diameters and areas than Angus steers. It is worth noting that Afrikaner bulls have a smaller mean fibre diameter than Holstein bulls (Dreyer et al., 1977).

Between-breed differences in muscle characteristics likely result from metabolic and physiological characteristics that have evolved in response to production purposes.

## 4.3 Sex

Testosterone is likely to contribute to the gender differences in specific fibre type percentages and sizes (Pette and Staron, 2000; Seideman and Crouse, 1986). Cattle can be considered as three different sexes – entire males, females (heifers and cull cows) and steers (castrated males).

Sex is unlikely to have an effect on muscle fibre type numbers in meat-producing animals (Dwyer et al., 1994; Seideman and Crouse, 1986; Staun, 1963), but it can strongly influence muscle fibre type composition (Lefaucheur and Gerrard, 2000) (Fig. 2).

Differences in fibre types are small between females and entire males (Chriki et al., 2012; Schreurs et al., 2008), but strong with castrated males. Steers tend to have a higher percentage of type IIA and a lower percentage of type IIX fibres than heifers (Chriki et al., 2012; Johnston et al., 1981).

Several studies have reported that muscle from bulls has a higher percentage of red oxidative muscle fibres (type I and IIA) and a lower percentage of white glycolytic muscle fibres (type IIX) than muscle from steers (Brandstetter et al., 1998b; Dreyer et al., 1977; Picard et al., 1995; Young and Bass, 1984).

Meta-analyses (Chriki et al., 2012; Schreurs et al., 2008) spanning a large number of data sets from several experiments report a lower proportion of type IIX fibres and a higher proportion of type I and IIA fibres in castrated males than females and entire males. The fact that this result contradicts previous studies may be explained by high variability in the samples, which came from animals of different breeds with varying ages and from different muscles. In the meta-analysis by Chriki et al. (2012), the average age for steers and bulls was 21 and 16 months, respectively, and it is known that oxidative activity increases while glycolytic activity decreases with increasing age after puberty (Jurie et al., 1999). The larger proportion of oxidative fibres in steers found by Chriki et al. (2012) could, therefore, be partially explained by age differences between steers and bulls.

There are also sex effects on fibre size – there are no significant differences between males and females, but in cattle, castration reduces fibre size as measured by cross-sectional area or diameter (Clancy et al., 1986; Dreyer et al., 1977; Picard et al., 1995; Young and Bass, 1984).

## 4.4 Muscle location and function

By far the greatest variability in fibre type composition is in differences between individual muscles related to body location and physiological function. The respiratory muscles that

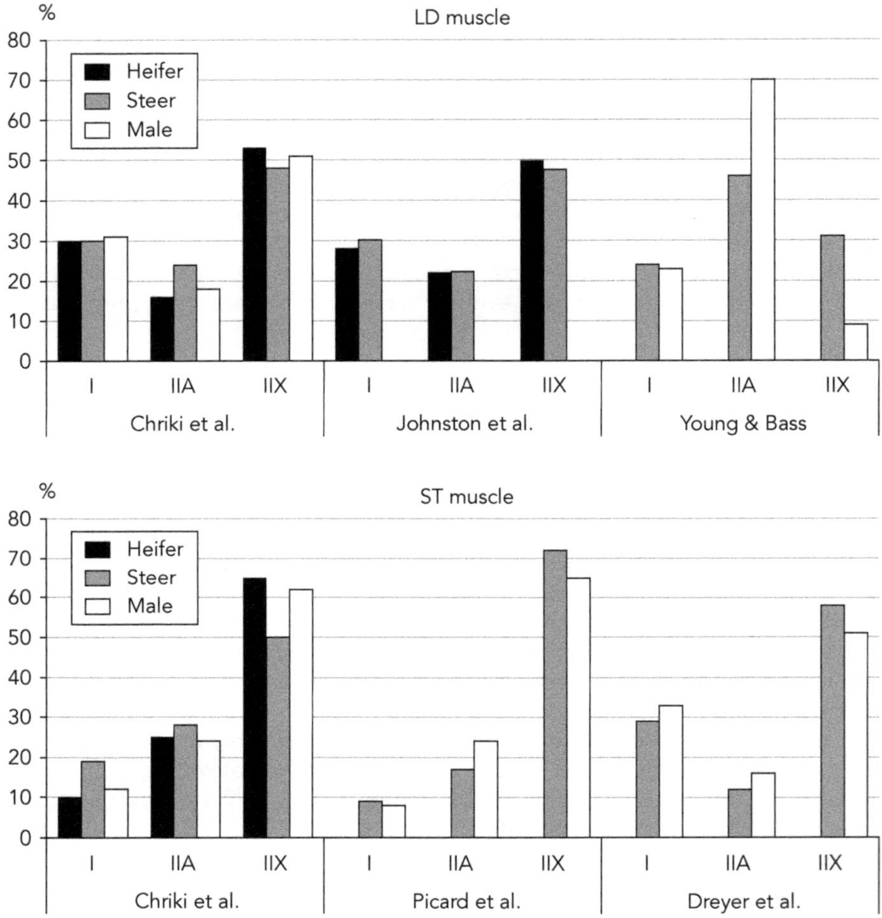


**Figure 2** Bovine muscle fibre characteristics according to gender. Source: From Chriki et al. (2012); Johnston et al. (1981); Young and Bass (1984); Picard et al. (1995) and Dreyer et al. (1977).

LD: Longissimus dorsi; ST: m. semitendinosus

are continuously working are mainly composed of red oxidative fibres. Muscles involved in maintenance of the upright standing position are rich in type I fibres, while their antagonist muscles usually have glycolytic fibres as the major component (Totland and Kryvi, 1991).

The many comparative studies of bovine muscles published (Hunt and Hedrick, 1977; Hwang et al., 2010; Kirchofer et al., 2002; Talmant et al., 1986; Totland and Kryvi, 1991) show significant compositional differences in fibre type and in fibre size according to muscles studied (see Fig. 3 and 4). Compositional variations in fibre type can also appear within the same muscle, as shown by Hunt and Hedrick (1977) on the intramuscular heterogeneity of bovine SM and ST muscles, where the inner part of the SM was richer in glycolytic fibres than the outer part and the inner part of the ST was richer in oxidative fibres than the outer part.

# 5 Variation in muscle fibre type proportions: additional factors

## 5.1 Feeding

There has been intensive research into the effect of diet on fibre type. At similar energy intakes, type of feed has no effect on proportion of different types of fibres (Listrat et al., 1999), but effects start to emerge at different energy intakes. Indeed, most relevant studies report that percentage of oxidative fibres decreased and percentage of glycolytic fibres increased with increasing energy level in the diet (Brandstetter et al., 1998b; Johnston et al., 1981; Jurie et al., 2006; Seideman and Crouse, 1986; Shibata et al., 2014; Suzuki et al., 1976). Moody et al. (1980) postulated that the available source of energy in diets appeared to cause a physiological shift from intermediate muscle fibres to white muscle fibres. However, Maltin et al. (2001) showed that feed regimen had little effect on muscle fibre proportion, but led to genotype-dependent variations in muscle fibre size. Most studies suggest that feeding restriction may lead to a higher frequency of type I or IIA

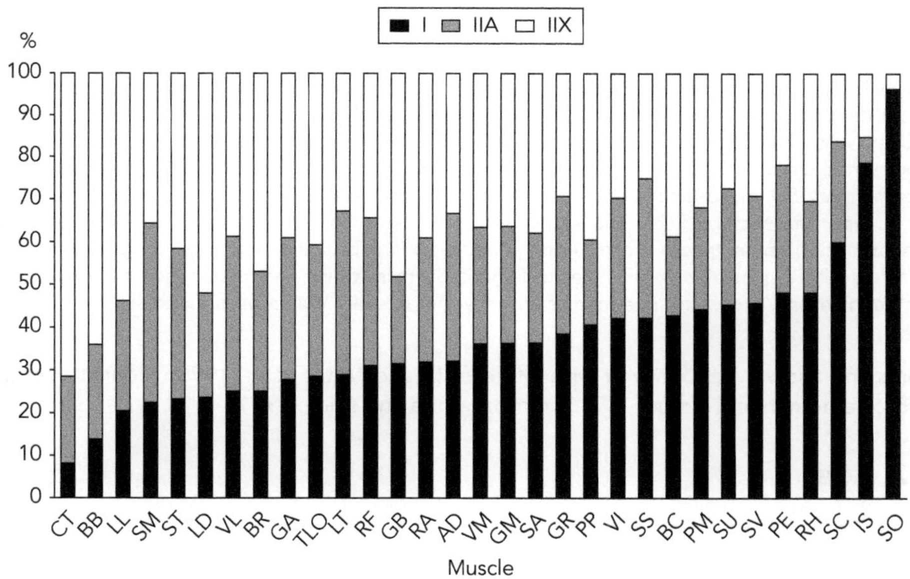


**Figure 3** Bovine muscle fibre proportion according to muscle.

Source: From Totland and Kryvi, 1991.

CT: m. cutaneus trunci; LT: m. longissimus thoracicus; SC: m. semispinalis capitis; BC: m. brachiocephalicus; SV: m. serratus ventralis; RH: m. rhomboideus; LD: m. latissimus dorsi; PP: m. pectoralis profundus; SS: m. supraspinatus; IS: m. infraspinatus; SU: m. subscapularis; TLO: m. triceps brachii caput longum; BB: m. biceps brachii; BR: m. brachialis; LL: m. longissimus lumborum; RA: m. rectus abdominis; PM: m. psoas major; GM: m. gluteus medius; PE: m. pectineus; AD: m. adductor; GB: m. gluteobiceps; SM: m. semimembranosus; ST: m. semitendinosus; SA: m. sartorius; RF: m rectus femoris; GR: m. gracilis; VL: vastus lateralis; VM: vastus mediatis; VI: vastus intermedius; GA: m. gastrocnemius; SO: m. soleus.

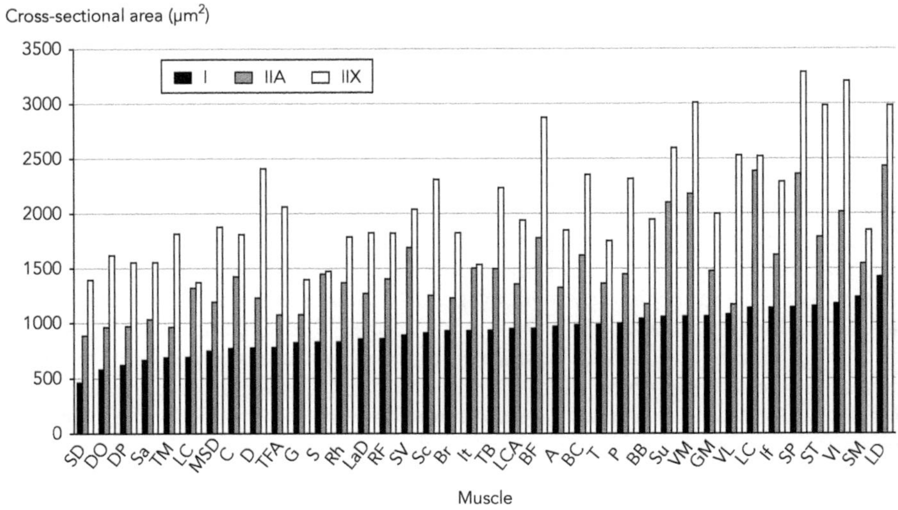


**Figure 4** Bovine muscle fibre cross-sectional area according to muscle.

Source: From Kirchofer et al. (2002).

BB: m. biceps brachii; Br: m. brachialis; BC: m. brachiocephalicus/omotransversarius; C: m. complexus; DP: m. deep pectoral; D: m. deltoidus; DO: m. dorsalis oblique; If: m. infraspinatus; It: m. intertransversarii; LaD: m. latissimus dorsi; LC: m. levatores costarum; LCA: m. longissimus capitis et atlantis; LC: m. longissimus costarum; LD: m. longissimus dorsi; MSD: m. multifidos and spinalus dorsi; Rh: m. rhomboideus; SD: m. scalenus dorsalis; SV: m. serratus ventralis; S: m. splenius; Sc: m. subscapularis; SP: m. superficial pectoral; Su: m. supraspinatus; TFA: m. tensor fasciae antebrachii; TM: m. teres major; T: m. trapezius; TB: m. triceps brachii; A: m. adductor; BF: m. biceps femoris; GM: m. gluteus medius; G: m. gracilis; P: m. pectineus; RF: m. rectus femoris; Sa: m. sartorius; SM: m. semimembranosus; ST: m. semitendinosus; VI: m. vastus intermedius; VL: m. vastus lateralis; VM: m. vastus medialis.

fibres and a lower frequency of type IIX fibres (Brandstetter et al., 1998b; Johnston et al., 1981; Seideman and Crouse, 1986; Yambayamba and Price, 1991).

## 5.2 Age

For a given muscle, muscle fibre composition is known to evolve with age (Fig. 5). The amount of muscle fibres is fixed at birth for most vertebrate species (Pearson, 1990), including bovines (Robelin et al., 1991; Wegner et al., 2000). While the number of muscle fibres varies little during growth, fibre size increases dramatically whatever the fibre type (Fig. 5). For example, the cross-sectional area of bovine ST muscle fibres grows tenfold between birth and 20 months old (Wegner et al., 2000).

Generally speaking, changes in the distribution of cattle muscle fibre types between 1 and 16 months of age mainly follow a pattern of conversion from oxidative IIA fibres into glycolytic fibres (Ashmore et al., 1972; Dreyer et al., 1977; Girard et al., 2011; Jurie et al., 1995; Klosowski et al., 1992; Seideman and Crouse, 1986; Solomon et al., 1985; Spindler et al., 1980). Some experiments have not detected any significant change in type I fibre rates during animal growth (Jurie et al., 1995; Micol et al., 1993), whereas other studies

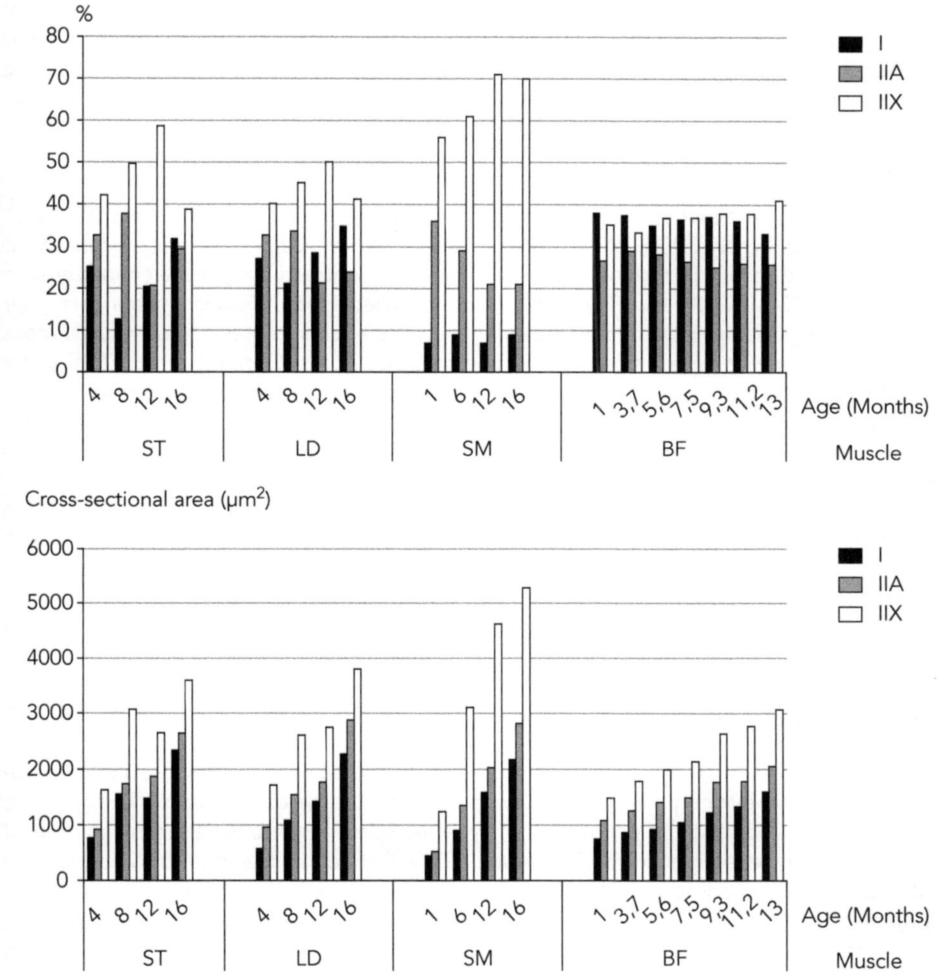


**Figure 5** Bovine muscle fibre characteristics according to age.

Source: From Brandstetter et al. (1998); Jurie et al. (1995) and Spindler et al. (1980).

ST: m. semitendinosus; LD: m. longissimus dorsi; SM: m. semimembranosus; BF: m. biceps femoris.

using several muscles of the carcass have shown animal age-related variations in type I fibre proportions (Brandstetter et al., 1998a), where the percentage of slow Myosin Heavy Chain (MHC1) decreased from 26% to 19% between 4 and 8 months and then steadily increased until 16 months to reach 34%. These changes were muscle dependent. Despite the fact that ST and LD muscles showed similar patterns of developmental change, ST muscle contained less type I fibres and more type IIX fibres than LD at 8 and 12 months of age (Brandstetter et al., 1998). Above 19 months of age, muscles become slower and more oxidative (Jurie et al., 2005). In older animals (from 4 to 9 years of age), fibre proportion and metabolic enzyme activities remain stable whatever the muscle considered (Jurie et al.,

2006). As the number of muscle fibres is fixed at the end of foetal life in cattle, postnatal muscle growth is due to an increase in fibre diameter from 30 μm at birth to about 60 μm at adulthood. The increase in fibre size happens irrespective of the contractile and metabolic features of the muscle fibres (Brandstetter et al., 1998a; Jurie et al., 1999; Jurie et al., 1995).

## 5.3 Exercise

Level of contractile activity likely influences myofibre composition of muscles (Pearson, 1990; Pette and Staron, 2000, 2001; Schiaffino and Reggiani, 2011). There is evidence that exercise training results in a fast-to-slower transition in contractile myofibre characteristics in the order IIB→IIX→IIA→I. Therefore, animals raised in conditions enabling regular physical activity, that is, outdoors or in a sufficiently big enclosure, tend to develop a more oxidative muscle with type II fibres converting to type I fibres (Gondret et al., 2005; Jurie et al., 1998; Vestergaard et al., 2000). In a study comparing tying-type housing and loose-housing young bulls, Jurie et al. (1998) observed that loose-housed animals had a higher percentage of type IIA and IIA-IIX hybrid fibres and a lower percentage of type IIX fibres in the ST muscle, while LD muscle showed no significant change. The authors concluded that the type of housing influenced contractile characteristics, especially in muscles involved in movement. Another experiment (Shibata et al., 2014) also evidenced a muscle fibre type conversion to slow-twitch (type I) from fast-twitch (type II) in bovine outdoor grazing steers compared to steers reared in stall barns, but only in LD and not in ST. These conflicting results probably stem from differences in breed, sex and age of the animals used in each experiment.

## 5.4 Growth-promoting agents

Growth hormone (GH), β-agonists and steroids that have been or are still used to increase muscle content in farm animals can influence muscle fibre types and sizes. Giving GH to steers tends to decrease the proportion of type IIA fibres and increase the proportion of type IIX fibres in several muscles (Ono et al., 1996). GH administration led to an increase in the size of all or some muscle fibre types in some muscles, including psoas major, supraspinatus and ST (Ono et al., 1996). Ractopamine hydrochloride and cimaterol β-agonists cause a transition of type I fibres to type II fast-twitch muscle fibres in cattle (Gonzalez et al., 2007, 2008, 2009; Vestergaard et al., 1994) and increases type I fibre cross-sectional area (Gonzalez et al., 2007; Vestergaard et al., 1994) (Fig. 6).

Clenbuterol and zilpaterol significantly increased the proportion of type IIX fibres in the LD steers and moreover their cross-sectional area was increased by 25%, 20% and 17% following a treatment with clenbuterol, zilpaterol and ractopamine, respectively (Strydom et al., 2009).

In a recent meta-analysis based on about 60 different studies (Lean et al., 2014), both zilpaterol and ractopamine highlighted a large weight gain and longissimus muscle area compared to controls, and these effects were particularly large for zilpaterol.

β-Agonists could increase meat toughness (for review see, Dikeman, 2007 ; Parr et al. 2016) . Zilpaterol causes meat toughness higher than ractopamine (Lean et al., 2014; Strydom et al., 2009). It is worth noting that these promoting agents are illegal in the European Union and that clenbuterol and cimaterol are illegal in numerous countries.

Steroid implants can produce an increase in type IIA fibres and a decrease in type IIX fibres in steers and bulls (Clancy et al., 1986). Steroids increase the size of both type I and

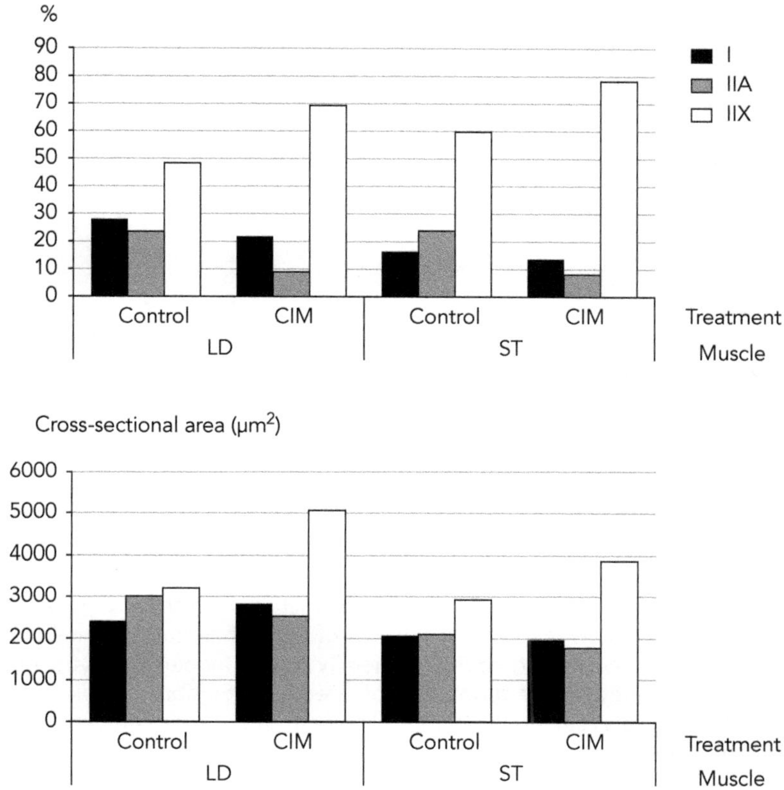


**Figure 6** Effects of cimaterol growth-promoting agent on bovine muscle fibre characteristics.

Source: From Vestergaard et al. (1994). CIM: cimaterol treatment; ST: m. semitendinosus; LD: m. longissimus dorsi.

type II fibres (Clancy et al., 1986; Hikim et al., 2002). β-Agonists and/or steroids increase the cross-sectional area of type I fibres in beef (Gonzalez et al., 2007; Vestergaard et al., 1994) and fast-twitch type II muscle fibres in young bulls and heifers (Vestergaard et al., 1994; Wu et al., 1986). Growth-promoting agents thus result in an increase in muscle volume due to an increase in muscle fibre size, sometimes following a muscle fibre type transition.

## 5.5 Individuals

Significant inter-individual variations have been detected in the fibre type composition of human skeletal muscles (Simoneau and Bouchard, 1989). Although the human population is very heterogeneous in terms of age, sex, physical activity and other factors, analyses in twins indicate that almost 50% of this variance is associated with genetic factors (Simoneau and Bouchard, 1995). There is scant literature data on inter-individual variability in fibre type composition in farm animals. However, a comparison on biochemical traits in 147 male Limousin cattle highlighted a coefficient of variation of 30% in the MHC1 content in both ST and LT muscles (Jurie et al., 1995). There is evidence that this coefficient of variation would be much higher after factoring in the whole heterogeneity of the bovine population, including cattle from different breeds, age and sex.

# 6 Impact of muscle fibre type on carcass and meat quality

Several reviews have been published establishing the links between fibre type proportions and meat quality (see Astruc, 2014; Choi and Kim, 2009; Klont et al., 1998; Lee et al., 2010; Lefaucheur, 2010; Lefaucheur and Gerrard, 2000 among others).

## 6.1 Muscle fibre types and carcass quality

Carcass quality is essentially related to its conformation reflecting its meat yield. In general, meat breeds have the best meat yield, with a slightly higher number of fibres than dairy breeds, and their better conformation is mainly explained by their larger-sized muscle fibres (Rehfeldt et al., 1984; Schreurs et al., 2008; Wegner et al., 2000). Double-muscled animals are an exception to the rule, as they are characterized both by large muscle fibres and by an approximately twofold higher number of fibres than conventional cattle. Beef breeds including double-muscled animals generally have a higher glycolytic metabolism associated with a slightly higher proportion of fast-glycolytic fibres (Chriki et al., 2012; Schreurs et al., 2008; Stavaux et al., 1994) in line with their lower content in myoglobin (Clinquart et al., 1998).

## 6.2 Muscle fibre types and meat colour

Colour has a critical influence on beef purchase decisions by consumers and is thus a critical factor for the beef industry (Cornforth, 1994). Consumers consider the red colour of beef as a guarantee of quality. Beef redness mainly depends on its myoglobin content that pigments the muscle red, but myoglobin is mainly found in oxidative red fibres (type I and IIA), so the red colour of meat is usually related to the proportion of oxidative fibres in the muscle (Hunt and Hedrick, 1977). As steady physical activity is likely to increase the proportion of oxidative fibre types, grazing animals that have access to large field areas usually show redder meat (Dunne et al., 2011; Vestergaard et al., 2000).

To a lesser extent, residual haemoglobin can also contribute to meat colour (Hunt and Hedrick, 1977). It is usually located in the red muscles that are heavily irrigated to carry oxygen to the oxidative fibres and, therefore, tends to strengthen muscle pigmentation.

In terms of colour intensity, meat from oxidative muscle has the advantage. However, myoglobin is very sensitive to oxidation. In the absence of oxygen, inside the muscle or when it is kept under vacuum, myoglobin gets reduced and exhibits a purple colour. In the presence of oxygen, it is transformed into bright red oxymyoglobin, which consumers appreciate, but after prolonged exposure to the air it oxidizes into dark brown metmyoglobin, which consumers reject (Renerre, 1984; Renerre and Labas, 1987; Suman and Poulson, 2013). Meat that is less rich in oxidative fibres presents a lower risk of oxidation during storage and thus better colour stability (Lefaucheur and Gerrard, 2000).

The market supply chain needs to take measures to preserve the bright red colour of meat by minimizing exposure to air and using suitable packaging.

Although to a lesser extent, meat colour can also be deteriorated by a low amplitude of pH drop in the muscle after slaughter (Macdougall, 1982). The emotional stress, fighting and unusual physical activity experienced by animals before slaughter causes an overconsumption of muscle glycogen (Mcveigh et al., 1982), and muscle glycogen content at exsanguination determines the ultimate pH of meat. If glycogen content at exsanguination is low, the pH of the meat will stabilize at values higher than 6.2 (against 5.6–5.8 on average for an animal slaughtered under good conditions) and the meat colour will be dark and rejected by consumers. 'Dark firm dry' (DFD) meat is mainly found in cattle whose muscles are more highly oxidative (and therefore rich in type I and IIA fibres). Because pre-slaughter stress can dramatically increase incidence of DFD, good-quality transport and housing of animals at slaughter can limit the production of the DFD meat that is considered bad quality.

## 6.3 Muscle fibre types and meat tenderness

Meat tenderness, especially beef intended for short cooking (grilled, roasted etc.), is a very important quality criterion for consumers. The first factor involved in meat tenderness is the intramuscular collagen content and cross-linking, but muscle fibre size is also thought to impact meat tenderness, even if to a lesser extent (Seideman and Crouse, 1986). Indeed, tenderness was negatively related to muscle fibre area in several studies (Crouse et al., 1991; Renand et al., 2001; Seideman et al., 1986, 1987). In another study, however, Seideman and Crouse (1986) found a positive relation between fibre size and tenderness and other scientists did not find any relationship between fibre size and meat tenderness (Geesink et al., 1995; Lewis et al., 1977; Ozawa et al., 2000). Overall, it seems that muscles composed of large fibres, especially type IIX, give tougher meat than muscles composed of small fibres. These results suggest that selecting for more muscular, leaner livestock can result in large and often type IIX muscle fibres, which may be associated with less tender meat (Gentry et al., 2002; Ryu and Kim, 2006). Conversely, Choi and Kim (2009) suggested in their review that overall, muscles with higher numbers of small- or medium-sized fibres tend to give meat with better tenderness without significant differences in muscle mass. Post-mortem proteolysis leads to a natural tenderizing of meat, and glycolytic fibres are more susceptible to proteolysis than oxidative fibres (Ouali, 1990). Indeed, the calpain/calpastatin ratio is higher in fast-twitch glycolytic (mostly type IIX and IIB fibres) than slow-twitch oxidative muscles (mostly type I fibres) (Ouali and Talmant, 1990), which could partly explain the faster rate of maturation in glycolytic muscles. Cattle muscles that are rich in oxidative (type I and IIA) fibres need to be stored in the cold room for several days to optimize their tenderness, but note that this only applies to collagen-poor muscles intended for steak or grilled meat.

Unlike glycolytic muscles, oxidative red muscles contract as they cool fast after slaughter. This phenomenon, called 'cold shortening' (Locker and Hagyard, 1963), could be linked to progressive low temperature-driven inactivation of the calcium ATPase pump and an increase in sarcoplasmic reticulum membrane phospholipid porosity that together are likely to increase the calcium content in the sarcoplasm and thus activate fibre contraction (Bendall, 1973; Davey and Gilbert, 1974). The build-up of calcium in myofibrils is less inclined to occur in fast-acting muscles rich in calcium-absorbing reticulum than in slow-acting muscles where the reticulum is not so abundant (Davey and Gilbert, 1974).

## 6.4 Muscle fibre types and meat flavour

Meat flavour is largely related to its content in lipids that release aroma as they cook (Mottram and Edwards, 1983). Intramuscular lipids are composed of triacylglycerols (TAG), which are stored in adipocytes localized in the perimysium and, to a lesser extent, as lipid droplets in the muscle fibre cytoplasm. TAG content ranges from 0.2% to 5% (Gandemer, 1999). In most farm animal species, oxidative muscles have a higher TAG content than glycolytic muscles (Gandemer, 1999). Phospholipids, which are components of biological membranes, account for 0.5 to 1% of total intramuscular fat (Gandemer, 1999). Phospholipid content is higher in oxidative (type I and IIA) than glycolytic (type IIX) fibres, partly because oxidative fibres have a higher proportion of mitochondria. As a rule, intramuscular fat content is positively correlated with oxidative fibre content (Gandemer, 1999), especially type I fibres in cattle (Calkins et al., 1981). Meats rich in type I fibres are, therefore, assumed to have a more developed flavour than white meat. In addition, fibre type composition-related differences have an influence on instrumental beef taste traits (Chikuni et al., 2010). In the experiment performed by Chikuni et al. (2010), the slow-type muscles (masseter and diaphragm) showed signals lower for sour and higher for bitter and astringent tastes than did the fast-type muscles (psoas major, longissimus thoracis and semitendinosus). Lactate and iron, whose content depends on fibre type, are also believed to influence meat taste. Therefore, muscle fibre type probably affects the beef taste characteristics.

## 6.5 Muscle fibre types and nutritional quality

Meat is rich in high nutritional quality proteins and contains all the essential amino acids that humans need from diet. It is also rich in vitamins and micronutrients. Beef with high proportions of oxidative fibres (type I and IIA fibres) has more iron and vitamin B12 than meats with high proportions of glycolytic fibres (type IIX and IIB fibres). Iron is an essential micronutrient, and heme iron, which is found mainly in type I and IIA fibres, is one of the nutrients that is easiest for humans to assimilate. Indeed, heme iron, which represents 50–80% of meat-source iron depending on species, is much better absorbed than the non-heme iron found in vegetables and cereals (Lopez and Martos, 2004). Moreover, meat doubles or even triples the absorption of non-heme iron from other foods that accompany it during the meal (Lopez and Martos, 2004). Vitamin B3 and carnosine are found in higher concentrations in muscle that has a higher proportion of glycolytic (type IIX and glycolytic) fibres. Carnosine is a bioactive dipeptide of β-alanine and histidine that probably helps buffer the intracellular pH variations induced by metabolic activity in glycolytic muscle. Carnosine is an antioxidant that protects against protein glycation and cross-linking and is believed to prevent Alzheimer's disease.

## 6.6 Muscle fibre types and water holding capacity

The water holding capacity of meat is an important factor affecting both the juiciness and technological yields of meat products during cold storage and cooking. Pre-slaughter stress is often the cause of 'pale soft exudative' (PSE) meats that have excessive drip loss and high cooking losses. Metabolism accelerates post-mortem, causing a rapid drop in pH. The combination of low pH and high carcass temperature causes protein denaturation and lateral contraction of myofibrils (Klont et al., 1998). The water ends up being expelled to the outside of the muscle. This phenomenon has mainly been observed in pork meat but is also observed in other species, including cattle (Guignot et al., 1993; Kim et al., 2014). Nevertheless, it is white meat rich in glycolytic fibres that are most susceptible to the PSE defect. Glycolytic fibres (type IIX and IIB) have the ability to quickly degrade their endogenous glycogen and produce a rapid pH drop. They generally have a high cross-sectional area, and high glycolytic fibre content muscles tend to be very thick, all of which reduces the cooling rate of the heart of the muscle. Thus, muscles with high glycolytic fibres (IIX and IIB) are more likely to produce PSE meat.

DFD meats, which are considered low quality due to their dark colour and higher susceptibility to bacterial growth, paradoxically have a high water holding capacity. Their high pH favours repulsion between the myofilaments and, therefore, water retention in the myofibrils. DFD meat is often from young cattle that have suffered stress and physical activity in the hours preceding slaughter. The DFD defect essentially affects oxidative muscle (with a high content of type I and IIA fibres).

# 7 Conclusion

The world's cattle population is a very heterogeneous collection of individuals. Genetic selection to better adapt animals to their environment and improve their milk and/or meat output has led to the emergence of many genetic types with different characteristics. The evolution and diversity of rearing methods and feed consumption habits are combined to further reinforce the variability of livestock at world scale. Muscle composition governs meat quality. In particular, the composition in oxidative and glycolytic fibres enhances certain qualities of meat but can deteriorate others. For example, an increase in the proportion of type I and IIA fibres improves the intensity of red beef colour but decreases its storability. An increase in the proportion of type IIX fibres improves meat yield but increases the risk of producing paler and possibly exudative meat. It is possible to change the proportion of different fibre types by genetic selection, farm system conditions and slaughter age. Means of action for variations in fibre type proportions, remain limited mainly due to biological reasons and also to preserve the breed characteristics that have been developed to meet specific objectives such as meat or milk productivity and rusticity to ensure good adaptation to sometimes harsh (feed resource, climate, topography, diseases) environments.

Routes to market and production strategies will be guided by economic market forces and opportunities.

# 8 Where to look for further information

Two of the greatest experts of the composition and physiology of skeletal muscle are Pr Stefano Schiaffino and Pr Carlo Reggiani from the University of Padova, Italy. They have published numerous articles on the types of muscle fibres in mammals and drafted a comprehensive review of the subject, such as the one published in Physiological Reviews in 2011, which is freely available at the following address: http://physrev.physiology.org/content/91/4/1447.long

In addition to the scientific review cited in the text which establish links between fibre types and meat quality (Lee et al., 2010, Lefaucheur, 2010 Klont et al., 1998, Choi and Kim, 2009), readers will find comprehensive information in the field by browsing:

- *Encyclopedia of Meat Science*, 2nd edition. In C. Devine and M. Dikeman (Eds). Oxford: Elsevier.
- *Applied Muscle Biology*. In M. Du and R. J. McCormick (Eds). CRC press, Taylor and Francis group.
- *Muscle Development of Livestock Animals, Physiology, Genetics and Meat Quality*. In M. F. W. Te Pas, M. E. Everts and H. P. Haagsman (Eds). Wallingford, UK, CABI Publishing.

The different chapters of these books were written by the domain expert whose details are available in the in-chapter headings.

# 9 References

Ashmore, C. R. and Doerr, L. (1971). Comparative aspects of muscle fiber types in different species. *Experimental Neurology, 31*, 408–18.

Ashmore, C. R., Tompkins, G. and Doerr, L. (1972). Postnatal development of muscle fiber types in domestic animals. *Journal of Animal Science, 34*(1), 37–41.

Astruc, T. (2014). Muscle fiber types and meat quality. In C. Devine and M. Dikeman (Eds), *Encyclopedia of Meat Science*, 2nd ed., Vol. 2, pp. 442–8. Oxford: Elsevier.

Bendall, J. R. (1973). Post mortem changes in muscle. In G. H. Bourne (Ed.), *Structure and Function of Muscle*, pp. 243–309. New York: Academic Press.

Brandstetter, A. M., Picard, B. and Geay, Y. (1998a). Muscle fibre characteristics in four muscles of growing bulls – I. Postnatal differentiation. *Livestock Production Science, 53*(1), 15–23.

Brandstetter, A. M., Picard, B. and Geay, Y. (1998b). Muscle fibre characteristics in four muscles of growing male cattle – II. Effect of castration and feeding level. *Livestock Production Science, 53*(1), 25–36.

Brooke, M. H. and Kaiser, K. K. (1970). Three 'myosin adenosine triphosphatase' systems: the nature of their pH lability and sulfhydryl dependence. *Journal of Histochemistry and Cytochemistry, 18*(9), 670–2.

Calkins, C. R., Dutson, T. R., Smith, G. C., Carpenter, Z. L. and Davis, G. W. (1981). Relationship of fiber type composition to marbling and tenderness of bovine muscle. *Journal of Food Science, 46*(3), 708–10.

Chagnot, C., Venien, A., Peyrin, F., Jamme, F., Refregiers, M., Desvaux, M. and Astruc, T. (2015). Deep UV excited muscle cell autofluorescence varies with the fibre type. *Analyst, 140*(12), 4189–96.

Chikuni, K., Oe, M., Sasaki, K., Shibata, M., Nakajima, I., Ojima, K. and Muroya, S. (2010). Effects of muscle type on beef taste-traits assessed by an electric sensing system. *Animal Science Journal, 81*(5), 600–5.

Choi, Y. M. and Kim, B. C. (2009). Muscle fiber characteristics, myofibrillar protein isoforms, and meat quality. *Livestock Science, 122*(2–3), 105–18.

Chriki, S., Picard, B., Jurie, C., Reichstadt, M., Micol, D., Brun, J. P., Journaux, L. and Hocquette, J. F. (2012). Meta-analysis of the comparison of the metabolic and contractile characteristics of two bovine muscles: Longissimus thoracis and semitendinosus. *Meat Science, 91*(4), 423–9.

Clancy, M. J., Lester, J. M. and Roche, J. F. (1986). The effects of anabolic agents and breed on the fibers of the longissimus muscle of male cattle. *Journal of Animal Science, 63*(1), 83–91.

Clinquart, A., Hornick, J. L., Van Eenaeme, C. and Istasse, L. (1998). The influence of double muscling on production and quality of meat in Belgian Blue cattle. *Productions Animales, 11*(4), 285–97.

Cole, J. M., Ramsey, C. B., Hobbs, C. S. and Temple, R. S. (1964). Effects of type and breed of british Zebu and dairy cattle on production, palatibility and composition. III – Percent wholesale cuts and yield of edible portion as determined by physical and chemical analysis. *Journal of Animal Science, 23*, 71–7.

Cornforth, D. (1994). Colour – its basis and importance. In A. M. Pearson and T. R. Dutson (Eds), *Quality Attributes and their Measurement in Meat, Poultry and Fish Products. Advances in Meat Research Series*, Vol. 9, pp. 34–78. London: Chapman & Hall.

Craig, R. W. and Padron, R. (2003). Molecular structure of the sarcomere. In A. G. Engel and C. Franzini-Armstrong (Eds), *Myology*, pp. 129–66. New York: McGraw-Hill.

Crouse, J. D., Koohmaraie, M. and Seideman, S. D. (1991). The relationship of muscle-fiber size to tenderness of beef. *Meat Science, 30*(4), 295–302.

Damon, R. A., Crown, R. M., Singletary, C. B. and McCraine, S. E. (1960). Carcass characteristics of purebred and crossbred beef steers in the gulf coast region. *Journal of Animal Science, 19*, 820–44.

Davey, C. L. and Gilbert, K. V. (1974). The mechanism of cold-induced shortening in beef muscle. *Journal of Food Technology, 9*, 51–8.

Dikeman, M. E., (2007). Effects of metabolic modifiers on carcass traits and meat quality. *Meat Science, 7*, 121–35

Dreyer, J. H., Naude, R. T., Henning, J. W. N. and Rossouw, E. (1977). Influence of breed, castration and age on muscle-fiber type and diameter in Friesland and Afrikaner cattle. *South African Journal of Animal Science-Suid-Afrikaanse Tydskrif Vir Veekunde, 7*(3), 171–80.

Dunne, P. G., Monahan, F. J. and Moloney, A. P. (2011). Current perspectives on the darker beef often reported from extensively-managed cattle: Does physical activity play a significant role? *Livestock Science, 142*(1–3), 1–22.

Dwyer, C. M., Stickland, N. C. and Fletcher, J. M. (1994). The influence of maternal nutrition on muscle-fiber number development in the porcine fetus and on subsequent postnatal-growth. *Journal of Animal Science, 72*(4), 911–17.

Fiems, L. O., De Campeneere, S., Van Caelenbergh, W., De Boever, J. L. and Vanacker, J. M. (2003). Carcass and meat quality in double-muscled Belgian Blue bulls and cows. *Meat Science, 63*(3), 345–52.

Fiems, L. O., Van Caelenbergh, W., Vanacker, J. M., De Campeneere, S. and Seynaeve, M. (2005). Prediction of empty body composition of double-muscled beef cows. *Livestock Production Science, 92*(3), 249–59.

Francisco, C. L., Jorge, A. M., Dal-Pai-Silva, M., Carani, F. R., Cabeco, L. C. and Silva, S. R. (2011). Muscle fiber type characterization and myosin heavy chain (MyHC) isoform expression in Mediterranean buffaloes. *Meat Science, 88*(3), 535–41.

Gandemer, G. (1999). Lipids and meat quality: lipolysis, oxidation, Maillard reaction and flavour. *Sciences Des Aliments, 19*(3–4), 439–58.

Geay, Y. and Robelin, J. (1979). Variation of meat production capacity in cattle due to genotype and level of feeding – Genotype-nutrition interaction. *Livestock Production Science, 6*(3), 263–76.

Geesink, G. H., Koolmees, P. A., Vanlaack, H. L. J. M. and Smulders, F. J. M. (1995). Determinants of tenderization in beef longissimus-dorsi and triceps brachii muscles. *Meat Science, 41*(1), 7–17.

Gentry, J. G., McGlone, J. J., Blanton, J. R. and Miller, M. F. (2002). Impact of spontaneous exercise on performance, meat quality, and muscle fiber characteristics of growing/finishing pigs. *Journal of Animal Science, 80*(11), 2833–9.

Girard, I., Aalhus, J. L., Basarab, J. A., Larsen, I. L. and Bruce, H. L. (2011). Modification of muscle inherent properties through age at slaughter, growth promotants and breed crosses. *Canadian Journal of Animal Science, 91*(4), 635–48.

Gondret, F., Combes, S., Lefaucheur, L. and Lebret, B. (2005). Effects of exercise during growth and alternative rearing systems on muscle fibers and collagen properties. *Reproduction Nutrition Development, 45*(1), 69–86.

Gonzalez, J. M., Carter, J. N., Johnson, D. D., Ouellette, S. E. and Johnson, S. E. (2007). Effect of ractopamine-hydrochloride and trenbolone acetate on longissimus muscle fiber area, diameter, and satellite cell numbers in cull beef cows. *Journal of Animal Science, 85*(8), 1893–901.

Gonzalez, J. M., Dijkhuis, R. D., Johnson, D. D., Carter, J. N. and Johnson, S. E. (2008). Differential response of cull cow muscles to the hypertrophic actions of ractopamine-hydrogen chloride. *Journal of Animal Science, 86*(12), 3568–74.

Gonzalez, J. M., Johnson, S. E., Thrift, T. A., Savell, J. D., Ouellette, S. E. and Johnson, D. D. (2009). Effect of ractopamine-hydrochloride on the fiber type distribution and shelf-life of six muscles of steers. *Journal of Animal Science, 87*(5), 1764–71.

Guignot, F., Vignon, X. and Monin, G. (1993). Postmortem evolution of myofilament spacing and extracellular-space in veal muscle. *Meat Science, 33*(3), 333–47.

Hikim, I. S., Artaza, J., Woodhouse, L., Gonzalez-Cadavid, N., Singh, A. B., Lee, M. I., Storer, T. W., Casaburi, R., Shen, R. Q. and Bhasin, S. (2002). Testosterone-induced increase in muscle size in healthy young men is associated with muscle fiber hypertrophy. *American Journal of Physiology-Endocrinology and Metabolism, 283*(1), E154–64.

Hunt, M. C. and Hedrick, H. B. (1977). Profile of fiber types and related properties of 5 bovine muscles. *Journal of Food Science, 42*(2), 513–17.

Hwang, Y. H., Kim, G. D., Jeong, J. Y., Hur, S. J. and Joo, S. T. (2010). The relationship between muscle fiber characteristics and meat quality traits of highly marbled Hanwoo (Korean native cattle) steers. *Meat Science, 86*(2), 456–61.

Johnston, D. M., Moody, W. G., Boling, J. A. and Bradley, N. W. (1981). Influence of breed type, sex, feeding systems, and muscle bundle size on bovine fiber type characteristics. *Journal of Food Science, 46*(6), 1760–5.

Johnston, D. M., Stewart, D. F., Moody, W. G., Boling, J. and Kemp, J. D. (1975). Effect of breed and time on feed on size and distribution of beef muscle-fiber types. *Journal of Animal Science, 40*(4), 613–20.

Jurie, C., Martin, J. F., Listrat, A., Jailler, R., Culioli, J. and Picard, B. (2005). Effects of age and breed of beef bulls on growth parameters, carcass and muscle characteristics. *Animal Science, 80*, 257–63.

Jurie, C., Martin, J. F., Listrat, A., Jailler, R., Culioli, J. and Picard, B. (2006). Carcass and muscle characteristics of beef cull cows between 4 and 9 years of age. *Animal Science, 82*, 415–21.

Jurie, C., Ortigues-Marty, I., Picard, B., Micol, D. and Hocquette, J. F. (2006). The separate effects of the nature of diet and grazing mobility on metabolic potential of muscles from Charolais steers. *Livestock Science, 104*(1–2), 182–92.

Jurie, C., Picard, B. and Geay, Y. (1998). Influences of the method of housing bulls on their body composition and muscle fibre types. *Meat Science, 50*(4), 457–69.

Jurie, C., Picard, B. and Geay, Y. (1999). Changes in the metabolic and contractile characteristics of muscle in male cattle between 10 and 16 months of age. *Histochemical Journal, 31*(2), 117–22.

Jurie, C., Picard, B., Hocquette, J. F., Dransfield, E., Micol, D. and Listrat, A. (2007). Muscle and meat quality characteristics of Holstein and Salers cull cows. *Meat Science, 77*(4), 459–66.

Jurie, C., Robelin, J., Picard, B. and Geay, Y. (1995). Postnatal changes in the biological characteristics of semitendinosus muscle in male limousin cattle. *Meat Science, 41*(2), 125–35.

Jurie, C., Robelin, J., Picard, B., Renand, G. and Geay, Y. (1995). Inter-animal variation in the biological characteristics of muscle-tissue in male limousin cattle. *Meat Science, 39*(3), 415–25.

Kim, Y. H. B., Warner, R. D. and Rosenvold, K. (2014). Influence of high pre-rigor temperature and fast pH fall on muscle proteins and meat quality: a review. *Animal Production Science, 54*(4), 375–95.

Kirchofer, K. S., Calkins, C. R. and Gwartney, B. L. (2002). Fiber-type composition of muscles of the beef chuck and round. *Journal of Animal Science, 80*(11), 2872–8.

Klont, R. E., Brocks, L. and Eikelenboom, G. (1998). Muscle fibre type and meat quality. *Meat Science, 49*, S219–29.

Klosowski, B., Bidwellporebska, K., Klosowska, D. and Piotrowski, J. (1992). Microstructure of skeletal-muscles of growing calves fed silage-based vs hay-based diets. 2. fiber type distribution. *Reproduction Nutrition Development, 32*(3), 257–63.

Laville, E., Astruc, T., Martin, V. and Bousset, J. (1995). Estimation of the muscle to bone ratio of the bovine pelvic limb using a morphometric method. *Meat Science, 41*(3), 345–55.

Lean, I. J., Thompson, J. M. and Dunshea, F. R. (2014). A meta-analysis of zilpaterol and ractopamine effects on feedlot performance, carcass traits and shear strength of meat in cattle. *PLoS ONE, 9*(12).

Lee, S. H., Joo, S. T. and Ryu, Y. C. (2010). Skeletal muscle fiber type and myofibrillar proteins in relation to meat quality. *Meat Science, 86*(1), 166–70.

Lefaucheur, L. (2010). A second look into fibre typing – Relation to meat quality. *Meat Science, 84*(2), 257–70.

Lefaucheur, L. and Gerrard, D. (2000). Muscle fiber plasticity in farm mammals. *Journal of Animal Science, 77*(E-Suppl), 1–19.

Lewis, P. K., Brown, C. J. and Heck, M. C. (1977). Fiber diameter, sarcomere length and tenderness of certain muscles of crossbred beef steers. *Journal of Animal Science, 45*(2), 254–60.

Listrat, A., Rakadjiyski, N., Jurie, C., Picard, B., Touraille, C. and Geay, Y. (1999). Effect of the type of diet on muscle characteristics and meat palatability of growing Salers bulls. *Meat Science, 53*(2), 115–24.

Locker, R. H. and Hagyard, C. J. (1963). A cold shortening effect in beef muscles. *Journal of the Science of Food and Agriculture, 14*, 787–93.

Lopez, M. A. A. and Martos, F. C. (2004). Iron availability: An updated review. *International Journal of Food Sciences and Nutrition, 55*(8), 597–606.

Macdougall, D. B. (1982). Changes in the color and opacity of meat. *Food Chemistry, 9*(1–2), 75–88.

Mcveigh, J. M., Tarrant, P. V. and Harrington, M. G. (1982). Behavioral stress and skeletal-muscle glycogen-metabolism in young bulls. *Journal of Animal Science, 54*(4), 790–5.

Micol, D., Robelin, J. and Geay, Y. (1993). Composition corporelle et caractéristiques biologiques des muscles chez les bovins en croissance et à l'engrais. *INRA Productions Animales, 6*(1), 61–9.

Moody, W. G., Kemp, J. D., Mahyuddin, M., Johnston, D. M. and Ely, D. G. (1980). Effect of feeding systems, slaughter weight and sex on histological properties of lamb carcasses. *Journal of Animal Science, 50*(2), 249–56.

Mottram, D. S. and Edwards, R. A. (1983). The role of triglycerides and phospholipids in the aroma of cooked beef. *Journal of the Science of Food and Agriculture, 34*(5), 517–22.

Ono, Y., Solomon, M. B., Elsasser, T. H., Rumsey, T. S. and Moseley, W. M. (1996). Effects of Synovex-S(R) and recombinant bovine growth hormone (Somavubove(R)) on growth responses of steers. 2. Muscle morphology and proximate composition of muscles. *Journal of Animal Science, 74*(12), 2929–34.

Ouali, A. (1990). Meat tenderization: possible causes and mechanisms. A review. *Journal of Muscle Foods, 1*, 129–65.

Ouali, A. and Talmant, A. (1990). Calpains and calpastatin distribution in bovine, porcine and ovine skeletal-muscles. *Meat Science, 28*(4), 331–48.

Ozawa, S., Mitsuhashi, T., Mitsumoto, M., Matsumoto, S., Itoh, N., Itagaki, K., Kohno, Y. and Dohgo, T. (2000). The characteristics of muscle fiber types of longissimus thoracis muscle and their influences on the quantity and quality of meat from Japanese Black steers. *Meat Science, 54*(1), 65–70.

Parr, T., Mareko, M. H. D., Ryan, K. J. P., Hemmings, K. M., Brown, D. M. and Brameld, J. M. (2016). The impact of growth promoters on muscle growth and the potential consequences for meat quality. *Meat Science, 120*, 93–99.

Pearson, A. M. (1990). Muscle growth and exercise. *Critical Reviews in Food Science and Nutrition, 29*(3), 167–96.

Peter, J. B., Barnard, R. J., Edgerton, V. R., Gillespie, C. A. and Stempel, K. E. (1972). Metabolic profiles of three fibre types of skeletal muscle in guinea pigs and rabbits. *Biochemistry, 11*, 2627–33.

Pette, D. and Staron, R. S. (1990). Cellular and molecular diversities of mammalian skeletal muscle fibers. *Reviews of Physiology, Biochemistry and Pharmacology, 116*, 1–76.

Pette, D. and Staron, R. S. (2000). Myosin isoforms, muscle fiber types, and transitions. *Microscopy Research and Technique, 50*(6), 500–9.

Pette, D. and Staron, R. S. (2001). Transitions of muscle fiber phenotypic profiles. *Histochemistry and Cell Biology, 115*(5), 359–72.

Picard, B. and Cassar-Malek, I. (2009). Evidence for expression of IIb myosin heavy chain isoform in some skeletal muscles of Blonde d'Aquitaine bulls. *Meat Science, 82*(1), 30–6.

Picard, B., Robelin, J. and Geay, Y. (1995). Influence of castration and postnatal energy restriction on the contractile and metabolic characteristics of bovine muscle. *Annales De Zootechnie, 44*(4), 347–57.

Rehfeldt, C., Fiedler, I. and Stickland, N. C. (1984). Number and size of muscle fibres in relation to meat production. In M. F. W. te Pas, M. E. Everts and H. P. Haagsman (Eds), *Muscle Development of Livestock Animals. Physiology, Genetics and Meat Quality*. Wallingford, UK: CABI Publishing.

Renand, G., Picard, B., Touraille, C., Berge, P. and Lepetit, J. (2001). Relationships between muscle characteristics and meat quality traits of young Charolais bulls. *Meat Science, 59*(1), 49–60.

Renerre, M. (1984). Variability between muscles and between animals of the color stability of beef meats. *Sciences Des Aliments, 4*(4), 567–84.

Renerre, M. and Labas, R. (1987). Biochemical factors influencing metmyoglobin formation in beef muscles. *Meat Science, 19*(2), 151–65.

Robelin, J., Lacourt, A., Bechet, D., Ferrara, M., Briand, Y. and Geay, Y. (1991). Muscle differentiation in the bovine fetus – a histological and histochemical approach. *Growth Development and Aging, 55*(3), 151–60.

Ryu, Y. C. and Kim, B. C. (2006). Comparison of histochemical characteristics in various pork groups categorized by postmortem metabolic rate and pork quality. *Journal of Animal Science, 84*(4), 894–901.

Schiaffino, S., Gorza, L., Sartore, S., Saggin, L., Ausoni, S., Vianello, M., Gundersen, K. and Lomo, T. (1989). 3 Myosin Heavy-Chain Isoforms in Type-2 Skeletal-Muscle Fibers. *Journal of Muscle Research and Cell Motility, 10*(3), 197–205.

Schiaffino, S. and Reggiani, C. (2011). Fiber types in mammalian skeletal muscles. *Physiological Reviews, 91*(4), 1447–531.

Schreurs, N. M., Garcia, F., Jurie, C., Agabriel, J., Micol, D., Bauchart, D., Listrat, A. and Picard, B. (2008). Meta-analysis of the effect of animal maturity on muscle characteristics in different muscles, breeds, and sexes of cattle. *Journal of Animal Science, 86*(11), 2872–87.

Seideman, S. C. and Crouse, J. D. (1986). The effects of sex condition, genotype and diet on bovine muscle-fiber characteristics. *Meat Science, 17*(1), 55–72.

Seideman, S. C., Crouse, J. D. and Cross, H. R. (1986). The effect of sex condition and growth implants on bovine muscle-fiber characteristics. *Meat Science, 17*(2), 79–95.

Seideman, S. C., Koohmaraie, M. and Crouse, J. D. (1987). Factors associated with tenderness in young beef. *Meat Science, 20*(4), 281–91.

Shibata, M., Matsumoto, K., Hikino, Y. and Yamamoto, N. (2014). Effect of indoor concentrate feeding vs. outdoor grazing on the expression of genes involved in muscle growth and nutrient content in Japanese black steer muscle. *Open Journal of Animal Sciences, 4*, 297–304.

Simoneau, J. A. and Bouchard, C. (1989). Human variation in skeletal-muscle fiber-type proportion and enzyme-activities. *American Journal of Physiology, 257*(4), E567–72.

Simoneau, J. A. and Bouchard, C. (1995). Genetic determinism of fiber-type proportion in human skeletal-muscle. *Faseb Journal, 9*(11), 1091–5.

Solomon, M. B., West, R. L. and Carpenter, J. W. (1985). Fiber types in the longissimus muscle from water buffalo and selected domestic beef breeds. *Meat Science, 13*(3), 129–35.

Spindler, A. A., Mathias, M. M. and Cramer, D. A. (1980). Growth changes in bovine muscle-fiber types as influenced by breed and sex. *Journal of Food Science, 45*(1), 29–31.

Staun, H. (1963). Various factors affecting number and size of muscle fiber in the pig. *Acta Agricultural Scandinavia, 13*, 293–322.

Stavaux, D., Art, T., Mcentee, K., Reznick, M. and Lekeux, P. (1994). Muscle-fiber type and size, and muscle capillary density in young double-muscled blue belgian cattle. *Journal of Veterinary*

*Medicine Series a-Zentralblatt Fur Veterinarmedizin Reihe a-Physiology Pathology Clinical Medicine*, *41*(3), 229–36.

Strydom, P. E., Frylinck, L., Montgomery, J. L. and Smith, M. F. (2009). The comparison of three beta-agonists for growth performance, carcass characteristics and meat quality of feedlot cattle. *Meat Science*, *81*(3), 557–64.

Suman, S. P. and Poulson, J. (2013). Myoglobin chemistry and meat color. *Annual Review of Food Science and Technology*, *4*(4), 79–99.

Suzuki, A., Tamate, H. and Okada, M. (1976). The effect of a high plane of nutrition during a given period of growth on size and proportion of skeletal muscle fiber types in the cattle. *Tohoku Journal of Agricultural Research*, *27*(1), 20–5.

Talmant, A., Monin, G., Briand, M., Dadet, M. and Briand, Y. (1986). Activities of metabolic and contractile enzymes in 18 bovine muscles. *Meat Science*, *18*(1), 23–40.

Tanabe, R., Muroya, S. and Chikuni, K. (1998). Sequencing of the 2a, 2x, and slow isoforms of the bovine myosin heavy chain and the different expression among muscles. *Mammalian Genome*, *9*(12), 1056–8.

Totland, G. K. and Kryvi, H. (1991). Distribution patterns of muscle-fiber types in major muscles of the bull (Bos-Taurus). *Anatomy and Embryology*, *184*(5), 441–50.

Vestergaard, M., Henckel, P., Oksbjerg, N. and Sejrsen, K. (1994). The effect of cimaterol on muscle-fiber characteristics, capillary supply, and metabolic potentials of longissimus and semitendinosus muscles from young friesian bulls. *Journal of Animal Science*, *72*(9), 2298–306.

Vestergaard, M., Oksbjerg, N. and Henckel, P. (2000). Influence of feeding intensity, grazing and finishing feeding on muscle fibre characteristics and meat colour of semitendinosus, longissimus dorsi and supraspinatus muscles of young bulls. *Meat Science*, *54*(2), 177–85.

Waritthitham, A., Lambertz, C., Langholz, H. J., Wicke, M. and Gauly, M. (2010). Muscle fiber characteristics and their relationship to water holding capacity of longissimus dorsi muscle in brahman and charolais crossbred bulls. *Asian-Australasian Journal of Animal Sciences*, *23*(5), 665–71.

Wegner, J., Albrecht, E., Fiedler, I., Teuscher, F., Papstein, H. J. and Ender, K. (2000). Growth-and breed-related changes of muscle fiber characteristics in cattle. *Journal of Animal Science*, *78*(6), 1485–96.

Wu, F. Y., Young, C. R., Coleman, M. E. and Smith, S. B. (1986). Total RNA and translatable mRNA levels in longissimus muscle from heifers fed diets containing a beta-agonist. *Journal of Animal Science*, *63*(Suppl. 1), 237.

Yambayamba, E., and Price, M. A. (1991). Fiber-type proportions and diameters in the longissimus muscle of beef heifers undergoing catch-up (compensatory) growth. *Canadian Journal of Animal Science*, *71*(4), 1031–5.

Young, O. A. (1984). The biochemical basis of fiber types in bovine muscle. *Meat Science*, *11*(2), 123–37.

Young, O. A., and Bass, J. J. (1984). Effect of castration on bovine muscle composition. *Meat Science*, *11*(2), 139–56.

Young, O. A., and Foote, D. M. (1984). Further-studies on bovine muscle composition. *Meat Science*, *11*(3), 159–70.

Chapter 4

# Factors affecting fat content and distribution of fat in cattle and carcasses

*Stephen B. Smith, Texas A&M University, USA*



## 1 Introduction

Like most animals, beef cattle are highly efficient in depositing fat. Estimates of efficiency rates for fat and protein deposition are approximately 0.75 and 0.25, respectively. Growth promotants such as β-adrenergic agonists can partition energy from fat growth to lean growth, but do not reduce the efficiency of the biochemical processes involved in fatty acid and triacylglycerol biosynthesis. Any ingested energy not used for maintenance and growth of lean tissues ultimately is deposited as carcass fat (adipose tissue). It has been known for decades that the various fat depots in cattle are deposited in a specific order. Abdominal (kidney, pelvic and heart) fat is deposited during the third trimester of gestation. This is followed by seam (intermuscular) fat, which is laid down between individual muscle groups, which is in turn followed by subcutaneous fat. The last fat depot to develop, and paradoxically the economically most important in many countries, is marbling (intramuscular, interfascicular) fat. Subcutaneous and seam fats also are of high value in the United States and in many other countries, because they are used extensively in the production of ground beef (mince), sausages and other processed meats. Excess fat trim is shipped to rendering facilities, where the extracted lipid is used to produce beef tallow.

The development of all fat depots is delayed by pasture or forage feeding, and is strongly promoted by grain feeding. Soon after adaptation to a grain-based diet, either in weaned calves or following backgrounding on pasture or forage, there is a pronounced increase in activities of enzymes associated with *de novo* fatty acid biosynthesis, such as acetyl-CoA carboxylase, ATP-citrate lyase and NADP-malate dehydrogenase. Additionally, the gene expression and activities of fatty acid elongase, and the Δ9 desaturase named stearoyl-CoA desaturase (SCD) increase four- to fivefold. These enzymes are responsible for the

http://dx.doi.org/10.19103/AS.2016.0009.04

elongation of shorter-chain saturated and monounsaturated fatty acids and desaturation of saturated fatty acids and some *trans*-unsaturated fatty acids.

In addition to changes in lipogenic enzymes with growth, fatty acid composition of fat depots changes with time on feed. In grain-fed cattle, the monounsaturated fatty acid, oleic acid (18:1n-9) and other *cis*-9 fatty acids accumulate over time on feed at the expense of stearic acid (18:0) and other saturated fatty acids. Additionally, grain feeding promotes the conversion of *trans*-vaccenic acid to *cis*-9, *trans*-11 conjugated linoleic acid (CLA). In grass-fed cattle, much less oleic acid accumulates in fat depots over time; instead, the fat depots become relatively enriched in saturated fatty acids and the n-3 fatty acid $\alpha$-linolenic acid (18:3n-3).

In addition to marked differences in carcass fat development between grain- and grass-fed cattle, there are remarkable differences in fatty acid composition across fat depots. Fat overlying the brisket is unusually enriched with monounsaturated fatty acids, while fat overlying the plate (lower rib cage) is very high in saturated fatty acids.

This chapter will address (1) the origin of adipocytes, (2) the ontogeny of fat development, (3) factors influencing fat development and (4) factors influencing fatty acid composition of fat depots.

# 2 Origin of adipocytes in beef fat depots

Much of what we know about the differentiation of adipocytes (fat cells) within fat depots comes from studies of secondary cell lines, such as 3T3-L1 pre-adipocytes. Adipocyte development in culture can be described simply. Pre-adipocytes are plated in culture dishes and are allowed to divide until the cells reach confluence, that is, the culture dishes are nearly completely filled with pre-adipocytes. Subsequently, hormones such as insulin and dexamethasone are added to initiate differentiation (lipid filling). Once lipid filling begins, proliferation ceases in secondary cell lines. The cessation of proliferation occurs even before detectable lipids have accumulated.

The processes of cell division and lipid filling in fat depots of growing cattle also occur generally as observed for pre-adipocytes in culture. Mesenchymal cells within connective beds such as the perimysium and the overlying epimyseum proliferate and then differentiate into pre-adipocytes. Initial lipid filling is characterized by the appearance of small lipid droplets, forming pre-adipocytes with a multilocular appearance. These coalesce as the adipocytes enlarge, leading to the unilocular adipocytes seen in mature fat depots. Unlike the situation in pre-adipocyte cell cultures, pre-adipocytes within fat depots of cattle can divide even after accumulation of some lipid, while they still are multilocular. The ability of the multilocular pre-adipocytes to divide increases the potential for the marked adiposity observed in long-fed cattle.

# 3 Ontogeny of fat development

## 3.1 Brown adipose tissue

Brown adipose tissue (BAT) is the first fat depot to develop in calves, and its development begins around the beginning of the third trimester of gestation. The largest BAT depots

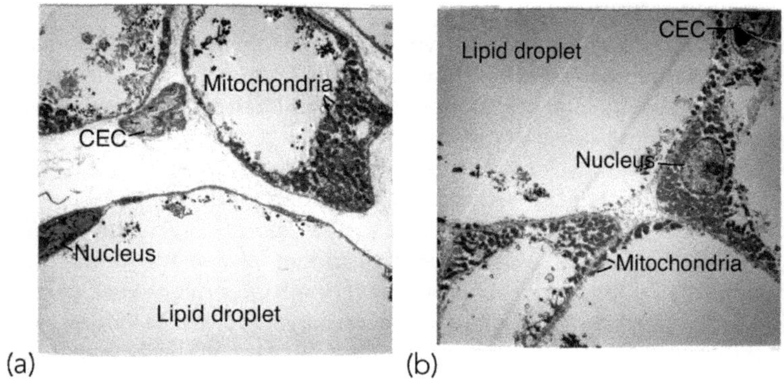


**Figure 1** Transmission electron micrographs of tailhead subcutaneous (a) and perirenal (b) fat depots from newborn Brahman calves. Large lipid droplets, adipocyte nuclei and tightly packed mitochondria are evident in the perirenal fat sample. In subcutaneous fat (a), a brown adipocyte (upper part of micrograph) sits adjacent to a white adipocyte (lower part of the micrograph). Abundant brown adipocytes are apparent in perirenal fat (b). CEC, capillary endothelial cell. The numerous mitochondria in the brown adipocytes are indicated in both micrographs.

are located as fat caps over the kidneys and in the pericardial cavity, where they provide heat for the kidneys and heart via non-shivering thermogenesis (heat production) during cold exposure of the newborn calves. Brown adipocytes are easily distinguishable from white adipocytes in that they contain numerous, highly differentiated mitochondria. In rodents, BAT adipocytes are small (5–10 μm diameter) and the lipid droplets within the adipocytes are small and multilocular. In newborn calves, the adipocytes are larger than those in rodents (20–50 μm diameter), and because the lipid droplets coalesce, the adipocytes are primarily unilocular (Fig. 1).

Uncoupling protein-1 (UCP1) gene expression in foetal calf perirenal BAT is virtually undetectable at the beginning of the third trimester, but reaches peak expression immediately before birth. UCP1 is unique to BAT, and as it is essential for the process of non-shivering thermogenesis, its expression level is taken as an index of the extent of BAT differentiation. UCP1 mRNA disappears within 4 weeks after calves are born, and it may disappear even sooner if calves are not exposed to cold ambient temperatures.

Historically, it was thought that subcutaneous fat did not develop initially as BAT, but more recently, adipocytes containing multiple, differentiated mitochondria and UCP1 mRNA have been demonstrated in subcutaneous fat of newborn calves. Unlike perirenal BAT, UCP1 gene expression in subcutaneous fat declines markedly prior to birth, and much of the mitochondrial capacity for thermogenesis is lost. During the last one to two weeks before birth, many of the brown adipocytes present in subcutaneous fat convert to white adipocytes. To date, neither abundant mitochondria nor UCP1 mRNA have been detected in marbling fat at any stage of development, indicating that marbling fat represents a unique fat depot.

## 3.2 White adipose tissue

Fat depots in adult cattle, which are completely devoid of brown adipocytes, are classified as white adipose tissue (WAT). The primary function of WAT is the storage of lipid in times of excess energy intake, to be called upon during times when food is less abundant. As subcutaneous adipose tissue also serves the important function of insulation against the cold, cattle adapted to the upper latitudes (*Bos taurus* breed types) typically have a greater capacity to accumulate subcutaneous fat cover than tropically adapted cattle (*Bos indicus* breed types). It is likely that marbling fat serves as an energy reservoir for adjacent muscle fibres, and those muscles that effectively accumulate marbling also typically have the capacity to oxidize fatty acids released from marbling fat.

Most fat depots in beef cattle use acetate as the primary precursor for *de novo* fatty acid biosynthesis. Acetate, a two-carbon by-product of ruminal fermentation, is the most abundant volatile fatty acid absorbed from the rumen. Very little glucose escapes fermentation in the rumen, and ruminant species have evolved to conserve glucose for essential physiological functions. Bovine subcutaneous fat has a low abundance of insulin receptors and glucose transporters (GLUT4), and the use of any glucose that enters subcutaneous adipocytes is strictly regulated by hexokinase and 6-phosphofructokinase. These enzymes occur early in the glycolytic pathway, and inhibition of 6-phosphofructokinase can divert glucose carbon to the pentose phosphate pathway for the generation of a portion of the reducing equivalents necessary for fatty acid synthesis.

# 4 Factors influencing fat development

## 4.1 Metabolic pathways in bovine fat

In the 1960s, it was thought that the incorporation of glucose carbon into fatty acids was limited by the activities of ATP-citrate lyase and NADP-malate dehydrogenase. These enzymes convert cytoplasmic citrate to oxaloacetate and acetyl-CoA, and malate to pyruvate and $CO_2$, respectively. Acetyl-CoA is the substrate for acetyl-CoA carboxylase, the first step in fatty acid synthesis, and the decarboxylation of malate by NADP-malate dehydrogenase reduces $NADP^+$ to $NADPH + H^+$, which thus produces a portion of the reducing equivalents necessary for the fatty acid synthase reactions. In the 1980s, researchers documented that lactate is incorporated into fatty acids at rates that exceed acetate incorporation into fatty acids *in vitro* because lactate conversion to fatty acids requires the activities of ATP-citrate lyase and NADP-malate dehydrogenase, and their activities *in situ* were sufficient to account for the high capacity of fat tissues to incorporate lactate carbon into fatty acids. The activities of acetyl-CoA carboxylase, ATP-citrate lyase and NADP-malate dehydrogenase increase in concert with total carcass fatness, and their activities are similarly depressed by high-roughage diets (Fig. 2). Thus, it is clear that ATP-citrate lyase and NADP-malate dehydrogenase could not be responsible for the low rates of glucose incorporation into fatty acids, and subsequently it was shown that hexokinase and 6-phosphofructokinase are responsible for limiting the use of glucose carbon for *de novo* fatty acid biosynthesis.

## 4.2 Marbling fat

As is the situation for subcutaneous and seam fat, the amount of marbling fat can vary remarkably. Lean beef, in which all subcutaneous fat and seam fat have been removed, contains approximately 1% extractable lipid. At the other extreme, closely trimmed beef from Japanese Black (or Wagyu) cattle can contain over 50% extractable lipid.

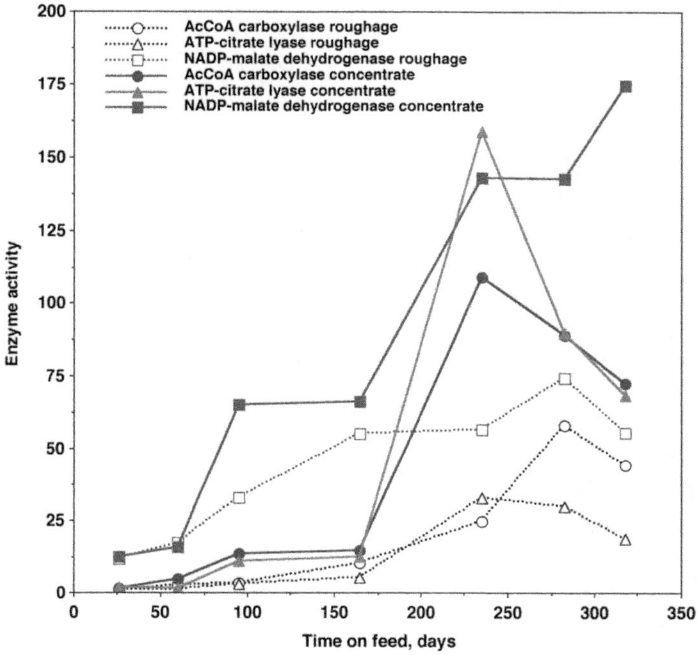


**Figure 2** Influence of diet and time on feed on subcutaneous fat enzyme activities. Angus × Hereford and Red Poll steers (initial age 265 days; initial weight 260 ± 4 kg) were fed pelleted sun-cured alfalfa hay (roughage) or a pelleted, high-corn concentrate diet (concentrate). Enzyme activities are nmol substrate converted to product per minute per $10^5$ adipocytes. For each line, the data are the means for 7–10 animals; subcutaneous fat samples were taken by biopsy at each sampling point, so each steer was sampled 7 times.

This extraordinary concentration of lipid can be attributed solely to lipid in marbling fat. Marbling fat can be distinguished from other fat depots by its location within perimysial connective tissues alongside myofibres (Fig. 3). Although some scientists have provided evidence for *trans*-differentiation of muscle satellite cells to pre-adipocytes, the localization of marbling adipocytes to the perimysium in most breed types supports the argument that marbling arises primarily from fibroblasts associated with perimysial connective tissues.

Marbling fat has several metabolic features that distinguish it from other depots. Marbling adipocytes typically display rates of fatty acid biosynthesis that are 5 to 10% of the rates observed in subcutaneous fat adipocytes, and are smaller (25–50 μm diameter) than subcutaneous adipocytes (100 to as much as 500 μm diameter). Triacylglycerol biosynthesis from most fatty acids also occurs at lower rates in marbling than in subcutaneous fat, and glucose contributes a greater proportion of carbon to *de novo* fatty acid biosynthesis in marbling fat than in subcutaneous fat (Fig. 4). Conversely, subcutaneous fat, and other fat depots besides marbling, primarily use acetate and lactate as the carbon source for *de novo* fatty acid biosynthesis. However, increasing free glucose uptake or infusion of glucose into the circulatory system of beef cattle promotes the development of marbling, subcutaneous and seam fat, possibly by providing carbon for the glycerol backbone of triacylglyerols.

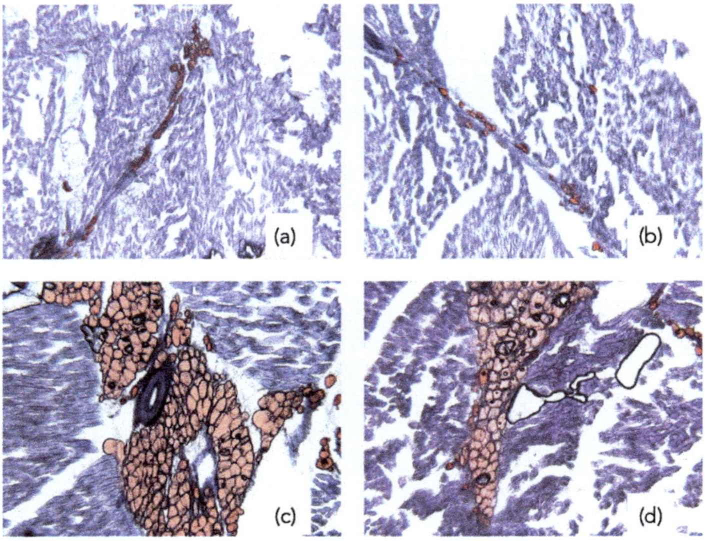


**Figure 3** Intramuscular adipocyte development in *M. longissimus thoracis* of Angus steers. Cells were stained with Oil Red O to stain adipocytes and haematoxylin solution to stain for nuclei. (a) Sample taken at weaning (8 months of age). (b) Sample taken after 4 months on native Texas pasture (12 months of age). (c) Sample taken in steers fed a corn-based, grain finishing diet for 4 months after weaning (12 months of age). (d) Sample taken in steers fed a corn-based, grain finishing diet for 4 months after grazing native pasture for 4 months (16 months of age). Intramuscular adipocytes are visible as thin threads of adipocytes (a and b) or as clusters of roughly spherical adipocytes (c and d). Magnification approximately × 100.

## 4.3 Distribution of carcass fat

In addition to having an enormous genetic potential to deposit marbling fat, Japanese Black cattle (Wagyu) are unusual in that they deposit much less backfat than British breeds. This represents one of the rare instances in which marbling fat is deposited without excessive backfat, and indicates that there is genetic potential to produce well-marbled beef without concomitant excessive carcass adiposity. The exceptional marbling ability of Wagyu cattle is not phenotypically expressed until the cattle achieve large body weights (generally over 600 kg). Thus, when raised to a typical body weight for small-framed cattle (around 525 kg), marbling in Wagyu steers is similar to that in Angus steers, although the Wagyu carcasses contain less fat trim. Similarly, carcasses of Angus crossbred cattle have higher marbling scores than Brahman crossbred cattle when raised a constant backfat level and age. In contrast, at a similar carcass weight, carcasses from British breeds (Angus–Hereford crosses) have higher marbling scores but also more fat trim than carcasses from Continental breeds (Charolais). Not including Asian breed types (Japanese Wagyu and Korean Hanwoo), heritability estimates for backfat thickness and marbling scores are both approximately 0.36. Heritabilities for backfat thickness and marbling scores of Wagyu cattle are 0.38 and over 0.50, respectively. Although Wagyu and Hanwoo cattle originally were derived from the same Asian breed type, heritabilities for backfat thickness and marbling scores in Hanwoo cattle are lower (0.20 and 0.28, respectively). The moderate-to-high

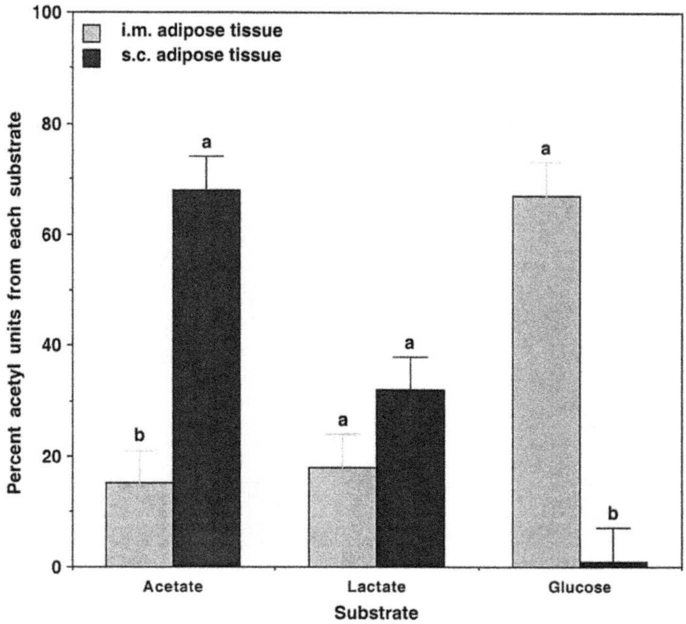


**Figure 4** Relative contributions of carbon precursors to *de novo* fatty acid biosynthesis in marbling fat (i.m.) and subcutaneous fat (s.c.). Total acetyl unit incorporation into fatty acids was the sum of acetyl units incorporated from acetate, lactate and glucose in *in vitro* incubations of intramuscular and subcutaneous fat. Each bar is the mean ± SEM of marbling or subcutaneous fat from four 18-month-old Angus steers fed high concentrate diets from weaning at 8 months of age.

heritabilites for backfat thickness and marbling scores indicate that the genetic potential exists to select for cattle with less trimmable fat, but for most breed types this would be at the expense of marbling fat. Thus, more genetic selection is required to produce cattle with adequate marbling without excessive carcass adiposity.

The fatty acid composition of subcutaneous fat does not differ between Angus and Brahman crossbred cattle when raised to the same body weight and subcutaneous fat thickness. Subcutaneous fat of Wagyu steers contains less saturated fatty acids and more monounsaturated fatty acids than subcutaneous fat of Angus steers when raised to the same body weight. This indicates that genetic potential exists to select for breed types or lines within breed types that differ in beef fatty acid composition.

# 5 Factors influencing fatty acid composition of fat depots

## 5.1 Cellular factors affecting fat content

As indicated in Fig. 2, the capacity to synthesize fatty acids increases with time on feed, and is greater in fat depots from concentrate-fed steers than in roughage-fed steers. Prior

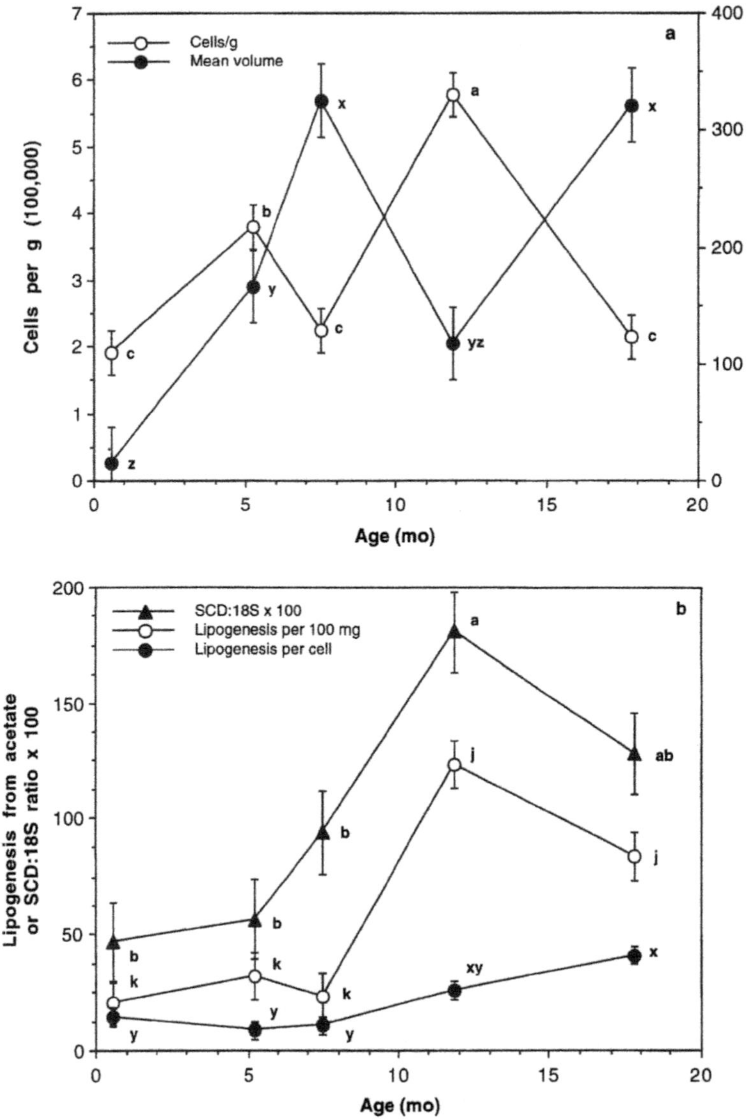


**Figure 5** Changes in cellularity, lipogenesis and stearoyl-CoA desaturase (*SCD*) gene expression during growth in preweaning (2.5 weeks to 7.5 months of age) and postweaning Angus steers. Top: Mean values for number of adipocytes per gram of adipose tissue and adipocyte volumes. Bottom: Lipogenesis and *SCD* gene expression.

to weaning, subcutaneous fat cells increase in size from approximately 20 pL at two weeks of age to nearly 300 pL by 5 months of age (Fig. 5a). As the enzymes responsible for *de novo* fatty acid biosynthesis are not yet expressed in pre-weaned calves, adipocyte lipid

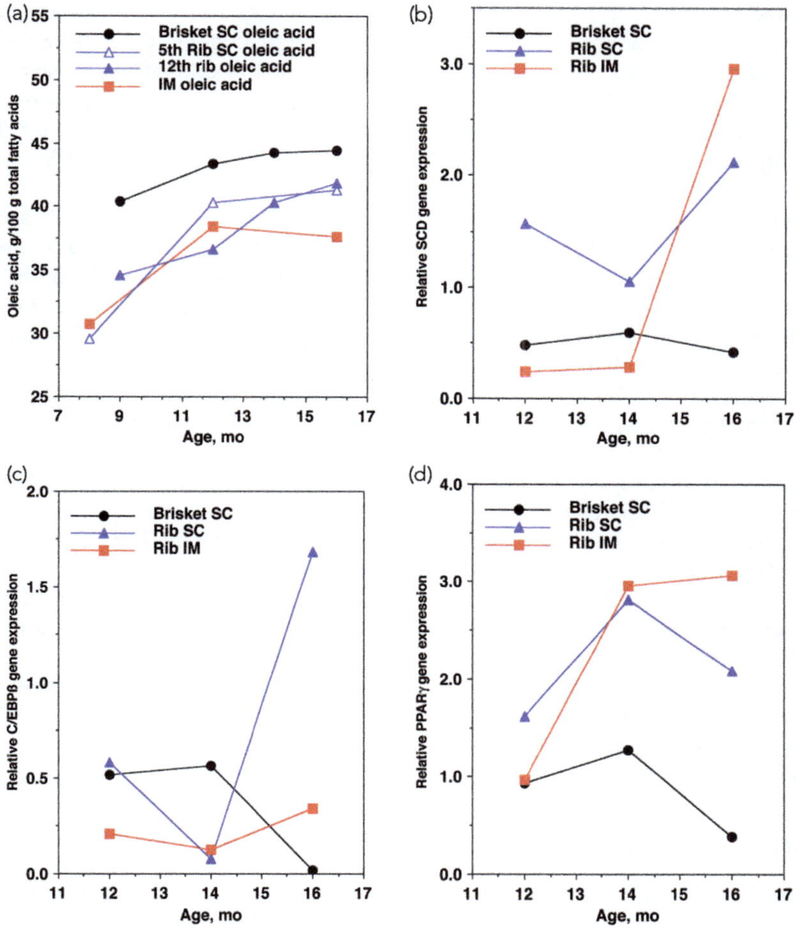


**Figure 6** Adipose tissue oleic acid and specific gene expression in post-weaned Angus steers fed grain-based diets. (a) Concentration of oleic in brisket fat, 5th rib subcutaneous (s.c.) fat, 12th rib s.c. fat and marbling (IM) fat (b) *SCD* gene expression (c) *C/EBPβ* gene expression and (d) *PPARγ* gene expression in brisket fat, rib fat and marbling fat.

filling during suckling is the result of absorption of milk fatty acids from the small intestine and conversion of circulating fatty acids to triacylglycerols in fat depots. Because marbling fat does not develop until after weaning, marbling adipocytes (or pre-adipocytes) must lack the ability to form lipids from fatty acids in the circulation. If weaned calves are placed on poor-quality pasture, subcutaneous fat cells can decrease in size as lipid stores are depleted. However, once calves are transitioned from pasture to a grain-based, finishing diet, adipocytes once again fill with lipid.

In the 1990s, researchers demonstrated that monounsaturated fatty acids increase with time on feed in subcutaneous fat of beef cattle. The enzyme responsible for the conversion of saturated fatty acids to monounsaturated fatty acids (e.g. stearic → oleic acid) is SCD, the Δ9 desaturase of animal tissues. Expression of the *SCD* gene increases with time in cattle fed a grain-based diet, but declines markedly at the end of the finishing phase (Fig. 5b). The decline in *SCD* gene expression at the end of the finishing phase is similar to depression in the activities of acetyl-CoA carboxylase and ATP-citrate lyase (Fig. 2), suggesting that subcutaneous fat loses the capacity for *de novo* fatty acid biosynthesis after extended periods of grain-based diets.

Monounsaturated fatty acids increase in fat depots until 16–18 months of age, after which they reach a fairly stable plateau (Fig. 6a). Brisket fat is unique among fat depots in that it contains a high proportion of oleic acid, which is elevated even at 9 months of age. In contrast, oleic acid is low in rib fat and marbling fat at 8 months of age, but increases substantially from 8 months of age to typical beef cattle slaughter age. The expression of the *SCD*, CCAAT enhancer-binding protein-β (*C/EBPβ*) and peroxisome proliferator-activated receptor-γ (*PPARγ*) has been documented as genetic markers of fat development and lipid metabolism. As seen for fatty acid composition, there are profound differences in *C/EBPβ*, *PPARγ* and *SCD* gene expression among adipose tissue depots (Fig. 6b–d).

Although oleic acid is highest in brisket fat at all ages tested, paradoxically *SCD* gene expression is low in brisket fat and does not change with age (Fig. 6b). *C/EBPβ* and *PPARγ* gene expressions decline to nearly undetectable levels in brisket fat, but increase in rib fat and marbling fat between 14 and 16 months of age (Fig. 6c and 6d). The low level of *SCD* gene expression and a decline in *C/EBPβ* and *PPARγ* gene expression suggest that brisket fat is a more mature adipose tissue than rib fat or marbling. What is unusual is that oleic acid is elevated in brisket adipose tissue at 9 months of age. This suggests that brisket fat differentiates early prenatally as BAT. It is possible that brisket fat originally differentiates as BAT with high *SCD* activity, and thus is enriched with oleic acid at birth. Because brisket fat is higher in oleic acid than rib fat, it is possible that brisket fat more fully maintains its BAT phenotype at birth than other subcutaneous fat depots.

## 5.2 Grass feeding and fatty acid composition

The United States and Pacific Rim countries (Australia, Japan and Korea) specialize in the production of grain-fed beef. In the 1990s, closely trimmed beef from the highest grading (A5) Wagyu cattle contained up to 35% total fat, but in response to competition from imported beef, Japanese farmers have increased total intramuscular fat to over 50%. This is in stark contrast to beef produced in Europe and Central and South America, which typically comes from grass-fed cattle that are fed little or no grain before processing. The United States and Australia have a small but growing niche market for grass-fed beef, primarily because grass feeding is perceived as more sustainable. In reality, recent studies have documented that grain feeding uses less land and less water and produces fewer greenhouse gases than does pasture/grass feeding, although this is not widely recognized.

As indicated above, grass feeding increases the concentration of α-linolenic acid, and this observation has led many scientists and producers to promote and produce grass-fed beef. However, very little α-linolenic acid is converted to the biologically important n-3 fatty acids, eicosapentaenoic acid (EPA) or, especially, docosahexaenoic acid (DHA). Additionally, the actual amount of α-linolenic acid in grass-fed beef remains biologically insignificant, and even the few randomized controlled studies comparing grass-fed and grain-fed beef have demonstrated no additional benefits of grass-fed beef over grain-fed beef.

# 6 Summary and future trends

## 6.1 How research can contribute to enhanced and sustainable animal production

Grass feeding clearly delays fat development but also increases the concentration of α-linolenic acid in muscle and fat. In beef steers fed pasture from weaning at 8 months to 12 months of age, marbling adipocytes are small and are not shown to have accumulated lipid beyond that seen at 8 months of age. Marbling adipocytes of calves fed immediately after weaning are large and numerous. Similarly, feeding a corn-based diet to calves previously grazed solely on pasture initiates lipid filling, so that marbling adipocytes become large and numerous. As backgrounding calves on pasture strongly depresses *SCD* gene expression (Fig. 7), fat depots in grass-fed cattle are higher in saturated fatty acids

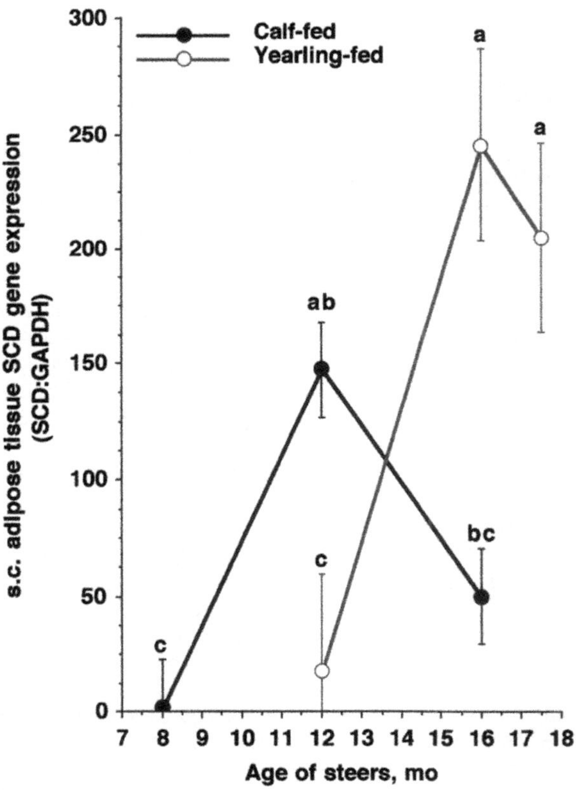


**Figure 7** Changes in *SCD* gene expression over time in subcutaneous fat (s.c. adipose tissue) from calf-fed (closed circles) and yearling-fed (open circles) steers. Steers were weaned at 8 months of age. The CF steers were harvested at 8, 12 or 16 months of age (0, 4 or 8 months free access to the corn-based diet). The YF steers were harvested at 12, 16 and 17.5 months of age (0, 4 or 5.5 months free access to the corn-based diet). From 10 to 12 months of age, the YF steers were supplemented with the corn-based diet to achieve a growth rate of 0.9 kg/d.

and correspondingly lower in monounsaturated fatty acids. Achieving the appropriate balance between grass forage feeding and finishing on a high concentrate diet would optimize the quantity and quality of fat deposition in beef cattle.

## 6.2 Future trends in research

What is not well understood is the mechanism by which grass/forage feeding depresses fat development. Is the delay in fat development in grass/forage-fed cattle caused by a limitation in caloric intake, or do the n-3 fatty acids in grasses and pastures specifically depress adipogenic gene expression? Alternatively, does the increased availability of glucose for absorption in grain-fed cattle specifically stimulate adipogenic gene expression? These questions will be difficult to answer in whole-animal models, due to the difficulty in limiting the ruminal fermentation of dietary starch and fatty acids in beef cattle. However, research with primary cell cultures of bovine pre-adipocytes should provide important information about the regulation of fat development in beef cattle.

# 7 References and further reading

Aberle, E. D., Forrest, J. C., Gerrard, D. E. and Mills, E. W. 2012. *Principles of Meat Science*. 5th edn. Kendall Hunt Publishing, Dubuque, IA.

Allen, C. E., Beitz, D. C., Kramer, D. A. and Kauffman, R. G. 1976. Biology of Fat in Meat Animal. North Central Regional Publication No. 234. Research Division, College of Agriculture and Life Sciences, University of Wisconsin, WI.

Campbell, E. M.G., Sanders, J. O., Lunt, D. K., Gill, C. A., Taylor, J. F., Davis, S. K., Riley, D. G. and Smith, S. B. 2016. Adiposity, lipogenesis, and fatty acid composition of subcutaneous and intramuscular adipose tissues of Brahman and Angus crossbred cattle. *J. Anim. Sci.* 94: 1415–25.

Gilmore, L. A., Walzem, R. L., Crouse, J. D. Smith, D. R., Adams, T. H., Vaidyananthan, V., Cao, X. and Smith, S. B. 2011. Consumption of high-oleic acid ground beef increases HDL cholesterol concentration but both high- and low-oleic acid ground beef decrease HDL particle diameter in normocholesterolemic men. *J. Nutr.* 141: 1188–94.

Mersmann, H. J. and Smith, S. B. 2005. Development of white adipose tissue lipid metabolism. In D. Burrin and H. J. Mersmann (eds), *Biology of Metabolism in Growing Animals*, pp. 275–302. Elsevier Science Publishers, Oxford.

Moody, W. B. and Cassens, R. G. 1968. A quantitative and morphological study of bovine longissimus fat cells. *J. Food Sci.* 33: 47–55.

Rule, D. C., Smith, S. B. and Romans, J. R. 1995. Fatty acid composition of muscle and adipose tissue of meat animals. In S. B. Smith and D. R. Smith (eds), *The Biology of Fat in Meat Animals: Current Advances*, pp. 144–65. American Society of Animal Science, Savoy, IL.

Smith, S. B. 1991. Dietary modification for altering fat composition of meat. In C. Haberstroh and C. E. Morris (eds), *Advances in Applied Biotechnology Series: Reduced Fat and Cholesterol Foods*, vol. 12, pp. 75–98. Gulf Publishing Company, Houston, TX.

Smith, S. B. 1995. Substrate utilization in ruminant adipose tissues. In S. B. Smith and D. R. Smith (eds), *The Biology of Fat in Meat Animals: Current Advances*, pp. 166–88. American Society of Animal Science, Savoy, IL.

Smith, S. B. and Carstens, G. E. 2005. Ontogeny and metabolism of brown adipose tissue in livestock species. In D. Burrin and H. J. Mersmann (eds), *Biology of Metabolism in Growing Animals*, pp. 303–22. Elsevier Science Publishers, Oxford.

Smith, S. B. and Smith, D. R. 2014. Chemical and physical characteristics of meat: Adipose tissue. In C. Devine and M. Dikemann (eds), *Encyclopedia of Meat Sciences 2e*, vol. 1, pp. 222–34. Elsevier Science Publishers, Oxford.

Smith, S. B. and Smith, D. R. 2016. Fats: production and uses of animal fats. In B. Caballero, P. Finglas and F. Toldra (eds), *The Encyclopedia of Food and Health*, vol. 2, pp. 604–8. Academic Press, Oxford.

Utera, A. R. and Van Vleck, LD. 2004. Heritability estimates for carcass traits in cattle: a review. *Genet. Mol. Res.* 30: 380–94.

Valin, C., Cross, H. R. and Smith, S. B. 1992. Tissue development related to meat quality. In R. Jarrige and C. Béranger (eds), *World Animal Science, Beef Cattle Production*, pp. 131–49. Elsevier Science Publishers.

Wheeler, T. L., Cundiff, L. V., Koch, R. M., Dikeman, M. E. and Crouse, J. D. 1997. Characterization of different biological types of steers (Cycle IV): Wholesale, subprimal, and retail product yields. *J. Anim. Sci.* 75: 2389–403.

# Part 2

# Management of cattle

Chapter 5

# Beef cattle nutrition and its effects on beef quality

*Christopher J. Richards, Oklahoma State University, USA; and Michael E. Dikeman, Kansas State University, USA*



## 1 Introduction

Beef quality is often related to the amount of intramuscular fat or marbling deposited in the longissimus muscle as an indicator of consumer appeal and palatability and as a component of carcass value. The evaluation of nutrition and its effect on beef quality involves the nutrition of cattle throughout their lifetime. Beef cattle are raised in diverse production systems with diverse resources. Beef production in the United States can be divided into three major segments, the cow-calf, stocker/backgrounding and finishing phases. Around the globe, the cow-calf and stocker segments of the industry exist nutritionally on a forage base. The finishing segment will primarily utilize either a grass/forage base, forage with supplemental nutrients from grain sources or energy-dense diets based on grains and grain by-products from other industries.

The Food and Agriculture Organization estimates that the 26% of the world land areas (FAO, 2012) are grasslands in environments or on lands not suitable for grain-crop production or human food production. The forage base also includes forage from residues of grain production or cover crops grown with grain crops. In the United States alone, crop residues are estimated at 4.5 million metric tons every year (USDA, 2006). The diversity of the forage base in terms of forage types, productivity and seasonality greatly impact the characteristics of the production systems and nutrition provided. Nutrition has been

http://dx.doi.org/10.19103/AS.2016.0009.05

**Figure 1** Cattle eating a high grain diet in a feedlot.

implicated to begin impacting beef quality as early as the prenatal and pre-weaning stages. As the stages of production get closer to the harvest point, the nutrition of animals has a more defined impact on the final beef product.

Current US conventional finishing/production systems are grain-finishing systems that utilize technologies such as implants, feed additives and antimicrobials, which have been extensively tested and found to be safe and efficacious when used according to label directions, to produce a safe, wholesome beef product in an efficient manner that minimizes the impact on the environment (NASEM, 2016) (Fig. 1).

Other beef production systems are used to market beef on the basis of characteristics of the previous management system(s) under which the animals were produced. Examples of product marketing with previous management factors include grass-fed, non-hormone treated, organic and natural. Beef products from these systems are designed to fit specific market niches. These niches exist because a small component of consumers have a preference for specific characteristics of the cattle or production systems for which they are willing to pay higher prices as they deem these as safer, higher quality, healthier, or more adequately aligned with their social beliefs. They may have focus on environmental benefits, taste, quality, sustainability, locally grown, safety, healthfulness, animal well-being, or other advantages that may be claimed as a result of a particular beef production system (Johnson and Mathews, 2010). Even though some consumers desire beef from niche production systems, they frequently desire the end product quality and flavour to be similar to conventional beef production systems (NASEM, 2016). Consequently, an understanding of niche market components that affect nutrition and traditional finishing technologies is necessary. The structured programmes through associations, government, harvesters and retailers that impact nutrition and use of conventional technologies include grass-fed, grass pastured, organic and natural.

The American Grassfed Association (AGA, 2016) has certification standards of Grassfed and Grass Pastured production systems (Fig. 2 and 3). The Grassfed certification requires grazing of grass during the growing season and use of harvested forages with a small allowance for supplementation. The AGA Grass Pastured program allows feeding approved supplements at a rate of up to 0.5% of body weight during the growth stage and up to

**Figure 2** Photo of cattle being finished on grass.

**Figure 3** Photo of carcasses and ground beef from grass-fed cattle.

0.75% of body weight during the last 200 pounds of gain. Neither programme allows hormone use, antibiotic use, feeding of animal by-products, use of organophosphates or supplement feeding 14 days prior to harvest.

Organic beef is certified and labelled if it meets the criteria set by the USDA National Organic Program (http://www.ams.usda.gov/AMSv1.0/nop). For a beef production system to be certified organic by the USDA, it must meet several criteria. Animals must be given no antibiotics or growth hormones. If animals become ill, they must be treated and removed from the National Organic Program. All grain and forage fed to the cattle must be from 100 per cent organic production systems for at least three years before harvest. The cattle must have year-round access to the outdoors, have access to pasture during the forage growing season and receive 30 per cent or more of their dry matter intake from pasture during the grazing season.

Natural programmes have evolved out of a desire by consumers for a product that has some of the components of the organic programmes, but are less costly. This has resulted in 'naturally raised' labels in the United States that are products of individual associations, companies or processors. While not required, most of these programmes follow the USDA Process Verified Program standards for the 'Never Ever 3' naturally raised programme. That programme label can be applied to programmes that are third-party verified not to use antibiotics, administer hormone growth promotants or feed animal-sourced proteins.

Industry experts estimate that approximately 400 000 head of cattle in the United States are marketed under natural product labels (Natural Beef, 2012). Estimates from retail sale recordings suggest that natural and organic beef sales account for only 3.3% of the US market according to a survey commissioned by the National Cattleman's Beef Association (2016). These programmes have experienced periods of growth, but have always remained under 5% of beef production. Recently, USDA Market News has worked with State Departments of Agriculture and other food systems to increase availability of quantity and price information for alternative beef production and beef sold through local and regional markets through their Local & Regional Food Market News (USDA, 2016).

Growth in all of these alternative beef production systems in the near term will be limited only to the extent that consumers are willing to pay price premiums for the alternative products because costs of production often are increased. Furthermore, if demand and willingness to pay premiums is reduced or non-existent for lower-end products and by-products, additional price pressure is placed on the premium products to maintain a profitable system (Johnson and Mathews, 2010). Further growth of the alternative systems may be limited by constraints such as pasture availability, reduced feed efficiency resulting in higher feed utilization and costs associated with programme compliance.

This chapter will focus on nutrition of growing cattle post-weaning through the backgrounding or stocker phase and the finishing phase. The beginning point will be at a traditional weaning age of approximately eight months of age with an additional focus on the nutritional impacts on calves that are weaned earlier. Production practices that fit within both niche markets and conventional production will be evaluated for each segment of the industry. The focus will be primarily on how the nutrition in a single segment impacts marbling or quality grade of the final products, but some other aspects such as tenderness, flavour, lipid or nutrient profiles, and shelf-life will be included when appropriate.

## 2 Early weaning nutrition

It has been suggested that there is a window of opportunity for improving future carcass marbling that is open until shortly after a traditional calf weaning age or by the time a calf is 250 days old. Du et al. (2013) suggested that there is a transition from adipocyte growth due to hyperplasia (increasing cell numbers) to adipocyte growth due to hypertrophy. Early weaning has been proposed as an opportunity to impact adipocyte growth through nutrition. The adipocyte growth could be impacted through total energy supply or specific substrates to the calf. The lipogenesis process for intramuscular fat deposition utilizes two substrates that can be affected by nutrition: fatty acids and glucose. Fatty acids are produced by bacterial fermentation of feed in the rumen to volatile fatty acids (acetate, propionate and butyrate) and glycerol 3-phosphate, which is commonly supplied by glucose from absorption through the small intestine or production from glycogen stored in the liver.

The total quantity of volatile fatty acids can be impacted by total energy intake. The proportion of acetate and propionate produced can be shifted on the basis of the diet consumed with grain diets having greater proportions of propionate. Smith et al. (1984) conducted a study that compared growing cattle that were consuming similar energy intakes from a grain-based or forage-based diet. Despite the relatively close levels of total energy intake, the grain-fed cattle had greater quantities of backfat and lipogenic enzyme activity. This study provides evidence that glucose has more of a role in the regulation of fat deposition than acetate and that it is independent of energy availability. Glucose supplies to the animal can be impacted by the supply of digestible starch to the small intestine (Richards and Hicks, 2007) and by protein supplies to the rumen and intestines (Richards et al., 2002).

Early weaning of calves can be done as early as 45 days of age versus traditional weaning at approximately 205 days of age. Early weaning has been considered a management tool to reduce nutritional demands of cows during times of limited nutrient availability from the forage base such as during a drought. Early weaned calves are highly efficient at converting feed to live weight gain but require a high-protein, energy-dense diet. It has been reported that intramuscular fat is deposited at a faster rate relative to subcutaneous fat in calves compared with yearlings (Wertz et al., 2001; Ahola, 2014). If intramuscular fat is preferentially deposited in early weaned calves, there may be opportunities to manage the nutrition of calves early in their lives to positively impact the quality of beef at harvest.

When like genetics and finishing systems have been compared, it appears that early weaning management can improve quality grades. Myers et al. (1999a) compared two calf crops of early weaned steers (approximately 160 days of age) that were placed directly in the feedlot to normal weaned calves weaned at 213 days of age and placed on a finishing diet. This study reported a 40% improvement in Average Choice or higher USDA quality grades (more marbling) when calves were early weaned. Steers were harvested on the basis of a common backfat at two and four endpoints in year one and two, respectively. Year two resulted in early weaned calves having heavier carcass weights. In both years there was a tendency for the early weaned calves to have higher yield grades (lower meat yields). In other studies, these authors (Myers et al., 1999c) weaned steers at 117 days of age on a 77% corn diet or rotated calves through tall fescue, orchard grass and smooth brome grass pastures with 0.9 kg of corn supplemented daily for 81 days. After the 81-day period, grazing steers were placed on the corn diet until finished. Ultrasound was used to determine a common backfat harvest point. Forage fed steers were in the feedlot 46 days less, but were 36 days older when harvested. In this study, grazed steers compensated in the feedlot to the point that there were no differences in carcass weights, yield grades or quality grades. Overall, the amount of concentrate fed in the two systems was not different. Therefore, the grazing system merely lengthened the time cattle were owned before marketing. In another study, Myers et al. (1999b) studied weaning ages of 90, 152 and 215 days of age. When finished to a constant backfat thickness, the number of days on feed decreased with increased weaning age, but no differences were determined in carcass weight, yield grade or marbling score. In this study, all groups graded over 90% USDA Choice.

While it has been most common for early weaned steers to receive a high-energy diet or grazed with large quantities of supplementation, it is possible that high-quality forages could be utilized. Arthington et al. (2005) weaned 20 crossbred steers at 89 days of age and 20 at an age of 300 days old. Early weaned steers were placed on annual ryegrass followed by perennial star grass with 1% of body weight supplementation of a commercial feed that was approximately 65% total digestible nutrients and 14% protein. When the 300-day-old steers were weaned, all calves were placed in the feedlot. The early weaned calves were 48 kg lighter at feedlot arrival, but had greater average daily gains over the first 112 days at the feedlot. Over the entire finishing period, there were no differences in feed efficiency, final backfat thickness, yield grade, marbling score or hot carcass weight. This study suggests that forage-based nutrition programmes during the early weaning period did experience improved gain at feedlot arrival. The improvements in early gains at the feedlot could be from lighter entry or reduced stress as they had been previously weaned.

The rate of gain between early weaning and traditional weaning can have an impact on subsequent finishing. Studies were conducted at two locations that compared early weaning at approximately 80 days to normal weaning at approximately 213 days of age (Waterman et al., 2012). At both locations, 32% forage diets were fed to early weaned steers until the normal weaned steers were weaned. All calves from one location were sent to a commercial feedlot and the second location to a university facility. Early weaned steers from both locations gained faster during the period from early weaning to traditional weaning time, but there was also a large difference in the gains between the two locations. Steers at the commercial facility gained an average of 0.98 kg per day and steers at the university facility gained 1.48 kg per day during the period from early weaning to weaning. At the commercial feedlot, early weaned steers averaged 36 kg heavier initially. All steers were fed the same number of days. Early weaned calves had 8.5 kg heavier carcasses but no difference in marbling score or quality grade distribution. These steers did have an average of 24% yield grade 4 (excessively fat) carcasses. At the second feedlot, early weaned steers averaged 42 kg heavier initially. Early weaned steers were on feed an average of 138 days, while normal weaned steers were on feed for 153 days. Early weaned steers averaged 18 kg heavier carcasses, greater backfat and higher marbling scores. All steers in the experiment were USDA Choice or above, with a greater percentage of early weaned calves grading in the upper 2/3 of the Choice Grade ('premium Choice'). The more rapid gain during the post weaning period of the early weaned steers in the second location decreased the number of days in the feedlot and resulted in greater marbling scores.

These studies suggest that there is potential to develop systems with early weaned calves that can result in very high-quality grades. A stronger understanding of how factors such as weaning age, growth rates and nutrient substrates fed during the early weaning period interact with subsequent production phases will be helpful in identifying systems that produce high-quality beef to receive premiums while maintaining efficiency and carcass weights to produce the greatest economic returns.

# 3 Nutrition in stocker or backgrounding systems for steers and heifers

'Stocker' or 'backgrounding' are terms used in the United States to describe systems to grow steers and heifers between the cow/calf and feedlot phases. These systems can be used to distribute highly seasonal production of calves from the cow/calf industry into a more constant supply of cattle for finishing and beef production. Stocker or backgrounding systems can also be used to increase the entry weight of cattle in the feedlot to 300 to 400 kg by growing them at rates of gain from 0.35 to 1.15 kg/day until they are 12 to 15 months of age (NASEM, 2016). These gains usually are very economical. In the United States, it is estimated that 76% of annual calf crop enters a stocker or backgrounding programme prior to finishing (Peel, 2003). Growing programmes are diverse and regionally can be based on annual forages, perennial forages, crop residues, harvested forages or silages, or limit-fed, high-grain diets to attain the desired growth rates.

The low-to-moderate growth rates that have been targeted during the stocker or backgrounding phases are to increase frame size as the body weight increases before entering the feedlot. The extent of the growth phase can be related to the biological type of the cattle where smaller framed or early maturing calves are grown longer when compared to larger framed calves. This results in a more constant weight at feedlot arrival. This extensive use of backgrounding indicates that any nutritional factors that carry over through the finishing period and affect final beef quality could have a significant impact on the industry. A number of recent summaries evaluating the impact of stocker and backgrounding factors have been published (Drouillard and Kuhl, 1999; Berger and Faulkner, 2003; Anderson and Gleghorn, 2007; Reuter and Beck, 2013; and Lancaster et al., 2014).

## 3.1 Concentration and quantity of starch intake: complete diets

A meta-analysis comparing the impacts of growing phase complete diets with high (79.4 or 72.8%), medium (45.2%) and low (16.4%) average levels of starch from grains was conducted by Lancaster et al. (2014). Comparisons were made between high and medium and high and low separately, with comparisons that had similar total energy intakes resulting in similar average daily gains. The only difference in feedlot performance or carcass characteristics in the analysis was an increased rate of feedlot gain for steers on high-starch growing phase diets compared to steers on low-starch diets during the growing phase. The diets fed in the original studies included corn (Wagner, 1988; Loerch, 1990), corn and sorghum (Sip and Pritchard, 1991; Coleman et al. 1995; and McCurdy et al., 2010), or corn and wheat middlings (Vasconcelos et al., 2009) as starch sources. The analysis consistently showed little to no difference in marbling scores due to starch level in the growing phase diets. The study of Sainz et al. (1995) compared an alfalfa hay-based diet to a corn and wheat-based diet. Despite the variation in starch level of growing diets fed, when energy intake was held constant, no impact was reported on the finishing parameters of steers. Collectively, these studies indicate that the level of starch in the backgrounding diet can have beneficial implications on finishing performance, but does not have an impact separate from energy level on marbling scores.

**Figure 4** Grain mix for cattle.

## 3.2 Forage system

In forage-only systems, impacts on quality of beef could be from the forage type or rates of gains attained on the grazing systems. Extensive diversity in soils, precipitation and forages that affect forage production and quality make vast comparisons difficult. In two studies looking at differing rates of gains on grass pastures (Klopfenstein et al., 2000) and wheat pastures (Hersom et al., 2004), quality grades after a finishing period were not impacted by the stocker programme when adjusted to a common backfat. A comparison of wheat and native range systems was conducted by Choat et al. (2003). They found that grazing wheat pasture resulted in heavier feedlot entry weights (28 kg) and increased carcass weights, longissimus muscle areas and quality grades compared to native range; there was no difference in fat thickness or yield grade. These studies suggest that grazing forages with divergent rates of gain may have an impact on subsequent quality when cattle are finished to a common endpoint.

## 3.3 Starch versus fibre energy supplementation on pasture

Providing supplemental energy to grazing cattle will supply a comparatively low percentage of the total dietary nutrients compared to complete diets. This lowered level of supplement consumption could affect responses subsequent to the grazing phase. Comparisons can be made for benefits of providing supplements and also the form of supplemental energy supplied.

To look at the effect of providing different levels of supplementation, Lake et al. (1974) supplemented corn at 0, 0.23, 0.45, 0.91 and 1.82 kg per day to steers grazing irrigated pastures containing orchard grass and smooth brome grass. As expected, weight gains improved as supplementation increased. The two higher levels of supplementation increased quality grade slightly. In a similar study, supplementing grain sorghum (Lomas et al., 2009), supplementation levels of 1.64 kg per day also increased quality grades. However, in a second study by Lake et al. (1974), supplementation at 6 levels up to 2.72 kg per day resulted in no difference in 12th rib fat thickness or marbling. A study by Carter et al. (2002) that evaluated intramuscular fat, stocker phase body weight and feedlot placement weights suggested that marbling was occurring after steers were at 66% of their final weight.

Studies have compared supplements on the basis of starch- or non-starch-based energy sources with steers grazing winter wheat pasture (Horn et al., 1995), ryegrass (Bumps, 2006) or dormant winter native range (Sharman et al., 2009). These studies consistently demonstrated that when supplements of up to 1% of body weight are supplied to grazing cattle, the form of energy supplement had no impact on marbling scores after a finishing phase.

In a regression analysis of three publications, Reuter and Beck (2013) determined that body weight during the stocker phase had no impact on quality grades when cattle were harvested at a common fat thickness. This concept was furthered in the meta-analysis of Lancaster et al. (2014), who concluded that stocker-phase rate of gain can impact feedlot performance, but stocker rate of gain and supplementation have minimal impact on quality grades. These are particularly true when cattle are finished to a common fat thickness. Assuming that fat thickness and marbling are not always strongly correlated, opportunities might exist for targeting common marbling points in coordinated systems.

For systems that are not focused on marbling, increasing energy supplied through supplementation or using high-quality forages will increase rates of gain. If you combine the conclusion of the early weaning work that suggests adipocyte hyperplasia is reduced after 250 days of age and the calculation that minimal marbling occurs before calves reach 66%, it could be concluded that there is a period of time that management could be focused for lean growth. This period could be evaluated for growth rates that maximize muscle growth without excessive fat deposition.

# 4 Nutrition in feedlots/finishing systems

Finishing of cattle on high-energy diets started as an opportunity to increase food production and consistency of beef products, while utilizing grains that were an economical source of energy for growing and finishing cattle. Grain-fed cattle are the predominant form of beef marketed in the US and Canadian marketing systems. These cattle are marketed on the basis of quality grades consisting of (USDA Select, Choice, Prime (Fig. 5); or Canadian A, AA, AAA, and Prime) with yield grades categories of 1 to 5 in the United States and 1 to 3 in Canada. Yield grades 5 and 3, respectively, have the lowest meat yields.

**Figure 5** USDA Choice Quality Grade steaks.

## 4.1 Calf-feds versus yearlings

Calves can be placed directly into the feedlot after weaning or go through a backgrounding system before feedlot entry as yearlings (Fig. 6). In general, smaller framed calves often go through a backgrounding programme before feedlot entry. Calves that are larger framed and heavier at weaning may go through a 45–60-day weaning period before being placed on finishing diets in the feedlot. The feeding of lighter weight calves in a feedlot after weaning is more common when the cost of putting the gain on the calves is more cost-effective in the feedlot than in forage-based grazing systems. This could be due to high grain availability or short-term or long-term limited forage availability. The balancing of age, weight, biological type and economic opportunities ultimately determines the path that calves take in the beef production chain.

**Figure 6** Calves weaned at a traditional age of about 200 days.

There has been a perception that, on average, yearling cattle will have higher-quality grades after the finishing period than calves placed directly in the feedlot (Berger and Pyatt, 2005). They suggest that much of that perception comes from calves that were 600 pounds being placed in the feedlot on grower diets producing moderate gains for less than 100 days and then finished for 100–150 days. Often, cattle in those management systems grade 40–50% low Choice (Berger and Pyatt, 2005). Winterholler et al. (2008) conducted research to compare fall born calves either finished as calves or yearlings. The calf feds were preconditioned for a short period and placed directly into the feedlot, while yearlings were grazed on wheat pasture for 164 days before entering the feedlot. Yearlings were 217 kg heavier at feedlot entry and fed for 81 days fewer, while calf feds had more backfat. With these systems, there were differences in final carcass weights and calf feds were fatter, but there was no difference in marbling scores.

Griffin et al. (2007) used eight years of data with purchased calves that were separated by initial weight for placement on calf-fed or yearling systems. The heaviest calves (293 kg) were placed directly into the feedlot, while lighter calves were grazed on corn residue and then on summer grass. Calf feds were finished for an average of 168 days, while yearlings were finished for an average of 90 days. Yearlings had 24 kg heavier carcasses and calves were slightly fatter with 0.15 cm more fat cover. However, there was no difference in quality grades.

These two studies demonstrate that the use of weight or random splitting of calves for feeding in calf or yearling production systems can result in carcass and fat thickness differences between systems. In a review of industry data, Anderson and Gleghorn (2007) used initial feedlot entry weight to compare carcass quality endpoints. They concede that feedlot entry weight is not always equal to age, but that on average, lighter cattle will be younger going into the feedlot than heavier cattle. Their data summary suggests that calves placed in the feedlot at approximately 250 kg have lighter carcasses, but have a slight advantage in percent of cattle grading USDA Choice or higher than cattle placed at approximately 385 kg. They note that the quality grade spread increases when cattle enter the feedlot at greater than 410 kg. The data from a number of sources agree that quality grades are not commonly lower for calf-fed cattle. The management of growth rates, time on finishing diets, genetic makeup and management of harvest endpoints may be more indicative of quality grades rather than simply age when placed on finishing diets.

## 4.2 Grain processing and type

Grain type and processing methods impact on quality grade have been extensively evaluated. Grains can have different energy levels and rates of digestion in the rumen (Britton and Stock, 1987). Wheat and barley are the most rapidly fermented, with high-moisture corn, steam-flaked corn, rolled corn and flaked milo intermediate. Whole corn and dry rolled milo are the slowest to break down in the rumen. In a review of processing methods and grain types, Owens and Gardner (2000) presented data demonstrating that steam flaking of grains reduces marbling compared to dry-rolling grains. They did not report a difference in marbling between corn, milo and wheat diets when processed in the same manner. However, a later study directly comparing steam-flaked corn and steam-flaked sorghum did not result in differences between the two grains (Duff et al., 2002).

Processing methods such as grinding and steam flaking not only change the rate at which the grain is degraded, but also tend to improve energy availability of the feed (Owens et al., 1997). As energy availability is increased, it is commonly associated with reductions in feed intake that improve feed efficiency, but do not dramatically impact the energy intake of the cattle. Krehbiel et al. (2006) found that increased metabolizable energy in feedlot diets increased average daily gains, improved feed efficiency and increased subcutaneous fat, but not muscling or marbling. Anderson and Glehorn (2007) conducted a database analysis on gain, feed intake, feed conversion, final weight and yield grade 4's of cattle that started in the feedlot at 700 to 750 pounds. They found that none of the factors had a significant relationship with quality grades.

Different grain sources and processing methods can also affect the proportion or quantity of starch digested in the rumen and small intestine. Changing the location of digestion can affect the energy substrates available from digestion for further metabolism (Richards and Hicks, 2007). Steam flaking of corn will shift digestion to the rumen as compared to whole or dry-rolled corn. The shifting of digestion would result in more starch being digested to volatile fatty acids in the rumen, which would reduce the quantity and potentially the availability of starch reaching the intestines to be digested to glucose. The quality grade advantages in marbling reported by Owens and Gardner (2000) for dry rolled grains over flaked grains could have been because of altered sites of starch digestion with dry-rolled grain diets increasing the amount of starch reaching the intestinal tract.

Generally, different grain sources and processing methods have significant impacts on rates and sites of starch digestion, diet intake levels and energy density of the diets. These changes can impact rates of grain and feed efficiency, but marbling responses have been more variable when finished cattle are evaluated at similar physiological endpoints.

## 4.3 By-product feeds

Growth of the grain milling industries for ethanol and sweeteners expanded rapidly within the last decade. In the early 1980s, ethanol production was below 200 million gallons and increased to 25 billion gallons worldwide with 14 8 billion gallons produced in the United States according to the Renewable Fuel Association. The sweetener industry in that country has been less dynamic with an estimated growth from 6 634 000 short tons (dry weight) in 1992 to 8 500 000 in 2015 (USDA Economic Research Service). With the growth of the grain milling industries, the by-products termed 'distillers grains' and 'corn gluten' feed have had increased availability for cattle finishing as both energy and protein sources (Fig. 7). Both dried and wet forms of products, with some variation in modified moisture or fat level, have become common feedlot ingredients. A survey of industry consultants (Samuelson et al., 2016) reported that 97.1% of their clients use grain by-products in their feedlot diets, which was up approximately 17% from a survey ten years earlier. Consultants reported that distillers grains were used extensively as wet (70.8%) and dry (16.7%) products. Corn gluten feed inclusion was reported as 8.3% wet and 4.2% dry of the diets. These numbers show that grain milling by-products have a large role in the diets of feedlot cattle. Such a large shift in dietary ingredients to a higher fibre product reduces the amount of starch fed and could impact beef quality.

As inclusion of milling by-products increased, concerns of reduced quality grades emerged. The largest part of the industry is using distillers grains that are characterized as high fibre, high protein and moderate fat. All three nutrients can have negative impacts on quality grades related to changes in both rumen fermentation profiles of volatile fatty acids and starch flow to the intestinal tract. Corrigan et al. (2006) reviewed distillers grain feeding data and concluded that optimum inclusion levels for feed efficiency were affected

**Figure 7** Dry distillers grains used in finishing cattle.

by grain processing methods. They recommended an inclusion level of 15–20% in steam-flaked grain diets and 27–40% in dry-rolled and high-moisture grain diets.

A pooled analysis of 21 studies showed that inclusion level of distillers grains (both wet and dry) in the diet affected quality grades (Reinhardt et al., 2007). At levels of inclusion of 16%, distillers grains increased marbling scores as it replaced grain in the diet. When levels of distillers grains inclusion exceeded 33%, quality grades declined below the control diets. In steam-flaked corn diets, levels of 0, 15, 30, 45, 60 and 75% dried corn distillers grains were fed to 330 kg heifers (Depenbusch et al., 2009). Increasing inclusion of distillers grains tended to increase marbling at the 15 and 45% inclusion levels and declined to a level similar to the control at 60% inclusion and were lowest at 75% distillers inclusion. The changes in marbling did result in a linear increase in the percent heifers grading Select with a rise starting at 45% distillers grains inclusion. These responses were associated with a linear decrease in fat thickness of 0.3 cm across the inclusion range. There is consistency that inclusion levels above 40% of distillers grains in feedlot diets will reduce feed efficiency and quality grades. Feeding with more rapidly digestible grain sources can result in negative effects at lower inclusion levels of distillers grains.

High levels of protein and fat in feedlot diets have been associated with reductions in marbling scores. In the study of Depenbusch et al. (2009), protein and fat levels increased from 14.0% to 19.7% and 3.7% to 8.0%, respectively, as distillers grains inclusion increased. Gunn et al. (2009) evaluated the effects of protein and fat from distillers. They had a dry-rolled corn control diet with 25% distillers grains, a 50% distillers grains diet and 25% distillers grains diets with the protein content, fat content or both balanced to the level of the 50% distillers grains diet using corn protein and vegetable oil. Their work showed that either high protein, fat or the combination decreased marbling scores and quality grades.

The feeding of distillers grains has had variable reports related to shelf life, tenderness and fatty acid content of meat. Shelf life was considered moderately unacceptable when distillers grains reached 40% of the diet (Roeber et al., 2005). Others have reported no difference in shelf life (Kinman et al., 2011). Depenbush et al. (2009) reported decreased tenderness as distillers grains levels increased, but others found no difference (Roeber et al., 2005; Kinman et al., 2011). Distillers grains feeding has not affected saturated or monounsaturated fatty acids, but has increased polyunsaturated fatty acids (Depenbusch et al., 2009; Kinman et al., 2011).

Corn gluten feed is lower in both protein and fat than traditional distillers grains. It does not tend to have the impact on marbling at higher feeding levels that distillers grains does. In steam-flaked corn diets, Block et al. (2005) fed 0, 20, 30 and 40% wet corn gluten feed. The increasing levels did not impact fat thickness or marbling scores. In a study feeding up to 50% wet corn gluten feed in dry-rolled corn diets, no differences in marbling scores were detected (Richards et al., 1998). In a second report (Richards et al., 1996), wet corn gluten feed was included to replace 50 or 100% of the dry-rolled corn in the diet. Even at the 100% inclusion level, quality grades were not impacted.

# 5 Vitamins A, E and growth promotors

Proper supply of minerals is a critical part of beef cattle nutrition programmes. Low levels have had a positive effect on marbling in cattle, but when supplies reach high levels, negative impacts have been recorded. It is thought that Vitamin A and D levels may impact marbling through adipocyte development (Kawada et al., 1996). Vitamin A is not produced by cattle, so requirements must be met through the diet or injections. Fresh-growing forages are good sources, while the concentration of Vitamin A is lower in mature forages, harvested forages or grains.

Practices involving vitamin A in finishing diets is highly variable. Feedlot consultants report that they supplement over two times the 2200 IU/kg recommended level of Vitamin A, but provide no scientific basis for doing so; most did not report supplementing Vitamin D (NRC, 1996; Samuelson et al., 2016). Bryant et al. (2010) evaluated vitamin A levels up to approximately 4 times the recommended levels. They concluded that the requirement level was adequate for finishing beef cattle, while higher levels had minimal impact on performance or marbling. Similar conclusions were derived by others (Pickworth et al., 2012a), although a second study by the same researchers had a tendency for reduced vitamin A to increase marbling (Pickworth et al., 2012b). Others have suggested that removal of vitamin A from finishing diets increases the marbling to 12th rib fat ratio (Arnett et al., 2009). Some studies suggest that Vitamin A removal from finishing diets for cattle under 15 months of age is more effective than that for older cattle (Oka et al., 1998). Most of these studies utilized Japanese black or Waygu cattle that have a very high propensity for marbling deposition. One study has suggested that vitamin A removal from feedlot diets has potential to increase colour display life and reduce lipid oxidation (Daniel et al., 1998).

It has been suggested that Vitamin A could be at least partially involved in seasonality of quality grades due to forages grazed before placement into the feedlot (Pyatt and Berger, 2005). These authors reviewed the potential of vitamin A in relation to seasonal quality grade lows associated with cattle that graze growing forage with high concentrations of Vitamin A in the spring and are finished during the summer with long daylight hours that could increase circulating Vitamin D. They suggest a possibility of a reduction in adipocyte hyperplasia and reduced marbling scores related to high vitamin levels. In contrast, Knock et al. (2007) supplemented 500 mg of 25-hydroxyvitamin $D_3$ to grain and grass finished steers and did not detect differences in beef quality. While further investigation is needed, there is certainly enough evidence that evaluations of stocker system impacts on quality grades should include an evaluation of mineral levels along with other factors.

Extensive research has been conducted to evaluate vitamin E supplementation for receiving and feedlot cattle. Experiments conducted have analysed feeding very high levels (supranutritional) of vitamin E for a short duration during receiving at the feedlot, feeding supranutritional levels for a short period prior to harvest, and feeding moderate levels for longer durations prior to harvest. While vitamin E supplementation does not have a consistent effect on animal health and performance (Secrist et al., 1997; Cusack et al., 2009), it has been shown that dietary vitamin E can delay metmyoglobin formation in fresh meat cuts. Sanders et al. (1997) evaluated feeding levels of 0, 1000, and 2000 IU per day of vitamin E supplementation for 100 days followed by freezing vacuum packaged stakes for up to 100 days. They concluded that for steaks stored less than 100 days, 1000 IU resulted in an approximately 60% increase in tissue alpha-Tocopherol concentration. Steaks from steers fed 1000 IU exhibited superior lean colour, less surface discoloration and less lipid oxidation during retail display. Japanese consumers preferred steaks from cattle supplemented with vitamin E 91% of the time with 58% of participants indicating muscle colour as the most important selection criteria. From a series of three experiments, Arnold et al. (1992) concluded that vitamin E supplementation, either by long, low dosage or by short, high dosage, extended the colour stability of fresh beef displayed during simulated retail display.

It has been demonstrated that feeding of distillers grains products in finishing diets can increase polyunsaturated fatty acids that make beef more susceptible to oxidative rancidity and reduce the shelf life (Kinman et al., 2011). Steers fed vitamin E for approximately 100 days prior to harvest have shown to have improved colour stability and increased shelf-life. de Mello Jr (2010) conducted an experiment with diets containing 35% distillers grains and levels of 0, 1000 and 2000 IU per day of vitamin E and concluded that 1000 IU resulted in improvements that were not increased by feeding 2000 IU. In a subsequent study, levels of 250 to 500 IU were determined to be sufficient to improve colour and shelf-life in finishing diets containing 35% distillers grains (Bloomberg et al., 2011). Economic estimations of supranutritional levels of vitamin E have indicated a benefit ratio of 8 to 10:1 (Liu et al., 1995; Smith et al., 1996), but industry estimations suggest that the practice is followed only in a low percentage of finishing cattle (Shaefer 2002).

Implanting growth promotors can affect the fat deposition and particularly marbling of cattle. Bruns et al. (2005) compared early versus delayed implanting with estradiol–trenbolone acetate implants on marbling scores. In their work, implanting time did not affect overall performance. Early implanting did result in reduced marbling scores when compared to no use of implants or delayed implanting. Roeber et al. (2000) compared marbling scores and consumer tenderness ratings of steers that received no implant or 7 combinations of implants. Control steers had higher marbling scores than all but 2 of the 7 combinations. Consumer-rated tenderness was greater for non-implanted steers than all but one implant combination. Maxwell et al. (2014) compared carcasses of steers that were produced under natural guidelines with no implants during the backgrounding or finishing phases of production to steers that received implants during backgrounding and finishing along with a beta agonist prior to harvest. The natural calves had lighter carcasses, smaller longissimus muscles and tended to have a greater percentage of carcasses grading Choice or greater. The use of the beta-agonist zilpaterol hydrochloride and ractopamine has been the review of numerous studies that have been subjected to a meta-analysis (Lean et al., 2014). They summarized that both beta-agonists increased weight gain, carcass weight and longissimus muscle area. Fat thickness was not affected by ractopamine but was decreased by zilpaterol hydrochloride. Both beta-agonists resulted in decreased tenderness, as estimated by Waren-Bratzler shear force measures, with zilpaterol hydrochloride having a greater impact.

# 6 Grass-fed and/or forage-based finishing systems

Discussion of forage-based systems requires definitions of both the production parameters for producing the product as well as the targeted endpoint. The name forage-based system has been applied to systems that utilize only forages, systems that use supplementation during parts or throughout the production cycle, and high levels of supplementation or feeding of cattle managed on pasture. As noted before, the grass finishing term as defined by USDA only applied to providing a forage system with little additional inputs. This would be similar to the American Grassfed Association Grassfed label. However, the Grassfed label also requires that no hormones, antibiotics, animal by-products or organophosphates are used. The Grass Pastured label by the American Grassfed Association allows supplementation levels that would be close to one third of the dry dietary intake at the end of the feeding period. A major draw to grass-fed beef has been a higher omega-3 fatty acid concentration per unit of total fat and higher ratios of omega-3 to other fatty acids (Mathews and Johnson, 2013). The more recent draw of consumers to these products has been a belief that forage-based feeding systems result in a healthier product by having a more desirable fatty acid profile and being produced under systems that would follow the 'Never Ever 3' programme for hormones, antibiotics and animal products. These products are often believed to be associated with local production by small producers. The challenge for consumers is that there is no broad consistency in these types of programmes and that these are common descriptors used in local marketing or direct marketing programmes, which can have great diversity in their true practices.

Another component of increased interest in the grass finished systems stemmed from beliefs that not feeding grain to cattle is better for the environment. The reality is that environmental impact is complex (Brocks et al., 2015). Grain-finishing cattle results in a lower carbon footprint than grass finishing due to a more efficient utilization of feed during the finishing period. Grass-finishing may have sustainable advantages in terms of reduced competition with humans when raised by grazing forage resources that are grown on lands not reasonably capable for human food production or on residues of cropping systems.

Endpoint targets also vary in forage-based feeding systems. Some grassfed systems may target a product that is very lean and has a 'grass' flavour where carcasses are typically lighter, have a yellow colour to the fat, darker lean and are more advanced in their physiological maturity than grain-fed systems. Other systems will target a product that has a higher level of finish to increase marbling sufficient to reach a Select or even Choice quality grade. There is a long history of forage-based system research including a focus on product quality and consumer acceptability. A book, *Forage Fed Beef: Production and Marketing Alternatives in the South*, was published from papers presented at a symposium and demonstrates a history of more than 40 years of research on forage management and feedstuff supplementation to produce consumer desirable beef products. Despite the long history of research focused in this area, it is challenging to sort out consistent effects of nutrition on beef quality due to variability of the systems compared and the variability of endpoints.

Supplementation of calves grazing forage allows for greater rates of gain that reduces the length of time before calves are ready for market and allows more opportunity to include lower-quality forages. A study of supplementation effects was conducted by Pugh (2003) where all weaned calves were either grown on cool season perennial pastures with winter annual grazing or cool season perennial pastures with 1% of bodyweight soybean hull supplementation. The calves were grazed through the following summer just prior to the weaning of the subsequent calf crop. This was conducted on the same pastures for two consecutive years with two calf crops. At harvest, the supplemented calves had increased carcasses weights, longissimus muscle areas, fat thicknesses and marbling scores (Baublits et al., 2004). Supplementation did decrease polyunsaturated and n-3 fatty acids, but the n-6 to n-3 ratio remained below 4:1 and the conjugated linoleic acid did not decline (Baublits et al., 2006a). In these 19-month-old steers and heifers, tenderness was not affected, but the grassy flavour was decreased by supplementation. Comparison of the steaks to USDA Choice and Select steaks purchased at retail resulted in the forage fed steaks having greater proportions of conjugated linoleic acid and lower n-6 to n-3 ratios than the retail steaks (Baublits et al., 2006b). Grassy flavour was the most intense in the non-supplemented steaks. The use of supplements in this perennial forage system allowed for the harvest of calves within 12 months after weaning with carcass weights and quality characteristics similar to conventionally raised cattle.

Along with the differences created from different feeding systems, use of technologies such as implants can play a role in performance and beef quality. Berthiaume et al. (2006) compared a grass silage-based growing system with graded levels of protein supplementation and with or without implants during a 98-day growing phase. These same programmes were carried through a finishing period where protein supplementation was switched to grain feeding levels. Both implants and feeding increased carcass weights and reduced days on feed. Grain feeding reduced days on feed by more than 20% and increased backfat, marbling scores and quality grades, while there was no difference in lean or fat colour. In this case, fat colour was evaluated on a five-point scale with 5 as white and treatments averaged 4.0. The steaks from the forage only treatment were more tender, had increased n-3 fatty acids and improved n-6 to n-3 ratio (Faucitano et al., 2008). Grass finishing without implants did have carcass weights that were 15% below the grain supplemented and implanted treatments. This study attempted to target common end points rather than similar days in the system, but the silage-only treatments fell short on carcass weight and marbling, which suggests that more than an additional 20% of time would be required to reach a common end point.

When evaluating quality differences between systems, bringing animals to a common finishing body weight and similar fat thickness usually makes the best comparison of systems. Muir et al. (1998) conducted a review by combining 15 studies with grain and forage-based feeding systems where all animals were taken to a common end weight or degree of fatness. At equal endpoints, calves had no difference in tenderness, juiciness or lean colour. When flavour was measured, two thirds of the time panellists could not detect a difference in flavour and no difference was noted in fat colour. These authors did not evaluate shelf life, which can sometimes be reduced in forage systems because of more lipid and lean oxidation.

Characteristics of individual forages could have an impact on subsequent beef quality. A comparison of alfalfa, pearl millet and mixed pastures was conducted where those forages were grazed the last 40 days prior to harvest (Duckett et al., 2013). There was no difference in carcass weights or marbling scores that averaged low Select quality grade.

Alfalfa and pearl millet grazing resulted in greater beef flavour than the mixed pastures. Alfalfa grazing increased linoleic acid compared to the other forages. When the forage systems were compared to a grain-based finishing system, grain finished steers did have higher marbling scores, but increased Omega-3 polyunsaturated fatty acid concentration and decreased n-6 to n-3 ratio.

Schmidt et al. (2013) finished steers on five different forages. There was some variation in fat thickness and quality grades, but no difference was detected in marbling scores. The legume forages (alfalfa and Cowpeas) improved estimates of tenderness and consumers' preference. A comparison of spring grazing of three warm season forages was compared for carcass characteristics after a common wintering on baleage and spring finishing period grazing tall fescue (Kurve et al., 2015). Forages evaluated included Indian grass, Bermuda grass and a mix of big bluestem, little blue stem and Indian grass. Grazing Indian grass resulted in 67% of the carcasses grading Select (approximately 3% intramuscular fat), whereas Bermuda grass grazing resulted in only 17% of the carcasses grading Select, with the rest grading lower (Standard). In these grass-based finishing systems that resulted in relatively low marbling scores, grazing immediately before harvest resulted in some detectable flavour differences, whereas forages grazed before a long common finishing period resulted in differences in marbling scores but no consistent responses in fatty acid profiles or consumer acceptability. These results suggest that forages fed immediately before harvest can influence the characteristics of the final product. Differences from forages fed earlier in the system are most likely related to animal performance.

The effects of different systems and targeted endpoints can have significant impacts on the tenderness, marbling, flavour, and colour of beef. Forage-based systems can be used to produce either lean beef with grassy flavours or carcasses that are comparable to those from conventional beef production. Systems can include intensively managed, high-quality forages, supplementation to increase rates of weight gain from a forage base, or some combination and produce high-quality beef. In many cases, management systems can retain the healthy fatty acid profiles associated with forage-fed beef. Forage-based systems usually require longer periods of time to reach an endpoint similar to conventionally finished beef due to slower rates of gain. If rates of gain in the systems are substantially different, carcass differences associated with age, such as tenderness and colour, may be impacted. If systems are compared at similar endpoint in terms of age and fat thickness, differences between systems are reduced.

# 7 Summary

Previously, it had been suggested that the finishing phase was all that affected quality (NRC, 1996), but Owens and Gardner (2000) suggested that other factors can affect body and meat composition. There is evidence that nutrition at all stages of growth has the potential to impact final beef quality. It is clear that marbling can be developed before the finishing phase by providing sufficient energy intake for muscle growth and marbling development. The form of nutrients supplied, especially starch, can have a positive influence on marbling by theoretically altering total energy supplied, by an impact on rumen metabolism, by quantity of substrate digested in the rumen, and/or by the proportion of starch digested in the intestine. While energy is a major driver of marbling, other nutrients can influence marbling development and should be considered.

The beef industry is becoming more integrated through retained ownership and coordinated supply chains, which creates opportunities to manage quality throughout the life of a calf. A better understanding of lifetime management on quality could lead to systems in which marbling can be developed with reduced production of waste fat. An improved understanding of marbling development could be used to improve production on the basis of parameters such as weight, maturity, initial fat thickness and initial marbling that can be estimated at feedlot arrival. Furthermore, consumers are increasingly having interest in the production of their food products and applying that to their product choices. Regardless of their choices, they maintain that the products must be safe and consistent and meet their expectations. An improved understanding of development of quality in beef systems will allow producers to develop systems focusing on consumers' expectations.

# 8 Where to look for further information

Additional scientific studies can be found in trade journals such as the *Journal of Animal Science*, *Professional Animal Scientist* and *Meat Science*. Extension web sites such as BeefExtension.com and the IowaBeefCenter.org have many materials related to beef cattle production. Oklahoma State University Extension Service offers a Beef Cattle Manual that addresses many areas of beef cattle production and can be found at http://www.agecon.okstate.edu/cattleman/order.asp. Many United States programme information and statistics can be found at https://www.ams.usda.gov/market-news/livestock-poultry-grain.

# 9 References

AGA (American Grassfed Association). 2016. Grassfed and Grass Pastured Ruminant Standards. Available online at http://www.americangrassfed.org/wp-content/uploads/2010/06/AGA-Grassfed-Standards-2016.pdf.

Ahola, J. K. 2014. Effects of the 60-d window around the time of weaning on subsequent quality grade and eating quality of beef. Available online http://www.cabpartners.com/articles/news/2709/60-d_window_at_weaning_Ahola_CAB_FINAL-to-CAB_8-15-14.pdf. Accessed 14 August 2016.

Anderson, P., and J. Gleghorn. 2007. Non-genetic factors that affect quality grade of fed cattle. In Proceedings of Beef Improvement Federation 39th Annual Research Symposium and Annual Meeting, Fort Collins, CO, pp. 31–43.

Arnett, A. M., M. E. Dikeman, M. J. Daniel, K. C. Olson, J. Jaeger and J. Perrett. 2009. Effects of vitamin A supplementation and weaning age on serum and liver retinol concentrations, carcass traits, and lipid composition in market beef cattle. *Meat Sci*. 81: 596–606.

Arnold, R. N., K. K. Scheller, S. C. Arp, S. N. Williams, D. R. Buege and D. M. Schaefer. 1992. Effect of long- or short-term feeding of alpha-tocopheryl acetate to Holstein and crossbred beef steers on performance, carcass characteristics, and beef color stability. *J. Anim. Sci*. 70: 3055–65.

Arthington, J. D., J. W. Spears and D. C. Miller. 2005. The effect of early weaning on feedlot performance and measures of stress in beef calves12. *J. Anim. Sci*. 83: 933–9.

Baublits, R. T., A. H. Brown Jr., F. W. Pohlman, D. C. Rule, Z. B. Johnson, D. O. Onks, C. M. Murrieta, C. J. Richards, H. D. Loveday, B. A. Sandelin and R. B. Pugh. 2006a. Fatty acid and sensory characteristics of beef from three biological types of cattle grazing cool-season forages supplemented with soyhulls. *Meat Sci*. 72: 100–7.

Baublits, R. T., F. W. Pohlman, A. H. Brown, D. C. Rule, Z. B. Johnson, D. O. Onks, C. M. Murrieta, B. A. Sandelin, C. J. Richards, H. D. Loveday and R. B. Pugh. 2006b. Comparison of fatty acid and sensory profiles of beef from forage-fed cattle with retail United States Department of Agriculture Choice and Select Beef. *J. Muscle Foods.* 17: 311–29.

Baublits, R. T., A. H Brown Jr., F. W Pohlman, Z. B Johnson, D. O Onks, H. D Loveday, R. E Morrow, B. A Sandelin, W. K Coblentz, C. J Richards and R. B Pugh. 2004. Carcass and beef color characteristics of three biological types of cattle grazing cool-season forages supplemented with soyhulls. *Meat Sci.* 68: 297–303.

Berger, L. L., and N. A. Pyatt. 2005. Nutritional and Management Factors Affecting Marbling Deposition. Available online at http://www.cabpartners.com/articles/news/241/berger_nutrition.pdf. Accessed 25 September 2016.

Berger, L. L., and D. B. Faulkner. 2003. Lifetime nutritional effects on feedlot performance and carcass merit. In *Plains Nutrition Council Spring Conference Proceedings*. Publication No. AREC 03–13. Texas A&M Research and Extension Center, Amarillo, pp. 63–78.

Berthiaume, R., I. Mandell, L. Faucitano and C. Lafrenière. 2006. Comparison of alternative beef production systems based on forage finishing or grain-forage diets with or without growth promotants: Feedlot performance, carcass quality, and production costs. *J. Anim. Sci.* 84: 2168–77.

Block, H. C., C. N. Macken, T. J. Klopfenstein, G. E. Erickson and R. A. Stock. 2005. Optimal wet corn gluten and protein levels in steam-flaked corn-based finishing diets for steer calves. *J. Anim. Sci.* 83: 2798–805.

Bloomberg, B. D., G. G. Hilton, K. G. Hanger, C. J. Richards, J. B. Morgan and D. L. VanOverbeke. 2011. Effects of vitamin E on color stability and palatability of strip loin steaks from cattle fed distillers grains. *J. Anim. Sci.* 89: 3769–82.

Britton, R. A., and R. A. Stock. 1987. Acidosis, rate of starch digestion and intake. Symposium Proceedings: Feed intake by beef cattle. Available online at http://www.beefextension.com/proceedings/feed_intake86/feed_intake86_8.pdf

Brocks, A., E. Andreini, M. Rolf and S. Place. 2015. Carbon footprint comparison between grass- and grain-finished beef. Accessed online at http://pods.dasnr.okstate.edu/docushare/dsweb/Get/Document-10189/ANSI-3292.pdf.

Bruns, K. W., R. H. Pritchard and D. L. Boggs. 2005. The effect of stage of growth and implant exposure on performance and carcass composition in steers1. *J. Anim. Sci.* 83: 108–16.

Bryant, T. C., J. J. Wagner, J. D. Tatum, M. L. Galyean, R. V. Anthony and T. E. Engle. 2010. Effect of dietary supplemental vitamin A concentration on performance, carcass merit, serum metabolites, and lipogenic enzyme activity in yearling beef steers. *J. Anim. Sci.* 88: 1463–78.

Bumpus, E. K. 2006. Influence of acetogenic versus propiogenic supplements on adipose tissue accretion in stocker steers grazing ryegrass pasture. MS Thesis. Texas A&M University, College Station.

Carter, J. N., P. A. Ludden, M. S. Kerley, M. Ellersieck, W. O. Herring and E. Berg. 2002. Intramuscular fat deposition in steers is accelerated at a set body weight. *Prof. Anim. Sci.* 18: 135–40.

Choat, W. T., C. R. Krehbiel, G. C. Duff, R. E. Kirksey, L. M. Lauriault, J. D. Rivera, B. M. Capitan, D. A. Walker, G. B. Donart and C. L. Goad. 2003. Influence of grazing dormant native range or winter wheat pasture on subsequent finishing cattle performance, carcass characteristics, and ruminal metabolism. *J. Anim. Sci.* 81: 3191–201.

Coleman, S. W., R. H. Gallavan, C. B. Williams, W. A. Phillips, J. D. Volesky, S. Rodriguez and G. L. Bennett. 1995. Silage or limit-fed grain growing diets for steers: I. Growth and carcass quality. *J. Anim. Sci.* 73: 2609–20.

Corrigan, M. E., G. E. Erickson, V. R. Bremer and T. J. Klopfenstein. 2006. Effect of corn processing in finishing diets containing grain milling byproducts. Proceedings of the OSU Cattle Grain Processing Symposium, pp. 173–82. Accessed online at http://beefextension.com/proceedings/cattle_grains06/06–24.pdf.

Cusack, P., N. McMeniman, A. Rabiee and I. Lean. 2009. Assessment of the effects of supplementation with vitamin E on health and production of feedlot cattle using meta-analysis. *Prev. Vet. Med.* 88: 229–46.

Daniel, M. J., M. E. Dikeman, A. M. Arnett and M. C. Hunt. 2009. Effects of dietary vitamin A restriction during finishing on color display life, lipid oxidation, and sensory traits of longissimus and triceps brachii steaks from early and traditionally weaned steers. *Meat Sci.* 81: 15–21.

de Mello, A. S. Jr. 2010. Effects of feeding distillers grains on beef quality. Ph.D. Diss. Univ. of Nebraska, Lincoln.

Depenbusch, B. E., C. M. Coleman, J. J. Higgins and J. S. Drouillard. 2009. Effects of increasing levels of dried corn distillers grains with solubles on growth performance, carcass characteristics, and meat quality of yearling heifers. *J. Anim. Sci.* 87: 2653–63.

Drouillard, J. S., and G. L. Kuhl. 1999. Effects of previous grazing nutrition and management on feedlot performance of cattle. *J. Anim. Sci.* 77: 136–46.

Du, M., Y. Huang, A. K. Das, Q. Yang, M. S. Duarte, M. V. Dodson and M.-J. Zhu. 2013. Meat Science and Muscle biology Symposium: Manipulating mesenchymal progenitor cell differentiation to optimize performance and carcass value of beef cattle. *J. Anim. Sci.* 91: 1419–27.

Duckett, S. K., J. P. S. Neel, R. M. Lewis, J. P. Fontenot and W. M. Clapham. 2013. Effects of forage species or concentrate finishing on animal performance, carcass and meat quality. *J. Anim. Sci.* 91: 1454–67.

Duff, G. C., K. J. Malcolm-Callis and M. L. Galyean. 2002. Effects of proportions of steam-flaked corn and grain sorghum and switching grain sources during the finishing period on performance and carcass characteristics of finishing beef steers. *Prof. Anim. Sci.*18: 387–94.

Faucitano, L., P. Y. Chouinard, J. Fortin, I. B. Mandell, C. Lafrenière, C. L. Girard and R. Berthiaume. 2008. Comparison of alternative beef production systems based on forage finishing or grain-forage diets with or without growth promotants: 2. Meat quality, fatty acid composition, and overall palatability. *J. Anim. Sci.* 86: 1678–89.

FAO. 2012. Plant Production and Protection Division: Grasslands, Rangelands and Forage Crops. Rome, Italy: Food and Agriculture Organization. Available online at http://www.fao.org/agriculture/crops/thematic-sitemap/theme/spi/grasslands-rangelands-and-forage-crops/en/. Accessed 23 September 2016.

Forage-Fed Beef, Production and marketing alternatives in the south. 1977. Bulletin 220. Available online at https://www.ars.usda.gov/research/software/?badid=1.

Griffin, W. A., T. J. Klopfenstein, G. E. Erickson, D. M. Feuz, J. C. MacDonald and D. J. Jordon. 2007. Comparison of performance and economics of a long-yearling and calf-fed system. *Prof. Anim. Sci.* 23: 490–9.

Gunn, P. J., A. D. Weaver, R. P. Lemenager, D. E. Gerrard, M. C. Claeys and S. L. Lake. 2009. Effects of dietary fat and crude protein on feedlot performance, carcass characteristics, and meat quality in finishing steers fed differing levels of dried distillers grains with solubles. *J. Anim. Sci.* 87: 2882–90.

Hersom, M. J., G. W. Horn, C. R. Krehbiel and W. A. Phillips. 2004. Effect of live weight gain of steers during winter grazing: I. Feedlot performance, carcass characteristics, and body composition of beef steers. *J. Anim. Sci.* 82: 262–72.

Horn, G. W., M. D. Cravey, F. T. McCollum, C. A. Strasia, E. G. Krenzer and P. L. Claypool. 1995. Influence of high-starch vs high-fiber energy supplements on performance of stocker cattle grazing wheat pasture and subsequent feedlot performance. *J. Anim. Sci.* 73: 45–54.

Johnson, R. J., and K. H. Mathews, Jr. 2010. Alternative Beef Production Systems – What's in a Name?. USDA LDP-M-193. Available online at http://usda.mannlib.cornell.edu/usda/ers/LDP-M/2010s/2010/LDP-M-07-21-2010.pdf. Accessed September 25, 2016.

Kawada, T., Y. Kamei and E. Sugimoto. 1996. The possibility of active form of vitamins A and D as suppressors on adipocyte development via ligand-dependant transcriptional regulators. *Int. J. Obes. Relat. Metab. Disord.* 20(Suppl 3): S52.

Kinman, L. A., G. G. Hilton, C. J. Richards, J. B. Morgan, C. R. Krehbiel, R. B. Hicks, J. W. Dillwith and D. L. VanOverbeke. 2011. Impact of feeding various amounts of wet and dry distillers grains to yearling steers on palatability, fatty acid profile, and retail case life of longissimus muscle. *J. Anim. Sci.* 89: 179–84.

Klopfenstein, T., R. Cooper, D. J. Jordon, D. Shain, T. Milton, C. Calkins and C. Rassi. 2000. Effects of backgrounding and growing programs on beef carcass quality and yield. *J. Anim. Sci.* 77(E-Suppl.): 1–11.

Knock, R. C. 2007. Carcass and meat quality characteristics of pasture- and feedlot-finished beef steers supplemented with 25-hydroxyvitamin $D_3$ Ph.D. Dissertation. Iowa State University.
Krehbiel, C. R., J. J. Cranston and M. P. McCurdy. 2006. An upper limit for caloric density of finishing diets. *J. Anim. Sci.* 84: E34–49.
Kurve, V., P. Joseph, J. B. Williams, H. T. Boland, S. K. Riffell, T. Kim and M. W. Schilling. 2015. The effect of feeding native warm-season grasses during the stocker phase on meat composition, quality characteristics, and sensory properties of loin steaks from forage-finished cattle. *J. Anim. Sci.* 93: 2576–86.
Lake, R. P., R. L. Hildebrand, D. C. Clanton and L. E. Jones. 1974. Limited Energy Supplementation of Yearling Steers Grazing Irrigated Pasture and Subsequent Feedlot Performance. *J. Anim. Sci.* 39: 827–33.
Lancaster, P. A., C. R. Krehbiel and G. W. Horn. 2014. A meta-analysis of effects of nutrition and management during the stocker and backgrounding phase on subsequent finishing performance and carcass characteristics. *Prof. Anim. Sci.* 30: 602–12.
Lean, I. J., Thompson, J. M. and Dunshea, F. R. 2014. A Meta-analysis of Zilpaterol and Ractopamine effects on feedlot performance, carcass traits and shear strength of meat in cattle. *PLoS ONE* 9(12): e115904.
Lomas, L. W., J. L. Moyer and G. A. Milliken. 2009. Effect of energy supplementation of stocker cattle grazing smooth bromegrass pastures on grazing and subsequent finishing performance and carcass traits. *Prof. Anim. Sci.* 25: 65–73.
Loerch, S. C. 1990. Effects of feeding growing cattle high-concentrate diets at a restricted intake on feedlot performance. *J. Anim. Sci.* 68: 3086–95.
Liu, Q., M. C. Lanari and D. M. Schaefer. 1995. A review of dietary vitamin E supplementation for improvement of beef quality. *J. Anim Sci.* 73: 3131–40.
Mathews, K. H., and R. J. Johnson. 2013. Alternative Beef Production Systems: Issues and Implications/ LDPM-218-01. Economic Research Service/USDA.
Maxwell, C. L., C. R. Krehbiel, B. K. Wilson, B. T. Johnson, B. C. Bernhard, C. F. O'Neill, D. L. VanOverbeke, G. G. Mafi, D. L. Step and C. J. Richards. 2014. Effects of beef production system on animal performance and carcass characteristics. *J. Anim. Sci.* 92: 5727–38.
McCurdy, M. P., G. W. Horn, J. J. Wagner, P. A. Lancaster and C. R. Krehbiel. 2010. Effects of winter growing programs on subsequent feedlot performance, carcass characteristics, body composition, and energy requirements of beef steers. *J. Anim. Sci.* 88: 1564–76.
Myers, S. E., D. B. Faulkner, T. G. Nash, L. L. Berger, D. F. Parrett and F. K. McKeith. 1999a. Performance and carcass traits of early-weaned steers receiving either a pasture growing period or a finishing diet at weaning. *J. Anim. Sci.* 77: 311–22.
Myers, S. E., D. B. Faulkner, F. A. Ireland and D. F. Parrett. 1999b. Comparison of three weaning ages on cow-calf performance and steer carcass traits. *J. Anim. Sci.* 77: 323–9.
Myers, S. E., D. B. Faulkner, T. G. Nash, L. L. Berger, D. F. Parrett and F. K. McKeith. 1999c. Performance and carcass traits of early-weaned steers receiving either a pasture growing period or a finishing diet at weaning. *J. Anim. Sci.* 77: 311–22.
Muir, P. D., Deaker. J. M. and Brown, M. D. (1998). Effect of forage- and grain-based feeding systems on beef quality: A review. *NZ J. Ag. Res.* 41: 623–35.
NASEM (National Academies of Sciences, Engineering, and Medicine). 2016. *Nutrient Requirements of Beef Cattle, Eighth Revised Edition*. Washington, DC: The National Academies Press.
National Cattleman's Beef Association. 2016. FreshLook Marketing Data for Total US Total Beef & Natural/Organic Pound Sales for 13 weeks ending 6/26/16. Available online at http://www.beefretail.org/natural-organicshareoftotalbeefdollarandpound.aspx
Natural Beef. 2012. Ag Marketing Resource Center. Available online at http://www.agmrc.org/commodities-products/livestock/beef/natural-beef/
NRC. 1996. Nutrient Requirements of Beef Cattle. 7th Ed. National Academy Press. Washington, D.C.
Oka, A., Y. Maruo, T. Miki, T. Yamasaki and T. Saito. 1998. Influence of vitamin A on the quality of beef from the Tajima strain of Japanese Black cattle. *Meat Sci.* 48: 159–67.

Owens, F. N., D. S. Secrist, W. J. Hill and D. R. Gill. 1997. The effect of grain source and grain processing on performance of feedlot cattle: a review. *J. Anim. Sci.* 75: 868–79.
Owens, F. N., and B. A. Gardner. 2000. A review of the impact of feedlot management and nutrition on carcass measurements of feedlot cattle. *J. Anim. Sci.* 77: 1–18.
Peel, D. S. 2003. Beef cattle growing and backgrounding programs. *Vet. Clin. North Am. Food Anim. Prac.* 19: 365–85.
Pickworth, C. L., S. C. Loerch and F. L. Fluharty. 2012a. Effects of timing and duration of dietary vitamin A reduction on carcass quality of finishing beef cattle. *J. Anim. Sci.* 90: 2677–91.
Pickworth, C. L., S. C. Loerch and F. L. Fluharty. 2012b. Restriction of vitamin A and D in beef cattle finishing diets on feedlot performance and adipose accretion. *J. Anim. Sci.* 90: 1866–78.
Pyatt, N. A., and L. L. Berger. 2005. Review: Potential effects of vitamins A and D on marbling deposition in beef cattle. *Prof. Anim. Sci.* 21: 174–81.
Pugh, R. B. 2003. Soybean hull supplementation to pasture-based beef cattle finishing diets. University of Tennessee Thesis.
Reinhardt, C. D., A. DiCostanzo and G. Milliken. 2007. Distillers byproducts alters carcass fat distribution of feedlot cattle. *J. Anim. Sci.* 85(Suppl. 2): 191.
Reuter, R. R., and P. A. Beck. 2013. Southern section interdisciplinary beef cattle symposium: Carryover effects of stocker cattle systems on feedlot performance and carcass characteristics. *J. Anim. Sci.* 91: 508–15.
Richards, C. J., and R. B. Hicks. 2007. Processing of corn and sorghum for feedlot cattle. In L. C. Hollis and K. C. Olson (eds), *Veterinary Clinics of North America Food Animal Practice: Topics in Nutritional Management of Feedlot Cattle*. Elsevier. Philadelphia, PA, pp. 207–22.
Richards, C. J., R. A. Stock, T. J. Klopfenstein and D. H. Shain. 1998. Effect of wet corn gluten feed, supplemental protein, and tallow on steer finishing performance. *J. Anim. Sci.* 76: 421–8.
Richards, C., R. Stock and T. J. Klopfenstein. 1996. Evaluation of levels of wet corn gluten feed and addition of tallow. Nebraska Beef Cattle Research Report, p. 61.
Richards, C. J., A. F. Branco, D. W. Bohnert, G. B. Huntington, M. Macari and D. L. Harmon. 2002. Intestinal starch disappearance increased in steers abomasally infused with starch and protein. *J. Anim. Sci.* 80: 3361–8.
Roeber, D. L., R. C. Cannell, K. E. Belk, R. K. Miller, J. D. Tatum and G. C. Smith. 2000. Implant strategies during feeding: impact on carcass grades and consumer acceptability. *J. Anim. Sci.* 78: 1867–74.
Roeber, D. L., R. K. Gill and A. DiCostanzo. 2005. Meat quality responses to feeding distiller's grains to finishing Holstein steers. *J. Anim. Sci.* 83: 2455–60.
Sainz, R. D., F. De la Torre and J. W. Oltjen. 1995. Compensatory growth and carcass quality in growth-restricted and refed beef steers. *J. Anim. Sci.* 73: 2971–9.
Samuelson, K. L., M. E. Hubbert, M. L. Galyean and C. A. Löest. 2016. Nutritional recommendations of feedlot consulting nutritionists: The 2015 New Mexico State and Texas Tech University survey1. *J. Anim. Sci.* 94: 2648–63.
Sanders, S. K., J. B. Morgan, D. M. Wulf, J. D. Tatum, S. N. Williams and G. C. Smith. 1997. Vitamin E supplementation of cattle and shelf-life of beef for the Japanese market. *J. Anim. Sci.* 75: 2634–40.
Schaefer, D. M. 2002. Fresh beef marketing opportunities due to dietary vitamin E. National Cattlemen's Beef Association. Available: http://www.beef.org/uDocs/Vitamin%20E.pdf. Accessed 12 October 2016.
Schmidt, J. R., M. C. Miller, J. G. Andrae, S. E. Ellis and S. K. Duckett. 2013. Effect of summer forage species grazed during finishing on animal performance, carcass quality, and meat quality. *J. Anim. Sci.* 91: 4451–61.
Secrist, D. S., F. N. Owens and D. R. Gill. 1997. Effects of vitamin E on performance of feedlot cattle: A review. *Prof. Anim. Sci.* 13: 47–54.
Sharman, E. D., P. A. Lancaster, G. G. Hilton, C. R. Krehbiel and G. W. Horn. 2009. Effects of starch- versus fiber-based supplements during winter grazing on growth rate of stocker cattle and final

carcass characteristics. http://www.ansi.okstate.edu/research/researchreports-1/2009/Sharman_Research%20Report%202009%20.pdf.

Sip, M. L., and R. H. Pritchard. 1991. Nitrogen utilization by ruminants during restricted intake of high-concentrate diets. *J. Anim. Sci.* 69: 2655–62.

Smith, S. B., R. L. Prior, C. L. Ferrell and H. J. Mersmann. 1984. Interrelationships among diet, age, fat deposition, and lipid metabolism in growing steers. *J. Nutr.* 114: 153–62.

Smith, G. C., J. B. Morgan, J. N. Sofos and J. D. Tatum. 1996. Supplemental vitamin E in beef cattle diets to improve shelf-life of beef. *Anim. Feed Sci. Technol.* 59(1–3): 207–14.

Smith, S. B., and J. D. Crouse. 1984. Relative contributions of acetate, lactate and glucose to lipogenesis in bovine intramuscular and subcutaneous adipose tissue. *J. Nutr.* 114: 792–800.

United States Standards for Livestock and Meat Marketing Claims, Grass (Forage) Fed Claim for Ruminant Livestock and the Meat Products Derived from Such Livestock. Federal Register p. 58631 Vol. 72, No. 199.

USDA. 2016. Local & Regional Food Market News. Available online at https://www.ams.usda.gov/market-news/local-regional-food. Accessed 26 September 2016.

USDA. 2006. Crop Residue Removal for Biomass Energy Production: Effects on soils and recommendations. Soil Quality-Agronomy Technical note No. 19. USDA Natural Resources Conservation Service. Available online at http://www.nrcs.usda.gov/Internet/FSE_DOCUMENTS/nrcs142p2_053253.pdf. Accessed 23 September 2016.

Vasconcelos, J. T., J. E. Sawyer, L. O. Tedeschi, F. T. McCollum and L. W. Greene. 2009. Effects of different growing diets on performance, carcass characteristics, insulin sensitivity, and accretion of intramuscular and subcutaneous adipose tissue of feedlot cattle. *J. Anim. Sci.* 87: 1540–7.

Wagner, J. J. 1988. Effect of previous growing rogram on the benefits of restricting feed intake during the finishing phase.In South Dakota State Univ. South Dakota Beef Rep. Accessed 15 August 2011, pp. 13–15. http://www.sdstate.edu/ars/species/beef/beef-reports/upload/Cattle_88-5_Wagner.pdf

Waterman, R. C., T. W. Geary, J. A. Paterson, R. J. Lipsey, W. R. Shafer, L. L. Berger, D. B. Faulkner and J. W. Homm. 2012. Early weaning in Northern Great Plains beef cattle production systems: III. Steer weaning, finishing and carcass characteristics. *Livestock Sci.* 148: 282–90.

Wertz, A. E., L. L. Berger, P. M. Walker, D. B. Faulkner, F. K. McKeith and S. L. Rodriguez-Zas. 2001. Early weaning and post weaning nutritional management affect feedlot performance of Angus X Simmental heifers and the relationship of 12th rib fat and marbling score to feed efficiency. *J. Anim. Sci.* 79: 1660–9.

Winterholler, S. J., D. L. Lalman, M. D. Hudson, C. E. Ward, C. R. Krehbiel and G. W. Horn. 2008. Performance, carcass characteristics, and economic analysis of calf-fed and wheat pasture yearling systems in the southern great plains. *Prof. Anim. Sci.* 24: 232–8.

Chapter 6

# Effects of metabolic modifiers on beef carcass composition and meat quality

*John M. Gonzalez, Sara M. Ebarb, Kelsey J. Phelps and Michael E. Dikeman, Kansas State University, USA.*



## 1 Introduction

Metabolic modifiers are compounds that are either fed to animals or implanted to improve rate of gain, improve feed efficiency, dressing per cent, carcass meat yield, visual meat quality, shelf-life or meat palatability. Most metabolic modifiers have been developed and researched to improve growth performance and carcass composition, with fewer of them developed or researched specifically to improve meat quality. Several in-depth review papers have been written on specific categories or types of metabolic modifiers. This chapter presents a general review of the effects of metabolic modifiers on carcass composition, fresh meat quality and cooked meat palatability. Emphasis will be on those metabolic modifiers that are approved for use in cattle in the United States and other developed countries. The following are the main types of modifier discussed in this chapter:

http://dx.doi.org/10.19103/AS.2016.0009.06

- Anabolic steroid implants
- Beta-adrenergic agonists
- Vitamins $D_3$, E and A
- Somatotropin

# 2 Anabolic steroid implants: introduction and the effect of single-implant programmes on carcass composition

## 2.1 Introduction to anabolic steroid implants

Anabolic implants in beef production systems have been studied extensively since their approval for use by the Federal Drug Administration in the 1950s (FDA, 2014). Several products are currently available for use in the United States and several developed countries, each consisting of various compounds and concentrations for use at different stages of animal growth. Most steroids are administered as subcutaneous implants placed in the middle third of the ear, usually as compressed pellets, but some as silastic rods. Release time for most implants range from 50 to 200 days. Approximately 97% of cattle in the United States entering the feedlot have received at least one anabolic implant during their production lives (NAHMS, 2000). Anabolic steroid implants are characterized as oestrogenic, androgenic or a combination of both oestrogenic and androgenic compounds (combination implants). Additionally, these compounds may also be categorized as natural hormones and xenobiotics, or combinations of these. Products available for use in beef cattle production, including doses and sex-specific usage, have been summarized by others (Montgomery et al., 2001; Duckett and Pratt, 2014); see Table 1, which includes the abbreviations used in this chapter. If trade names are not declared, steroid and concentrations used will be reported.

Implant programmes are used widely by the beef cattle industry in the United States and a few other countries in growing and finishing cattle because of economic incentives associated with improved growth rates and efficiency of feed utilization (Dikeman, 1997). In 1996, Duckett et al. (1996) estimated a return on investment ranging from $22.39 for a single oestrogenic implant to $112.53 for administration of two (reimplant) combination implants, with the average return being $45.36. These values were more recently adjusted for inflation by Duckett and Pratt (2014) to yield a return on investment from $54.02 to $218.58, with an average return of $102.62. Because the return on investment is heavily dependent on the implant programme used, producers utilize multiple regimens that cause the effects of anabolic implants on carcass composition and meat quality to vary.

## 2.2 The effect of single-implant programmes on carcass composition

Depending on the status of the animals and implant strategy, carcass alterations commonly include increased carcass muscling at the expense fat deposition, which results in alterations in economically important measures associated with those tissues. The impacts of single-implant programmes on carcass characteristics of females are somewhat inconsistent in terms of affecting the same attributes from study to study.

**Table 1** Anabolic implants utilized in commercial beef production and research discussed in this chapter[1]

| Trade name | Active ingredients | Sex utilization | Manufacturer | Chapter abbreviation |
|---|---|---|---|---|
| Oestrogenic | | | | |
| Component E-S[2] | 20 mg ES[3] benzoate + 200 mg progesterone | Steers | Elanco | Not needed |
| Component E-C[2] | 10 mg ES benzoate + 100 mg progesterone | Suckling calves | Elanco | Not needed |
| Compudose 200 | 25.7 mg ES | Heifers and steers | Elanco | Com-200 |
| Compudose 400 (Encore) | 43.9 mg ES | Heifers and steers | Elanco | Not needed |
| Ralgro | 36 mg zeranol | Heifers and steers | Merck | RAL |
| Ralgro Magnum | 72 mg zeranol | Steers | Merck | Not needed |
| Synovex-S | 20 mg ES benzoate + 200 mg progesterone | Steers | Zoetis | SYN-S |
| Synovex-C | 10 mg ES benzoate + 100 mg progesterone | Heifers and steers | Zoetis | SYN-C |
| Androgens | | | | |
| Component T-H[2] | 200 mg TBA[4] | Heifers | Elanco | COM-TH |
| Component T-S[2] | 200 mg TBA + 43.9 mg ES | Steers | Elanco | COM-TS |
| Finaplix-H | 200 mg TBA | Heifers | Merck | FIN-H |
| Finaplix-S | 140 mg TBA | Steers | Merck | FIN-S |
| Combination | | | | |
| Component E-H[1] | 20 mg ES benzoate + 100 mg testosterone propionate | Heifer | Elanco | Not needed |
| Component TE-G[2] | 8 mg ES + 80 mg TBA | Heifer and steers | Elanco | Not needed |
| Component TE-IH[2] | 8 mg ES + 80 mg testosterone propionate | Heifer | Elanco | Not needed |
| Component TE-IS[2] | 16 mg ES + 80 mg TBA | Steers | Elanco | Not needed |
| Component TE-S[2] | 24 mg ES + 120 mg TBA | Steers | Elanco | COM-TES |
| Component TE-200 | 20 mg ES + 200 mg TBA | Heifers and steers | Elanco | Not needed |

*(Continued)*

**Table 1** *(Continued)*

| Trade name | Active ingredients | Sex utilization | Manufacturer | Chapter abbreviation |
|---|---|---|---|---|
| Revalor-G | 4 mg ES + 40 mg TBA | Heifers and steers | Merck | Not needed |
| Revalor-H | 14 mg ES + 140 mg TBA | Heifers | Merck | REV-H |
| Revalor-IH | 8 mg ES + 80 mg TBA | Heifers | Merck | REV-IH |
| Revalor-S | 24 mg ES + 120 mg TBA | Steers | Merck | REV-S |
| Revalor-IS | 16 mg ES + 80 mg TBA | Steers | Merck | REV-IS |
| Revalor-200 | 20 mg ES + 200 mg TBA | Heifers and steers | Merck | REV-200 |
| Revalor-XS | 40 mg ES + 200 mg TBA | Steers | Merck | REV-XS |
| Synovex-Choice | 14 mg ES benzoate + 100 mg TBA | Heifers and steers | Zoetis | SYN-C |
| Synovex-H | 20 mg ES benzoate + 200 mg testosterone propionate | Heifers | Zoetis | SYN-H |
| Synovex Plus | 28 mg ES benzoate + 200 mg TBA | Heifers and steers | Zoetis | SYN-Plus |
| Synovex-T40 | 8 mg ES + 40 mg TBA | Heifers and steers | Zoetis | Not needed |
| Synovex-T80 | 16 mg ES + 80 mg TBA | Steers | Zoetis | Not needed |
| Synovex-T120 | 24 mg ES + 120 mg TBA | Steers | Zoetis | Not needed |

[1]Table adapted from Montgomery et al. (2001) and Duckett and Pratt (2014).
[2]Available with 29 mg tylosin.
[3]ES = estradiol.
[4]TBA = trenbolone acetate.

In early studies utilizing combination implants, studies reported approximately 12 kg increases in hot carcass weight (HCW), up to 4.1 cm$^2$ increases in *Longissimus* muscle (LM) area (Crouse et al., 1987; Moran et al., 1991; Cranwell et al., 1996; Cook et al., 2000), but only Cranwell et al. (1996) reported a 0.25 cm decrease in 12th rib fat. Henricks et al. (1997) was the only study that observed no difference in carcass characteristics when heifers were implanted with REV-H or FIN-H.

As combination implants were developed, Adams et al. (1990) reported that heifers implanted with SYN-H had a 13.1 kg increase in HCW, with no effect on other carcass measures. Garber et al. (1990) observed SYN-H increased LM area by 7.5 cm$^2$; however, marbling score was reduced by 18%, which resulted in a 16% reduction in USDA Quality Grade. Kreikemeier and Mader (2004) reported no independent COM-200 or FIN-H effects, but found when they were simultaneously implanted, they produced an 11.9 kg increase in HCW and a 44 % reduction in marbling score. Cleale et al. (2013) observed a 12.1 kg increase in HCW, 3.8 cm$^2$ increase in LM area and a 25% reduction in marbling score for heifers that received a SYN-Plus implant. In summary, single-implant programmes tend to affect female HCW and LM area most consistently of all carcass measures. Of the fat depots, marbling is most often negatively affected.

The majority of single-implant studies focus on steers. Additionally, the majority of these studies utilize TBA-based combination implants. Early work by Perry et al. (1991), Cecava and Hancock (1994), Gerken (1995) and Herschler et al. (1995) with TBA/estradiol (ES) implants (SYS-S, REV-S, FIN-S or TBA/ES benzoate [in ratios of 1:5 or 1:10]) found increases in HCW between 8 and 21 kg, LM area between 3 and 4.8 cm$^2$, with Herschler et al. (1995) reporting decreases in marbling and percentage of carcasses grading Choice. Hermesmeyer et al. (2000) examined the effects of REV-S and SYN-Plus at two feed levels (full vs restricted) and resulting differences in subcutaneous fat thicknesses. Implants modestly increased HCW by 4 kg, DP by 0.4%, and LM area by 2 cm$^2$, and the implant response in restricted steers was nearly equal to full-fed steers. Bruns et al. (2005) examined the effect of early (day 1 of finishing period) and late (day 57 of finishing period) administration of REV-S and reported early administration decreased marbling score by 45% and late administration decreased YG by 0.3 units. Timing of implants did not differentially influence HCW, DP or LM area; however, implanting did increase these measures by 12 kg, 0.17% and 4.95 cm$^2$, respectively.

In recent years, there have been larger impacts on carcass characteristics for single administration implant regimens. Boles et al. (2009) pooled data from studies conducted over two years where steers were implanted with SYN-C in year 1 and steers and heifers were implanted with 24 mg ES benzoate and 120 mg TBA in year 2. Data were pooled and implants increased HCW by 22.5 kg, LM by 4.7 cm$^2$, and decreased marbling score by 15%. In two small pen studies, Winterholler et al. (2008) only observed an increase in HCW of 22.2 kg in only one of the trials when steers were administered with REV-S. Kellermeier et al. (2009) reported that REV-S increased LM area by 10.55 cm$^2$, decreased 12th rib fat by 0.22 cm, YG by 0.7 units and marbling score by 5%. Bass et al. (2009), found no effect of timing of various TBA/ES implants (80 mg TBA and 16 mg ES or 120 mg TBA and 24 mg ES) in Holstein steers, but implanting increased HCW by 11.4 kg and LM area by 5.2 cm$^2$. The authors also reported decreased 12th rib fat thickness and marbling score for steers that received the moderate strength implant, with no effect by the aggressive implant. Finally, USDA Quality Grade for all implanted steers was decreased to an average of mid-Select. Baxa et al. (2010) reported dramatically increased HCW, DP and LM area by 20.9 kg, 2.2% and 11.9 cm$^2$, respectively, when steers received a REV-S implant 91 days before slaughter. Marbling score was also reduced by 8%. These studies demonstrate that anabolic implants administered to cattle in the feedlot phase can increase HCW and increase DP, which primarily occurs through increased muscling. Fat measures seems to be affected more in recent studies, which is most likely due to the increase in potency of the implants administered.

# 3 Anabolic steroid implants: the effect of multiple implant (reimplant) programmes on carcass composition

The majority of multiple implant regimen studies have been conducted in steers. Initial studies described below do not demonstrate large advantages on carcass characteristics when compared to single-implant programmes, which is most likely due to the small number of replications within each study. Initial small replication studies with zeranol, RAL and SYN-S found that implanting increased HCW up to 16 kg, and reduced marbling score by 5%; however, reimplanting had no significant advantage over the single-implant treatment (Loy et al., 1988; Simms et al., 1988).

Hoechst-Roussel Agri-Vet Company (1991) evaluated five implant strategies and reported that REV-S/REV-S and SYN-S/FIN-S significantly reduced marbling compared to the other implant treatments and, consequently, reduced the percentage of carcasses grading USDA Choice (Table 2). The SYN-S/FIN-S treatment also had the largest LM and lowest YG. Samber et al. (1996) reported that steers from 6 multiple implant programmes differed from non-implanted steers in HCW, DP, fat thickness and per cent KPH. LM area of steers receiving 3 REV-S implants was increased (maximum difference, 5.6 $cm^2$) when compared to non-implanted, twice-implanted, and RAL/SYN-S/REV-S (implanted

**Table 2** Effect of implant treatment on carcass traits

| | Day 1: | Revalor-S | Synovex-S | Revalor-S | Synovex-S | Synovex-S |
|---|---|---|---|---|---|---|
| Carcass data | Day 67–70: | — | Revalor-S | Revalor-S | Synovex-S | Finaplix-S |
| Hot carcass wt. | | 777[c] | 783[b] | 792[a] | 770[d] | 787[ab] |
| Marbling[d] | | 4.29[a] | 4.22[a] | 4.10[b] | 4.33[a] | 4.10[b] |
| % Choice | | 68[a] | 70[a] | 61[b] | 69[a] | 59[b] |
| KPH fat, %[e] | | 2.39 | 2.45 | 2.39 | 2.40 | 2.41 |
| Backfat thickness, in. | | 0.51[b] | 0.54[a] | 0.51[b] | 0.52[b] | 0.50[b] |
| Ribeye area, $in^2$[f] | | 12.49[b] | 12.66[b] | 13.04[b] | 12.65[b] | 13.53[a] |
| Yield grade | | 3.09[a] | 3.13[a] | 2.99[a] | 3.00[a] | 2.75[b] |

[a,b,c] = Means in same row with different superscripts differ ($P < 0.05$).
[d] 3 = slight, 4 = small, 5 = modest, 6 = moderate.
[e] Kidney, pelvic and heart fat as a % of hot carcass weight – estimated visually.
[f] Ribeye muscle area between the 12th and 13th ribs.
From Hoechst-Roussel Agri-Vet Company (1991).

**Table 3** Carcass traits and shear force values as affected by implant treatment and dietary protein level

| | Experimental treatment group | | | | | | |
|---|---|---|---|---|---|---|---|
| Trait[a] | CON | RAL/ SYN/REV | RAL/ REV/REV | SYN/ REV | REV/ REV | REV/ 3X-12.5% | REV/ 3X-14.5% |
| Carcass weight, kg | 381[c] | 391[c] | 397[c] | 396[c] | 394[c] | 400[c] | 396[c] |
| FT, cm | 1.36[cd] | 1.26[d] | 1.43[c] | 1.44[c] | 1.49[c] | 1.26[d] | 1.36[cd] |
| YG | 3.35[de] | 3.23[ef] | 3.40[cde] | 3.47[cd] | 3.62[c] | 3.12[f] | 3.23[ef] |
| Marbling[b] | 457[c] | 421[de] | 420[de] | 458[c] | 443[cd] | 400[e] | 435[cd] |
| % Choice and Prime | 85.9[c] | 62.2[fg] | 65.3[efg] | 80.2[cd] | 79.5[cde] | 54.9[g] | 73.1[def] |
| Shear force, kg | 2.58[f] | 2.74[def] | 2.75[def] | 2.64[ef] | 3.01[c] | 2.92[cd] | 2.86[cde] |

[a] Abbreviations used: FT, fat thickness; YG, calculated USDA yield grade.
[b] 300 = slight, 400 = small, 500 = modest.
[c,d,e,f,g] Means in the same row lacking a common superscript letter differ ($P < 0.05$).
From Samber et al. (1996).

separately over 150 days) steers (Table 3). When Holstein steers received implants with combinations of 36 mg zeranol; 20 mg ES benzoate/progesterone; 140 mg TBA/20 mg ES benzoate/200 mg progesterone; or 140 mg TBA/36 mg zeranol on days 0, 56, 112 and 168 of a 249-day finishing period, implanted steers had increased HCW by 19.3 kg, 23.9 kg, 27.9 kg and 16.8 kg, respectively, compared to non-implanted steers. Additionally, the latter 3 treatments had increases in LM area of 5.8 cm$^2$, 11 cm$^2$ and 7.09 cm$^2$, respectively (Apple et al., 1991). These data indicate that different implant protocols perform differently when implanted multiple times.

More recent research has used larger studies with more consistent results. Roeber et al. (1999) examined various steer implant strategies consisting of an initial implant on day 0 of COM-TS, RAL, REV-S, SYN-S or no implant, followed by a second implant of SYN-Plus, REV-S, or no implant (Table 4). There was no clear advantage in any carcass measure when comparing the 7 implant programmes to one another. All implant combinations except the COM-TS regimen differed from non-implanted carcasses in HCW, with an average increase of 27.3 kg. Per cent KPH was affected by 5 of the 7 protocols, with a maximum difference of 0.36%. Marbling score (maximum reduction of 14%) and quality grade (maximum reduction of 84%) were affected by 5 and 4, respectively, of the 7 implant regimens. indicating some combinations of implants can have a greater detrimental effect on marbling. Scheffler et al. (2003) reported dramatically increased HCW of 36 kg and LM area by 8.8 cm$^2$ for carcasses from steers implanted three times with Component-TS compared to carcasses from non-implanted steers. Once-implanted steers had similar HCW and LM area compared to non-implanted ones, and both these implant programmes had reduced HCW and LM areas than steers administered implants three times during finishing. Twice-implanted steers did not differ in HCW from those implanted three times, and these two treatments averaged 5.7 cm$^2$ larger LM area compared to non-implanted carcasses. Bryant et al. (2010) observed a reimplanting effect when carcasses from twice-implanted steers (REV-IS/REV-S) had 12.6 kg heavier HCW, 0.3% greater DP, 3.3 cm$^2$ larger LM area, but similar YG compared to the once-implanted steers (REV-S/no implant).

**Table 4** Carcass traits stratified by implant strategy groups ($n$ = 403)

| Implant strategy | Hot Car. Wt. (lb.) | Adj. Fat (in.) | YG | Marbling* | QG** | % Prime/ Choice | % Y4/ Y5 |
|---|---|---|---|---|---|---|---|
| No Implant/No Implant | 737.4[b] | 0.66 | 3.51[ab] | 524.3[a] | 523.8[a] | 94.4[a] | 24.1 |
| Encore & Component T-S/ No implant | 777.2[ab] | 0.65 | 3.47[ab] | 511.6[ab] | 515.1[ab] | 93.2[ab] | 18.2 |
| Ralgro/Synovex Plus | 798.8[a] | 0.69 | 3.59[ab] | 459.2[bc] | 460.7[c] | 77.4[c] | 28.3 |
| Ralgro/Synovex-S | 796.8[a] | 0.64 | 3.40[ab] | 482.7[abc] | 487.5[abc] | 81.3[abc] | 6.3 |
| Revalor-S/Revalor-S | 809.5[a] | 0.67 | 3.45[ab] | 449.6[c] | 454.1[c] | 66.7[c] | 18.8 |
| Revalor-S/No Implant | 795.9[a] | 0.71 | 3.70[a] | 467.0[bc] | 474.0[abc] | 76.9[c] | 34.6 |
| No Implant/Synovex Plus | 791.0[a] | 0.64 | 3.35[b] | 458.3[c] | 463.4[bc] | 78.9[bc] | 7.8 |
| Synovex Plus/No Implant | 794.1[a] | 0.71 | 3.61[ab] | 470.4[bc] | 471.4[bc] | 75.0[c] | 25.5 |

[a,b,c] Means in the same column with different superscript letters differ ($P < 0.05$).
* Marbling scores are coded as 300 = slight, 400 = small, 500 = modest and 600 = moderate.
** USDA quality grades are coded as 100 to 299 = Standard, 300 to 349 = Low Select, 350 to 399 = High Select, 400 to 499 = Low Choice, 500 to 599 = Average Choice and 600 to 699 = High Choice.
From Roeber et al. (1999).

In the past decade, the focus of multiple implant research has shifted more to time of implant administration. DeHaan et al. (1990) were the first group to indicate time of second implant administration is important in maximizing carcass response. In experiment 1 of their study, the authors reported steers implanted with 200 mg progesterone plus 20 mg ES benzonate at feedlot day 1 and 85 had a 0.7% reduction in KPH and 13% decrease in YG. In experiment 2 when the same implants were administered at feedlot day 1 and 112, HCW increased by a dramatic 40.9 kg, YG decreased 9%, and marbling score decreased by 49%. Parr et al. (2011b) conducted 4 studies to examine the effect of a longer lasting implant product, REV-XS. In experiment 1, implanting with REV-XS at day 0 was compared to no-implant, single REV-S at day 0 of feeding, and REV-IS at day 0 of feeding followed by reimplanting with REV-S at day 68 to 74 of feeding. Implanting increased HCW and LM area, and reduced 12th rib fat and marbling compared to non-implanted carcasses; however, REV-XS had greater impacts on these measures than single implanted steers, but did not differ from the double implant treatment. During the remainder of the experiments, REV-XS was compared to various double implant programmes, with the double implant programme varying in time of second implant. Rarely were there differences in carcass responses to the REV-XS and double implant regimen. When differences were found, the REV-XS carcasses had greater HCW, LM area and reduced KPH. The authors concluded that time of initial and terminal implant administration are keys to controlling overall carcass response.

Moran et al. (1991) explored various TBA, ES and zeranol programmes in heifer carcasses and found that carcasses from heifers implanted with TBA on 1, 84, 168 and 252 of feeding averaged 12 kg heavier HCW compared to non-implanted carcasses, with no effect on fat score. In a multiple trial study, Mader and Lechtenberg (2000) observed increased HCW of 8 kg for heifers that received a RAL implant at day 0, followed by a RAL/Finnaplix-H implant at day 70, compared to heifers given a RAL/Finnaplix-H implant only at d 70. Carcass measures including DP, LM area and YG were similar between the two treatments. Sissom et al. (2007) reported when carcasses of heifers implanted with REV-200 at trial

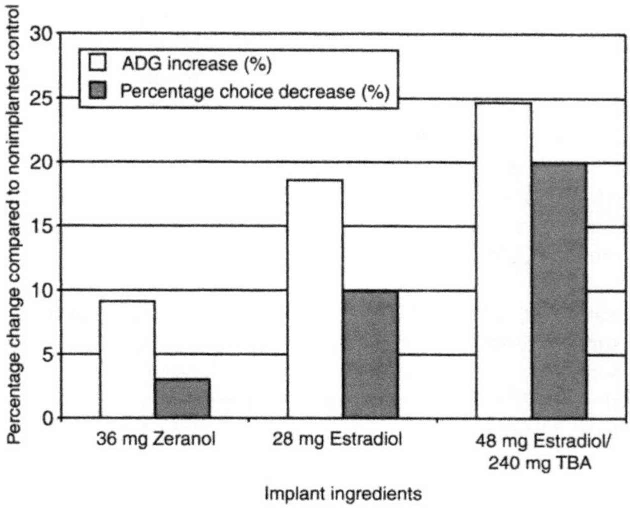


**Figure 1** Typical results from commonly used feedlot implant programmes.
Adapted from Anderson et al. (2000).

initiation were compared to carcasses from heifers initially implanted with REV-IH and reimplanted on day 58 with FIN-H, carcass characteristics were unaffected except for a tendency for marbling score to be decreased, which resulted in a 13% decrease in the percentage of carcasses grading Choice. Schneider et al. (2007) examined the effect of implanting with different combinations and concentrations of TBA plus ES administered once or twice during finishing on carcass characteristics of feedlot heifers. Carcasses from heifers implanted twice had an additional 6.0 kg of HCW, 5 $cm^2$ of LM area and improved YG of 8% when compared to single implanted carcasses. Marbling score was unaffected, but the percentage of carcasses grading Choice or Prime was reduced by 13.5%.

In summary, carcass responses to anabolic implants are dependent on the implant regimen employed. Overall, average carcass responses to anabolic implants include 10 to 20 kg increases in HCW, 5 to 12 $cm^2$ increases in LM area, but almost no improvement in YG. Anderson et al. (2014) summarized the effects of mild/single implants with strong combination implants and showed that ADG increased and percentage of Choice carcasses decreased with the more potent (stronger) implants (Fig. 1). Carcasses from implanted cattle may be leaner, but this is due to increased quantity of muscle, not due to decreased quantity of fat deposition; however, there may be some fat depot specificity because some studies have shown reduced marbling, especially in males administered aggressive implants.

# 4 Anabolic steroid implants: the effect of anabolic steroid implants on meat quality

Meat quality is a multifaceted aspect of meat that can be defined by consumers by different attributes; this depends on when they are evaluating the meat item in the product life cycle. At retail, colour is the single most important quality attribute consumers consider in their purchasing decision (Mancini and Hunt, 2005). While there are few literature reports examining the impact of implants on meat colour, combination implants administered to steers and heifers less than 100 days before harvest have been shown to increase the incidence of dark cutters as demonstrated by Scanga et al. (1998). At time of consumption, the 2010 National Beef Quality Audit (NBQA) reported that consumers evaluate two attributes to determine eating satisfaction: tenderness and flavour (Igo et al., 2011). Because tenderness has long been considered the most important quality attribute (Beermann, 2009), the majority of the research described below will focus on that trait.

Much of the early work with anabolic implants found little to no effect on objective and subjective measures of meat quality. Early research with zeranol and TBA implants reported no effects on sensory panel tenderness, juiciness, flavour or acceptability scores, and WBSF (Borger et al., 1973; Crouse et al., 1987; Apple et al., 1991). Only Apple et al. (1991) demonstrated that 24-hour lean colour scores for steaks from 140 mg-TBA implanted carcasses were greater than five other implant treatments by an average of 0.5 points on an eight-point scale. Perry et al. (1991) also reported no difference in all sensory panel scores for LM steaks from Holstein and beef breed steers implanted once with 128 mg of TBA and 28 mg of ES. Gerken et al. (1995) utilized cloned steers to determine the effect of one-time administration of three implant types (none, SYN-S, FIN-S, or REV-S) on objective and subjective tenderness measures of strip loin, top sirloin and top round steaks. Authors reported no implant effects on WBSF or trained sensory panellist scores of tenderness of strip loin and top round steaks aged 7, 14 or 21 days postmortem; however, panellists' scores for SYN-S top sirloin steaks were 0.37 and 1.12 points (8-point scale) lower than control and FIN-S steaks, respectively. Overall, many of the early implant studies do not indicate substantial influences on meat quality attributes, possibly due to lack of use of aggressive strategies described below.

**Table 5** Effect of implant treatment on ribeye steak peak shear force values and sensory panel tenderness scores at 14 days of ageing

| | Implant treatment[a] | | | | | |
|---|---|---|---|---|---|---|
| Trait | CONT | MGA | FINMGA | REV | REVMGA | Effect[b] |
| Shear force, lb. | 7.76[c] | 7.95[cd] | 7.93[cd] | 8.27[de] | 8.34[e] | CI<br>RF |
| Very tender steaks[f], % | 70.70 | 66.70 | 72.60 | 61.20 | 59.10 | |
| Tender steaks[f], % | 18.10 | 20.00 | 12.80 | 23.30 | 26.10 | |
| Tough steaks[f], % | 11.20 | 13.30 | 14.60 | 15.50 | 14.80 | |
| Sensory panel[g] | 5.96[c] | 5.88[cd] | 5.87[cde] | 5.72[e] | 5.76de | CI |

[a] Implant treatments: CONT = non-implanted, MGA = melangestrol acetate, FINMGA = Finaplix-H® plus MGA, REV = Revalor-H®, REVMGA = Revalor-H® plus MGA.

[b] Contrast effects: CI ($P < 0.05$) = control vs all implants; RF ($P < 0.05$) = REVMGA vs FINMGA.

[c,d,e] Means in the same row with different superscripts differ ($P < 0.05$).

[f] Based on shear force: very tender = <8.5 lb., tender = 8.6 to 10 lb., tough = >0 lb.

[g] 8 = extremely tender; 6 = moderately tender; 4 = slightly tough.

From Nichols et al. (1996).

In contrast to the early studies described above, research in the mid-1990s began to detect an implant effect on meat quality. Specifically, two factors contributed to this change in findings: use of various commercial products currently available that are more powerful and employment of multiple implant programmes. We will first present the data documenting the effects of two (initial and terminal) implant uses. Nichols et al. (1996) evaluated five implant treatments and reported that the REV and REV plus feeding of MGA treatments had the highest shear force values and these treatments plus the FIN plus feeding of MGA treatment had the highest percentages of tough steaks (15%) determined by a sensory panel (Table 5). Foutz et al. (1997) reported that implanted cattle (SYN-S, REV, SYN-S plus FIN-S, or SYN-S plus FIN-S/reimplanted with FIN-S) had increased WBSF of 7-d aged steaks by an average of 0.32 kg compared to non-implanted steers. Roeber et al. (2000) examined the effects of seven implant regimens involving different combinations of non-implanting, COM-TS, SYN-Plus, RAL and REV-S, on 14-day aged WBSF and consumer panellists' scores. Results indicated when steers only received an initial REV-S implant with no second implant, WBSF was increased by 0.54 kg compared to the controls, but did not differ from the other implant programmes. Unlike the WBSF results, steaks from all implant regimens except the COM-TS treatment were rated less tender compared to controls by a consumer panel, and steaks from all implant regimens, except the COM-TS and initial REV-S implant with no second implant, were rated less tender.

Schoonmaker et al. (2001) tested aggressive (SYN-C at day 163 followed by REV-S at day 204 and 295, or SYN-C at day 107 followed by REV-S at day 170 and 198) versus non-aggressive (SYN-S at day 204 and 295 or SYN-S at day 198 and 275) implant programmes in steers. They found no effect of implant strategy on WBSF or trained panel scores of LM steaks. Kerth et al. (2003) reported an initial REV-H implant, followed by another REV-H implant increased heifer 14-day aged steak WBSF by 0.31 kg, but no other differences were found for myofibrillar fragmentation index or trained panel scores. Barham et al. (2003) found a SYN-S initial implant followed by either a SYN-S or REV-S implant did not affect WBSF or trained sensory panel juiciness, beef flavour, flavour intensity or overall mouth feel scores when steaks were aged an average of 19 days. However, trained panellists did detect that implant protocols produced steaks with reduced tenderness scores (0.35 point on 8 point scale), but consumers were unable to detect any differences. Platter et al. (2003) examined the effect of two, three, four and five implants (combinations of SYN-C, RAL,

**Table 6** Consumer sensory ratings of steaks from implant strategies differing by number of implants administered[a]

| | Means by number of implants administered | | | | | |
|---|---|---|---|---|---|---|
| Consumer sensory response | 0 | 2 | 3 | 4 | 5 | SEM |
| Tenderness[b] | 3.15[z] | 3.79[y] | 3.97[y] | 3.88[y] | 3.93[y] | 0.16 |
| Flavour[b] | 3.34[z] | 3.62[y] | 3.76[y] | 3.77[y] | 3.82[y] | 0.09 |
| Juiciness[b] | 3.54[z] | 3.91[y] | 4.10[y] | 4.10[y] | 4.13[y] | 0.12 |
| Satisfaction with overall eating quality, % | 73.6[y] | 65.0[z] | 63.2[z] | 63.5[z] | 63.5[z] | 3.08 |

[a] Adapted from Platter et al. (2003).
[b] Tenderness, flavour, and juiciness like to dislike ratings by consumers where: 1 = liked extremely and 9 = disliked extremely.
[y,z] Means in the same row that do not have a common superscript differ ($P < 0.05$).

**Table 7** Experimental design outlining implant strategy, marbling scores, USDA quality grade distribution, Warner–Bratzler shear force values (WBS) for striploin steaks and consumer sensory panel scores[a]

| | Experimental treatment group[b] | | | | | | | | | | |
|---|---|---|---|---|---|---|---|---|---|---|---|
| Item | 1 | 2 | 3 | 4 | 5 | 6 | 7 | 8 | 9 | 10 | 11 |
| Experimental design | | | | | | | | | | | |
| Implant at branding | NO | NO | NO | NO | C | NO | C | C | C | C | C |
| Implant at weaning | NO | NO | NO | NO | NO | RA | NO | NO | RA | RA | RA |
| Implant at backgrounding | NO | NO | RA | S | S | S | RA | S | RA | S | S |
| Implant at feedlot entry | NO | S | S | S | NO | S | S | S | S | S | REV |
| Reimplant in feedlot | NO | REV | REV | REV | REV | REV | REV | REV | REV | REV | REV |
| Meat quality data | | | | | | | | | | | |
| Overall mean WBS, kg | $3.54^{z}$ | $3.95^{y}$ | $4.46^{w}$ | $41.9^{wxy}$ | $4.19^{wxy}$ | $4.15^{wxy}$ | $4.12^{wxy}$ | $4.05^{xy}$ | $4.05^{xy}$ | $4.14^{wxy}$ | $4.38^{wx}$ |
| Steaks ≤4.5 kg (14-d), % | $82^{w}$ | $66^{wx}$ | $44^{wx}$ | $56^{wx}$ | $54^{wx}$ | $58^{wx}$ | $50^{wx}$ | $70^{wx}$ | $62^{wx}$ | 58wx | $38^{x}$ |
| Steaks ≤4.5 kg (21-d), % | 94 | 88 | 76 | 74 | 76 | 80 | 78 | 84 | 82 | 80 | 64 |
| Tenderness[e] | $3.15^{z}$ | $3.79^{y}$ | $4.05^{xy}$ | $4.00^{xy}$ | $3.87^{xy}$ | $3.91^{xy}$ | $3.78^{y}$ | $3.96^{xy}$ | $3.71^{y}$ | $3.80^{xy}$ | $4.25^{x}$ |
| Flavour[e] | $3.34^{z}$ | $3.62^{yz}$ | $3.81^{xy}$ | $3.76^{xy}$ | $3.71^{xy}$ | $3.73^{xy}$ | $3.74^{xy}$ | $3.83^{xy}$ | $3.70^{xy}$ | $3.82^{xy}$ | $3.92^{x}$ |
| Juiciness[e] | $3.54^{z}$ | $3.91^{y}$ | $4.17^{xy}$ | $4.11^{xy}$ | $4.00^{xy}$ | $4.12^{xy}$ | $4.02^{xy}$ | $4.17^{xy}$ | $4.06^{xy}$ | $4.02^{xy}$ | $4.30^{x}$ |

[a] Adapted from Platter et al. (2003).
[b] NO = no implant; C = Synovex-C (10 mg of estradiol benzoate, 100 mg of progesterone); RA = Ralgro (36 mg of zeranol); S = Synovex-S (20 mg of estradiol benzoate, 200 mg of progesterone); REV = Revalor-S (24 mg of 17-ß estradiol, 120 mg of trenbolone acetate).
[c] Adjusted to a common fat thickness.
[d] 300 = slight, 400 = small, 500 = modest.
[e] Tenderness, flavour and juiciness like to dislike ratings by consumers where: 1 = like extremely and 9 = disliked extremely.
[w,x,y] Means in the same row lacking a common superscript letter differ ($P < 0.05$).

SYN-S, REV-S) administered throughout the production life of steers. Results indicated that implanting two or more times decreased consumer panel tenderness and percentage of steaks with overall eating quality satisfaction (Table 6). Interestingly, implanting steers three times produced steaks that had 0.30 kg and 0.73 kg greater WBSF than steaks from non-implanted and twice-implanted steers, respectively (Table 7). Steaks from steers that were implanted four and five times did not differ in WBSF from steaks from twice-implanted steers, possibly indicating that three implants produce the maximum negative effect on tenderness.

In agreement with Platter et al. (2003), other literature indicates that implanting cattle with three implants (an initial and two boosters) has the greatest effect on meat quality, specifically tenderness. Samber et al. (1996) examined the effect of six multiple implant strategies involving RAL, REV-S and SYN-S during finishing on WBSF of LM steaks aged 14 days. Indicating that implant number and combinations can have a differential effect on tenderness compared to non-implanted steaks, only steers that received two successive REV-S implants had 0.43 and 0.31 kg greater WBSF, respectively, than controls. While Scheffler et al. (2003) reported no implant effect on objective meat colour, steers that received three COM-TES implants produced 14-day aged steaks with 0.5 and 0.75 kg greater WBSF compared to steers subjected to none or one implant, respectively.

Most literature tested meat quality attributes at one fixed postmortem ageing time. Schneider et al. (2007) was one of the first groups to demonstrate postmortem ageing could alleviate the negative implant effects on tenderness. The authors tested four single-implant and seven two-implant programmes of various TBA and ES concentrations. Authors reported that single-implant programmes did not differ from no-implant programmes in WBSF at days 3, 7, 14, 21 and 28 of postmortem ageing. Two-implant programmes had greater WBSF than single-implant programmes on all days of ageing; however, when evaluating consumer panel data, the authors reported at least 14 days of ageing are required to negate the effects of implants and provide consumers with an acceptable product. In contrast, Smith et al. (2007) reported no day of ageing time by implant treatment (two SYN-Plus implants) interaction on WBSF values of steaks from implanted cattle; implanted steaks on average had WBSF that were 0.34 kg greater than non-implanted steaks.

In summary, most studies focusing on the effect of implants on meat quality target tenderness as the main variable of interest. Colour effects are rarely reported and when they are, there is little effect on 24 to 72-h subjective lean colour scores. When implants affect tenderness, there is typically a 0.3–0.5 kg increase in WBSF and less than half a point reduction in trained panel scores. When trained panellists detect a decrease in tenderness, consumers often do not detect any differences. This is especially true when meat is aged longer than 14 days postmortem.

## 5 Beta-adrenergic agonists: ractopamine hydrochloride (RH)

Beta-adrenergic agonist (BAA) compounds are similar in structure to the catecholamines norepinephrine and epinephrine. They act on tissues through the β-1, -2 or -3 adrenergic receptors as nutrient repartitioning agents by directing nutrients towards protein accumulation (lean muscle growth), indirectly decreasing lipogenesis (fat synthesis and storage), and increasing lipolysis (fat mobilization and hydrolysis). Cattle proportionately have more β-2 than β-1 receptors, and the two options available for use in the United States target a specific receptor type. Ractopamine hydrochloride (RH; Optaflexx™, Elanco Animal Health; FDA approved in 2003) is a β-1 BAA (binds to β-2 at a lower affinity) for use in steers or market heifers during the last 28 to 42 days on feed. Zilpaterol hydrochloride (ZH; Zilmax™, Merck Animal Health; FDA approved in 2006) is a β-2 BAA approved to be fed to cattle the last 20 to 40 days of the finishing period. Due primarily to perceived animal welfare concerns in the United States, Merck Animal Health has ceased selling ZH. Although the use of BAA greatly improves efficiency of beef production and carcass muscling, there are data indicating use of these compounds can have negative impacts on marbling, meat tenderness and overall palatability.

## 5.1 Ractopamine hydrochloride (RH): effects on carcass composition

As RH approached approval and widespread acceptance in the industry, numerous studies documenting its effect on carcass characteristics were published. Gruber et al. (2007) was one of the first studies to demonstrate 200 mg·head$^{-1}$·d$^{-1}$ of RH increased HCW by 5.5 kg and LM area by 2.3 cm$^2$, but did not affect fat when fed to two crossbred classes and British cattle. Winterholler et al. (2007) and Sissom et al. (2007) both fed 200 mg·head$^{-1}$·d$^{-1}$ of RH for the final 28 days of feeding to steers and heifers, respectively, and reported up to a 8 kg increase in HCW and 2.6 cm$^2$ increase in LM area. Bryant et al. (2010) found that anabolic implant and RH effects on steer and heifer carcass characteristics occurred independently. Additionally, the authors demonstrated a dose-dependent carcass response when steers fed 200 mg·head$^{-1}$·d$^{-1}$ had 6.3 kg greater HCW, 0.46 % greater DP, and 2.3 cm$^2$ larger LM area, whereas carcasses from steers fed 100 mg·head$^{-1}$·d$^{-1}$ were similar to control cattle. Fat measures were also not altered by feeding RH. In contrast, Gonzalez et al. (2010) did not find increases in HCW or LM area, but reported marbling score was slightly reduced by 3.5% when steers were fed 200 mg·head$^{-1}$·d$^{-1}$.

Boler et al. (2012) also conducted a dose titration study (0, 200, or 300 mg·head$^{-1}$·d$^{-1}$) when RH was fed to crossbred steers 28 days prior to harvest. Steers fed RH had a 4% increase in HCW, a 1% increase in DP, and 4 cm$^2$ increase in LM area; however, there were no differences between 200 and 300 mg·head$^{-1}$·d$^{-1}$ RH treatments. Bohrer et al. (2014) observed a 2% increase in HCW, but no changes in marbling score, 12th rib fat thickness, LMA, or DP when steers were fed 300 mg·head$^{-1}$·d$^{-1}$ of RH for 35 d prior to harvest. Kononoff et al. (2014) fed the same level of RH for 28 d and found HCW increased by 1.5% and DP slightly increased, but fat measures, YG and quality grade were not significantly affected. Arp et al. (2014) fed 200, 300 and 400 mg·head$^{-1}$·d$^{-1}$ of RH for the final 30 days of feeding and reported increased HCW by 3.9 and 6.3 kg for steers fed 300 and 400 levels, respectively, when compared to controls. Feeding RH at the 200 and 300 levels did not increase LM area over controls, but carcasses from 400 level steers had 2.4 and 2.2 cm$^2$ larger LM areas compared to control and 200 level carcasses, respectively. Only carcasses from the 300 level had 4% lower marbling scores than control carcasses, although all other fat measures were unaffected.

Holstein cattle are often discounted due to producing carcasses with lower muscle to bone ratios; therefore, these cattle would benefit from the repartitioning effects associated with β-AA supplementation. Bass et al. (2009) demonstrated these effects when 200 mg·head$^{-1}$·d$^{-1}$ of RH was fed for 36 days and HCW increased 5.7 kg and LM area increased 1.9 $cm^2$; fat measures were unaffected. Vogel et al. (2009) compared 200 and 300 $mg{\cdot}head^{-1}{\cdot}d^{-1}$ of RH in calf-fed and yearling Holsteins. For calf feeding, 200 and 300 $mg{\cdot}head^{-1}{\cdot}d^{-1}$ of RH increased HCW by 4.7 and 5.1 kg and LM area by 1.74 and 2.87 $cm^2$, respectively. Dressing percentage and KPH were not affected by RH, but the 300 level decreased fat thickness compared to controls and 200 level, and the 200 level had lower marbling scores than the controls. In contrast, when fed to yearlings, RH had no effect on carcass characteristics.

In summary, most studies demonstrate that RH fed at 200 $mg{\cdot}head^{-1}{\cdot}d^{-1}$ or greater elicit positive effects on HCW and LM area, with no to little effect on fat measures. Supplementing with RH at a level of 200 $mg{\cdot}head^{-1}{\cdot}d^{-1}$ does not affect carcass composition and feeding greater levels for longer time does not always stimulate maximum benefit. This is most likely due to desensitization to the product, which serves as a defence mechanism against prolonged BAA receptor stimulation.

## 5.2 Ractopamine hydrochloride (RH): effects on meat quality

Early research with non-approved BAA reported steak tenderness decreased and that elevated calpastatin activity was hypothesized as the cause for this negative impact. Therefore, the focus of RH meat quality research focused on establishing its impact on steak tenderness and will be the focus of this and the ZH meat quality section. To date, no clear conclusions have been accepted due to inconsistent results between studies. Avendaño-Reyes et al. (2006) reported steers fed 300 $mg{\cdot}head^{-1}{\cdot}d^{-1}$ RH for 33 days produced steaks that were 0.44 kg tougher than control steaks through 14 days of ageing. After 14 days of ageing, Quinn et al. (2008) observed LM steaks from heifers fed 200 $mg{\cdot}head^{-1}{\cdot}d^{-1}$ for 28 days had similar WBSF as steaks from control heifers. Bohrer et al. (2014) found a 0.39 kg increase in WBSF from 14-day aged steaks fabricated from cattle fed 300 $mg{\cdot}head^{-1}{\cdot}d^{-1}$ for 35 days.

Because of the apparent decrease in tenderness resulting from RH supplementation, researchers began studying postmortem ageing past 14 days as a means to advert the negative effects. In addition to finding a breed dependent RH increase in WBSF, Gruber et al. (2008) reported ageing up to 21 days did not alleviate the negative effects of RH on WBSF (+0.38 average) or slice shear force (+1.4 kg average). Authors also reported trained sensory panels rated RH steaks lower for tenderness and juiciness. After supplementing crossbred steers with 200 $mg{\cdot}head^{-1}{\cdot}d^{-1}$ RH for 33 days and ageing loins for 3, 7, 14 or 21 days postmortem, Scramlin et al. (2010) reported that RH steaks aged 3 and 7 days had 0.70 and 0.48 kg greater WBSF, respectively, when compared to control steaks. By day 14 of ageing, however, RH steaks did not differ from control steaks. Woerner et al. (2011) observed LM steaks from steers and heifers supplemented with 200 $mg{\cdot}head^{-1}{\cdot}d^{-1}$ were not affected by postmortem ageing up to 28 days, but on average produced steaks that were 0.23 kg tougher when compared to the control fed cattle. The authors also used WBSF values to predict consumer acceptability and found RH fed steers were predicted to have less consumer acceptance. Boler et al. (2012) fed steers 200 or 300 $mg{\cdot}head^{-1}{\cdot}d^{-1}$ RH for 28 days and aged loins 4, 7, 14, 21 and 28 days postmortem. On days 7 and 14, steaks from 300 level cattle had greater WBSF than controls (+0.62 kg) and 200 level (+0.38 kg), which did not differ from each other. By day 21, steaks from the 300 level had greater WBSF than controls (+0.43), but not different than 200 steaks. By 28 days, WBSF values were not different, indicating longer ageing periods may be required to alleviate the negative impacts of RH on tenderness. In contrast, Martin et al. (2014) fed

300 mg·head$^{-1}$·d$^{-1}$ for 28 days and aged meat for 16 and 23 days postmortem. Steaks from RH carcasses were 0.41 and 0.42 kg greater in WBSF after 16 and 23 days of ageing, respectively. The same trend was seen for slice shear force, where RH steaks were 2.22 and 1.53 kg greater, respectively.

In conclusion, RH can have a negative effect on tenderness, but this negative effect is not always found and little research has been conducted to explore if consumers can detect differences. In step with WBSF data, ageing can compensate for the somewhat negative effect on tenderness, but advantages of extended ageing past 14 days are not always demonstrated. More research is needed to explore the biological mechanisms associated with reduced tenderness.

# 6 Beta-adrenergic agonists: zilpaterol hydrochloride (ZH)

## 6.1 Zilpaterol hydrochloride (ZH): effects on carcass composition

Because ZH is a β-2 agonist and cattle contain proportionately more β-2 receptors than other meat animal species, this compound elicits a more potent effect in growth, carcass weight and meat yields than ZH. As mentioned earlier, this product has been prohibited for sale in the United States by the manufacturer due to perceived animal welfare concerns and decreased tenderness; therefore, we will briefly summarize its effects on carcass characteristics and meat quality. Typically, in conventional beef steers and heifers, HCW and LM area are the two carcass characteristics dramatically altered by ZH supplementation. Feeding ZH at levels around 8.3 mg/kg produced carcasses with HCW gains between 17.2 and 27 kg and LM area gains between 5.6 and 15.2 cm$^2$. Gains in these measures caused 1.2 to 2.2% increases in DP and 17.6 and 27% reductions in YG (Plascencia et al., 1999; Vasconcelos et al., 2008; Kellermeier et al., 2009; Montgomery et al., 2009a; Montgomery et al., 2009b; Parr et al., 2011b; Rathmann et al., 2012; Rodas-González et al., 2012). When compared to the RH results described above, ZH has a distinctly greater effect on muscling of steer and heifer carcasses.

As nutrients are being deposited towards lean muscle accretion described above, the amount of fat being deposited can be decreased for ZH supplemented cattle resulting in inferior meat quality. Robles-Estrada et al. (2009) and Rodas-González et al. (2012) reported that ZH supplementation does not affect 12th rib fat thickness or marbling score. Other authors also reported no change in marbling score for cattle subjected to ZH (Neill et al., 2009; Parr et al., 2011a; Rathmann et al., 2012; Hales et al., 2014). Claus et al. (2010) were the only group to objectively measure intramuscular fat percentage within the *Longissimus* muscle for cattle subjected to ZH supplementation up to 40 days prior to harvest and found no effect.

Rathmann et al. (2012) was the lone study to report a 0.08 cm decrease in 12th rib fat, with no effect on marbling. Montgomery et al. (2009b), Baxa et al. (2010) and Lean et al. (2014) found ZH decreased both 12th rib fat thickness and marbling score between 0.9 and 0.14 cm, and 4.4% and 23%, respectively. Vasconcelos et al. (2008) reported a linear decline in 12th rib fat thickness and tendency for a decline in marbling as days of feeding ZH increased. Elam et al. (2009) found the same trend when feeding ZH for 20, 30 or 40 days. When compared to the 0 supplemented group, steers supplemented with

ZH for 20, 30 or 40 days had modest decreases in marbling score by 3.3, 6.5 and 7.4%, respectively, and a tendency for a linear decrease in FT by 0.1, 0.13 and 0.14 cm. Finally, Montgomery et al. (2009a) and Holland et al. (2010) reported marbling score reductions between 10% and 4% with ZH supplementation.

Similar to conventional steer and heifer beef breeds, ZH elicits greater Holstein carcass effects than RH by increasing HCW between 11.6 and 15.5 kg and LM area between 5.1 and 7 $cm^2$. This resulted in some studies reporting increases in DP between 1.5 and 1.6%, and decreases in YG between 4.9 and 7% (Beckett et al., 2009; Brown et al., 2014; Martin et al., 2014). Beckett et al. (2009) also demonstrated a day of ZH supplementation linear effect when they reported greater advantages in HCW, LM area, DP and YG when steers were fed ZH for 30 and 40 days versus 20 days. Garmyn et al. (2010) was the only group that observed no difference for all carcass measures between Holstein steers fed 0 and 8.3 mg/kg of ZH for 20 days.

As demonstrated above, ZH elicits potent effects on carcass characteristics of steers and heifers, especially attributes related to muscling. These responses are regularly more than double the response seen when using RH. Fat deposition will be negatively affected through reductions in 12th rib fat thickness and marbling, which usually is not seen with RH.

## 6.2 Zilpaterol hydrochloride (ZH): effects on meat quality

In line with the effects seen with carcass characteristics, ZH drastically negatively impacts meat tenderness, but to what extent has been a topic of spirited discussion throughout the past decade. Studies have focused on the impact that days of supplementation and postmortem ageing elicit on tenderness. Supplementing with ZH for greater than 20 days can have lasting impacts on tenderness attributes for steaks subjected to postmortem ageing.

Brooks et al. (2009) observed LM, *gluteus medius* and *triceps brachii* steaks from USDA Select carcasses had linear increases in WBSF as days fed ZH increased, which did not occur in Choice steaks. Feeding ZH also reduced the percentage of steaks with WBSF values less than 4.5 kg; however, the authors suggested these negative effects could be alleviated by ageing for 21 days postmortem. In another study, Brooks et al. (2010) found ZH fed for 20, 30 or 40 days all increased WBSF following 7 and 14 days of ageing, but were not different from each other. By day 21 of ageing, only steaks from cattle fed ZH for 30 days had greater WBSF than controls. After 21 days of ageing, Claus et al. (2010) reported feeding ZH at a concentration of 7.6 ppm for 0, 20, 30 or 40 days caused a linear type increase in WBSF in steer LM steaks, and a quadratic type response in heifer LM and *Triceps brachii* steaks, where WBSF peaked and levelled at 30 days of feeding (Fig. 2). Steaks from the *gluteus medius* muscle of heifers demonstrated the detrimental effect of increased ZH supplementation period, as there was an increase in the percentage of steaks greater than 4.6 and 5.0 kg in WBSF after 21 days of ageing (Fig. 3).

Possibly learning from the RH data, most of the studies conducted involve postmortem ageing as part of the design. Strydom et al. (2009) reported a day of ageing by ZH interaction for LM steaks in which at 2, 7 and 14 days of ageing, steaks were approximately 1.4, 1.8 and 1.9 kg (WBSF) tougher than the control steaks, respectively. The authors also looked at the response of the *Semitendinosus* muscle and found a lack of an ageing by ZH interaction, but ZH increased WBSF by 0.8 kg during the ageing study. Finally, Strydom et al. (2009) reported calpastatin activity was elevated in ZH LM and *Semitendinosus* by 67 and 55%,

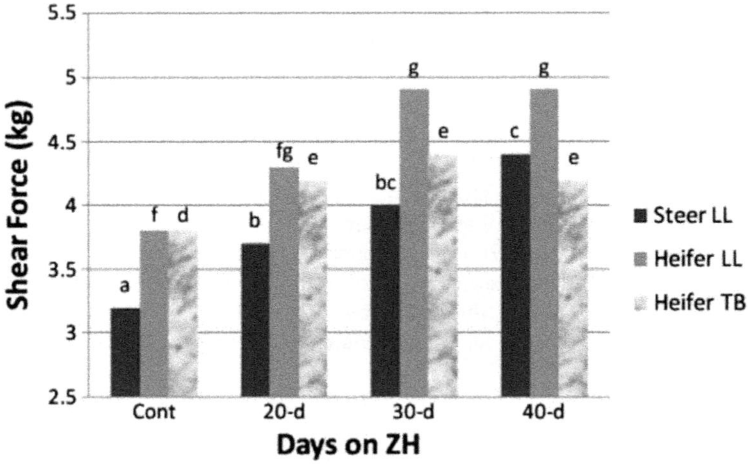


**Figure 2** Effects of zilpaterol hydrochloride treatment on mean Warner–Bratzler shear force of steer and heifer longissimus lumborum and heifer triceps brachii (TB) muscles. Within muscle, means lacking a common superscript letter differ ($P < 0.05$). Adapted from Claus et al. (2010).

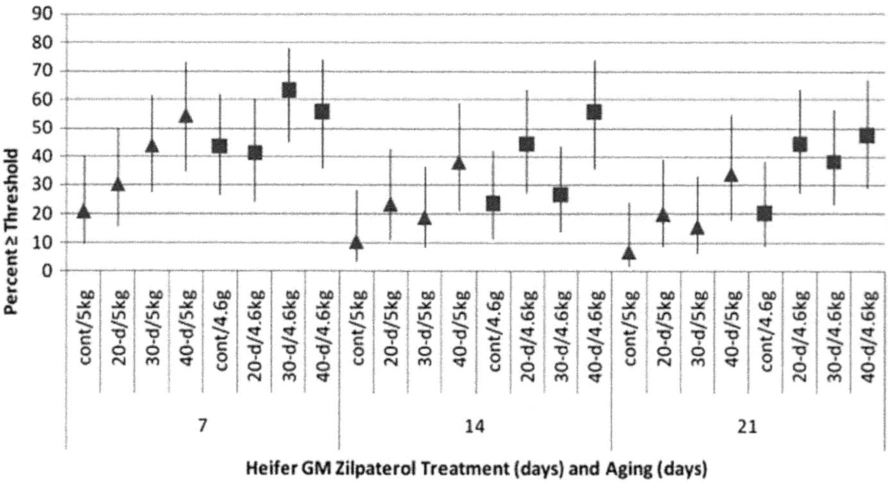


**Figure 3** Percentages of heifer gluteus medius (GM) steaks exceeding thresholds of either 5.0 (▲) or 4.6 (■) kg from zilpaterol hydrochloride treatments aged 7, 14 and 21 days. Bars represent 95% confidence intervals. Adapted from Claus et al. (2010).

respectively, indicating this was a reason for decreased tenderness. Rathmann et al. (2009) also observed ZH strip steaks were tougher than control steaks at days 7, 14 and 21 days of ageing. Additionally, the authors found the longer ZH was supplemented, the greater the increase in WBSF at day 21 of ageing. Scramlin et al. (2010) found similar results when ZH steaks had 2.23, 2.15, 1.73 and 1.24 kg greater WBSF at 3, 7, 14 or 21 days of ageing, respectively, indicting ageing through 21 days does not alleviate the negative effect of ZH on tenderness. Kellermeier et al. (2009) and Korn et al. (2013) also found 21 days does not completely alleviate the ZH negative impact on WBSF when steaks had 1.4 and 1.0 kg greater values compared to controls, respectively. Shook et al. (2009), Rathmann et al. (2012) and McEvers et al. (2012) found smaller differences with ZH steaks having 0.4, 0.5 and 0.34 kg greater WBSF, respectively.

Because of the lack of ability of 21 days of ageing to completely alleviate the negative effects of ZH on WBSF, researchers examined ageing meat for longer times. Rodas-González et al. (2012) reported that calcium chloride injection of steaks from cattle fed ZH did not improve tenderness after 28 days of ageing because ZH steaks had 0.55 kg greater WBSF, indicating that calcium chloride injection did not alleviate the toughness problem. Martin et al. (2014) observed 0.43 kg greater WBSF for LM steaks aged 16 days from ZH-fed steers compared to the control steers and, after 23 days of ageing, ZH steaks remained 0.42 kg greater. Two studies have examined extended ageing to 35 days postmortem. Garmyn et al. (2011) reported following 28 days of ageing, ZH steaks remained tougher by 0.71 kg when compared to the control steaks, and extending ageing to 35 days reduced the difference to 0.29 kg, but did not completely alleviate the differences. Ebarb et al. (2016) found after 35 day of ageing, ZH steak WBSF values were 0.70 kg greater than control steaks (Fig. 4) and authors indicated the reduced tenderness may have been due to enlargement of type IIX muscle fibre cross-sectional area (Fig. 5). The role of ZH and other growth-promoting technologies influence on cell enlargement needs to be further explored and quantified.

Zilpaterol hydrochloride has a major negative effect on meat tenderness. This negative effect on tenderness are improved by ageing for more than 21 days postmortem; however, significant differences are still found between ZH and non-ZH steaks. Although not extensively covered in this section, Garmyn and Miller (2014) concluded that while

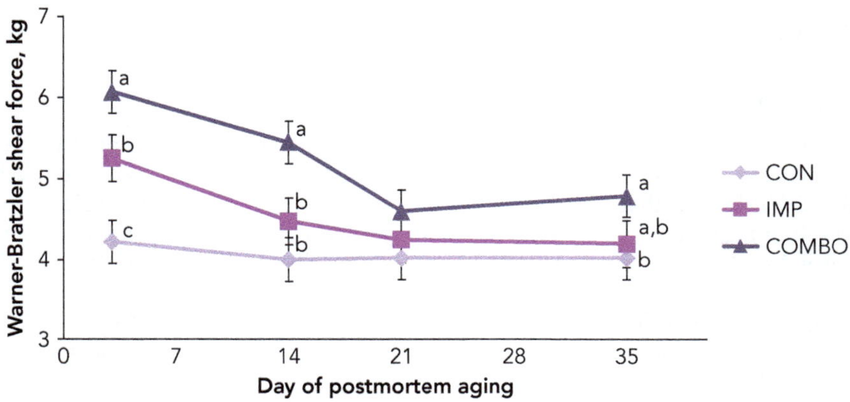


**Figure 4** Effect of growth technologies on Warner–Bratzler shear force after 35 days of ageing. Crossbred feedlot heifers ($n$ = 32) were subjected to 1 of 3 treatments: no implant or zilpaterol hydrochloride (CON); Component TE-200 implant (Elanco Animal Health, Greenfield, IN) on d 0 of study, no zilpaterol hydrochloride (IMP); and Component TE-200 implant on d 0 of study and zilpaterol hydrochloride (Merck Animal Health, Summit, NJ) at 8.3 mg/kg of ZH from d 50 to d 71, followed by a 3-d withdrawal (COMBO). [a,b,c]Points within a day of ageing lacking common superscripts differ ($P < 0.05$). Adapted from Ebarb et al. (2016).

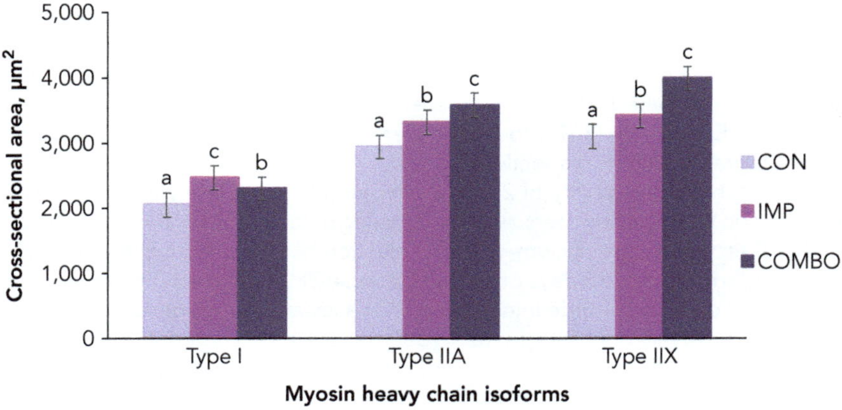


**Figure 5** Effect of growth technologies on myosin heavy chain type I, IIA and IIX cross-sectional area. Crossbred feedlot heifers ($n$ = 32) were subjected to 1 of 3 treatments: no implant or zilpaterol hydrochloride (CON); Component TE-200 implant (Elanco Animal Health, Greenfield, IN) on day 0 of study, no zilpaterol hydrochloride (IMP); and Component TE-200 implant on day 0 of study and zilpaterol hydrochloride (Merck Animal Health, Summit, NJ) at 8.3 mg/kg of ZH from d 50 to d 71, followed by a 3-d withdrawal (COMBO). [a,b,c]Points within a fibre type lacking common superscripts differ ($P < 0.05$). Adapted from Ebarb et al. (2016).

consumers can detect toughness in ZH steaks, they label these steaks acceptable for tenderness. If ZH is ever sold again in the United States, the exact mechanisms responsible for decreased tenderness need to be further explored.

# 7 Vitamins $D_3$, A and E

## 7.1 Vitamin $D_3$

Feeding vitamin $D_3$ has received attention for its potential to improve meat tenderness. Dietary vitamin $D_3$ has been shown to increase blood and muscle calcium levels. Wheeler et al. (1997) showed that elevated levels of calcium in muscle increases calpain enzyme activity, thus promoting proteolysis. Feeding high levels of vitamin $D_3$ ($0.5 \times 10^6$ to $7.5 \times 10^6$ IU/hd/d) 4 to 10 days before slaughter improved tenderness of the *longissimus thoracis* at 7 days postmortem, but there was no advantage at 14 or 21 days (Swanek et al., 1999) They reported more calpain enzyme activity for cattle fed $D_3$. However, Karges et al. (1999) reported that calpastatin activity, the inhibitor of calpain activity, was not altered in cattle fed vitamin $D_3$ for 4 or 6 days.

Montgomery et al. (1999) reported that control carcasses had temperatures 15.6°C higher at 24 hour postmortem than carcasses from cattle fed vitamin $D_3$, but did not state whether this had any effect on their results. They reported that vitamin $D_3$ only reduced the variation and not the mean shear force values in *longissimus* steaks. These authors concluded that feeding vitamin $D_3$ will effectively improve tenderness if cattle tend to be tough and would have no impact on cattle that produce tender beef. Karges et al. (1999) reported a decrease in WBSF of both *LM* and *gluteus medius* steaks at 14 and 21 days postmortem. However, they also reported a decrease in HCW from feeding 6 × $10^6$ IU/hd/d for 4 or 6 days because of decreased feed intake by the cattle. On the other hand, Vargas et al. (1999) fed steers vitamin $D_3$ alone (6 × $10^6$ IU/hd/d), a combination of vitamin $D_3$ and vitamin E (1000 IU/hd/d), or neither vitamin and found that steaks from both vitamin $D_3$ treatments required less ageing time than steaks from control cattle to reach a WBSF value of <3.86 kg, which they considered to be 'very tender'.

Karges et al. (1999) found that dry matter intake in cattle was reduced from feeding vitamin $D_3$ for 2 d at 7.5 × $10^6$ IU/d; 4 d at 15 × $10^6$ IU/d; 5 d at 7.5 × $10^6$ IU/d; and 6 d at 5 × $10^6$ IU/d. In the study by Montgomery et al. (2002), feeding vitamin $D_3$ at 2.5, 5.0 and 7.5 × $10^6$ IU/d for 9 days before harvest resulted in negative ADG during the last 21 days of feeding, but not for those fed 0.5 or 1.0 × $10^6$ IU/d. This decreased gain was partially due to decreased feed intake on days 7 and 8 of feeding vitamin $D_3$. Although not statistically significant, final weight was 5, 11 and 16 kg lighter for cattle fed the 2.5, 5.0 and 7.5 × $10^6$ IU/d, respectively. These findings raise important questions about the practicality of feeding vitamin $D_3$ to improve tenderness because livestock producers are not likely to use this practice if it reduces performance and efficiency.

Puls (1994) suggested that supplementing cattle with at least 2 × $10^6$ IU/d vitamin $D_3$ could result in cattle toxicity. Furthermore, high levels of vitamin $D_3$ can be toxic to humans. The U.S. Recommended Dietary Allowance is 400 IU/d (10 μg) for infants, children, adults and pregnant/lactating women. An intake of 10,000 IU/d for several months resulted in marked disturbances in calcium metabolism (Council on Scientific Affairs, 1987). Montgomery et al. (1999) reported a four to seven times higher concentration of vitamin $D_3$ in liver and muscle than normal when cattle were fed 5 × $10^6$ IU/hd/d of vitamin $D_3$. Montgomery et al. (2000) further reported that concentrations of vitamin $D_3$ in liver were increased 71- and 114-fold by doses of 5 and 7.5 × $10^6$ IU/d and that these levels raise a caveat with regard to the commercial adoption of feeding vitamin $D_3$ to improve beef tenderness because consumption of as little as 45 μg vitamin $D_3$ per day has been associated with signs of hypervitaminosis D in young children (citing American Academy of Pediatrics, 1963). Montgomery et al. (2002) stated that the increase in residues in liver samples poses a serious toxicological hazard, requiring livers to be removed from the food chain. However, these authors further stated that the increase in beef muscle concentrations does not seem to pose a toxicological hazard. Because Montgomery et al. (2002) found that feeding as little as 0.5 × $10^6$ IU/d resulted in a significant improvement in tenderness of top round steaks at all times postmortem, an improvement in top loin steaks at 7 days postmortem and no negative effect on feedlot performance or tissue residues, they concluded that cattle should be supplemented with no more than 0.5 × $10^6$ IU/d vitamin $D_3$.

In summary, feeding vitamin $D_3$ likely will not be adapted by the industry if it reduces feed intake and performance and potentially could cause toxicity in humans

## 7.2 Vitamin A

Recently, the role of vitamin A in adipocyte differentiation and marbling development has been investigated. Steers finished on diets containing low levels of vitamin A have produced carcasses with increased marbling scores (Oka et al., 1998; Nade et al., 2003; Kruk et al., 2004; Gorocica-Buenfil, Fluharty and Loerch, 2006). Marbling scores have been negatively correlated with concentration of vitamin A in blood (Nakai et al., 1992; Torii et al., 1996; Matsuzaki et al., 1998; Adachi et al., 1999) and liver (Oka et al., 1998; Chae et al., 2003). Most of these studies utilized Japanese black or Waygu cattle that have a very high propensity for marbling deposition. Gorocica-Buenfil et al. (2006) demonstrated that marbling score and percentage of USDA Choice carcasses were 10% higher in steers fed diets with no supplemental vitamin A.

Arnett et al. (2009) utilized Angus steers that were either early weaned or traditionally weaned and superimposed a high supplement level of vitamin A or no supplemental vitamin A. Steers not supplemented with vitamin A had higher marbling/fat thickness ratios than steers supplemented with high vitamin A (Fig. 6). In addition, not supplementing with vitamin A resulted in a higher percentage of carcasses grading Choice and Prime. In all of these studies, there have been no significant effects of vitamin A supplemental level on growth rate, feed intake or carcass composition. However, retinoic acid, a form of vitamin A, regulates growth hormone gene expression (Bedo, et al., 1990) and has increased growth rates of cattle (Perry et al., 1968).

## 7.3 Vitamin E

Numerous studies have shown that feeding supra-nutritional levels of vitamin E improves meat colour and can extend shelf life of beef. There have been no reports of negative effects on feed intake or performance of cattle fed vitamin E as is reported for feeding vitamin $D_3$. Therefore, this metabolic modifier is practical and effective.

Several studies indicate that dietary vitamin E supplementation to steers results in accumulation of α-tocopherol in muscle tissue and that this antioxidant delays lipid and myoglobin oxidation. Consequently, colour stability and retail shelf life of beef are prolonged. Faustman et al. (1989) demonstrated that percent metmyoglobin was reduced (Fig. 7) in Holstein steers fed with high levels of vitamin E. Arnold et al. (1992) fed Holstein calves diets containing vitamin E at 300 IU/d for 266 d; 1,140 IU/d for 67 d; 1,200 IU/d

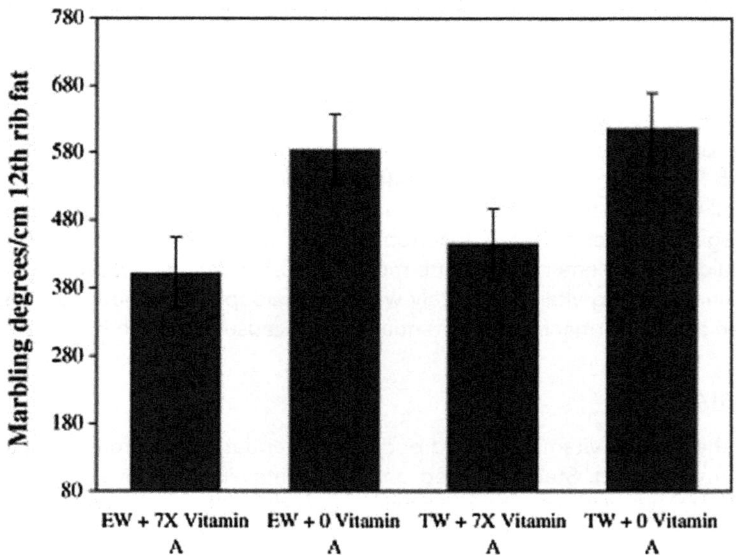


**Figure 6** Effects of supplementing diets with high or no vitamin A on the marbling/fat thickness ratio of early and traditionally weaned Angus calves. Adapted from Arnett et al. (2009).

for 38 d and a control diet in which no vitamin E was fed. They found that colour stability during retail display of *longissimus lumborum* steaks was extended by 2.5 to 4.8 d (Fig. 8). *Gluteus medius* steaks had an extended colour display life of 1.6 to 3.8 d. In addition, the accumulation of lipid oxidation products was suppressed for muscles from vitamin E-supplemented steers.

Liu et al. (1996) fed Holstein steers 0, 250, 500 or 2,000 IU/d vitamin E for durations of 42 or 126 d and evaluated the effects on meat quality. Colour display of fresh beef was extended from 0.9 to 1.8 d from the lowest to highest feeding level when using the 'metmyoglobin threshold method' for determining when beef reached the end of its shelf life. In addition, increases in TBARS were delayed in beef from cattle fed vitamin E. Collectively, supplementation of 500 mg of α-tocopherol acetate per steer daily improved the mean colour display life of the three muscles by 2.3 days, or essentially 100%. However, when muscles were vacuum aged for 56 days, colour display life was decreased to 56% of that for meat aged 14 days, even though α-tocopherol concentrations after 56 days were still 96% of 14-day values.

Lynch et al. (1998) compared diets containing 20 (basal diet) or 2,000 IU α-tocopherol acetate/kg feed/d for 50 days in Friesian steers. Supplemented fresh, frozen and vacuum packaged *longissimus lumborum, gluteus medius* and *psoas major* muscles showed greater colour and lipid oxidative stability than meat from the basal group after 7 days of retail display.

These research reports on supplementing cattle diets with supra-nutritional levels of vitamin E consistently show an advantage in colour display life and reduced oxidative deterioration of meat in various packages and chilled states. There do not appear to be any negative effects on feed intake, growth rate, feed efficiency, dressing per cent or meat yield percentage. Thus, it seems that the beef cattle industry should be routinely feeding supra-nutritional levels of vitamin E.

Williams et al. (1992) conducted a blind study of consumers regarding beef from cattle fed 500 IU/steer of vitamin E for 100 to 120 days and beef from cattle not fed vitamin E. They found 3.6 percentage points reduction in losses in retail value from the vitamin E fed

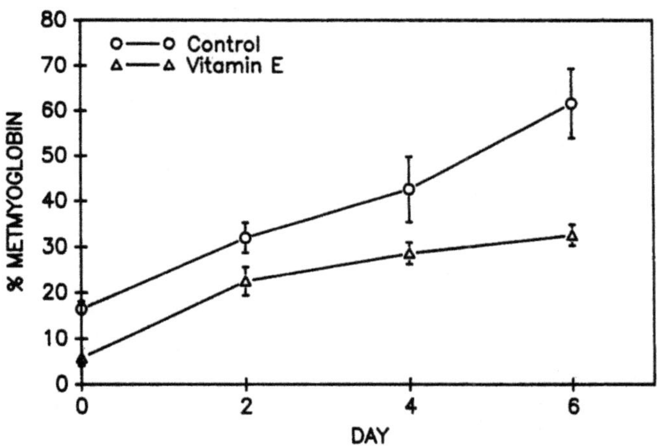


**Figure 7** Metmyoglobin accumulation during storage at 4°C for fresh ground sirloin patties from control and vitamin E-supplemented Holstein steers. $n$ = 11 for each group; standard error bars are indicated. Adapted from Faustman et al. (1989).

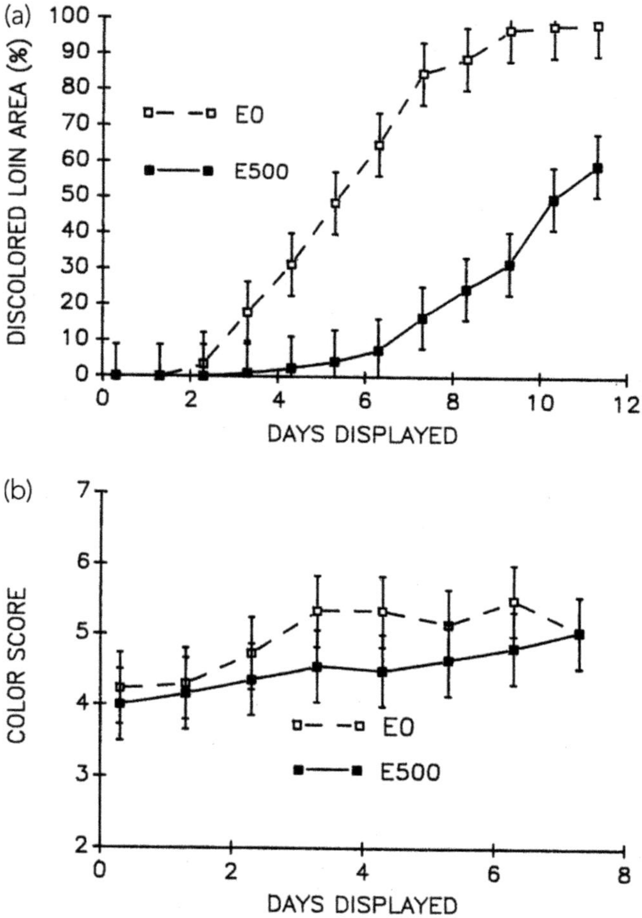


**Figure 8** Effect of level of vitamin E supplementation on area of discoloration (a) and colour score (b) of *longissimum lumborum* (loin) steaks ($n$ = 34 steers). Colour scores were associated with the following descriptive terms: 3 = moderately light cherry red, 4 = cherry red, 5 = slightly dark red, and 6 = moderately dark red. E0 = 0 IU/d of supplemental vitamin E; E500 = 300 IU/d of actual supplemental vitamin E. Adapted from Arnold et al. (1992).

cattle. Liu et al. (1995) described the cost/benefit ratio of supplementing 500 IU of vitamin E for 126 days is estimated to be $3 per animal. If retailers could improve their receipts by 3.6% (Williams et al., 1992), this suggests a financial gain of $792 million annually to the beef industry. The benefit/cost ratio for the packing, fabrication, distribution and retail marketing segments of the beef industry would be 10.4:1. The only issue that needs to be worked out is for cattle feeders to be compensated for the additional cost of feeding higher levels of vitamin E. It may also require a method to rapidly verify that cattle actually received adequate vitamin E supplementation when marketed with that guarantee.

The entire production, processing and retail segments of the meat industry would gain from this practice.

# 8 Somatotropin

There is no scientific data to prevent the approval of somatotropin for use in beef cattle production. Somatotropin is extremely effective in improving growth performance and meat yield percentage. However, it will decrease marbling significantly and decrease tenderness a majority of the time. It has a neutral to slightly negative effect on colour and firmness. It is not likely that somatotropin will be used for beef production in the near future, partly because of the labour involved in administering the product to cattle.

# 9 Summary

Anabolic implants are a very cost-effective technology that can improve growth rate and efficiency, increase carcass leanness, and have minimal or modest negative effects on marbling and meat palatability, depending on implant composition, and time and frequency of implanting. When used as recommended, marbling and meat palatability will only be subtly affected. Ractopamine hydrochloride is a cost-effective feed additive that improves animal performance and efficiency and can increase dressing percentage and carcass composition. It can have negative effects on marbling and tenderness if fed longer or in higher concentration than recommended. Zilpaterol improves cattle performance and efficiency and distinctly increases dressing percentage, carcass weight and carcass leanness. It negatively affects meat tenderness without 21 or more days of ageing and can decrease marbling. Feeding high levels of vitamin E clearly improves beef colour and extends retail shelf life but producers typically are not rewarded for the cost of feeding it. It has no negative effect on marbling or meat palatability. Restricting vitamin A late in the finishing phase can increase marbling and quality grade. Feeding vitamin D3 can increase tenderness early postmortem, but the advantage is lost by 14 days. And, it can decrease feed intake and cattle performance. Plus, there can be toxicity effects in certain age groups of humans that are frequent consumers of beef. It is not likely that somatotropin will be utilized by the industry, partly because of the challenging delivery system.

# 10 Where to look for further information

The objective of this chapter was to provide a more detailed, historical documentation of metabolic modifier research in beef cattle. Dikeman (2007) served as the template for the topics covered in this chapter. Recently, several review articles have been published that examine the use of anabolic steroids and BAA on feedlot performance and meat quality. Duckett and Pratt (2014) and Garmyn and Miller (2014) provide reviews on the impacts of implants and beta-agonists on meat quality and palatability. Lean et al. (2014) provides an excellent meta-analysis of approximately 62 studies that quantifies the effect of ZH and RH on feedlot performance, carcass characteristics and meat tenderness. Finally, Strydom

(2016), Dunshea et al. (2016) and Dilger et al. (2016) provide further reviews on growth-promoting technologies and global implications of their utilisation.

# 11 References


Adachi, K., H. Kawano, K. Tsuno, Y. Nomura, N. Yamamoto, A. Arikawa, A. Tsuji, M. Adachi, T. Onimaru and K. Ohwada. 1999. Relationship between serum biochemical values and marbling scores in Japanese black cattle. *J. Vet. Med. Sci.* 61:961–4.

Adams, T. E., J. R. Dunbar, S. L. Berry, W. N. Garrett, T. R. Famula and Y. B. Lee. 1990. Feedlot performance of beef heifers implanted with Synovex-H: effect of melengestrol acetate, ovariectomy or active immunization against GnRH. *J. Anim. Sci.* 68:3079–85. doi:/1990.682324x.

American Academy of Pediatrics. 1963. The prophylactic requirement and the toxicity of vitamin D. *Pediatrics* 31:512–25.

Anderson, P. T., B. J. Johnson and M. E. Dikeman. 2014. Metabolic modifiers. In M. E. Dikeman and C. Devine (eds), *Meat Science Encyclopedia*, 2nd ed. Academic Press, San Diego, CA. pp. 62–9.

Apple, J. K., M. E. Dikeman, D. D. Simms and G. Kuhl. 1991. Effects of synthetic hormone implants, singularly or in combinations, on performance, carcass traits, and *longissimus* muscle palatability of Holstein steers. *J. Anim. Sci.* 69:4437–48.

Arnett, A. M., M. E. Dikeman, M. J. Daniel, K. C. Olson, J. Jaeger and J. Perrett. 2009. Effects of vitamin A supplementation and weaning age on serum and liver retinol concentrations, carcass traits, and lipid composition in market beef cattle. *Meat Sci.* 81:596–606.

Arnold, R. N., K. K. Scheller, S. C. Arp, S. N. Williams, D. R. Buege and D. M. Schaefer. 1992. Effect of long- or short-term feeding of α-tocopherol acetate to Holstein and crossbred beef steers on performance, carcass characteristics, and beef color stability. *J. Anim. Sci.* 70:3055.

Arp, T. S., S. T. Howard, D. R. Woerner, J. A. Scanga, D. R. McKenna, W. H. Kolath, P. L. Chapman, J. D. Tatum and K. E. Belk. 2014. Effects of dietary ractopamine hydrochloride and zilpaterol hydrochloride supplementation on performance, carcass traits, and carcass cutability in beef steers. *J. Anim. Sci.* 92:836–43.

Avendaño-Reyes, L., T. Torres-Rodríguez, F. J. Meraz-Murillo, C. Pérez-Linars, F. Figueroa-Saavedra and P. H. Robinson. 2006. Effects of two β-adrenergic agonists on finishing performance, carcass characteristics, and meat quality of feedlot steers. *J. Anim. Sci.* 84:3259–65. doi:10.2527/jas.2006-173.

Barham, B. L., J. C. Brooks, J. R. Blanton, Jr, A. D. Herring, M. A. Carr, C. R. Kerth and M. F. Miller. 2003. Effects of growth implants on consumer perceptions of meat tenderness in beef steers. *J. Anim. Sci.* 81:3052–6. doi:/2003.81123052x.

Bass, P. D., J. L. Beckett and R. J. Delmore, Jr. 2009. Case study: Effects of ractopamine in combination with various hormone implant regimens on growth and carcass attributes in calf-fed Holstein steers. *Prof. Anim. Sci.* 25:195–201.

Baxa, T. J., J. P. Hutcheson, M. F. Miller, J. C. Brooks, W. T. Nichols, M. N. Streeter, D. A. Yates and B. J. Johnson. 2010. Additive effects of a steroidal implant and zilpaterol hydrochloride on feedlot performance, carcass characteristics, and skeletal muscle messenger ribonucleic acid abundance in finishing steers. *J. Anim. Sci.* 88:330–7. doi:10.2527/jas.2009-1797.

Beckett, J. L., R. J. Delmore, G. C. Duff, D. A. Yates, D. M. Allen, T. E. Lawrence and N. Elam. 2009. Effects of zilpaterol hydrochloride on growth rates, feed conversion, and carcass traits in calf-fed Holstein steers. *J. Anim. Sci.* 87:4092–4100.doi:10.2527/jas.2009-1808.

Bedo, G., P. Santisteban and A. Aranda. 1989. Retinoic acid regulates growth hormone gene expression. *Nature* 339: 231–4.

Beermann, D. H. 1995. Growth promotants – promise, problems, and perceptions. *Proc. Recip. Meat Conf.* 48:45–50.

Beermann, D. H. 2009. ASAS centennial paper: A century of pioneers and progress in meat science in the United States leads to new frontiers. *J. Anim. Sci.* 87:1192–8.

Bohrer, B. M., B. M Edenburn, D. D Boler, A. C. Dilger and T. L. Felix. 2014. Effect of feeding ractopamine hydrochloride (Optaflexx) with or without supplemental zinc and chromium propionate on growth performance, carcass characteristics, and meat quality of finishing steers. *J. Anim. Sci.* 92:3988–96. doi:10.2527/jas.2014-7824.

Boler, D. D., A. L. Shreck, D. B. Faulkner, J. Killefer, F. K. McKeith, J. W. Homm and J. A. Scanga. 2012. Effect of ractopamine hydrochloride (Optaflexx) dose on live performance, carcass characteristics and tenderness in early weaned beef steers. *Meat Sci.* 92:458–63. doi: 10.1016/j.meatsci.2012.05.011.

Boles, J. A., D. L. Boss, K. I. Neary, K. C. Davis and M. W. Tess. 2009. Growth implants reduced tenderness of steaks from steers and heifers with different genetic potentials for growth and marbling. *J. Anim. Sci.* 87:269–74. doi: 10.2527/jas.2008-1256.

Borger, M. L., L. L. Wilson, J. D. Sink, J. H. Ziegler and S. L. Davis. 1973. Zeranol and dietary protein level effects on live performance, carcass merit, certain endocrine factors and blood metabolite levels of steers. *J. Anim. Sci.* 36:706–11.

Brooks, J. C., H. C. Claus, M. E. Dikeman, J. Shook, G. G. Hilton, T. E. Lawrence, J. M. Mehaffey, B. J. Johnson, D. M. Allen, M. N. Streeter, W. T. Nichols, J. P. Hutcheson, D. A. Yates and M. F. Miller. 2009. Effects of zilpaterol hydrochloride feeding duration and postmortem aging on Warner-Bratzler shear force of three muscles from beef steers and heifers. *J. Anim. Sci.* 87:3764–9.

Brooks, J. C., J. M. Mehaffey, J. A. Collins, H. R. Rogers, J. Legako, B. J. Johnson, T. Lawrence, D. M. Allen, M. N. Streeter, W. T. Nichols, J. P. Hutcheson, D. A. Yates and M. F. Miller. 2010. Moisture enhancement and blade tenderization effects on the shear force and palatability of strip loin steaks from beef cattle fed zilpaterol hydrochloride. *J. Anim. Sci.* 88:1809–16.

Brown, T. R., A. K. Sexton, T. E. Lawrence, M. F. Miller, C. L. Thomas, D. A. Yates, J. P. Hutcheson, J. M. Hodgen and J. C. Brooks. 2014. Comparative effects of zilpaterol hydrochloride and ractopamine hydrochloride on live performance and carcass characteristics of calf-fed Holstein steers. *J. Anim. Sci.* 92:4217–22.

Bruns, K. W., R. H. Pritchard and D. L. Boggs. 2005. The effect of stage of growth and implant exposure on performance and carcass composition in steers. *J. Anim. Sci.* 83:108–16.

Bryant, K. W., T. E. Engle, M. L. Galyean, J. J. Wagner, J. D. Tatum, R. V. Anthony and S. B. Laudert. 2010. Effects of ractopamine and trenbolone acetate implants with or without estradiol on growth performance, carcass characteristics, adipogenic enzyme activity, and blood metabolites in feedlot steers and heifers. *J. Anim. Sci.* 88:4102–19. doi.10.25271.jas.2010.2901.

Cecava, M. J., and D. L. Hancock. 1994. Effects of anabolic steroids on nitrogen metabolism and growth of steers fed corn silage and corn-based diets supplemented with urea or combinations of soybean meal and feathermeal. *J. Anim. Sci.* 72:515–22. doi:/1994.722515x.

Chae, S. H., K. K. Jung and C. B. Choi. 2003. Correlated relationships between vitamin A concentration in serum and liver and carcass characteristics of Hanwoo steers. *J. Anim. Sci. Tech.* 45:585–92.

Claus, H. L., M. E. Dikeman, L. Murray, J. C. Brooks, J. Shook, G. G. Hilton, T. E. Lawrence, J. M. Mchaffey, B. J. Johnson, D. M. Allen, M. N. Streeter, W. T. Nichols, J. P. Hutcheson, D. A. Yates, M. F. Miller, M. C. Hunt and J. Killefer. 2010. Effects of supplementing feedlot steers and heifers with zilpaterol hydrochloride on Warner-Bratzler shear force interrelationships of steer and heifer longissumus lumborum and heifer triceps brachii and gluteus medius muscles aged for 7, 14, and 21 d. *Meat Sci.* 88:347–55.

Cleale, R. M., D. Amodie, D. T. Bechtol, J. S. Drouillard, J. D. Edmonds, M. Edmonds, B. D. Hunsaker, L. A. Kraft, T. E. Lawrence, R. D. Rulli and A. R. Waite. 2013. Effects of estradiol benzoate and trenbolone acetate, alone or in combination at dose levels present in Synovex Choice, on performance by feedlot heifers. *J. Anim. Sci.* 91:970–7.

Cook, R. B., J. D. Popp, T. A. McAllister, J. P. Kastelic and R. Harland. 2000. Effects of immunization against GnRH, melengestrol acetate, and a trenbolone acetate/estradiol implant on growth and carcass characteristics of beef heifers. *Theriogenology*. 55:973–81. doi:10.1016/S0093-69X(01)00458-7.

Cranwell, C. D., J. A. Unruh, J. R. Brethour, D. D. Simms and R. E. Campbell. 1996. Influence of steroid implants and concentrate feeding on performance and carcass composition of cull beef cows. *J. Anim. Sci.* 74:1770–6. doi:/1996.7481770x.

Council on Scientific Affairs, American Medical Association. 1987. Vitamin preparations as dietary supplements and as therapeutic agents. *J. Amer. Med. Assoc.* 257:1929–36.

Crouse, J. D., B. D. Schanbacher, H. R. Cross, S. C. Seideman and S. B. Smith. 1987. Growth and carcass traits of heifers as affected by hormonal treatment. *J. Anim. Sci.* 64:1434–40.

DeHaan, K. C., L. L. Berger, D. J. Kesler, F. K. McKeith, D. B. Faulkner, G. F. Cmarik and R. J. Favero. 1990. Effects of prenatal testosterone treatment and postnatal steroid implantation on growth performance and carcass traits of heifers and steers. *J. Anim. Sci.* 68:2198–2207. doi:/1990.6882198x.

Dikeman, M. E. 1997. Reducing the fat content by production practices. In A. M. Pearson and T. R. Dutson (eds), Advances in Meat Research Series, Vol. 11. *Production and Processing of Healthy Meat, Poultry and Fish Products.* Blackie Academic & Professional, Chapman & Hall, London, pp. 150–90.

Duckett, S. K., D. G. Wagner, F. N. Owens, H. G. Dolezal and D. R. Gill. 1996. Effects of estrogenic and androgenic implants on performance, carcass traits, and meat tenderness in feedlot steers: A review. *Prof. Anim. Sci.* 12:205–14.

Duckett, S. K., and S. L. Pratt. 2014. Meat science and muscle biology symposium: anabolic implants and meat quality. *J. Anim. Sci.* 92:3–9. doi:10.2527/jas.2013-7088.

Ebarb, S. M., J. S. Drouillard, K. R. Maddock-Carlin, K. J. Phelps, M. A. Vaughn, D. D. Burnett, C. B. Paulk, D. M. Grieger and J. M. Gonzalez. 2016. Effect of growth-promoting technologies on *Longissimus lumborum* muscle fiber morphometrics, collagen solubility, and cooked meat tenderness. *J. Anim. Sci* . 94:869–81.

Elam, N. A., J. T. Vasconcelos, G. Hilton, D. L. VanOverbeke, T. E. Lawrence, T. H. Montgomery, W. T. Nichols, M. N. Streeter, J. P. Hutcheson, D. A. Yates and M. L. Galyean. 2009. Effect of zilpaterol hydrochloride duration of feeding on performance and carcass characteristics of feedlot cattle. *J. Anim. Sci.* 87:2133–41. doi:10.2527/jas.2008-1563.

Faustman, C., R. G. Cassens, D. M. Schaefer, D. R. Buege, S. N. Williams and K. K. Scheller. 1989. Improvement of pigment and lipid stability in Holstein steer beef by dietary supplementation with vitamin E. *J. Food Sci.* 54:858.

FDA. 2014. Product safety information. http://www.fda.gov/AnimalVeterinary/SafetyHealth/ProductSafetyInformation/ucm055436.htm (Accessed 4 November 2016.)

Foutz, C. P, H. G. Dolezal, T. L. Gardner, D. R. Gill, J. L. Hensley and J. B. Morgan. 1997. Anabolic implant effects on steer performance, carcass traits, subprimal yields, and *longissimus* muscle properties. *J. Anim. Sci.* 75:1256–65.

Garber, M. J., R. A. Roeder, J. J. Combs, L. Eldridge, J. C. Miller, D. D. Hinman and J. J. Ney. 1990. Efficacy of vaginal spaying and anabolic implants on growth and carcass characteristics in beef heifers. *J. Anim. Sci.* 68:1469–75.

Gruber, S. L., J. D. Tatum, T. E. Engle, K. J. Prusa, S. B. Laudert, A. L. Schroeder and W. J. Platter. 2008. Effects of ractopamine supplementation and postmortem aging on *Longissimus* muscle palatability of beef steers differing in biological type. *J. Anim. Sci.* 86:205–10.

Garmyn, A. J., J. N. Shook, D. L. VanOVerbeke, J. L. Beckett, R. J. Delmore, D. A. Yates, D. M. Allen and G. G. Hilton. 2010. The effects of zilpaterol hydrochloride on carcass cutability and tenderness of calf-fed Holstein steers. *J. Anim. Sci.* 88:2476–85.

Garmyn, A. J., S. M. Knobel, K. S. Spivey, L. F. Hightower, J. C. Brooks, B. J. Johnson, S. L. Parr, R. J. Rathmann, J. D. Starkey, D. A. Yates, J. M. Hodgen, J. P Hutcheson and M. F. Miller. 2011. Warner-Bratzler and slice shear force measurements of 3 beef muscles in response to various aging periods after trenbolone acetate and estradiol implants and zilpaterol hydrochloride supplementation of finishing beef steers. *J. Anim. Sci.* 89:3783–91.

Garmyn, A. J., and M. F. Miller. 2014. Meat science and Muscle Biology Symposium-Implant and beta agonist impacts on beef palatability. *J. Anim. Sci.* 92:10–20.

Gerken, C. L., J. D. Tatum, J. B. Morgan and G. C. Smith. 1995. Use of genetically identical (clone) steers to determine the effects of estrogenic and androgenic implants on beef quality and palatability characteristics. *J. Anim. Sci.* 73:3317–24.

Gonzalez, J. M., S. E. Johnson, A. M. Stelzleni, T. A. Thrift, J. D. Savell, T. M. Warnock and D. D. Johnson. 2010. Effect of ractopamine-HCl supplementation for 28 days on carcass characteristics, muscle fiber morphometrics, and whole muscle yields of six distinct muscles of the loin and round. *Meat Sci.* 85:379–84.

Gorocica-Buenfil, M. A., F. L. Fluharty and S. C. Loerch. 2005. Effect of dietary vitamin A intake on marbling. *J. Anim. Sci.* 83(Suppl. 1):370 (Abstr.)

Gruber, S. L., J. D. Tatum, T. E. Engle, M. A. Mitchell, S. B. Laudert, A. L. Schroeder and W. J. Platter. 2007. Effects of ractopamine supplementation on growth performance and carcass characteristics of feedlot steers differing in biological type. *J. Anim. Sci.* 85:1809–15. doi:10.2527/jas.2006-634.

Hales, K. E., S. D. Shackelford, J. E. Wells, D. A. King, M. D. Hayes, T. M. Brown-Brandl, L. A. Kuehn, H. C. Freetly and T. L. Wheeler. 2014. Effects of feeding dry-rolled corn-based diets with and without wet distillers grains with solubles and zilpaterol hydrochloride on performance, carcass characteristics, and heat stress in finishing beef steers. *J. Anim. Sci.* 92:4023–33. doi:10.2527/jas.2014-7638.

Henricks, D. M., R. T. Brandt, Jr., E. C. Titgemeyer and C. T. Milton. 1997. Serum concentrations of trenbolone-17 beta and estradiol-17 beta and performance of heifers treated with trenbolone acetate, melengestrol acetate, or estradiol-17 beta. *J. Anim. Sci.* 75:2627–33. doi:/1997.7510627x.

Hermesmeyer, G. N., L. L. Berger, T. G. Nash and R. T. Brandt, Jr. 2000. Effects of energy intake, implantation, and subcutaneous fat end point on feedlot steer performance and carcass composition. *J. Anim. Sci.* 78:825–31.

Herschler, R. C., A. W. Olmsted, A. J. Edwards, R. L. Hale, T. Montgomery, R. L. Preston, S. J. Bartle and J. J. Sheldon. 1995. Production responses to various doses and ratios of estradiol benzoate and trenbolone acetate implants in steers and heifers. *J. Anim. Sci.* 73:2873–81. doi:/1995.73102873x.

Holland, B. P., C. R. Krehbiel, G. G. Hilton, M. N. Streeter, D. L. VanOverbeke, J. N. Shook, D. L. Step, L. O. Burciaga-Robles, D. R. Stein, D. A. Yates, J. P. Hutcheson, W. T. Nichols and J. L. Montgomery. 2010. Effect of extended withdrawal of zilpaterol hydrochloride on performance and carcass traits in finishing beef steers. *J. Anim. Sci.* 88:338–48.

Hoechst-Roussel Agri-Vet Company. 1991. Pooled summary of five feedyard trials with Revalor-S®, Finaplix-S® and/or Synovex-S® administered to yearling steers during the finishing period. Tech. Bull. 1, Hoechst-Roussel Agri-Vet Co., Sommerville, NJ.

Igo, J. L., J. C. Brooks, B. J. Johnson, J. Starkey, R. J. Rathmann, A. J. Garmyn, W. T. Nichols, J. P. Hutcheson and M. F. Miller. 2011. Characterization of estrogen-trenbolone acetate implants on tenderness and consumer acceptability of beef under the effect of 2 aging times. *J. Anim. Sci.* 89:792–7.

Karges, K., J. B. Morgan, F. N. Owens and D. R. Gill. 1999. Effects of feeding vitamin $D_3$ on carcass characteristics of beef cattle. *J. Anim. Sci.* 77 (Suppl. 1): 172 (Abstr.).

Karges, K., F. N. Owens, D. R. Gill and J. B. Morgan. 1998. Effects of vitamin $D_3$ levels on feed intake by yearling steers. 1998 Plains Nutrition Council Spring Conference, 16–17 April, San Antonio, TX.

Kellermeier, J. D., A. W. Titlor, J. C. Brooks, M. L. Galyean, D. A. Yates, J. P. Hutcheson. W. T. Nichols, M. N. Streeter, B. J. Johnson and M. F. Miller. 2009. Effects of zilpaterol hydrochloride with or without an estrogen-trenbolone acetate terminal implant on carcass traits, retail cutout, tenderness, and muscle fiber diameter in finishing steers. *J. Anim. Sci.* 87:3702–11. doi:10.2527/jas.2009-1823.

Kerth, C. R., J. L. Montgomery, K. J. Morrow, M. L. Galyean and M. F. Miller. 2003. Protein turnover and sensory traits of *Longissimus* muscle from implanted and nonimplanted heifers. *J. Anim. Sci.* 81:1728–35. doi:/2003.8171728x.

Kononoff, P. J., P. J. Defoor, M. J. Engler, R. S. Swingle, S. T. James and F. L. S. Marquess. 2014. Impact of a leptin single nucleotide polymorphism and ractopamine hydrochloride on growth and carcass characteristics in finishing steers. *J. Anim. Sci.* 92:2702–7. doi:10.2527/jas.2013-7355.

Korn, K. T., R. P. Lemenager, M. C. Claeys, M. Engstrom and J. P. Schoonmaker. 2013. Supplemental vitamin $D_3$ and zilpaterol hydrochloride. I. Effect on performance, carcass traits, tenderness, and vitamin D metabolites of feedlot steers. *J. Anim. Sci.* 91:3322–31.

Kreikemier, W. M., and T. L. Mader. 2004. Effects of growth-promoting agents and season on yearling feedlot heifer performance. *J. Anim. Sci.* 82:2481–8.

Kruk, Z. A., B. D. Siebert, W. S. Pitchford, J. Davis, G. S. Harper and C. D. K. Bottema. 2004. The effect of vitamin A on fatness in lot-fed cattle: (I) Fat Quantity. Proceedings 50th International Congress of Meat Science and Technology, Helsinki, Finland. E. Puolanne, ed. University of Helsinski.

Lean, I. J., J. M. Thompson and F. R. Dunshea. 2014. A meta-analysis of zilpaterol and ractopamine effects on feedlot performance, carcass traits, and shear strength of meat in cattle. *PLoS ONE.* doi:101371/journal.pone.0115904.

Liu, Q., M. C. Lanari and D. M. Schaefer. 1995. A review of dietary vitamin E supplementation for improvement of beef quality. *J. Anim. Sci.* 73:3131–40.

Liu, Q., K. K. Scheller, S. C. Arp, D. M. Schaefer and S. N. Williams. 1996. Titration of fresh meat color stability and malondaldehyde development with Holstein steers fed vitamin E-supplemented diets. *J. Anim. Sci.* 74:117–126.

Loy, D. D., H. W. Harpster and E. H. Cash. 1988. Rate, composition and efficiency of growth in feedlot steers reimplanted with growth stimulants. *J. Anim. Sci.* 66:2668–77.

Lynch, M. P., J. P. Kerry, D. J. Buckley, C. Faustman and P. A. Morrissey. 1998. Effect of dietary vitamin E supplementation on the colour and lipid stability of fresh, frozen and vacuum-packaged beef. *Meat Sci.* 52:95–9.

Mader, T. L., and K. F. Lechtenberg. 2000. Growth-promoting systems for heifer calves and yearlings Martin, J. N., A. J. Garmyn, M. F. Miller, J. M. Hodgen, K. D. Pfeiffer, C. L. Thomas, R. J. Rathmann, D. A. Yates, J. P. Hutcheson, and J. C. Brooks. 2014. Comparative effects of beta-adrenergic agonist supplementation on the yield and quality attributes of selected subprimals from calf-fed Holstein steers. *J. Anim. Sci.* 92:4204–16. doi:10.2527/jas2014-7881.

Mancini, R. A., and M. C. Hunt. 2005. Current research in meat color. *Meat Sci.* 71:100–21.

Martin, J. N., A. J. Garmyn, M. F. Miller, J. M. Hodgen, K. D. Pfeiffer, C. L. Thomas, R. J. Rathmann, D. A. Yates, J. P. Hutcheson and J. C. Brooks. 2014. Comparative effects of beta-adrenergic agonist supplementation on the yield and quality attributes of selected subprimals from calf-fed Holstein steers. *J. Anim. Sci.* 92:4204–16. doi:10.2527/jas2014-7881.

Matsuzaki, M., S. Hara, M. Ogawa, N. Shiba and E. Tsuneishi. 1998. Effects of vitamins A, D3 and E on growth performance and carcass characteristics, and on lipolytic responsiveness and insulin secretion in Japanese black steers. *Jpn. Anim. Sci. Tech.* 69:741.

McEvers, T. J., W. T. Nichols, J. P. Hutcheson, M. D. Edmonds and T. E. Lawrence. 2012. Feeding performance, carcass characteristics, and tenderness attributes of steers sorted by the Igenity tenderness panel and fed zilpaterol hydrochloride. *J. Anim. Sci.* 90:4140–7.

Montgomery, J. L., M. A. Carr, C. R. Kerth, G. G. Hilton, L. L. Behrends and M. F. Miller. 1999. Dietary modifications using vitamin $D_3$ to improve beef tenderness. *J. Anim. Sci.* 77(Suppl. 1):45 (Abstr.).

Montgomery, T. H., P. F. Dew and M. S. Brown. 2001. Optimizing carcass value and the use of anabolic implants in beef cattle. *J. Anim. Sci.* 79:E296–E306.

Montgomery, J. L., M. A. Carr, C. R. Kerth, G. G. Hilton, B. P. Price, M. L. Galyean, R. L. Horst and M. F. Miller. 2002. Effect of vitamin $D_3$ supplementation level on the postmortem tenderization of beef from steers. *J. Anim. Sci.* 80:971–81.

Montgomery, J. L., F. C. Parrish, Jr., D. C. Beitz, R. L. Horst, E. J. Huff-Lonergan and A. H. Trenkle. 2000. The use of vitamin $D_3$ to improve beef tenderness. *J. Anim. Sci.* 78:2615–21.

Montgomery, J. L., C. R. Krehbiel, J. J. Cranston, D. A. Yates, J. P. Hutcheson, W. T. Nichols, M. N. Streeter, D. T. Bechtol, E. Johnson, T. TerHune and T. H. Montgomery. 2009a.

Dietary zilpaterol hydrochloride. I. Feedlot performance and carcass traits of steers and heifers. *J. Anim. Sci.* 87:1374–83. doi: 10.2527/jas.2008-1162.

Montgomery, J. L., C. R. Krehbiel, J. J. Cranston, D. A. Yates, J. P. Hutcheson, W. T. Nichols, M. N. Streeter, R. S. Swingle and T. H. Montgomery. 2009b. Effects of dietary zilpaterol hydrochloride on feedlot performance and carcass characteristics of beef steers fed with and without monensin and tylosin. *J. Anim. Sci.* 87:1013–23. doi:/10.2527/jas.2008-1169.

Moran, C., J. F. Quirke, D. J. Prendiville, S. Bourke and J. F. Roche. 1991. The effect of estradiol, trenbolone acetate, or zeranol on growth rate, mammary development, carcass traits, and plasma estradiol concentrations of beef heifers. *J. Anim. Sci.* 69:4249–58. doi:/1991.69114249x.

NAHMS. 2000. Implant usage by U. S. Feedlots. http://www.aphis.usda.gov/animal_health/nahms/feedlot/downloads/feedlot99/Feedlot99_is_ImplantUsage.pdf (Accessed 17 June 2015).

Nade, T., S. Hirabara, T. Okumura and K. Fujita 2003. Effects of vitamin A on carcass composition concerning younger steer fattening of Wagyu cattle. Asian-australas. *J. Anim. Sci.* 16:353–8.

Nakai, A., K. Kita, K. Hasegawa and M. Hiramitsu. 1992. Vitamin A deficiency in Japanese Black cattle. *J. Clinical Vet. Med.* 10:2143.

Neill, S., J. A. Unruh, T. T. Marston, J. R. Jaeger, M. C. Hunt and J. J. Higgins. 2009. Effects of implanting and feeding zilpaterol hydrochloride on performance, carcass characteristics, and subprimal beef yields of fed cows. *J. Anim. Sci.* 87:704–10. doi:10.2527/jas.2008-1254.

Nichols, W. T., M. I. Wray, T. H. Montgomery, B. Schutte, J. B. Morgan, H. G. Dolezal and D. P. Hutcheson. 1996. The effects of anabolic agents alone and in combination on feedyard performance, carcass characteristics, and meat quality of finishing heifers fed for 108, 131, or 143 days. Revalor-H Tech. Bull. 3, Hoechst-Roussel Agri-Vet Company, Sommerville, NJ.

Oka, A., Y. Maruo, T. Miki, T. Yamasaki and T. Saito. 1998. Influence of vitamin A on the quality of beef from the Tajmia strain of Japanese Black Cattle. *Meat Sci.* 48:159–67.

Parr, S. L., K. Y. Chung, J. P. Hutcheson, W. T. Nichols, D. A. Yates, M. N. Streeter, R. S. Swingle, M. L. Galyean and B. J. Johnson. 2011a. Dose and release pattern of anabolic implants affects growth of finishing beef steers across days on feed. *J. Anim. Sci.* 89:863–73. doi:10.2527/jas.2010-3447.

Parr, S. L., K. Y. Chung, M. L. Galyean, J. P. Hutcheson, N. DiLorenzo, K. E. Hales, M. L. May, M. J. Quinn, D. R. Smith and B. J. Johnson. 2011b. Performance of finishing beef steers in response to anabolic implant and zilpaterol hydrochloride supplementation. *J. Anim. Sci.* 89:560–70. doi:10.2527/jas.2010-3101.

Perry, T. W., Smith, W. H., Beeson, W. M. and Mohler, M. T. 1968. Value of supplemental vitamin A for fattening beef cattle on pasture. *J. Anim. Sci.* 27:190–4.

Perry, T. C., D. G. Fox and D. H. Beermann. 1991. Effect of an implant of trenbolone acetate and estradiol on growth, feed efficiency, and carcass composition of Holstein and beef steers. *J. Anim. Sci.* 69:4696–4702. doi:/1991.69124696x.

Plascencia, A., N. Torrentera and R. A. Zinn. 1999. Influence of the beta-agonist, zilpaterol, on growth performance and carcass characteristics of feedlot steers. *J. Anim. Sci.* 77(Suppl. 1):114(Abstr.).

Platter, W. J., J. D. Tatum, K. E. Belk, J. A. Scanga and G. C. Smith. 2003. Effects of repetitive use of hormonal implants on beef carcass quality, tenderness, and consumer ratings of beef palatability. *J. Anim. Sci.* 81:984–96.

Puls, R. 1994. *Vitamin Levels in Animal Health*, p. 84. Sherpa Int., Clearbrook, BC, Canada.

Quinn, M. J., C. D. Reinhardt, E. R. Loe, B. E. Depenbusch, M. E. Corrigan, M. L. May and J. S. Drouillard. 2008. The effects of ractopamine-hydrogen chloride (Optaflexx) on performance, carcass characteristics, and meat quality of finishing feedlot heifers. *J. Anim. Sci.* 86:902–8. doi:10.2527/jas.2007-0117.

Rathmann, R. J., J. M. Mehaffey, T. J. Baxa, W. T. Nichols, D. A. Yates, J. P. Hutcheson, J. C. Brooks, B. J. Johnson and M. F. Miller. 2009. Effects of duration of zilpaterol hydrochloride and days on the finishing diet on carcass cutability, composition, tenderness, and skeletal muscle gene expression in feedlot steers. *J. Anim. Sci.* 87:3686–3701. doi:10.2527/jas.2009-1818.

Rathmann, R. J., B. C. Bernhard, R. S. Swingle, T. E. Lawrence, W. T. Nichols, D. A. Yates, J. P. Hutcheson, M. N. Streeter, J. C. Brooks, M. F. Miller and B. J. Johnson. 2012. Effects of zilpaterol hydrochloride and days on the finishing diet on feedlot performance, carcass characteristics, and tenderness in beef heifers. *J. Anim. Sci.* 90:3301–11. doi:10.2527/jas.2011-4375.

Robles-Estrada, J. C., A. A. Arrizon, A. Barreras, J. F. Calderon, F. Figueroa-Saavedra, N. Torrentera, A. Plascencia and R. A. Zinn. 2009. Effects of preslaughter withdrawal period on response of feedlot heifers to zilpaterol hydrochloride supplementation: Growth performance and carcass characteristics. *J. Anim. Sci.* 87:1759–63.doi:10.2527/jas.2008-1071.

Rodas-González, A., S. B. Pflanzer, A. J. Garmyn, J. N. Martin, J. C. Brooks, S. M. Knobel, B. J. Johnson, J. D. Starkey, R. J. Rathmann, P. E. de Felicio, M. N. Streeter, D. A. Yates, J. M. Hodgen, J. P. Hutcheson, and M. F. Miller. 2012. Effects of postmortem calcium chloride injection on meat palatability traits of strip loin steaks from cattle supplemented with or without zilpaterol hydrochloride. J. Anim. Sci. 90:3584-3595. doi:10.2527/jas2012-5159

Roeber, D. L., R. C. Cannell, K. E. Belk, R. K. Miller, K. Dean, J. D. Tatum and G. C. Smith. 1999. Impact of feedlot growth promotant implant strategies on carcass grade characteristics and subsequent cooked beef palatability traits when applied to small/medium framed, 3-way British crossbred steers. Final Report to the Beef Quality Assurance Advisory Board, Nat. Cattlemen's Beef Assoc., Denver, CO.

Roeber, D. L., R. C. Cannell, K. E. Belk, R. K. Miller, J. D. Tatum and G. C. Smith. 2000. Implant strategies during feeding: impact on carcass grades and consumer acceptability. *J. Anim. Sci.* 78:1867–74. doi:/2000.7871867x.

Scramlin, S. M., W. J. Platter, R. A. Gomez, W. T. Choat, F. K. McKeith and J. Killefer. 2010. Comparative effects of ractopamine hydrochloride on growth performance, carcass traits, and *Longissimus* tenderness of finishing steers. *J. Anim. Sci.* 88:1823–9.doi:10.2527/jas.2009-.2405

Samber, J. A, J. D. Tatum, M. I. Wray, W. T. Nichols, J. B. Morgan and G. C. Smith. 1996. Implant program effects on performance and carcass quality of steer calves finished for 212 days. *J. Anim. Sci.* 74:1470–6.

Scanga, J. A, K. E. Belk, J. D. Tatum, T. Grandin and G. C. Smith. 1998. Factors contributing to the incidence of dark cutting beef. *J. Anim. Sci.* 76:2040–7.

Scheffler, J. M., D. D. Buskirk, S. R. Rust, J. D. Cowley and M. E. Doumit. 2003. Effect of repeated administration of combination trenbolone acetate and estradiol implants of growth, carcass traits, and beef quality of long-fed Holstein steers. *J. Anim. Sci.* 81:2395–2400.doi:/2003.81102395x.

Schneider, B. A., J. D. Tatum, T. E. Engle and T. C. Bryant. 2007. Effects of heifer finishing implants on beef carcass traits and *Longissimus* tenderness. *J. Anim. Sci.* 85:2019–30. doi:10.2527/jas.2007-0004.

Schoonmaker, J. P., F. L. Fluharty, S. C. Loerch, T. B. Turner, S. J. Moeller and D. M. Wulf. 2001. Effect of weaning status and implant regimen on growth, performance, and carcass characteristics of steers. *J. Anim. Sci.* 79:1074–84. doi:/2001.7951074x.

Shook, J. N., D. L. VanOverbeke, L. A. Kinman, C. R. Krehbiel, B. P. Holland, M. N. Streeter, D. A. Yates and G. G. Hilton. 2009. Effects of zilpaterol hydrochloride and zilpaterol hydrochloride withdrawal time on beef carcass cutability, composition, and tenderness. *J. Anim. Sci.* 87:3677–85.

Simms, D. D., T. B. Goehring, R. T. Brandt, Jr., G. L. Kuhl, J. J. Higgins, S. B. Laudert and R. W. Lee. 1988. Effect of sequential implanting with zeranol on steer lifetime performance. *J. Anim. Sci.* 66:2736–41.

Sissom, E. K., C. D. Reinhardt, J. P. Hutcheson, W. T. Nichols, D. A. Yates, R. S. Swingle and B. J. Johnson. 2007. Response to ractopamine-HCl is altered by implant strategy across days on feed. *J. Anim. Sci.* 85:2125–32. doi:10.2527/jas.2006-660.

Smith, K. R., S. K. Duckett, M. J. Azain, R. N. Sonon, Jr. and T. D. Pringle. 2007. The effect of anabolic implants on intramuscular lipid deposition in finished beef cattle. *J. Anim. Sci.* 85:430–40. doi:10.2527/jas.2006-280.

Strydom, P. E., L. Frylinck, J. L. Montgomery and M. F. Smith. 2009. The comparison of three β-agonists for growth performance, carcass characteristics and meat quality of feedlot cattle. *Meat Sci.* 81:557–64. doi:10.1016/j.meatsci.2008.10.011.

Swanek, S. S., J. B. Morgan, F. M. Owens, D. R. Gill, C. A. Strasia, H. G. Dolezal and F. K. Ray. 1999. Vitamin $D_3$ supplementation of beef steers increases *longissimus* tenderness. *J. Anim. Sci.* 77:874–81.

Vargas, D. N., A. E. Down, D. S. Webb, J. B. Morgan and H. G. Dolzal. 1999. Effect of supplementing feedlot steers with vitamin $D_3$ and E on carcass traits, shelf-life attributes and *longissimus* muscle tenderness. *J. Anim. Sci.* 77 (Suppl. 1):172 (Abstr.).

Vasconcelos, J. T., R. J. Rathmann, R. R. Reuter, J. Leibovich, J. P. McMeniman, K. E. Hales, T. L. Covey, M. F. Miller, W. T. Nichols and M. L. Galyean. 2008. Effects of duration of zilpaterol hydrochloride feeding and days on the finishing diet on feedlot cattle performance and carcass traits. *J. Anim. Sci.* 86:2005–15. doi:10.2527/jas.2008-1032.

Vogel, G. J., G. C. Duff, J. Lehmkuhler, J. L. Beckett, J. S. Drouillard, A. L. Schroeder, W. J. Platter, M. T. Van Koevering and S. B. Laudert. 2009. Effect of ractopamine hydrochloride on growth performance and carcass traits in calf-fed and yearling Holstein steers fed to slaughter. *Prof. Anim. Sci.* 25:26–32.

Wheeler, T. L., M. Koohmaraie and S. D. Shackelford. 1997. Effect of postmortem injection time and post-injection aging time on the calcium-activated tenderization process in beef. *J. Anim. Sci.* 75:2652–60.

Williams, S. N., T. M. Frye, M. Frigg, D. M. Schaefer, K. K. Scheller and Q. Liu.1992. Vitamin E as an *in situ* post-mortem pigment and lipid stabilizer in beef. Pacific Northwest Animal Nutrition Conference, Spokane, WA, p. 149.

Winterholler, S. J., G. L. Parsons, C. D. Reinhardt, J. P. Hutcheson, W. T. Nichols, D. A. Yates, R. S. Swingle and B. J. Johnson. 2007. Response to ractopamine-hydrogen chloride is similar in yearling steers across days on feed. *J. Anim. Sci.* 85:413–19. doi:10.2527//jas.2006-555.

Winterholler, S. J., G. L. Parsons, D. K. Walker, M. J. Quinn, J. S. Drouillard and B. J. Johnson. 2008. Effect of feedlot management system on response to ractopamine-HCl in yearling steers. *J. Anim. Sci.* 86:2401–14. doi.10.2527/jas.2007-0482.

Woerner, D. R., J. D. Tatum, T. E. Engle, K. E. Belk and D. W. Couch. 2011. Effects of sequential implanting and ractopamine hydrochloride supplementation on carcass characteristics and *Longissimus* muscle tenderness of calf-fed steers and heifers. *J. Anim. Sci.* 89:201–9. doi:10.2527/jas.2010-2857.

Chapter 7

# Understanding the effects of handling, transportation, lairage and slaughter on cattle welfare and beef quality

*Michael S. Cockram, University of Prince Edward Island, Canada*



## 1 Introduction

Handling, transport, lairage and slaughter are important components of beef production that can affect both animal welfare and meat quality. Each stage of pre-slaughter management can potentially present multiple negative welfare issues. These issues can include fear and distress, injury and pain, fatigue and discomfort, and thirst and hunger. The presence of multiple potential stressors and their interactions makes it difficult for commercial operations to completely standardise handling, transport, lairage and

http://dx.doi.org/10.19103/AS.2016.0009.07

slaughter. It is also difficult to replicate this complexity in a research environment to investigate factors affecting animal welfare and beef quality. Variations in each stage of pre-slaughter management, variations between cattle and differences in how humans interact with the animals can have significant effects on welfare and meat quality outcomes. Some of the risks of poor welfare can be mitigated by the provision of appropriate facilities and equipment for handling, transport and slaughter. However, their effectiveness is dependent on the training, experience, skill and supervision of the humans that operate the facilities and handle the cattle.

In many countries, the manner in which handling, transport and slaughter are conducted is highly regulated by detailed legal provisions (de Witte, 2009; Gavinelli et al., 2008; Thornber and Adams, 2008). On top of this legal framework, there are highly successful industry-based quality assurance schemes, protocols and performance targets that are audited internally by the meat industry and externally by the retail and restaurant sectors (Barnett et al., 2008; Edge and Barnett, 2009; Grandin, 2012).

It is not possible to avoid all of the stress associated with handling, transport and slaughter, and it is not possible to undertake these procedures without occasional difficulties. However, a consequence of investment to improve the manner in which handling, transport and slaughter is conducted is that, in many countries, it is possible for most cattle to be transported (Warren et al., 2010a) and stunned with a penetrating captive bolt and then exsanguinated, without experiencing severe welfare issues (Gregory and Shaw, 2000).

There are many different opinions and ethical positions on the issues under consideration, and there are many people who are concerned about the risk to cattle welfare during transport and slaughter. Research has an important role in identifying the major welfare issues. The onus is on both industry and research to work together to develop improvements to handling, transport and slaughter to optimise animal welfare and beef quality.

# 2 Beef quality

Handling, transport, lairage and slaughter can affect meat quality in several ways. If cattle are stressed, the colour and texture of the beef could be affected. Several factors can reduce muscle glycogen concentration, and this can affect post-mortem acidification of the meat and the prevalence of dark, firm, dry (DFD) meat. If cattle are injured during handling and transport, bruising can occur, and subsequent trimming affects the value of the carcase.

## 2.1 Stress and meat quality

Stress responses can be behavioural, such as arousal and excitement, and physiological, such as activation of the autonomic nervous system and neuroendocrine responses (Moberg, 2000). Acute physiological responses to stressors include cortisol and catecholamine (epinephrine/adrenaline and norepinephrine/noradrenaline) release, increased metabolism, heart rate and body temperature. In addition to the mobilisation of hepatic and muscle glycogen that results in reduced post-mortem acidification and increased ultimate muscle pH ($pH_u$), stress responses could potentially affect other aspects of muscle physiology and post-mortem changes important to beef quality (Ferguson and Warner, 2008).

Each stage of pre-slaughter management can be stressful to cattle (Grandin, 1997; Mitchell et al., 1988), and there are multiple factors that influence their behavioural and physiological responses to stressors. Stress responses have a psychological or emotional basis and occur when cattle perceive that their environment is threatening (Terlouw et al., 2008; Veissier and Boissy, 2007). Many of the stressors that cattle face during the pre-slaughter period are novel and have the potential to induce fear (Terlouw et al., 2012).

The way in which cattle respond to stressors depends on many factors, including breed, sex, temperament and experience. Cattle reared extensively have less frequent interaction with humans and are more likely to show more expressive behaviour during handling compared with cattle reared intensively (Fordyce et al., 1985). More excitable cattle have increased stress responses during handling compared with those with a calmer temperament (Cooke, 2014; Curley et al., 2006). Repeated early gentle physical human contact with calves can reduce plasma cortisol concentration at slaughter and improve muscle tenderness (Probst et al., 2012). However, this effect was not seen when the physical interaction was not started until 5 weeks before slaughter (Probst et al., 2013).

Calmer temperament as assessed by degree of agitation during confinement in a chute/race or crush and reduced speed during subsequent exit from confinement (Curley et al., 2006) can in some circumstances increase the rate of post-mortem muscle pH decline and improve tenderness, but these effects are inconsistent (Cafe et al., 2011). King et al. (2006) reported that in one out of three groups of steers, those categorised as excitable (based on exit velocity from restraint and behaviour when restrained) had tougher meat than in those categorised as calmer. Voisinet et al. (1997) found decreased tenderness and more borderline carcases with DFD meat from excitable than from calm *Bos indicus* steers and heifers, whereas Coombes et al. (2014) did not find a significant effect of temperament on muscle glycogen concentration or $pH_{u.}$

Although some studies have shown that relationships between pre-slaughter stress and aspects of beef quality can be demonstrated in some circumstances, multiple factors and their interactions likely affect whether the conditions are such that beef quality is affected. There are many useful studies that have identified stressors associated with various pre-slaughter procedures, but the relationship between stress and beef quality has proven difficult to demonstrate consistently. The evidence for relationships between pre-slaughter procedures, stress and beef quality was reviewed by Ferguson et al. (2001) and Ferguson and Warner (2008). Other than the clear relationships between pre-slaughter factors and muscle glycogen; $pH_u$ and DFD; and those between handling, transport and injury, the effects on other aspects of beef quality are not as clear.

There is considerable variation in the stress responses of cattle to pre-slaughter management (Gruber et al., 2010). The variation in aspects of beef quality between groups of cattle exposed to differing pre-slaughter management suggests that a relationship probably exists between pre-slaughter management and beef quality. For example, Gruber et al. (2010) found a positive correlation between the plasma epinephrine concentration of cattle during restraint before transport, muscle colour and tenderness. Cattle categorised by a higher magnitude of their physiological stress responses to restraint had tougher muscle compared with calmer cattle. However, muscle $pH_u$ did not differ between behavioural categories.

A clear relationship between pre-slaughter management and the amount of muscle glycogen available for post-mortem conversion to lactic acid has been demonstrated. When there is sufficient muscle glycogen concentration, the pH falls over a 24–48 h period after slaughter from about pH 7 to an ultimate pH ($pH_u$) of about 5.5–5.6. If some of the muscle glycogen has been utilised before the animal is slaughtered, the amount of lactic acid formed after slaughter is reduced, and the $pH_u$ might only decline to 5.8–6.0. Beef with a $pH_u \geq 6$ is described as DFD and has a darker colour, firmer texture, little or no exudate when cut and has an increased risk of bacterial spoilage. Although it can be more tender than normal beef, some flavour attributes can be adversely affected (Beltran et al., 1997; Grayson et al., 2016; Warriss, 1990) In bulls, there is a curvilinear relationship between $pH_{24}$ and Warner–Bratzler shear force values with a toughness peak at $pH_{24}$ 6.2–6.4 (Jelenikova et al., 2008). Young bulls are especially prone to DFD, but it can occur in all cattle (Warriss, 1990). Although stress hormones such as epinephrine are likely to cause some of the muscle glycogen depletion (McVeigh and Tarrant, 1983) that occurs during social mixing, and stress is a risk factor for DFD (Apple et al., 2005), blockage of the effects of epinephrine does not completely stop the reduction in muscle glycogen that occurs during social mixing of young bulls (McVeigh and Tarrant, 1983). Muchenje et al. (2009) did not find consistent correlations in all breeds of steers between urinary concentrations of norepinephrine and epinephrine sampled after slaughter and muscle $pH_u$ and tenderness. Certain types of exercise are a contributory factor to the occurrence of DFD (Gregory, 1996). However, exercise on a treadmill for 0.25 h at 8 km/h failed to affect muscle glycogen concentration or $pH_{24}$ (Apple et al., 2006).

## 2.2 Bruising

Bruising can occur at any point from handling on the farm, loading, during transport, unloading in the lairage, during handling before slaughter and during stunning (McCausland and Millar, 1982). Bruising is caused by trauma that ruptures blood vessels resulting in blood leaking into surrounding tissues. There is an increased risk of food spoilage from bruised meat and it is aesthetically unpleasing to consumers (Marshall, 1977). Bruising can result in considerable trim loss during processing. A positive relationship between bruising and muscle $pH_{24}$ has been demonstrated (McNally and Warriss, 1996; Strappini et al., 2010). The trauma that causes a bruise and subsequent swelling and inflammation are likely to cause pain and discomfort (Gregory, 1996). Cattle with no fat cover and those with a large amount of fat cover have an increased risk of bruising compared with those with

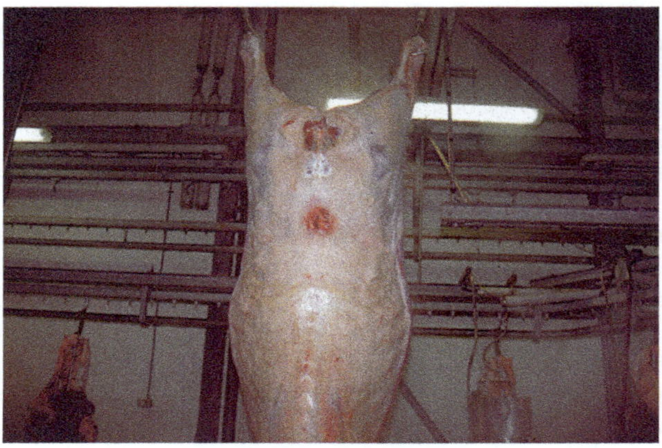

**Figure 1** Bruising on the back of a carcase.

intermediate fat cover (Hoffman and Lühl, 2012). In cattle, bruises inflicted antemortem are not visible in live animals and can only be detected post-mortem (Fig. 1). Bruises, therefore, provide information about the injuries that occurred before slaughter (Strappini et al., 2009).

Defects in the construction and maintenance of handling facilities (Strappini et al., 2013) and vehicles, for example, sharp edges, protruding bolts, posts and narrow entrances/exits, can cause injury and bruising (Grandin, 1980b). Poor handling, use of driving instruments, such as sticks (McNally and Warriss, 1996, 1997), and forceful impacts between cattle (Strappini et al., 2013) are also risk factors for bruising.

# 3 Fitness for transport, loading and responses to transportation

## 3.1 Fitness for transport

Fat cattle (steers, heifers and young bulls) sent for slaughter are normally fit and healthy for transport to the slaughter plant. However, cows are also culled and sent for slaughter from dairy and beef herds for production, breeding or health reasons. For economic reasons, there is the temptation to transport cattle that are not fully fit to slaughter. In addition, some cattle with health issues that might be fit at the start of a journey can deteriorate during the journey.

Cattle that are transported to slaughter with pathology (e.g. arthritis or pneumonia) have an increased risk of suffering during transportation, and many conditions will result in partial or total condemnation of carcases as unfit for human consumption (Dupey et al., 2013; Rezac et al., 2014; Vial et al., 2015; White and Moore, 2009). As non-ambulatory cattle are both unsuitable for transportation and also likely to be condemned as unfit for human consumption (Doonan et al., 2003; Stull et al., 2007), in many countries they cannot be transported to slaughter. In some situations, cattle that are not fit for transport due to painful conditions, such as fractures (Cullinane et al., 2012), and other types of injuries, in an emaciated body condition and showing signs of disease that would have made them less able to cope with transport and were likely to have been associated with suffering, are transported to slaughter (Ahsan et al., 2014).

In a survey in North America of over 200 000 fat cattle, 0.003% were lame at loading, 0.013% were lame at unloading, 0.020% were non-ambulatory and 0.007% were dead on arrival. Out of 1,704 cull cattle, 0.4% were lame at loading, 0.1% were lame at unloading, 0.3% were non-ambulatory and 0.06% were dead on arrival (González et al., 2012d). In a survey, conducted in the United States in 2007, of cattle on arrival at 23 slaughter plants, Nicholson et al. (2013) observed that 2.7% of cull dairy cows and 1.9% of beef bulls were severely lame (i.e. displaying an arched back at all times and refusing to bear weight on one leg), 4% of cattle were extremely emaciated and 4% had foot abnormalities.

Care is required to ensure that cattle, particularly cull dairy cows (Malena et al., 2007; Rezac et al., 2014; Večerek et al., 2006), are still fit for transport and handling, otherwise there is a risk of them experiencing severe welfare issues, for example, becoming non-ambulatory and dying during the journey (González et al., 2012a; Grandin, 2001b). However, mortality during transport is not common (González et al., 2012a). Post-mortem examination of cows that died during a 12-h journey found subcutaneous and muscle bruising, mainly on the lateral areas; lung emphysema, congestion and oedema; and renal congestion. Respiratory distress while dying was considered to have been a likely factor that could have caused the interstitial emphysema. Lateral bruising and lung congestion were considered to have been associated with recumbency during the journey, resulting in trampling and hypostatic lung congestion. Dehydration was considered to have contributed to hypovolemic shock. The post-mortem muscle pH was greater in cows that died during the journey than in those that arrived alive after a 2-h journey (Burns et al., 2014).

## 3.2 On-farm loading

Ramp angles >20° make loading more difficult and cause increased heart rate (Holleben et al., 2003). During loading, cattle can move along a level ramp smoothly, but if the ramp angle is >7°, some cattle will resist climbing the ramp, and some may fall or slip (Gebresenbet et al., 2012). Loading steers onto a vehicle via a 24° ramp can increase serum cortisol concentration and rectal temperature (Pettiford et al., 2008).

## 3.3 Behavioural and physiological responses to transport

Transport exposes cattle to a number of potential stressors, for example, novelty, mixing of groups, thermal stress, feed and water withdrawal, vehicular movement, responses to maintain stability and reduced opportunities for rest can result in variable responses, depending on the characteristics of the journey, the type of cattle and their fitness (Cockram, 2007). Although some cattle are transported to slaughter using other modes of transportation, only road transportation will be considered here. A number of physiological measurements are commonly used to assess the impact of a range of different potential welfare issues that might affect cattle during transport, including

- Fear and stress: cortisol, catecholamines, heart rate, respiration rate and rectal temperature
- Muscle damage/exertion: creatine kinase
- Dehydration: osmolality and total protein
- Food restriction: free fatty acids and β-hydroxybutyrate (Broom, 2003).

Considerable care is required to ensure that loading and transportation are undertaken appropriately and that cattle are not injured. In Nigeria, Minka and Ayo (2007) observed skin wounds, bruising, fractures and dislocations in *B. indicus* young bulls and heifers that were transported for 10–12 h. They attributed these injuries to horned cattle, actions by handlers and poor vehicle design. Although most cattle transported for slaughter do not have horns, one survey in the United States recorded horns on 16% of the cattle sent for slaughter (Nicholson et al., 2013). Transport of cattle with horns increases the amount of bruising compared with those transported without horns (Ramsay et al., 1976; Shaw et al., 1976; Wythes et al., 1979b).

In many countries, changes in practice because of research, advice, inspection and audits have reduced the risk of injury during transport (Nicholson et al., 2013).

Kenny and Tarrant (1987a,b) showed that stress responses of young bulls or steers to transportation were due to a combination of several factors: novelty of re-penning, mixing of social groups, confinement in a smaller area and vehicle movement.

- In steers, loaded onto a vehicle using a 34° ramp and immediately unloaded, the heart rate was increased (Kenny and Tarrant, 1987b).
- After young bulls from a stable social group were loaded on to a stationary vehicle at a stocking density of 355 kg/m$^2$ for 1 h, there was an increased frequency of butting, exploration, mounting and changes in position, and increases in heart rate and plasma cortisol concentration (Kenny and Tarrant, 1987a). Similar physiological responses were found after loading steers onto a stationary vehicle for 1 h, at a stocking density of 494 kg/m$^2$ (Kenny and Tarrant, 1987b). After loading young bulls from a mixed social group, the responses were greater and there was an increased plasma creatine kinase activity.
- When the young bulls were transported for 1 h, the behavioural and physiological responses were similar, but there were fewer social interactions and no increased mounting activity (Kenny and Tarrant, 1987a). In comparison with 1 h of stationary confinement on the vehicle, during 1 h of transport, steers showed fewer agonistic interactions, did not ruminate, but defecated more frequently (Kenny and Tarrant 1987b).

The rectal temperature and the plasma cortisol and epinephrine responses of calm Bos indicus to transport can be lower than in those categorised as excitable (Burdick et al., 2011). The plasma concentrations of ACTH and cortisol rise during the first part of a journey and then fall. This decrease is due to a decreased responsiveness of the anterior pituitary gland to stimulation by corticotrophin-releasing factor rather than a decrease in the perception of transportation as a stressor (Knights and Smith, 2007).

During transport, for example, at a stocking density of 327 kg/m$^2$ for 3 h, steers and heifers stood still for most of the journey and did not lie down (Cockram and Spence, 2012). Following transportation for 3 h, at stocking densities of either 270 or 337 kg/m$^2$, for a 3-h period after unloading, steers laid down more, spent slightly more time drinking, had a decreased frequency of interactions, but no change in eating or rumination in comparison with their pre-journey behaviour (Booth-McLean et al., 2007).

# 4 Road type, driving events, stocking density and thermal environment during transportation

## 4.1 Road type, driving events and stocking density

Road type can affect the vibration experienced by cattle during a journey. Vibration and the risk of bruising are greater on rough surfaces than on smooth surfaces (Gebresenbet et al., 2011; Huertas et al., 2010), and the heart rate is lower on highways than when transported on smaller, single carriageway roads (Eldridge et al., 1984).

Road type can also affect the behavioural responses of cattle to driving events. On small single carriageway roads, a driver will need to make frequent vehicular adjustments to respond to road features, thereby causing an increased frequency of accelerations (lateral acceleration from cornering, forward acceleration and deceleration from braking) to be experienced by cattle. These accelerations require the cattle to make more frequent postural, leg and feet adjustments to maintain balance than when on multi-lane carriageways with limited access points (Cockram and Spence, 2012).

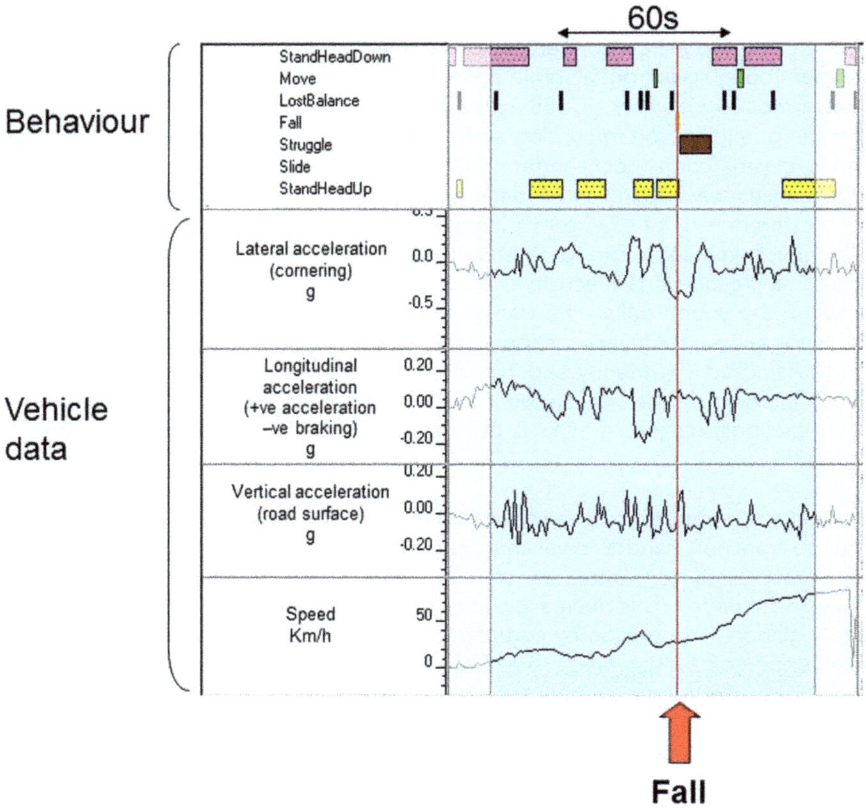


**Figure 2** The effects of lateral, longitudinal and vertical acceleration, and vehicle speed on cattle behaviour before and after a fall during a journey (Cockram and Spence, 2012, Copyright UFAW, reproduced with permission).

The driving events that precede most losses of balance by cattle are cornering, acceleration, braking, vertical acceleration and gear changes, especially when two driving events coincide (Kenny and Tarrant 1987b; Tarrant et al., 1988, 1992; Cockram and Spence, 2012). Therefore, driver skill and technique can affect how comfortable the cattle are during a journey. Falls are infrequent events. When they occur, they are preceded by a series of events (Fig. 2), but these are not necessarily consistent between each fall event (Cockram and Spence, 2012).

In situations, where there are rapid changes in vehicle motion, due to poor driving technique or an emergency response requiring rapid braking or swerving, there could be a benefit from a stocking density that provided mutual support (Tarrant, 1990). Heifers transported for 6 h had a lower heart rate at a stocking density of 409 kg/m$^2$ compared with those transported at 319 kg/m$^2$ (Eldridge et al., 1988). However, in steers transported for 6 h, some cattle fell over at a stocking density of 460 kg/m$^2$, but no animals fell at stocking densities of 345 or 288 kg/m$^2$. The severity of bruising was lower at a stocking density of 345 kg/m$^2$ compared with 460 kg/m$^2$ or 288 kg/m$^2$ (Eldridge and Winfield, 1988).

Tarrant et al. (1988, 1992) showed that cattle were at greater risk of stress and injury at high stocking density than at low stocking density. In steers transported for 4 h at stocking densities of 196, 312 and 591 kg/m$^2$, both plasma cortisol concentration and plasma creatine kinase activity increased with stocking density. The frequency of losses of balance and falls and the severity of bruising were greater in cattle transported at 591 kg/m$^2$ than in those transported at the lower stocking densities (Tarrant et al., 1988). In steers transported for 24 h, at a stocking density of 592 kg/m$^2$, the numbers of losses of balance and falls, the severity of bruising, the plasma cortisol concentration and the plasma creatine kinase activity were greater and the frequency of changes in position were lower than in those transported at 465 kg/m$^2$ or 499 kg/m$^2$ (Tarrant et al., 1992).

At the high stocking density, when an animal fell, it was sometimes trapped on the floor due to the remaining cattle 'closing over' the fallen animal. The fallen animal will cause others to lose their footing and they will trample the fallen animal resulting in the increased bruising at high stocking density (Tarrant et al., 1988; Tarrant, 1990). Romero et al. (2013) found that stocking densities ≥300 kg/m$^2$ increased the risk of bruising and the risk of DFD in cattle transported for 1–4 h (Romero et al., 2013).

Several factors can affect the stocking density used, including body size/weight, position/type of compartment used on the vehicle, type of vehicle (e.g. number of axles in relation to axle weight regulations), risk of heat distress (temperature and humidity within a vehicle increases with increased stocking density [Wikner et al., 2003]), journey duration (in some countries, the need to provide sufficient space to lie down and, on some vehicles, sufficient space to provide water and/or feed) and size of vehicle in relation to the number of cattle available to fill the load (Knowles, 1999; González et al., 2012b).

The floor areas required for cattle to stand or to lie down are related to the size and shape of the animals. Equations that are used to calculate space requirements are in the form of a constant (k) multiplied by the live weight to the power of $^{0.6667}$ (Petherick, 2007). Using published research, Petherick and Philips (2009) estimated that a space allowance for cattle during transport calculated using a constant of ≥0.02 would reduce the risk of the cattle experiencing stress, injury and fatigue. In one study in North America, an increased risk of death was found when the space allowance (m$^2$/animal) either exceeded a high or was lower than a low threshold (i.e. *k* values >0.035 and <0.015 respectively, where space allowance = k live weight $^{0.6667}$) (González et al., 2012d).

## 4.2 Thermal environment

Ventilation during transport is required to avoid the build-up of excessive heat and humidity within the vehicle. Although some vehicles have mechanical fans, most vehicles are passively ventilated through openings along each side of the vehicle (Fig. 3). Airflow in a vehicle is caused by (a) vehicle motion creating differential air pressure outside relative to inside of the vehicle, (b) wind or (c) metabolic heat produced by the cattle warming the air that then rises and draws cooler air in through the inlets (Randall and Patel, 1994). Too high a stocking density and too low a deck height relative to the height of the cattle can obstruct airflow by blocking ventilation openings and flow of air over the cattle (Randall, 1993). During extreme cold, or if the cattle are not fit and healthy, excessive airflow through the ventilation openings, especially if combined with rain or snow could result in thermal discomfort. Depending on the conditions, the ventilation openings can be closed or boarded to provide the cattle with protection from the elements (Goldhawk et al., 2015).

**Figure 3** Type of vehicle commonly used to transport cattle in North America.

Cattle are transported in a range of thermal conditions, for example –25 to 45°C. Conditions within the vehicle can result in combinations of temperature and humidity that can cause thermal distress (Randall, 1993; Goldhawk et al., 2014). In North America, one study showed that the risk of cattle becoming non-ambulatory was greater if the average air temperature was >20°C than when it was <20°C. However, the risk of dying was greater at average temperatures <0°C than when it was >20°C. The risk of death increased when the temperature was <−15°C. (González et al., 2012d). Although there is little research in this area (Tucker et al., 2015), animals that are transported in very cold conditions are at risk of frostbite, especially if they come into contact with metal surfaces.

Cattle can, to some extent, adjust to a wide range of thermal conditions, depending on their prior acclimatisation, but can experience difficulty if subjected to sudden changes in their thermal environment (Randall, 1993). The opportunities for cattle to thermoregulate by changes in behaviour are limited during transportation. In cold conditions, feed deprivation, air movement and a wet coat can increase the temperature at which cattle need to increase their metabolic heat production to maintain thermal balance and high stocking density can decrease this temperature. In hot conditions, feed deprivation and air movement can increase and high stocking density, solar radiation, and high relative humidity can decrease the temperature at which cattle need to increase evaporative heat loss (Schrama et al., 1996; Silanikove, 2000).

The ability of cattle to lose heat via the evaporation of water is dependent on a temperature and vapour pressure gradient (Silanikove, 2000). Cattle can increase heat loss by evaporation of water through sweating and by increasing their respiration rate (Caulfield et al., 2014; Warren et al., 2010a). In dehydrated steers, sweating in response to high temperature and humidity can be impaired (Bianca, 1965). There is limited research on heat stress during transport. Nanni Costa et al. (2003) transported bulls for 3 h at a stocking density of 333 kg/m$^2$ followed by lairage in conditions of high temperature and humidity (temperature humidity index was at times >75), but did not find significant effects of the environmental conditions on the plasma cortisol concentration or muscle $pH_u$, but plasma creatine kinase activity was increased. Heifers exposed to air temperatures >35°C for 1–2 d before slaughter have an increased the risk of DFD (Scanga et al., 1998).

# 5 Journey duration, feed and water provision, and the use of livestock markets

## 5.1 Journey duration

Journey duration has increased in many countries because the number of slaughter plants has decreased and they have become large and specialised facilities (Håkansson et al., 2016). In the EU, there is transport of cattle between countries for slaughter (Hardstaff et al., 2015).

If only animals fit for transport are selected, the environmental conditions (including driving style, road conditions, vehicle design and operation, space allowance, thermal conditions and ventilation) and the pre- and post-transport handling of the animals are optimal, it should be possible to transport certain types of cattle over long distances without major welfare problems (Cockram, 2007). There is, of course, a limit on how long cattle can be transported without experiencing suffering arising from the cumulative effects of factors such as feed and water restriction, fatigue, stress, discomfort and injury.

Some cattle journeys to slaughter can be long (e.g. up to 45 h, González et al., 2012a and up to 68 h, Warren et al., 2010a). Although it can be difficult to use the behavioural and physiological responses of cattle to transportation to identify unambiguously when a critical limit has been reached for journey duration (Cockram, 2007), the evidence that the risks of cattle becoming non-ambulatory or dying increases with journey duration, especially when the journey exceeds 30 h (González et al., 2012d; Malena et al., 2006), is compelling. Teke (2013) studied the transport of bulls from Hungary to Turkey, at a stocking density of 186 $kg/m^2$, during which the ambient temperatures ranged from −8 to 32°C, on journeys lasting 30 h, with a 2-h rest period after 14 h, during which the bulls were offered feed and water. The mortality rate for 121 journeys involving 3,874 bulls was 0.46% and the percentage live weight loss was 5.6%.

Towards the end of journeys lasting 24–31 h, some cattle will lie down (Knowles et al., 1999; Tarrant et al., 1992). This suggests that the cattle may have become fatigued during the last part of the journey. In cattle transported up to 31 h, at a stocking density of 369 $kg/m^2$ with a 1-h stationary period after 14 h when the cattle were provided with water, the plasma creatine kinase activity, the plasma concentration of total protein and the plasma osmolality increased during transportation and the cattle lost about 7% live weight. However, there were no consistent effects of journey duration on these variables, but in some replicates the plasma osmolality increased with journey duration (Knowles et al., 1999).

Immediately after a long journey, cattle eat and drink, then after several hours they can show increased resting (Knowles et al., 1999; Ross et al., 2016). The live weight recovers and the plasma concentration of total protein and the plasma osmolality returns to normal over a 72-h period. Although there were some small decreases in the glycogen concentration in some muscles, no effects of transportation for up to 31 h were found on muscle $pH_{24}$ or bruising (Knowles et al., 1999).

Some cattle are transported with access to feed and water and provided with a mid-journey break. Whether a mid-journey break is always required is not clear. After a 9-h journey, cattle did not spend the mid-journey period resting. However, access to feed and water during a 12-h mid-journey period can avoid signs of feed restriction and dehydration during a second 9-h journey (plasma concentrations of β-hydroxybutyrate, free fatty acids

**Table 1** Effects of increased journey duration on meat quality

| Type of cattle | Space/stocking density | Journey durations (h) | Lairage duration (h) | Muscle $pH_{24}$ | Tenderness | Reference |
|---|---|---|---|---|---|---|
| Young bulls | 1.8 $m^2$/animal | 0.6, 3.5 and 6.3 | Overnight | No effect | No effect | Maria et al. (2003) |
| Young bulls and heifers | 1.5 $m^2$/animal | 1.5 and 4.4 | | No effect | | Marenčić et al. (2012) |
| Young bulls | 389 kg/$m^2$ | 0.5, 3 and 6 | 12 | No effect | | Chacon et al. (2005) |
| Young bulls | 1.8 $m^2$/animal | 0.5, 3 and 6 | Overnight | No effect | Increased | Villarroel et al. (2003) |
| Steers | 465 to 592 kg/$m^2$ | 1 and 24 | 3 | Increased | | Tarrant et al. (1992) |
| Steers and heifers | 369 kg/$m^2$ | 1 and 14–31 | 72 | No effect | | Knowles et al. (1999) |
| Cows and bulls | 321–393 kg/$m^2$ | 4, 8 and 12 h | | No effect | | Bulitta et al. (2015) |

and total plasma protein were not raised during this second stage of the journey) (Earley et al., 2013).

Most studies have not found large effects of journey duration on meat quality (Table 1).

## 5.2 Feed and water

Feed is often withdrawn before transport to reduce the risk of faecal contamination at slaughter and to reduce the risk of spillage during carcase processing after slaughter (Pointon et al., 2012). Fasting steers for 24 h, before a 2-h journey, at a stocking density of 376 kg/$m^2$ reduced the weight of faeces produced during the journey and the faeces were less 'runny' compared with steers transported directly off pasture. After steers were kept in a lairage overnight with access to water, but no feed, steers without feed for 24 h before transport had rumen contents and digesta with a lower dry matter composition than steers that had been transported directly off pasture. There were no treatment effects on the plasma cortisol concentration and muscle $pH_u$ (Gregory et al., 2000). Long periods without feed before transport can increase the risk of bruising (Dodt et al., 1979).

On some long journeys, a mid-journey 'rest' period may be provided to allow the cattle to be offered feed and water either on or off the vehicle, and if the cattle are held in a lairage they might be offered feed. However, for many cattle from the time that they are prepared for loading on the farm until slaughter, they usually do not have access to feed. Some vehicles provide drinking water, but in most cases, water will not be normally available during the journey. However, water should be available during lairage. After transportation, provision of water gives cattle an opportunity to drink without detrimental effects on meat quality (Wythes et al., 1985).

The rumen and large intestines of cattle provide nutrients and water during periods without access to feed and water (Warriss, 1990). However, after 48 h of fasting, steers need to mobilise body energy reserves; plasma free fatty acids, β-hydroxybutyrate and urea concentrations increase and plasma glucose concentration decrease (Frohi and Blum, 1988; Rule et al., 1985).

Most of the live weight loss that occurs during transportation is a consequence of food and water deprivation. The weight of water and dry matter in the rumen decreases within the first 24 h without feed (Bass and Duganzich, 1980). In a study in North America, fat cattle lost 0.09% of live weight/h during transport and cull cattle lost 0.53% of live weight/h during transport. The rate of live weight loss (shrink) increased with both temperature and journey duration. The percentage live weight loss was also increased if the cattle were loaded during the afternoon/evening versus the night/morning and was affected by origin of loading, for example, auction market versus farm/feedlot (González et al., 2012c).

Extended periods without access to water will result in dehydration caused by a decrease in body fluids and an increase in osmolality. *B. indicus* steers, either deprived of feed and water for 60 h or deprived of feed and water for 12 h and then transported for 48 h, had increased plasma total protein concentrations (Parker et al., 2003). Steers offered a nutritional supplement containing amino acids, electrolytes and energy, 24 h before a 1 h journey that was followed by lairage for up to 12 h with access to water, had lower percentage live weight loss and greater carcase weight than controls (Grumpelt et al., 2015).

Heat stress that causes increased water loss through sweating and panting will increase the speed of dehydration during periods of water restriction. When the loss of body weight reaches a critical threshold, the ability of cattle to thermoregulate in response to heat stress and to maintain cardiovascular function is impaired (Silanikove, 1994). In North America, the risk of cattle dying during transportation was greater if the percentage live weight loss was >8% than when it is <8%. (González et al., 2012d). Steers can lose 7% of their live weight after 15–31 h of transportation (Knowles et al., 1999; Warriss et al., 1995). Non-transported steers can lose 8% of their live weight after 24 h without feed and water, 11% after 48 h and 14% after 72 h without feed and water (Wythes et al., 1980).

## 5.3 Livestock markets

Many slaughter cattle are sold via auction markets/sale yards/sale barns rather than directly from farm to slaughter (Iglesias and East, 2015; Warren et al., 2010a). Markets

**Figure 4** Holding pens in a cattle market.

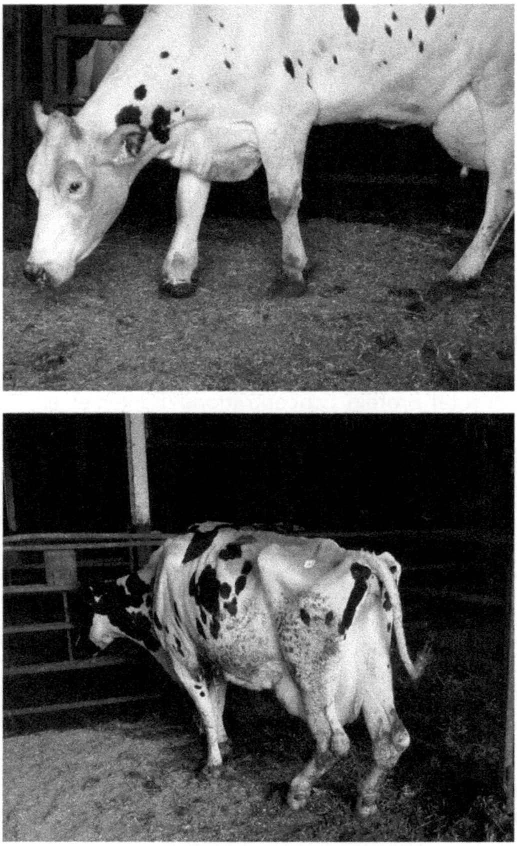

**Figure 5** Lame cull dairy cows that had been transported to an auction market even though they were not fit for transportation or sale at a market.

increase the duration between farm and slaughter, expose the cattle to more handling and social mixing is more likely (Fig. 4). The length of fasting is increased, the risks of cattle becoming fatigued and dehydrated, experiencing fear, distress and injury are greater than if the cattle were sent directly to slaughter.

In North America, some cull dairy and beef cattle are sold via markets, even though they are not in a suitable condition for transportation, for example, due to emaciation, severe lameness and showing signs of illness (Ahola et al., 2011) (Fig. 5). In some countries, the handling and transport of cattle to market and slaughter are not optimal because of poor facilities and poor handling (Adeyemo et al., 2009; Ahsan et al., 2014).

There is considerable variation between markets, but cattle from markets often have more bruising than those sent to slaughter directly from farms (Jarvis et al., 1995; McNally and Warriss, 1996, 1997; Strappini et al., 2010; Warner et al., 1998; Weeks et al., 2002).

Slips and falls, and contact with equipment can occur during unloading, grading, moving into and through the sale ring and during restraint (Blackshaw et al., 1987; Gregory, 2008).

Cattle that arrive at slaughter plants from markets can show greater signs of feed and water restriction effects (more drinking and greater plasma concentrations of total protein and free fatty acids) than in those direct from farms (Jarvis et al., 1996a,b). There is variation between studies on the effects of markets on meat quality. Warner et al. (1998) found increased plasma cortisol concentration, lower muscle glycogen concentration, greater muscle $pH_u$ and darker muscle after steers had been transported to a market, kept overnight with water, but no feed and then transported to slaughter compared with those sent directly to slaughter.

Ferguson et al. (2007) did not find major or consistent effects of the marketing method on the muscle glycogen concentration and $pH_u$ of cattle that had been transported to a market, kept in their social group overnight with access to water, moved to a sale pen adjacent to cattle that were exposed for sale, moved to another pen after the sale and then transported to the slaughter plant where they were slaughtered without overnight lairage compared with those that had been transported directly to a slaughter plant and slaughtered after overnight lairage with access to water. Warren et al. (2010b) found that cattle from markets tended to have an increased risk of DFD compared with those sent directly to slaughter.

# 6 Lairage

## 6.1 Unloading

During unloading and movement of cattle into lairage pens, slips and falls can occur and electric goads/prods are sometimes used (Miranda-de la Lama et al., 2012; Nicholson et al., 2013). Ideally, the unloading bay should be built at the level of the vehicle deck so that cattle can walk off a vehicle without having to go down a ramp (Fig. 6). Grandin (1990) recommends a maximum angle of 25° for adjustable ramps that are raised to the second or third deck of a vehicle and a maximum angle for non-adjustable livestock ramps of 20°. The surface on which the cattle step onto after leaving the vehicle should be flat, wide and straight (Grandin, 1980a). Stair steps are preferable to a slope with cleats (Grandin, 1980a). Grandin (2008) provided recommendations on dimensions and designs for stairs, cleat spacing for ramps and non-slip flooring for cattle. A large height difference between the vehicle and an unloading ramp can cause cattle to slip and fall, and cattle running into lairage pens rather than walking can increase the risk of bruising (Hoffman and Lühl, 2012).

## 6.2 Lairage duration

Cattle can be slaughtered immediately after arrival at the slaughterhouse or placed in lairage pens for various durations (Fig. 7). Lairage provides an opportunity for antemortem inspection; it allows recovery from transportation, provides time for wet cattle to dry before slaughter and provides a reserve of cattle to provide a constant supply of animals for slaughter, especially for the start of slaughter first thing in the morning. Cattle may also be placed in lairage if the slaughter plant has reached its daily slaughter capacity, there is a delay in slaughter (e.g. mechanical breakdown), or the load arrives after slaughtering

**Figure 6** Unloading ramp and unloading bay.

**Figure 7** Lairage pen.

**Table 2** Effects of increased lairage duration on meat quality

| Type of cattle | Space/stocking density during transport | Journey duration (h) | Lairage duration (h) | Provision of feed in lairage | Muscle $pH_{24}$ | Tenderness | Cooking losses/drip loss | Reference |
|---|---|---|---|---|---|---|---|---|
| Bulls | | 2 | 0 and 6 | | No effect | | | Ferreira et al. (2006) |
| | | 5 | 0 and 6 | | Decreased | | | |
| Young bulls | | 54 | 24–36 and 57–59 | Straw | Increased | Decreased | Increased | Liotta et al. (2007) |
| Bulls | 233 $kg/m^2$ | 30 | 24, 48 and 72 | Feed until 12 h before slaughter | Decreased | No effect | No effect | Teke et al. (2014) |
| Young bulls in mixed group | | | Overnight, 2 days and 4 days | | Decreased | | | Warriss et al. (1984) |
| Males and females | 0.23–1.83 $m^2$/animal | 1.2–15.7 | 0.9 to $\leq$ 15.7 $\geq$ 15.8 to 37 | | Increased | | | Mach et al. (2008) |
| Steers | | | 2.5 and 26.5 | | | Increased | | Wythes et al. (1988) |
| Steers | | 4 | 12, 24 and 48 | | | Decreased | | Jones et al. (1990) |

had finished for the day (Ljungberg et al., 2007). The logistics for ensuring that cattle are available when required for slaughter is a complex and variable process (Ljungberg et al., 2007). Cattle are collected from many different farms/markets, and these sources vary from one day to the next. The timing of deliveries of cattle to the slaughter plant can be affected by many factors that affect the speed of loading and journey duration.

If cattle are fatigued, for example, after sale through a market, but are provided with appropriate conditions, they will settle down and rest in a lairage (Cockram, 1990). However, the environment within a lairage may not always provide conditions suitable for rest (Cockram, 1991). There are many factors that can affect the ability of cattle to rest in a lairage, for example, provision of bedding; group size and social composition; floor, drinker and feeding space; ventilation to prevent heat distress; and noise (Cockram, 1990; Weeks, 2008). Lairages can be noisier than the conditions experienced on a farm and cattle can show signs of fear (increased heart rate and movement) in response to noises such as shouting and metal clanging (Waynert et al., 1999). There is noise from the movement of vehicles outside of the lairage and noise from the movement of cattle into and out of pens with associated opening and closing of gates (Weeks, 2008). Other sources of noise are the cleaning of pens and cattle vocalisations (Weeks et al., 2009). The effect of keeping young bulls for about 9 h in a lairage that also contained pigs was an increased muscle $pH_{24}$ and increased meat toughness compared with those not lairaged in the same area as pigs (Peña et al., 2014). Unless overnight lairage produces an improvement in the handling of cattle from the lairage to the point of slaughter, then keeping cattle in lairage overnight may not be beneficial for their welfare. Cattle kept in lairage overnight can show a greater stress response to slaughter than those lairaged for only a few hours (Lowe et al., 2004).

Lairage provides an opportunity for cattle to eat and drink and, therefore, to improve their muscle glycogen concentration and to rehydrate (Warriss et al., 1984). In bulls, transported for 4 h and then lairaged for 17–20 h, with either no water, water or electrolytes, they drank 30 L of water and 17 L of electrolytes. Plasma osmolality was lower and the extracellular fluid volume was greater in those offered water compared with cattle offered electrolytes or no water (Gortel et al., 1992). Whether there is a significant improvement in the meat quality of cattle kept overnight in a lairage is unclear (Table 2).

## 6.3 Social grouping

Disruption of social grouping as a result of separation from familiar or mixing with unfamiliar animals either during regrouping on farm before loading, at a market and after unloading at the slaughter plant or in lairage can cause behavioural and physiological signs of stress. In Hereford, but not in Angus steers, Colditz et al. (2007) found that regrouping 1 week before slaughter increased both beef compression values and taste panel appraisal of juiciness. Other aspects of meat quality were not affected. In young bulls, transported for 1.5 h and then lairaged 3–5 h in mixed social groups, the plasma cortisol concentration was not increased, but urinary norepinephrine concentration, muscle $pH_u$ and toughness were greater than in those kept in stable social groups (Lowe et al., 2004).

In young bulls, social mixing in the lairage increases physical activity, especially mounting and agonistic interactions, reduces muscle glycogen concentration, increases muscle $pH_{24}$ and the risk of DFD (Franc et al., 1988; McVeigh et al., 1982; Mohan Raj et al., 1992). Although DFD beef induced by social mixing can become more tender than normal beef (Beltran et al., 1997), taste can be affected (Grayson et al., 2016). If the frequency of mounting is reduced by an overhead electric grid and darkness, the muscle $pH_{24}$ and the occurrence of DFD are reduced (Bartoš et al., 1988). However, if bulls are kept in a stable social group, the muscle $pH_{24}$ and the occurrence of DFD are reduced (Bartoš et al., 1993).

Minimising the duration that young bulls spend in lairage is also beneficial in reducing the occurrence of DFD (Bartoš et al., 1993). In young bulls that had been tethered, mixing unfamiliar bulls in lairage overnight compared with those slaughtered without lairage, decreased muscle glycogen concentration and increased muscle $pH_u$ (Lahucky et al., 1998). Mixing social groups of steers does not always affect muscle $pH_{24}$ (Wythes et al., 1979a). Warren et al. (2010b) found an interaction between type of cattle, origin and lairage on the risk of DFD; steers and heifers kept in lairage overnight tended to have an increased risk of DFD compared with those that were slaughtered sooner.

# 7 Handling and movement of cattle through lairage and to the stunning pen

## 7.1 Facility design and handling methods

If facilities are designed well and handlers are trained and supervised, it is easier to handle and move the cattle; there are fewer injuries and less stress (Grandin, 1994, 2006; Tucker et al., 2015). Grandin has successfully encouraged handlers to use the natural behaviour of cattle to keep together and move as a group, improving the flow and reducing the need for intervention from a handler (Grandin, 1980a). An appreciation of how cattle perceive their environment (e.g. field of vision and responses to noise and movement), an understanding of how to use the flight zone and point of balance has improved animal handling by facilitating the flow of cattle (Grandin, 1980a; Lyles and Calvo-Lorenzo, 2014).

Grandin (1980a) described how the wide field of view of cattle and their poor depth perception influence appropriate designs for cattle handling facilities. When moving along a passageway or race, cattle will balk/hesitate less if sudden visual discontinuities in the floor level, texture and appearance are avoided. Solid sides to pens/races/chutes avoid peripheral distractions to the cattle. Curved races leading up to a restrainer or stunning pen will encourage cattle to move until they are almost into the pen (Grandin, 1980a).

Depending on the extent of the previous handling, most cattle are fearful of human handlers. Once a human enters the flight zone of an animal, it moves away, and leaving the flight zone causes the animal to stop moving. The flight distance varies with the type of cattle and their experience of handling. The most efficient flight distance to encourage the steady movement of cattle is on the boundary of the flight zone. Cattle can be handled easier if they can maintain visual contact with a handler who orients themselves at an angle from the animal's shoulder instead of standing directly behind it (Grandin, 1980a).

Cattle move easier from a dimly lit area to a brighter area (Grandin, 1990). Common causes of balking/hesitation of cattle during movement are shiny reflections on metal or water (Klinglmair et al., 2011); dangling objects, for example, chains; seeing humans in front of them; air movement towards them; and sudden intermittent noise (Grandin, 2006, 2012, 2013).

Handling facilities at large slaughter plants should be one-way and use curved and diagonal arrangements of pens and passageways. Transitions into a single race or chute need to avoid bunching and jamming, for example, by using a circular pen with a crowding

**Figure 8** Curved handling area with solid sides, non-slip flooring and a crowding gate that swings around a central post.

gate which swings around a central post (Fig. 8). One-way gates prevent cattle from backing up and non-slip flooring is essential to avoid falls (Grandin, 1980a, 1990, 1996).

Pre-slaughter handling methods identified in Australia to move cattle included tactile interactions: touching and pushing, slapping and hitting, and use of an electric goad/prod. Auditory interactions consisted of talking, whistling, shouting and shaking metallic rattles and banging on pen fittings. The only visual interaction was waving. The attitudes of the handlers affected how they handled the cattle. For example, compared with handlers who did not agree, handlers who believed that poor flooring makes cattle unsure of their footing used a goad less and talked and whistled less. Those that considered that goads

**Figure 9** Cattle in single file race/chute leading to the stunning pen.

are not stressful and that it is important to arouse cattle, used a goad more and talked more, but waved less (Coleman et al., 2012).

Examples of poor handling include inappropriate use of driving aids such as sticks, goads, whips, pipes and paddles (application to sensitive body regions; unnecessary and frequent use and forceful use resulting in injury); dragging of non-ambulatory animals; routine contact of cattle with sharp/hard structures; and forcing cattle to move too quickly resulting in falls (Lyles and Calvo-Lorenzo, 2014; Nicholson et al., 2013).

Jarvis et al. (1996b) observed more potentially traumatic events during handling of cattle in the race leading to the stunning pen (Fig. 9) than during unloading, movement into and out of lairage pens and while in a lairage pen. Observations in slaughter plants in Sweden of cattle in the race leading to the stunning box showed that 8% of cattle slipped, 0% fell, 7% refused to move forward, 14% vocalised, 18% urinated or defecated, 16% were hit with a driving instrument behind the hip bone, 9% were hit with a driving instrument in front of the hip bone, 1% were kicked, 2% had their tail twisted, 13% had an electric goad used on them and 19% had a hydraulic gate lowered onto them (Hultgren et al., 2014).

## 7.2 Handling/driving instruments

Cattle can experience frequent and routine use of driving instruments during handling at a slaughter plant. Bruises on cattle consisting of parallel red marks can readily be identified

**Figure 10** Handling instruments: flag and electric goad/prod.

as stick marks (Weeks et al., 2002). Frequent use of an electrical goad can be stressful (raised plasma cortisol concentration at slaughter) (Hemsworth et al., 2011) and cause cattle to vocalise (Grandin, 1998a, 2001a). In US slaughter plants, audit data showed that electric goads were used on 15% of beef cattle and 29% of cows and bulls. There was not a consistent effect of increasing line speed on goad use (Grandin, 2005).

In steers subjected to repeated electrical stimulation from a goad for 24 h before slaughter, in some muscle groups, the muscle pH and toughness were increased, but there were variable effects on meat quality taste panel determinations (Lewis et al., 1962). In cattle that were given six electric shocks from a goad in the race leading to the stunning box, there was no effect on muscle $pH_u$ or toughness, but plasma lactate increased and taste panel assessment was that the beef was less tender, less juicy and had a less acceptable flavour (Warner et al., 2007).

Examples of alternatives to driving instruments, such as goads and sticks (that can cause pain or injury) include various types of flags (Fig. 10), a plastic bag or plastic streamers taped to the end of a stick (Grandin, 2006) and an air-powered prod/goad that creates an intense vibration rather than an electric shock (Grandin, 2013).

## 7.3 Restraint in stunning pen

The duration that cattle are restrained in the race/chute leading up to the stunning pen and in the stunning pen should be minimised to reduce stress (Cockram and Corley, 1991; Probst et al., 2014). Although some cattle will walk into the stunning area, many require some encouragement, for example by the use of a goad (Bourguet et al., 2011; Grandin, 2005), even if they enter the stunning area using a restrainer conveyor (Grandin, 1988, 2013).

The most common design of stunning pen for cattle is a narrow pen with a guillotine entry gate with solid sides (Grandin, 1980a) (Fig. 11). The parts of the pen that make contact with the animal should move with a smooth, steady motion and not apply excessive pressure that can cause struggling or vocalisation. The floor should be non-slip (Grandin,

**Figure 11** Stunning pen with vertical guillotine entry gate.

2013). A light placed above the front of the stunning pen makes cattle raise their head to look up and this assists stunning (Grandin, 1980).

# 8 Overview of the slaughter process

## 8.1 Controversy associated with cattle slaughter

Slaughter is the stage in beef production associated with the most controversy. There are deeply held cultural and ethical views on the killing of cattle to produce meat for human consumption. These are reflected in the different terms used, for example, the word 'slaughter' for the 'killing' of animals intended for human consumption rather than just killing, and the expression 'harvesting of animals', that is, not distinguishing between animals and plants in their gathering and use for consumption. In most societies, there are individuals and groups, for example, vegetarians and vegans that hold ethical views that oppose the slaughter of cattle for human consumption. Many consumers of beef prefer not to think about how their beef is produced or place their trust in the meat industry to treat cattle in an appropriate, that is, humane manner. Others place reliance on detailed legal regulations to control the manner in which cattle are slaughtered. Increasingly, end users (consumers of beef), retailers and restaurants require the meat industry to undertake slaughter and the pre-slaughter management of live cattle according to an agreed set of procedures that are then monitored, audited and publicised.

## 8.2 Stunning and slaughter

Slaughter is normally a two-stage process in which the animal is first stunned to cause a temporary loss of consciousness or insensibility (inability to perceive stimuli) and then it is killed by exsanguination or cardiac arrest. Death can be caused by permanently damaging the brain by anoxia (lack of oxygen supply) and/or permanent physical damage to vital centres in the brain responsible for respiration. The main method of causing anoxia of the brain is exsanguination (bleeding out) following cutting/sticking of the major blood vessels in the neck/chest (Blackmore, 1993). Using electricity to stop the heart and the flow of blood to the brain will also cause anoxia and death.

Stunning should produce immediate insensibility of sufficient duration to ensure that the animal remains insensible or unconscious until bleeding is initiated and has caused permanent insensibility by cerebral anoxia (Blackmore, 1989). If a method causes permanent insensibility that inevitably results in death with no possibility of recovery, the time taken for the onset of permanent insensibility is the relevant criteria to use when determining the welfare implications of a method rather than the time taken to die (death usually occurs in stages as the various tissues progressively cease to function). If the stunning method causes temporary insensibility, the time taken for the onset of insensibility, the duration of insensibility and the time taken to cause sufficient exsanguination to result in permanent insensibility and death are the relevant criteria to use.

Some concerns remain over the effectiveness of some stunning methods, the time to loss of sensibility and the subsequent duration of insensibility. Stunning and exsanguination is a technical operation that requires practical skill, effective equipment and good restraint of the animals. In addition, the behavioural and neurophysiological responses to different methods of slaughter are complex, and the interpretation of some of these responses in relation to consciousness or sensibility is to some extent unclear (Terlouw et al., 2015). There are several practical methods for assessing whether an animal is unconscious, but some methods may not be sensitive enough to identify the time when the onset of insensibility occurred.

The following signs that could indicate a return to consciousness after stunning should be absent before hoisting or any invasive procedures are performed (Grandin, 2013):

- Rhythmic breathing
- Spontaneous blinking or response to a hand waved in front of the eyes
- Nystagmus (vibrating eyelids)
- Vocalisation
- Righting reflex or attempt to get up or lift the head
- Response to a pinprick on the nose
- Eye tracking of moving objects.

If an animal did regain sensibility on the slaughter line, it might experience pain and distress from the ineffective stunning procedure, from the cut in the neck or chest, and from inversion and suspension of its body weight by one leg on the slaughter rail (Gouveia et al., 2009). This event should be extremely rare. If, before hoisting and exsanguination or when on the slaughter line, there is any indication that an animal has not been stunned correctly or is showing any signs of returning to sensibility it should be re-stunned immediately. A stunned and exsanguinated animal must be left to bleed out and die before any skinning, head removal or limb removal takes place (Grandin, 2002).

The methods used to slaughter cattle have evolved and developed with improvements in handling, restraint and equipment (Gregory, 1990). Most cattle are stunned using a penetrating captive bolt stunner. There is also some use of electrical stunning. Religious slaughter is often performed using exsanguination without prior stunning, but some Halal slaughter can involve some form of stunning.

The following factors influence the choice of slaughter method:

1. welfare considerations
   - ability to induce loss of consciousness and death without causing pain, fear or distress
   - ability to induce rapid loss of consciousness
   - reliability
2. practical considerations
   - safety of personnel
   - ability to maintain equipment
   - economics in relation to line speed and capacity
3. food safety and beef quality considerations
   - not pose a risk to food safety
   - not adversely affect meat quality
4. religious considerations.

At the time of slaughter, stress hormone concentrations in cattle are greater than the concentrations measured at the farm before transport (Bourguet et al., 2010). This stress is partly due to pre-slaughter handling and restraint, and partly due to the effects of stunning (Shaw and Tume, 1992; Tume and Shaw, 1992). If stunning is effective in causing immediate loss of consciousness, the cattle will not experience distress as a result of stunning (Grandin, 2014a), but meat quality might be affected if one method of stunning caused more stress than another (Grandin, 1980c).

# 9 Methods of stunning

## 9.1 Penetrating captive bolt stunning

Captive bolt stunning requires that the slaughter operator place the captive bolt stunner in a precise location and at the correct angle to ensure that the bolt causes a percussive blow to the skull that causes shock waves that induce concussion, haemorrhage and brain damage. In addition, the captive bolt penetrates the skull and causes variable degrees of brain damage (Finnie, 1995, 1997). The concussive effects of captive bolt stunning are not dependent on the direct damage caused by the bolt as it passes through the brain (Daly and Whittington, 1989) and damage to any one particular part of the brain is not required to cause loss of consciousness. However, damage to the brain stem is associated with a rapid onset of unconsciousness (Gregory and Shaw, 2000).

The captive bolt is fired either automatically when in contact with the head or when a trigger is pulled (Fig. 12). The bolt is propelled either by an explosive charge from a cartridge of the appropriate size for the thickness of the skull or by pneumatic force. The velocity of the bolt varies with the charge used. The velocity of the bolt affects the severity and duration of the disruption in brain activity (Daly et al., 1987).

The penetrating captive bolt creates a discrete hole in the skull that can, but does not always cause, radiating skull fracture lines. The bolt creates a well-defined haemorrhagic track through the brain, and there is severe destruction to the cerebellum, brainstem and caudal aspect of the cerebral hemispheres with subarachnoid and intraventricular haemorrhage (Finnie, 1997). However, the areas of the brain destroyed or damaged by captive bolt stunning are to some extent variable, and there are some differences between types of captive bolt devices and types of cattle in the damage caused (Schwenk et al., 2016). Common types of captive bolts have a penetrating depth of between 9 and 13.5 cm. The distance between the skin and the inside of the skull of a bull can be 4 cm and the distance between the skin and the thalamus in the forebrain can be 12 cm (Schwenk et al., 2016).

Neurophysiological research techniques have been used to assess the onset of insensibility. The electrical activity of the brain can be recorded from electrodes on the surface of the head (generating an electroencephalogram or EEG) or the brain (electrocorticography or ECoG). The pattern of EEG recordings is different when an animal is conscious, unconscious or dead. However, the differentiation between the characteristics of an EEG recorded during various states or degrees of consciousness can be unclear and are open to interpretation (Shaw, 2002). Severe reduction in spontaneous cortical activity is used to provide evidence of brain failure. A flat line or isoelectric EEG occurs when no brain electrical activity is recorded and is consistent with brain death.

The ability of the brain to perceive a discrete stimulus to the nervous system can be recorded as a discrete electrical activity (evoked potential) in the brain. The time when visual (light flashing) or somatosensory (nerve stimulation) evoked responses are abolished after stunning has been used to indicate that sufficient brain damage has occurred to

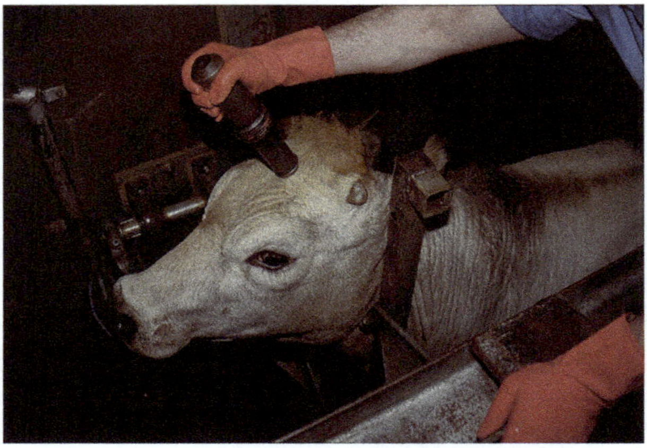

**Figure 12** Captive bolt stunning with head restraint.

cause failure of a primary sensory pathway and this degree of damage would normally be inconsistent with sensibility. However, the recovery of evoked potentials after stunning is more difficult to interpret (Daly et al., 1987, 1988).

After successful captive bolt stunning, both visual and somatosensory evoked responses are no longer recorded and the ECoG shows high-amplitude, low-frequency activity and it becomes isoelectric after about 44 s (Daly et al., 1988).

Following successful stunning, the animal should collapse, stop breathing and then become rigid with its head extended and its hind legs flexed (tonic activity). This is followed by a period of involuntary kicking movements (clonic activity). Respiratory arrest occurs due to damage to brain stem function (Gregory and Shaw, 2000). The absence of a corneal reflex indicates that the pathways from the optic nerve to the optic chiasma and brain stem and back to the eye have been damaged. Disruption of brain function will stop a corneal reflex, but it could also be stopped by damage to the optic nerve without loss of consciousness (Gregory and Shaw, 2000). However, in practice, absence of a corneal reflex is a useful indication of effective captive bolt stunning (Gregory and Shaw, 2000).

The main causes of the failure of a penetrating captive bolt stun to cause insensibility with one shot are lack of maintenance of the stunner, damp cartridges in cartridge-fired stunners, low air pressure in pneumatic stunners, movement of the head – making it difficult to position the stunner correctly – and inexperienced operators not placing the stunner in the ideal position (Atkinson et al., 2013; Grandin, 2002, 2012, 2013). To improve the performance of captive bolt stunners, many slaughter plants have a documented maintenance programme and use equipment to measure the bolt velocity (Grandin, 2006).

Accurate placement of the captive bolt stunner on the head requires skill. The location used to shoot cattle with a captive bolt stunner is the cross-point on the forehead between lines from the centre of the base of each horn/horn bud to the middle of the contralateral eye (Gregory and Shaw, 2000). In a survey of skulls from two slaughter plants in Germany, Fries et al. (2012) found that 65% of the skulls had a hole made by the captive bolt that was within 2.5 cm of the ideal position, 31% indicated medium precision and 4% had been shot >5 to 6 cm from the ideal location. In France, one study of the use of a captive bolt stunner with no head restraint did not find a relationship between the presence of ocular signs or breathing and the site of impact or morphometric characteristics of the skull (Terlouw et al., 2015).

Table 3 shows that, in practice, some cattle stunned using a penetrative captive bolt show signs indicative of inadequate stunning. There can be problems in effectively stunning bulls with a captive bolt. Standard practice is to shoot the animal a second time with a captive bolt stunner if it shows any signs of an ineffective stun or shows signs of recovery. However, cattle that are stunned with a captive bolt more than once have a greater plasma cortisol concentration than in those shot only once (Chulayo et al., 2016).

von Wenzlawowicz et al. (2012) showed that the use of either close head restraint with a back pusher or the provision of a concave table for head placement (with or without a back pusher) resulted in fewer inaccurate placements of the captive bolt stunner than when no head restraint was used. However, head restraint can increase the risk of some cattle showing signs of struggling and stress (Ewbank et al., 1992).

Although captive bolt stunning can result in permanent insensibility, unless the vital areas of the brain stem required to maintain respiration are damaged, the stun is likely to be only temporary. The animal must be killed by exsanguination before it has an opportunity to regain consciousness. The interval between stunning and exsanguination (sticking) should be as short as possible to reduce the chance of recovery. However, observations in slaughter plants show that there are often difficulties in keeping within the recommended maximum stun-to-stick interval of 60 s (Atkinson et al., 2013). It has been suggested that the maintenance of blood pressure due to a prolonged stun-to-stick interval can increase the risk of blood splash (Belk et al., 2002).

With captive bolt stunning, the brain is penetrated, and there is a risk of direct and indirect dissemination of brain tissue (Prendergast et al., 2003). Pithing, that is, physical disruption of the brain and rostral part of the spinal cord using a rod inserted through the hole in the skull made by a penetrating captive bolt was used as an alternative to exsanguination; however, its use has been discontinued due to the risk of dissemination of brain and spinal cord material (Appelt and Sperry, 2007).

## 9.2 Non-penetrating captive bolt stunning

Non-penetrating captive bolt stunning reduces the risk of the dissemination of brain material compared with the use of a penetrating captive bolt (Anil et al., 2002; Grandin, 2006). It causes a concussive impact on the forehead resulting in a depressed fracture of the skull with fracture lines radiating from the site of impact and subarachnoid haemorrhages

**Table 3** Examples from studies in slaughter plants in France, Sweden and the United Kingdom of the percentages of cattle stunned using a penetrating captive bolt that showed some selected signs of an inadequate response to stunning

| | Bulls | Steers | Heifers/cows |
|---|---|---|---|
| Failed to collapse immediately | 1–4 | 0–1 | 1–2 |
| Rhythmic breathing | 2–3 | 0–3 | 0.3–1 |
| Corneal reflex | 1 | 0–0.1 | 0–3 |

Sources: Atkinson et al., 2013; Bourguet et al., 2011; Gregory et al., 2007.

and some petechial haemorrhage. The responses to a successful stun are similar to those from a penetrating captive bolt (Neves et al., 2016). An isoelectric EEG has been recorded to occur after 60 s (Finnie, 1995; Gibson et al., 2009c).

Grandin (2014b) considers that non-penetrating captive bolt stunners are not suitable for some types of cattle, for example those with woolly heads and that they are more effective if the stunner has sufficient power to fracture the skull. Non-penetrating captive bolt guns that are sometimes used before Halal slaughter are used in a manner that attempts to avoid fracturing the skull.

## 9.3 Electrical stunning

If a current of sufficient strength and duration is passed through the head of an animal, there is a massive depolarisation of the neurons and a state similar to an epileptic seizure is produced. In head-to-body stunning, current is first passed through the brain for about 3 s via electrodes on the nose and the neck to cause insensibility and then current is passed for about 15 s through a brisket electrode. This stops the heart and the animal dies due to lack of supply of oxygenated blood to the brain. Head-to-body electrical stunning, therefore, causes immediate and permanent insensibility, even without exsanguination (Blackmore, 1989). Gasping, rhythmic breathing, eyeball rotation and corneal reflexes can occur after head-to-body electrical stunning that stops the heart. However, these responses disappear as brain death occurs without the animal regaining consciousness (Mpamhanga and Wotton, 2015; Wotton et al., 2000).

Sufficient current is required to overcome the impedance or electrical resistance of body tissue. Factors that reduce impedance are firm contact between the body and the electrodes, use of water and the absence of mud or faeces on the coat (Grandin, 2013; Wotton et al., 2000). A current of greater than 1.5 A is required to ensure cardiac arrest and this is sufficient to have caused an effective stun by inducing characteristic epileptiform brain activity (high-amplitude, low-frequency EEG) before cardiac arrest occurs (Wotton et al., 2000). If the animal is unconscious when the current flows through the heart, it will not experience pain. It takes less than 100 ms for the current to reach the threshold required to induce an epileptic seizure and unconsciousness (Mpamhanga and Wotton, 2015).

If a third step is included in which current is applied through the body to the rear of the stunning box for about 4 s, it will cause skeletal and respiratory muscles to contract, and the animal will be immobilised and not able to breathe. Immobilisation facilitates exsanguination, but if the animal is not fully unconscious or recovers consciousness, pain, fear and distress could be experienced (Wotton et al., 2000), but signs of recovery, such as vocalisation or struggling would not be apparent (Devine et al., 1986). Severe muscle contractions can cause femoral fractures (with potential for bone fragments in the meat) and petechial haemorrhage in muscles (blood splash) resulting in trimming and/or downgrading (Mpamhanga and Wotton, 2015).

Head-only electrical stunning produces an epileptiform seizure and no heart fibrillation. Although eye reflexes are initially absent they can return within 1–4 minutes (Cook et al., 1991). Therefore, head-only electrical stunning is only temporary. In cattle stunned and allowed to recover, the tonic phase ended after about 10 s, the clonic phase ended after 56 s, respiration started after 51 s (range 33–90 s) and cattle stood up after 144 s (range 47–319 s) (Wotton et al., 2000). Head-only electrical stunning is sometimes used before Halal slaughter because it is reversible and does not injure or kill the animal (Grandin, 2013).

# 10 Exsanguination

After stunning, cattle are killed by exsanguination, that is, bleeding or 'sticking' using either a cut across the neck severing the carotid arteries and jugular veins or by chest sticking. After the animal has rolled out of the stunning box, and as soon as post-stunning reflexes permit, a chain is attached around a hind leg, and the animal is hoisted and suspended in an inverted position (Fig. 13). 'Sticking' is undertaken when the animal is suspended to facilitate the drainage of blood from the carcase. To reduce the risk of cattle recovering consciousness, sticking needs to be undertaken as soon as possible.

In cattle, vertebral arteries can maintain cerebral blood flow after bilateral severance of carotid and jugular vessels (Baldwin and Bell, 1963; Blackman et al., 1986). Therefore, if a neck cut is used, some blood will continue to flow to the brain via the vertebral arteries for 1–2 minutes (Anil et al., 1995b). Chest sticking is preferable as this severs the major blood vessels in the brachiocephalic trunk near the heart. The cut is made through an incision in the thorax by inserting a knife in front of the sternum. This stops blood flow to the brain; there is less risk of occluded arteries, and the systemic blood pressure falls within 8 s (Anil et al., 1995b).

The welfare implications of religious slaughter, that is, Jewish (Shechita) and Muslim (Halal) slaughter are controversial. In Shechita slaughter, and in most Halal slaughter, the neck is cut without any prior stunning of the animal. However, some Halal slaughter is undertaken after stunning (head-only electrical stunning or non-penetrating captive bolt stunning). In Halal slaughter, stunning is not permitted to stop the heart, damage or kill the animal (Anil, 2012; Nakyinsige et al., 2013).

**Figure 13** Hoisting prior to exsanguination.

Research has identified that the welfare issues related to exsanguination without prior stunning are stress during handling/restraint before exsanguination, pain during and/or immediately following the neck incision, suffering arising from aspiration of blood into the respiratory tract and the delay before the animal loses consciousness after exsanguination (Gibson et al., 2009a,b; Gregory et al., 2009, 2010, 2012; Mellor and Littin, 2004).

After Shechita slaughter, similar ECoG changes occur to those that have been observed after captive bolt stunning, but the changes take longer to occur. Evoked responses gradually decreased in amplitude, and visually evoked responses can stop after 55 s and somatosensory evoked responses can stop after 77 s. The ECoG changes to high-amplitude, low-frequency activity and becomes isoelectric after about 72 s (Daly et al., 1988). In one study in France, the percentages of cattle showing the following signs 10 s and 30 s after exsanguination without stunning (Halal slaughter) and 30 s after hoisting were

- Rhythmic breathing 70%, 0% and 4%, respectively, and
- Corneal reflex 58%, 47% and 13%, respectively (Bourguet et al., 2011).

As Shechita is undertaken using a long sharp knife that makes one quick, large and deep incision across the neck, blood should readily flow out of the wound; blood pressure falls rapidly and cattle may not appear to suffer (Grandin and Regenstein, 1994; Rosen, 2004). The requirements for the type of knife used for Halal slaughter are less precise and in some countries, Halal slaughter is performed using a sawing motion to cut the blood vessels in the upper neck (Ahsan et al., 2014).

In about 8% of cattle slaughtered by Shechita and Halal methods (Gregory, et al., 2008), blockages can form in the severed carotid arteries and this can maintain arterial blood pressure (for about 50 s vs. 26 s when the arteries are not blocked) and delay the time when insensibility occurs (Anil et al., 1995a). One study showed that the duration before cattle slaughtered by the Halal method finally collapsed after their neck had been cut varied from 6 to 180 s (Gregory et al., 2009). A high neck cut, undertaken at the position of $\leq$2 tracheal rings, reduces (a) the time taken for cattle to collapse, (b) the duration when corneal and palpebral reflexes are still present and (c) the risk of carotid blockage compared with a conventional low ($\geq$2.5 tracheal rings) neck cut (Gibson et al., 2015).

Exsanguination without stunning compared with exsanguination after captive bolt stunning does not improve bleed out (rate of blood loss and total blood loss) and does not affect muscle $pH_{24}$ (Anil et al., 2006; Önenc and Kaya, 2004; Sazili et al., 2013).

Some of the methods used to restrain cattle before exsanguination without stunning can cause distress. In some countries, cattle are casted using ropes; the procedure takes 3–4 minutes and can involve struggling and trauma (Ahsan et al., 2014). In some countries, conscious cattle are hoisted by one leg before exsanguination resulting in vocalisation (Anil et al., 2006; Grandin, 2013).

Pens that tilt the animal to either a 45° angle, 90° angle or 180° (i.e. upside down) are used to aid exsanguination without stunning (Anil, 2012; Velarde et al., 2014). Upright and Facomia (45° angle) pen designs include features for extra restraint such as a belly lift, back push and chin lift (Anil, 2012; Dunn, 1990). These features should operate using a slow, steady movement and not cause discomfort from excessive pressure (Grandin, 2001b). If the animal is kept in an upright position rather than inverted, there is reduced stress and struggling (Dunn, 1990; Grandin, 2001b), and delays while the animal is turned are avoided (Novelli et al., 2016).

# 11 Case study: auditing systems

In North America, audits of slaughter plants by restaurant companies that have been adopted by the meat industry have improved the handling and stunning of cattle. Grandin introduced numerical scoring of a limited number of criteria to measure a number of different welfare issues. This proved to be an effective way of conducting audits and providing plant managers with a numerical goal to work towards and maintain (Grandin, 1998a,b; 2001a, 2006). The scoring system scores 100 cattle using five variables (Grandin, 1998a,b). The American Meat Institute auditing system criteria are as follows:

1. 95% of cattle rendered insensible with one application of the stunner
2. 100% of cattle must be rendered insensible before they are hoisted or invasive procedures are started
3. ≤5% of cattle vocalise in the restrainer, the stun box and while entering the stunning areas
4. ≤1% of cattle fall
5. ≤5% of cattle are moved with an electric goad
6. There are no acts of abuse, such as dragging non-ambulatory animals, deliberate slamming of gates on cattle or poking sensitive parts of cattle (Grandin, 2012).

The audits motivated the industry to improve their supervision of employees and their maintenance of equipment. In 1996, only three plants out of 10 passed the stunning audit. For a 5-year period from 1999 to 2003, 90% or more of the plants passed. In 2010, all 30 of the plants audited passed the stunning audit (Grandin, 2012). In 94% of the plants none of the animals fell during handling (Grandin, 2012). For electric goad use, the percentage of plants that passed by moving 75% or more of the cattle without using an electric goad/prod was 76% in 1999 and 90% in 2003 (Grandin, 2006).

Remote video auditing has been introduced by some large companies where auditors at a remote location observe unloading, handling and stunning at several slaughter plants (Grandin, 2013). Farm-to-fork audits that provide customers with the ability to trace beef products back to the feedlot or farm of origin to ensure compliance with health, welfare and food safety standards throughout the whole food production chain have been developed and will become increasingly important (Smith et al., 2005, 2015).

# 12 Summary

Each stage of pre-slaughter management can be stressful to cattle, and there are multiple factors that influence their behavioural and physiological responses to stressors. The way in which cattle respond to stressors depends on many factors, including breed, sex, temperament and experience. Research has identified factors affecting muscle $pH_u$ and the risk of bruising. However, consistent relationships between stressors and other aspects of meat quality such as meat tenderness and taste have proven more difficult to identify.

Transport exposes cattle to a number of potential stressors, for example, novelty, mixing of groups, thermal stress, feed and water withdrawal, vehicular movement, responses to maintain stability and reduced opportunities for rest. These factors can result in variable responses, depending on the characteristics of the journey, the type of cattle and their fitness. Most studies have not found large effects of journey duration on meat quality, and there are few meat quality and no welfare benefits of lairaging cattle before slaughter.

Studies in slaughter plants have identified that poor handling practices occur. However, the work of Temple Grandin has shown that if facilities are designed well and handlers are trained and supervised, it makes it easier to handle and move the cattle; there are also fewer injuries and less stress.

Conventional slaughter is a two-stage process in which animals are first stunned to cause a temporary loss of consciousness and then killed by exsanguination or cardiac arrest. Stunning should produce immediate insensibility of sufficient duration to ensure that the animal remains unconscious until bleeding is initiated and has caused permanent insensibility by cerebral anoxia.

The concussion arising from the use of a penetrating captive bolt can stun cattle effectively, but some concerns remain over the consistently of its application. There are some concerns over the effectiveness of some stunning methods, the time to loss of sensibility and the subsequent duration of insensibility. Stunning and exsanguination are technical operations that require practical skill, effective equipment and good restraint of the animal. In addition, the behavioural and neurophysiological responses to different methods of slaughter are complex and the interpretation of some of these responses in relation to consciousness is to some extent debatable.

In Shechita slaughter, and in most Halal slaughter, the neck is cut without any prior stunning. With this method of slaughter, there is stress during handling/restraint before exsanguination, pain during and/or immediately following the neck incision, aspiration of blood into the respiratory tract and a delay before the animal loses consciousness.

Audits of the handling and stunning of cattle have been demonstrated to produce significant improvements. For the sustainability of the beef industry, it is vital that the industry implement research on how to improve animal welfare and that consumers are provided with evidence that this has occurred, so that they can be reassured that cattle do not experience major welfare issues during handling, transport, lairage and slaughter.

# 13 Future trends in research

As handling, transport and slaughter is a complex process, and multiple factors influence whether specific outcomes occur, research in this area needs to consider how to examine the effects of various interactions. Large-scale epidemiological studies of industry practice and experimental studies with the resources to examine a range of factors and their interactions are required.

Owing to uncontrolled factors that occur when cattle are transported, replication of journey treatments and detailed specification of the journey characteristics are required. Studies on transportation that report relevant outcome measurements in response to a range of experimental treatments imposed during transportation, for example, different stocking densities can provide valuable information. Identifying the fitness of cattle in terms of their health status in relation to their potential journey requires further research.

Multidisciplinary approaches are advantageous, they could involve the following:

- engineering, for example, to characterise the physical effects that cattle experience, such as accelerations, temperature, humidity and air movement,
- physiology, for example, to characterise the degree to which cattle have to respond to challenges in their environment including remote monitoring of physiological variables during journeys,
- behaviour, for example, to characterise responses to factors such as comfort, rest, fatigue, thermal distress, driving events, thirst and hunger,
- health and pathology, for example, to identify fitness, injury and other pathological changes, and
- meat science, for example, to identify factors affecting meat quality,

These are required to further understand the effects of transport on cattle and to identify ways of mitigating factors that reduce animal welfare and meat quality.

Research on how to facilitate knowledge transfer to industry and consumers, and methods for encouraging the adoption of practices that improve animal welfare would be beneficial (Grandin, 2003).

One of the greatest challenges is to identify a consistent way of slaughtering cattle that minimises the risks of unnecessary suffering. Further research is required to provide a greater understanding of how to meet the needs of communities that currently require cattle to be slaughtered without stunning while ensuring that cattle do not suffer unnecessarily, and that cattle do not experience fear, distress and pain during slaughter. One associated research goal is to remove some of the uncertainty in identifying when various methods of stunning and slaughter are not fully effective in inducing immediate loss of consciousness and what signs indicate recovery of consciousness. This will require neurophysiology, neuropathology/anatomical pathology and behavioural approaches to be used in studies in experimental and commercial situations.

# 14 Where to look for further information

## 14.1 Websites

www.grandin.com is Temple Grandin's website that provides resources on cattle handling and slaughter.

www.animalhandling.org is a US meat industry website that provides information on voluntary guidelines including the American Meat Institute's Recommended Animal Handling Guidelines and Audit Guide.

www.hsa.org.uk is the website for The Humane Slaughter Association, a UK animal welfare charity that provides technical and practical information on handling transport and slaughter.

## 14.2 Books

Livestock Handling and Transport, 4th edition. 2014. CABI. Edited by Temple Grandin. This book contains definitive reviews on cattle handling and transport.

Animal Welfare and Meat Science. 1998. CABI. This book by Neville Gregory contains chapters on stunning, slaughter, animal welfare and meat quality.

Animal Welfare and Meat Production. 2007. CABI. This book by Neville Gregory is a 2nd edition of Animal Welfare and Meat Science that provides updates on stunning, slaughter, animal welfare and meat quality.

Humane Livestock Handling. 2008. Storey Publishing. This book by Temple Grandin contains information on cattle behaviour and handling and provides details on the planning and construction of facilities.

## 14.3 Reports

Opinion of the Scientific Panel on Animal Health and Welfare on a request from the Commission related to welfare aspects of the main systems of stunning and killing the main commercial species of animals, *The EFSA Journal* 2004; 45: 1–29. www.efsa.eu.int.

This is a review and interpretation of the welfare aspects of stunning and killing of cattle and other animals that was undertaken by the European Food Safety Authority Scientific Panel on Animal Health and Welfare.

EFSA AHAW Panel (EFSA Panel on Animal Health and Welfare), 2013. Guidance on the assessment criteria for studies evaluating the effectiveness of stunning methods regarding animal protection at the time of killing. *The EFSA Journal* 2013; 11(12): 3486, 41 pp. doi:10.2903/j.efsa.2013.3486.

This document provides guidance on how to conduct and report research on the welfare implications of stunning.

EFSA AHAW Panel (EFSA Panel on Animal Health and Welfare), 2013. Scientific opinion on monitoring procedures at slaughterhouses for bovines. *The EFSA Journal* 2013; 11(12): 3460, 65 pp. doi:10.2903/j.efsa.2013.3460.

Guidance on the assessment of consciousness in cattle after stunning with a penetrative captive bolt and the assessment of unconsciousness and death in cattle during slaughter without stunning.

OIE World Organisation for Animal Health. Terrestrial Animal Health Code. Section 7. Animal Welfare. 2016. www.oie.int.

International recommendations on handling, transport and slaughter.

## 14.4 Video material

'Responsible cattle handling and transport for optimum welfare and meat quality' Kansas State University Division of Continuing Education, DCE Facilitation Center, 13 College Court Building, Manhattan, KS 66506, USA.

'Careful Driving and Animal Welfare. The effect of driver behaviour on the behaviour of livestock in-transit' University of Edinburgh. www.youtube.com/watch?v=4sQ3Etj9K5I.

# 15 References

Adeyemo, O. K., Adeyemi, I. G. and Awosanya, E. J. (2009), 'Cattle cruelty and risks of meat contamination at Akinyele cattle market and slaughter slab in Oyo State, Nigeria', *Trop. Anim. Health Prod.*, 41(8), 1715–21.

Ahola, J. K., Foster, H. A., VanOverbeke, D. L., Jensen, K. S., Wilson, R. L., Glaze, J. B., Fife, T. E., Gray, C. W., Nash, S. A., Panting, R. R. and Rimbey, N. R. (2011), 'Survey of quality defects in market beef and dairy cows and bulls sold through livestock auction markets in the Western United States: I. Incidence rates', *J. Anim. Sci.*, 89(5), 1474–83.

Ahsan, M., Hasan, B., Algotsson, M. and Sarenbo, S. (2014), 'Handling and welfare of bovine livestock at local abattoirs in Bangladesh', *J. Appl. Anim. Welf. Sci.*, 17(4), 340–53.

Anil, M. H. (2012), 'Religious slaughter: A current controversial animal welfare issue', *Anim. Front.*, 2(3), 64–7.

Anil, M. H., McKinstry, J. L., Wotton, S. B. and Gregory, N. G. (1995a), 'Welfare of calves – 1. Investigations into some aspects of calf slaughter', *Meat Sci.*, 41(2), 101–12.

Anil, M. H., McKinstry, J. L., Gregory, N. G., Wotton, S. B. and Symonds, H. (1995b), 'Welfare of calves – 2. Increase in vertebral artery blood flow following exsanguination by neck sticking and evaluation of chest sticking as an alternative slaughter method', *Meat Sci.*, 41(2), 113–23.

Anil, M. H., Love, S., Helps, C. R. and Harbour, D. A. (2002), 'Potential for carcass contamination with brain tissue following stunning and slaughter in cattle and sheep', *Food Control*, 13, 431–6.

Anil, M. H., Yesildere, T., Aksu, H., Matur, E., McKinstry, J. L., Weaver, H. R., Erdogan, O., Hughes, S. and Mason, C. (2006), 'Comparison of Halal slaughter with captive bolt stunning and neck cutting in cattle: Exsanguination and quality parameters', *Anim. Welfare*, 15(4), 325–30.

Appelt, M. and Sperry, J. (2007), 'Stunning and killing cattle humanely and reliably in emergency situations - A comparison between a stunning-only and a stunning and pithing protocol', *Can. Vet. J.*, 48, 529–34.

Apple, J. K., Kegley, E. B., Galloway, D. L., Wistuba, T. J. and Rakes, L. K. (2005), 'Duration of restraint and isolation stress as a model to study the dark-cutting condition in cattle', *J. Anim. Sci.*, 83(5), 1202–14.

Apple, J. K., Kegley, E. B., Galloway, D. L., Wistuba, T. J., Rakes, L. K. and Yancey, J. W. S. (2006), 'Treadmill exercise is not an effective methodology for producing the dark-cutting condition in young cattle', *J. Anim. Sci.*, 84(11), 3079–88.

Atkinson, S., Velarde, A. and Algers, B. (2013), 'Assessment of stun quality at commercial slaughter in cattle shot with captive bolt', *Anim. Welfare*, 22(4), 473–81.

Baldwin, B. A. and Bell, F. R. (1963), 'The anatomy of the cerebral circulation of the sheep and ox. The dynamic distribution of the blood supplied by the carotid and vertebral arteries to cranial regions', *J. Anat.*, 97, 203–15.

Barnett, J. L., Edge, M. K. and Hemsworth, P. H. (2008), 'The place of quality assurance in managing animal welfare during long distance transport', *Vet. Ital.*, 44, 121–31.

Bartoš, L., Franc, Č., Albiston, G. and Beber, K. (1988), 'Prevention of dark-cutting (DFD) beef in penned bulls at the abattoir', *Meat Sci.*, 22(3), 213–20.

Bartoš, L., Franc, Č., Řehák, D. and Štípková (1993), 'A practical method to prevent dark-cutting (DFD) in beef', *Meat Sci.*, 34(3), 275–82.

Bass, J. J. and Duganzich, D. M. (1980), 'A note on effects of starvation on the bovine alimentary tract and its contents', *Anim. Prod.*, 31, 111–13.

Belk, K. E., Scanga, J. A., Smith, G. C. and Grandin, T. (2002), 'The relationship between good handling/stunning and meat quality in beef, pork, and lamb', http://www.grandin.com/meat/hand.stun.relate.quality.html (accessed 20 July 2016)

Beltrán, J. A., Jaime, I., Santolaria, P., Sañudo, C., Albertí, P. and Roncalés, P. (1997), 'Effect of stress-induced high post-mortem pH on protease activity and tenderness of beef', *Meat Sci.*, 45(2), 201–7.

Bianca, W. (1965), 'Sweating in dehydrated steers', *Res. Vet. Sci.*, 6, 33–7.

Blackman, N. L., Cheetham, K. and Blackmore, D. K. (1986), 'Differences in blood supply to the cerebral cortex between sheep and calves during slaughter, *Res. Vet. Sci.*, 40, 252–4.

Blackmore, D. K. (1989), 'Electrocardiograms as an aid in assessing the stunning of cattle and pigs', *N.Z. Vet. J.*, 37, 59–61.

Blackmore, D. K. (1993), 'Euthanasia – not always eu', *Aust. Vet. J.*, 70, 409–13.

Blackshaw, J. K., Blackshaw, A. W. and Kusano, T. (1987), 'Cattle behaviour in a saleyard and its potential to cause bruising', *Aust. J. Exp. Agric.*, 27, 753–7.

Booth-McLean, M. E., Schwartzkopf-Genswein, K. S., Brown, F. A., Holmes, C. L., Schaefer, A. L., McAllister, T. A. and Mears, G. J. (2007), 'Physiological and behavioural responses to short-haul transport by stock trailer in finished steers', *Can. J. Anim. Sci.*, 87(3), 291–7.

Bourguet, C., Deiss, V., Gobert, M., Durand, D., Boissy, A. and Terlouw, E. M. C. (2010), 'Characterising the emotional reactivity of cows to understand and predict their stress reactions to the slaughter procedure', *Appl. Anim. Behav. Sci.*, 125, 9–21.

Bourguet, C., Deiss, V., Tannugi, C. C. and Terlouw, E. M. C. (2011), 'Behavioural and physiological reactions of cattle in a commercial abattoir: Relationships with organisational aspects of the abattoir and animal characteristics', *Meat Sci.*, 88, 158–68.

Broom, D. M. (2003), 'Transport stress in cattle and sheep with details of physiological, ethological and other indicators.', *Dtsch. Tierarztl. Wochenschr.*, 110(3), 83–9.

Bulitta, F. S., Aradom, S. and Gebresenbet, G. (2015), 'Effect of transport time of up to 12 hours on welfare of cows and bulls', *J. Serv. Sci. Manage.*, 8(2), 161–82.

Burdick, N. C., Carroll, J. A., Randel, R. D., Willard, S. T., Vann, R. C., Chase, J., C. C., Lawhon, S. D., Hulbert, L. E. and Welsh, J. T. H. (2011), 'Influence of temperament and transportation on physiological and endocrinological parameters in bulls', *Livest. Sci.*, 139, 213–21.

Burns, L. V., Moron, S. E., Córdova, F. M.d., Sousa, L. F., Veiga, A. P. M., Barbosa, S. M., Cândido, F. R., Cruz, O. M., Prata, D. M., Feitosa, A. C. F., Stefanine, N. R., Monteiro, K. B., e Silva, A. K., Zimermann, F. C. and Ramos, A. T. (2014), 'Deaths of cows during long distance transportation to the slaughterhouse. Changes on histopathological examination, muscle and liver glycogen storage and muscle pH', *Braz. J. Vet. Pathol.*, 7(1), 7–13.

Cafe, L. M., Robinson, D. L., Ferguson, D. M., McIntyre, B. L., Geesink, G. H. and Greenwood, P. L. (2011), 'Cattle temperament: Persistence of assessments and associations with productivity, efficiency, carcass and meat quality traits', *J. Anim. Sci.*, 89(5), 1452–65.

Caulfield, M. P., Cambridge, H., Foster, S. F. and McGreevy, P. D. (2014), 'Review: Heat stress: A major contributor to poor animal welfare associated with long-haul live export voyages', *Vet. J.*, 199, 223–8.

Chacon, G., Garcia-Belenguer, S., Villarroel, M. and Maria, G. A. (2005), 'Effect of transport stress on physiological responses of male bovines', *Dtsch. Tierarztl. Wochenschr.*, 112(12), 465–9.

Chulayo, A. Y., Bradley, G. and Muchenje, V. (2016), 'Effects of transport distance, lairage time and stunning efficiency on cortisol, glucose, HSPA1A and how they relate with meat quality in cattle', *Meat Sci.*, 117, 89–96.

Cockram, M. S. (1990), 'Some factors influencing behaviour of cattle in a slaughterhouse lairage', *Anim. Prod.*, 50, 475–81.

Cockram, M. S. (1991), 'Resting behaviour of cattle in a slaughterhouse lairage', *Br. Vet. J.*, 147, 109–19.

Cockram, M. S. (2007), 'Criteria and potential reasons for maximum journey times for farm animals destined for slaughter', *Appl. Anim. Behav. Sci.*, 106(4 SPEC. ISS.), 234–43.

Cockram, M. S. and Corley, K. T. T. (1991), 'Effect of pre-slaughter handling on the behaviour and blood composition of beef-cattle', *Br. Vet. J.*, 147, 444–54.

Cockram, M. S. and Spence, J. Y. (2012), 'The effects of driving events on the stability and resting behaviour of cattle, young calves and pigs', *Anim. Welfare*, 21(3), 403–17.

Colditz, I. G., Ferguson, D. M., Greenwood, P. L., Doogan, V. J., Petherick, J. C. and Kilgour, R. J. (2007), 'Regrouping unfamiliar animals in the weeks prior to slaughter has few effects on physiology and meat quality in Bos taurus feedlot steers', *Aust. J. Exp. Agric.*, 47(7), 763–9.

Coleman, G. J., Rice, M. and Hemsworth, P. H. (2012), 'Human-animal relationships at sheep and cattle abattoirs', *Anim. Welfare*, 21(Suppl. 2), 15–21.

Cook, C. J., Devine, C. E., Gilbert, K. V., Tavener, A. and Day, A. M. (1991), 'Electroencephalograms and electrocardiograms in young bulls following upper cervical vertebrae-to-brisket stunning', *N.Z. Vet. J.*, 39, 121–5.

Cooke, R. F. (2014), 'Bill E. Kunkle Interdisciplinary Beef Symposium: temperament and acclimation to human handling influence growth, health, and reproductive responses in Bos taurus and Bos indicus cattle', *J. Anim. Sci.*, 92(12), 5325–33.

Coombes, S. V., Gardner, G. E., Pethick, D. W. and McGilchrist, P. (2014), 'The impact of beef cattle temperament assessed using flight speed on muscle glycogen, muscle lactate and plasma lactate concentrations at slaughter', *Meat. Sci.*, 98, 815–21.

Cullinane, M., O'Sullivan, E., Collins, G., Collins, D. M. and More, S. J. (2012), 'Veterinary certificates for emergency or casualty slaughter bovine animals in the Republic of Ireland: are the welfare needs of certified animals adequately protected?', *Anim. Welfare,* 21(Suppl. 2), 61–7.

Curley Jr., K. O., Paschal, J. C., Welsh Jr., T. H. and Randel, R. D. (2006), 'Technical note: Exit velocity as a measure of cattle temperament is repeatable and associated with serum concentration of cortisol in Brahman bulls', *J. Anim. Sci.*, 84(11), 3100–3.

Daly, C. C. and Whittington, P. E. (1989), 'Investigation into the principal determinants of effective captive bolt stunning of sheep.', *Res. Vet. Sci.*, 46(3), 406–8.

Daly, C. C., Gregory, N. G. and Wotton, S. B. (1987), 'Captive bolt stunning of cattle – effects on brain-function and role of bolt velocity', *Br. Vet. J.*, 143, 574–80.

Daly, C. C., Kallweit, E. and Ellendorf, F. (1988), 'Cortical function in cattle during slaughter – conventional captive bolt stunning followed by exsanguination compared with shechita slaughter', *Vet .Rec.*, 122, 325–9.

de Witte, K. (2009), 'Development of the Australian animal welfare standards and guidelines for the land transport of livestock: Process and philosophical considerations', *J. Vet. Behav.*, 4(4), 148–56.

Devine, C. E., Tavener, A., Gilbert, K. V. and Day, A. M. (1986), 'Electroencephalographic studies of adult cattle associated with electrical stunning, throat cutting and carcass electro-immobilization', *N.Z. Vet. J.*, 34, 210–13.

Dodt, R. M., Anderson, B. and Horder, J. C. (1979), 'Bruising in cattle fasted prior to transport for slaughter', *Aust. Vet. J.*, 55(11), 528–30.

Doonan, G., Appelt, M. and Corbin, A. (2003), 'Nonambulatory livestock transport: The need for consensus', *Can. Vet. J.*, 44(8), 667–72.

Dunn, C. S. (1990), 'Stress reactions of cattle undergoing ritual slaughter using two methods of restraint', *Vet. Rec.*, 126(21), 522–5.

Dupuy, C., Morignat, E., Maugey, X., Vinard, J., Hendrikx, P., Ducrot, C., Calavas, D. and Gay, E. (2013), 'Defining syndromes using cattle meat inspection data for syndromic surveillance purposes: a statistical approach with the 2005–2010 data from ten French slaughterhouses', *BMC. Vet. Res.*, 9, 88.

Earley, B., Drennan, M. and O'Riordan, E. G. (2013), 'The effect of road transport in comparison to a novel environment on the physiological, metabolic and behavioural responses of bulls', *Res. Vet. Sci.*, 95(2), 811–18.

Edge, M. K. and Barnett, J. L. (2009), 'Development of animal welfare standards for the livestock transport industry: process, challenges, and implementation', *J. Vet. Behav.*, 4(5), 187–92.

Eldridge, G. A. and Winfield, C. G. (1988), 'The behaviour and bruising of cattle during transport at different space allowances', *Aust. J. Exp. Agric.*, 28, 695–8.

Eldridge, G. A., Cahill, D. J. and Ballandis, R. (1984), 'The behaviour and heart rate of cattle at different loading densities during road transport', *Anim. Prod. Aust.*, 15, 676–.

Eldridge, G. A., Winfield, C. G. and Cahill, D. J. (1988), 'Responses of cattle to different space allowances, pen sizes and road conditions during transport', *Aust. J. Exp. Agric.*, 28, 155–9.

Ewbank, R., Parker, M. J. and Mason, C. W. (1992), 'Reactions of cattle to head restraint at stunning: a practical dilemma', *Anim. Welfare,* 1, 55–63.

Ferguson, D. M. and Warner, R. D. (2008), 'Have we underestimated the impact of pre-slaughter stress on meat quality in ruminants?', *Meat Sci.*, 80(1), 12–19.

Ferguson, D. M., Bruce, H. L., Thompson, J. M., Egan, A. F., Perry, D. and Shorthose, W. R. (2001), 'Factors affecting beef palatability – Farmgate to chilled carcass', *Aust. J. Exp. Agric.*, 41(7), 879–91.

Ferguson, D. M., Warner, R. D., Walker, P. J. and Knee, B. (2007), 'Effect of cattle marketing method on beef quality and palatability', *Aust. J. Exp. Agric.*, 47(7), 774–81.

Ferreira, G. B., Andrade, C. L., Costa, F., Freitas, M. Q., Silva, T. J. P. and Santos, I. F. (2006), 'Effects of transport time and rest period on the quality of electrically stimulated male cattle carcasses', *Mea Sci.*, 74(3), 459–66.

Finnie, J. W. (1995), 'Neuropathological changes produced by nonpenetrating percussive captive bolt stunning of cattle', *N.Z. Vet. J.*, 43, 183–5.

Finnie, J. W. (1997), 'Traumatic head injury in ruminant livestock', *Aust. Vet. J.*, 75(3), 204–8.

Fordyce, G., Goddard, M. E., Tyler, R., Williams, G. and Toleman, M. A. (1985), 'Temperament and bruising of *Bos indicus* cross cattle', *Aust. J. Exp. Agric.*, 25(2), 283–8.

Franc, C., Bartos, L., Hanys, Z. and Tomes, Z. (1988), 'Pre-slaughter social activity of young bulls relating to the occurrence of dark-cutting beef', *Anim. Prod.*, 46, 153–61.

Fries, R., Schrohe, K., Lotz, F. and Arndt, G. (2012), 'Application of captive bolt to cattle stunning – a survey of stunner placement under practical conditions', *Animal*, 6(7), 1124–8.

Frohli, D. and Blum, J. W. (1988), 'Effects of fasting on blood plasma levels, metabolism and metabolic effects of epinephrine and norepinephrine in steers', *Acta. Endocrinol-COP.*, 118, 254–9.

Gavinelli, A., Ferrara, M. and Simonin, D. (2008), 'Formulating policies for the welfare of animals during long distance transportation', *Vet. Ital.*, 44, 71–86.

Gebresenbet, G., Aradom, S., Bulitta, F. S. and Hjerpe, E. (2011), 'Vibration levels and frequencies on vehicle and animals during transport', *Biosyst. Eng.*, 110(1), 10–19.

Gebresenbet, G., Bosona, T., Feleke, M. and Bobobee, E. Y. H. (2012), 'Improving loading facilities and methods to minimize stress on animals during transport from farm to abattoir', *J. Agric. Sci. Tech. A*, 2(6), 784–99.

Gibson, T. J., Dadios, N. and Gregory, N. G. (2015), 'Effect of neck cut position on time to collapse in halal slaughtered cattle without stunning', *Meat Sci.*, 110, 310–14.

Gibson, T. J., Johnson, C. B., Murrell, J. C., Chambers, J. P., Stafford, K. J. and Mellor, D. J. (2009a), 'Components of electroencephalographic responses to slaughter in halothane - anaesthetised calves: effects of cutting neck tissues compared with major blood vessels', *N.Z. Vet. J.*, 57(2), 84–9.

Gibson, T. J., Johnson, C. B., Murrell, J. C., Hulls, C. M., Mitchinson, S. L., Stafford, K. J., Johnstone, A. C. and Mellor, D. J. (2009b), 'Electroencephalographic responses of halothane – anaesthetised calves to slaughter by ventral-neck incision without prior stunning', *N.Z. Vet. J.*, 57(2), 77–83.

Gibson, T. J., Johnson, C. B., Murrell, J. C., Mitchinson, S. L., Stafford, K. J. and Mellor, D. J. (2009c), 'Electroencephalographic responses to concussive non-penetrative captive-bolt stunning in halothane – anaesthetised calves', *N.Z. Vet. J.*, 57(2), 90–5.

Goldhawk, C., Crowe, T., Janzen, E., González, L. A., Kastelic, J., Pajor, E. and Schwartzkopf-Genswein, K. (2014), 'Trailer microclimate during commercial transportation of feeder cattle and relationship to indicators of cattle welfare', *J. Anim. Sci.*, 92(11), 5155–65.

Goldhawk, C., Janzen, E., Gonzalez, L. A., Crowe, T., Kastelic, J., Kehler, C., Siemens, M., Ominski, K., Pajor, E. and Schwartzkopf-Genswein, K. S. (2015), 'Trailer temperature and humidity during winter transport of cattle in Canada and evaluation of indicators used to assess the welfare of cull beef cows before and after transport', *J. Anim. Sci.*, 93(7), 3639–53.

González, L. A., Schwartzkopf-Genswein, K., Bryan, M., Silasi, R. and Brown, F. (2012a), 'Benchmarking study of industry practices during commercial long haul transport of cattle in Alberta, Canada', *J. Anim. Sci.*, 90(10), 3606–17.

González, L. A., Schwartzkopf-Genswein, K., Bryan, M., Silasi, R. and Brown, F. (2012b), 'Space allowance during commercial long distance transport of cattle in North America', *J. Anim. Sci.*, 90(10), 3618–29.

González, L. A., Schwartzkopf-Genswein, K., Bryan, M., Silasi, R. and Brown, F. (2012c), 'Factors affecting body weight loss during commercial long haul transport of cattle in North America', *J. Anim. Sci.*, 90(10), 3630–9.

González, L. A., Schwartzkopf-Genswein, K., Bryan, M., Silasi, R. and Brown, F. (2012d), 'Relationships between transport conditions and welfare outcomes during commercial long haul transport of cattle in North America', *J. Anim. Sci.*, 90(10), 3640–51.

Gortel, K., Schaefer, A. L., Young, B. A. and Kawamoto, S. C. (1992), 'Effects of transport stress and electrolyte supplementation on body-fluids and weights of bulls', *Can. J. Anim. Sci.*, 72, 547–53.

Gouveia, K. G., Ferreira, P. G., da Costa, J. C. R., Vaz-Pires, P. and da Costa, P. M. (2009), 'Assessment of the efficiency of captive-bolt stunning in cattle and feasibility of associated behavioural signs', *Anim. Welfare*, 18(2), 171–5.
Grandin, T. (1980a), Observations of cattle behavior applied to the design of cattle-handling facilities', *Appl. Anim. Ethol.*, 6, 19–31.
Grandin, T. (1980b), 'Bruises and carcass damage', *Int. J. Study Anim. Probl.*, 1(2), 121–37.
Grandin, T. (1980c), 'The effect of stress on livestock and meat quality prior to and during slaughter', *Int. J. Study Anim. Probl.*, 1, 313–37.
Grandin, T. (1988), 'Double rail restrainer conveyor for livestock handling', *J. Agric. Eng. Res.*, 41(4), 327–38.
Grandin, T. (1990), 'Design of loading facilities and holding pens', *Appl. Anim. Behav. Sci.*, 28(1–2), 187–201.
Grandin, T. (1994), 'Farm animal-welfare during handling, transport, and slaughter', *J. Am. Vet. Med. Ass.*, 204, 372–7.
Grandin, T. (1996), 'Factors that impede animal movement at slaughter plants', *J. Am. Vet. Med. Assoc.*, 209(4), 757–9.
Grandin, T. (1997), 'Assessment of stress during handling and transport', *J. Anim. Sci.*, 75(1), 249–57.
Grandin, T. (1998a), 'The feasibility of using vocalization scoring as an indicator of poor welfare during cattle slaughter', *Appl. Anim. Behav. Sci.*, 56(2–4), 121–8.
Grandin, T. (1998b), 'Objective scoring of animal handling and stunning practices at slaughter plants', *J. Am. Vet. Med. Assoc.*, 212(1), 36–9.
Grandin, T. (2001a), 'Cattle vocalizations are associated with handling and equipment problems at beef slaughter plants', *Appl. Anim. Behav. Sci.*, 71(3), 191–201.
Grandin, T. (2001b), 'Welfare of cattle during slaughter and the prevention of nonambulatory (downer) cattle', *J. Am. Vet. Med. Assoc.*, 219, 1377–82.
Grandin, T. (2002), 'Public veterinary medicine: food safety and handling. Return-to-sensibility problems after penetrating captive bolt stunning of cattle in commercial beef slaughter plants', *J. Am. Vet. Med. Assoc.*, 221(9), 1258–61.
Grandin, T. (2003), 'Transferring results of behavioral research to industry to improve animal welfare on the farm, ranch and the slaughter plant', *Appl. Anim. Behav. Sci.*, 81, 215–28.
Grandin, T. (2005), 'Maintenance of good animal welfare standards in beef slaughter plants by use of auditing programs', *J. Am. Vet. Med. Assoc.*, 226(3), 370–3.
Grandin, T. (2006), 'Progress and challenges in animal handling and slaughter in the U.S', *Appl. Anim. Behav. Sci.*, 100(1–2), 129–39.
Grandin, T. (2008), 'Engineering and design of holding yards, loading ramps and handling facilities for land and sea transport of livestock', *Vet. Ital.*, 44, 235–45.
Grandin, T. (2012), 'Auditing animal welfare and making practical improvements in beef-, pork- and sheep-slaughter plants', *Anim. Welfare*, 21(Suppl. 2), 29–34.
Grandin, T. (2013), 'Making slaughterhouses more humane for cattle, pigs, and sheep', *Annu. Rev. Anim. Biosci.*, 1, 491–512.
Grandin, T. (2014a), 'Animal welfare and society concerns finding the missing link', *Meat Sci.*, 98(3), 461–9.
Grandin, T. (2014b), 'Recommended captive bolt stunning techniques for cattle', http://www.grandin.com/humane/cap.bolt.tips.html (accessed 27 June 2016)
Grandin, T. and Regenstein, J. M. (1994), 'Religious slaughter and animal welfare: a discussion for meat scientists', *Meat Focus Int.*, 3, 115–23.
Grayson, A. L., Shackelford, S. D., King, D. A., McKeith, R. O., Miller, R. K. and Wheeler, T. L. (2016), 'The effects of degree of dark cutting on tenderness and sensory attributes of beef', *J. Anim. Sci.*, 94(6), 2583–91.
Gregory, N. G. (1990), 'Slaughtering methods and equipment', *Vet. Hist.*, 6, 73–84.
Gregory, N. G. (1996), 'Welfare and hygiene during preslaughter handling', *Meat Sci.*, 43, S 35-S 46.
Gregory, N. G. (2008), 'Review: Animal welfare at markets and during transport and slaughter', *Meat Sci.*, 80(10–15), 2–11.

Gregory, N. and Shaw, F. (2000), 'Penetrating captive bolt stunning and exsanguination of cattle in abattoirs', *J. Appl. Anim. Welfare Sci.*, 3(3), 215–30.
Gregory, N. G., Lee, C. J. and Widdicombe, J. P. (2007), 'Depth of concussion in cattle shot by penetrating captive bolt', *Meat Sci.*, 77(4), 499–503.
Gregory, N. G., Jacobson, L. H., Nagle, T. A., Muirhead, R. W. and Leroux, G. J. (2000), 'Effect of preslaughter feeding system on weight loss, gut bacteria, and the physico-chemical properties of digesta in cattle', *N.Z. J. Agric. Res.*, 43(3), 351–61.
Gregory, N. G., von Wenzlawowicz, M. and von Holleben, K.. (2009), 'Blood in the respiratory tract during slaughter with and without stunning in cattle', *Meat Sci.*, 82(1), 13–16.
Gregory, N. G., von Wenzlawowicz, M., Alam, R. M., Anil, H. M., Yeşildere, T. and Silva-Fletcher, A. (2008), 'False aneurysms in carotid arteries of cattle and water buffalo during shechita and halal slaughter', *Meat Sci.*, 79(2), 285–8.
Gregory, N. G., Fielding, H. R., von Wenzlawowicz, M. and von Holleben, K. (2010), 'Time to collapse following slaughter without stunning in cattle', *Meat Sci.*, 85, 66–9.
Gregory, N. G., von Wenzlawowicz, M., von Holleben, K., Fielding, H. R., Gibson, T. J., Mirabito, L. and Kolesar, R. (2012), 'Complications during shechita and halal slaughter without stunning in cattle', *Anim. Welfare*, 21, 81–6.
Gruber, S. L., Tatum, J. D., Engle, T. E., Chapman, P. L., Belk, K. E. and Smith, G. C. (2010), 'Relationships of behavioral and physiological symptoms of preslaughter stress to beef longissimus muscle tenderness', *J. Anim. Sci.*, 88(3), 1148–59.
Grumpelt, B., Hoffer, W., Curie, O., Jones, O., Jones, K., Kimmel, D., McDonald, B., Paterson, R. and Schaefer, A. (2015), 'Short communication: The Pre-transport management of antemortem stress in cattle: Impact on carcass yield', *Can. J. Anim. Sci.*, 95(4), 557–60.
Håkansson, N., Flisberg, P., Algers, B., Jonsson, A., Rönnqvist, M. and Wennergren, U. (2016), 'Improvement of animal welfare by strategic analysis and logistic optimisation of animal slaughter transportation', *Anim. Welfare*, 25(2), 255–63.
Hardstaff, J. L., Häsler, B. and Rushton, J. R. (2015), 'Livestock trade networks for guiding animal health surveillance', *BMC. Vet. Res.*, 11, 82–.
Hemsworth, P. H., Rice, M., Karlen, M. G., Calleja, L., Barnett, J. L., Nash, J. and Coleman, G. J. (2011), 'Human-animal interactions at abattoirs: relationships between handling and animal stress in sheep and cattle', *Appl. Anim. Behav. Sci.*, 135(1), 24–33.
Hoffman, L. C. and Lüxehl, J. (2012), 'Causes of cattle bruising during handling and transport in Namibia', *Meat Sci.*, 92(2), 115–24.
Holleben, K. V., Henke, S., Schmidt, T., Bostelmann, N., Wenzlawowicz, M. V. and Hartung, J. (2003), 'Handling of slaughter cattle in pre and post transport situations including loading and unloading on journeys up to 8 hours in Germany', *Dtsch. Tierarztl. Wochenschr.*, 110(3), 93–9.
Huertas, S. M., Gil, A. D., Piaggio, J. M. and Van Eerdenburg, F. J. C. M. (2010), 'Transportation of beef cattle to slaughterhouses and how this relates to animal welfare and carcase bruising in an extensive production system', *Anim .Welfare*, 19(3), 281–5.
Hultgren, J., Wiberg, S., Berg, C., Cvek, K. and Kolstrup, C. L. (2014), 'Cattle behaviours and stockperson actions related to impaired animal welfare at Swedish slaughter plants', *Appl. Anim. Behav. Sci.*, 152, 23–37.
Iglesias, R. M. and East, I. J. (2015), 'Cattle movement patterns in Australia: an analysis of the NLIS database 2008–2012', *Aust. Vet. J.*, 93(11), 394–403.
Jarvis, A. M., Selkirk, L. and Cockram, M. S. (1995), 'The influence of source, sex class and pre-slaughter handling on the bruising of cattle at two slaughterhouses', *Livest. Prod. Sci.*, 43(3), 215–24.
Jarvis, A. M., Harrington, D. W. J. and Cockram, M. S. (1996a), 'Effect of source and lairage on some behavioural and biochemical measurements of feed restriction and dehydration in cattle at a slaughterhouse', *Appl. Anim. Behav. Sci.*, 50(1), 83–94.
Jarvis, A. M., Messer, C. D. A. and Cockram, M. S. (1996b), 'Handling, bruising and dehydration of cattle at the time of slaughter', *Anim. Welfare*, 5(3), 259–70.

Jeleníková, J., Pipek, P. and Staruch, L. (2008), 'The influence of ante-mortem treatment on relationship between pH and tenderness of beef', *Meat Sci.*, 80(3), 870–4.

Jones, S. D. M., Schaefer, A. L., Robertson, W. M. and Vincent, B. C. (1990), 'The effects of withholding feed and water on carcass shrinkage and meat quality in beef cattle', *Meat Sci.*, 28, 131–9.

Kenny, F. J. and Tarrant, P. V. (1987a), 'The reaction of young bulls to short-haul road transport', *Appl. Anim. Behav. Sci.*, 17(3), 209–27.

Kenny, F. J. and Tarrant, P. V. (1987b), 'The physiological and behavioural responses of crossbred Friesian steers to short-haul transport by road', *Livest. Prod. Sci.*, 17, 63–75.

King, D. A., Schuehle Pfeiffer, C. E., Randel, R. D., Welsh, J., T. H., Oliphint, R. A., Baird, B. E., Curley, J., K. O., Vann, R. C., Hale, D. S. and Savell, J. W. (2006), 'Influence of animal temperament and stress responsiveness on the carcass quality and beef tenderness of feedlot cattle', *Meat Sci.*, 74, 546–56.

Klinglmair, K., Stevens, K. B. and Gregory, N. G. (2011), 'Luminance and glare in indoor cattle-handling facilities', *Anim. Welfare*, 20(2), 263–9.

Knights, M. and Smith, G. W. (2007), 'Decreased ACTH secretion during prolonged transportation stress is associated with reduced pituitary responsiveness to tropic hormone stimulation in cattle', *Domest. Anim. Endocrinol.*, 33, 442–50.

Knowles, T. G. (1999), 'A review of the road transport of cattle', *Vet. Rec.*, 144(8), 197–201.

Knowles, T. G., Warriss, P. D., Brown, S. N. and Edwards, J. E. (1999), 'Effects on cattle of transportation by road for up to 31 hours', *Vet. Rec.*, 145(20), 575–82.

Lahucky, R., Palanska, O., Mojto, J., Zaujec, K. and Huba, J. (1998), 'Effect of preslaughter handling on muscle glycogen level and selected meat quality traits in beef', *Meat Sci.*, 50(3), 389–93.

Lewis, P. K., Brown, C. J. and Heck, M. C. (1962), 'Effect of preslaughter treatments on certain chemical and physical characteristics of certain beef muscles', *J. Anim. Sci.*, 21, 433–8.

Liotta, L., Costa, L. N., Chiofalo, B., Ravarotto, L. and Chiofalo, V. (2007), 'Effect of lairage duration on some blood constituents and beef quality in bulls after long journey', *Ital. J. Anim. Sci.*, 6(4), 375–84.

Ljungberg, D., Gebresenbet, G. and Aradom, S. (2007), 'Logistics chain of animal transport and abattoir operations', *Biosyst. Eng.*, 96, 267–77.

Lowe, T. E., Devine, C. E., Wells, R. W. and Lynch, L. L. (2004), 'The relationship between postmortem urinary catecholamines, meat ultimate pH, and shear force in bulls and cows', *Meat Sci.*, 67, 251–60.

Lyles, J. L. and Calvo-Lorenzo, M. (2014), 'Practical developments in managing animal welfare in beef cattle: What does the future hold?', *J. Anim. Sci.*, (12), 5334.

Mach, N., Bach, A., Velarde, A. and Devant, M. (2008), 'Association between animal, transportation, slaughterhouse practices, and meat pH in beef', *Meat Sci.*, 78, 232–8.

Malena, M., Voslarova, E., Tomanova, P., Lepkova, R., Bedanova, I. and Vecerek, V. (2006), 'Influence of travel distance and the season upon transport-induced mortality in fattened cattle', *Acta Vet. Brno.*, 75, 619–24.

Malena, M., Voslářová, E., Kozák, A., Bělobrádek, P., Bedáňová, I., Steinhauser, L. and Večerek, V. (2007), 'Comparison of mortality rates in different categories of pigs and cattle during transport for slaughter', *Acta Vet. Brno.*, 76(SUPPL. 8), S109–S116.

Marenčic´, D., Ivankovic, A., Pintic, V., Kelava, N. and Jakopovic, T. (2012), 'Effect of the transport duration time and season on some physicochemical properties of beef meat', *Arch. Tierz.*, 55(2), 123–31.

Maria, G. A., Villarroel, M., Sanudo, C., Olleta, J. L. and Gebresenbet, G. (2003), 'Effect of transport time and ageing on aspects of beef quality', *Meat Sci.*, 65, 1335–40.

Marshall, B. L. (1977), 'Bruising in cattle presented for slaughter', *N.Z. Vet. J.*, 25(4), 83–6.

McCausland, I. P. and Millar, H. W. C. (1982), 'Time of occurrence of bruises in slaughtered cattle', *Aust. Vet. J.*, 58, 253–5.

McNally, P. W. and Warriss, P. D. (1996), 'Recent bruising in cattle at abattoirs', *Vet. Rec.*, 138(6), 126–8.

McNally, P. W. and Warriss, P. D. (1997), 'Prevalence of carcase bruising and stick-marking in cattle bought from different auction markets', *Vet .Rec.*, 140(9), 231–2.

McVeigh, J. M. and Tarrant, P. V. (1983), 'Effect of propranolol on muscle glycogen metabolism during social regrouping of young bulls', *J. Anim. Sci.*, 56(1), 71–80.

McVeigh, J. M., Tarrant, P. V. and Harrington, M. G. (1982), 'Behavioral stress and skeletal muscle glycogen metabolism in young bulls', *J. Anim. Sci.*, 54(4), 790–5.

Mellor, D. J. and Littin, K. E. (2004), 'Using science to support ethical decisions promoting humane livestock slaughter and vertebrate pest control' *Anim. Welfare*, 13, (Suppl. 1), 127–32.

Minka, N. S. and Ayo, J. O. (2007), 'Effects of loading behaviour and road transport stress on traumatic injuries in cattle transported by road during the hot-dry season', *Livest. Sci.*, 107(1), 91–5.

Miranda-de la Lama, G. C., Leyva, I. G., Barreras-Serrano, A., Pérez-Linares, C., Sánchez-López, E., María, G. A. and Figueroa-Saavedra, F. (2012), 'Assessment of cattle welfare at a commercial slaughter plant in the northwest of Mexico', *Trop. Anim. Health Prod.*, 44(3), 497–504.

Mitchell, G., Hattingh, J. and Ganhao, M. (1988), 'Stress in cattle assessed after handling, after transport and after slaughter', *Vet. Rec.*, 123(8), 201–5.

Moberg, G. P. (2000), 'Biological response to stress: Implications for animal welfare', in G. P. Moberg and J. A. Mench (Eds.), *The Biology of Animal Stress. Basic Principles and Implications for Animal Welfare*, CABI Publishing, Wallingford, pp. 1–21.

Mohan Raj, A. B., Moss, B. W., Rice, D. A., Kilpatrick, D. J., McCaughey, W. J. and McLauchlan, W. (1992), 'Effect of mixing male sex types of cattle on their meat quality and stress-related parameters', *Meat Sci.*, 32(4), 367–86.

Mpamhanga, C. J. and Wotton, S. B. (2015), 'The effects of pre-slaughter restraint (for the purpose of cattle identification) on post-slaughter responses and carcass quality following the electrical stun/killing of cattle in a Jarvis Beef stunner', *Meat Sci.*, 107, 104–8.

Muchenje, V., Dzama, K., Chimonyo, M., Strydom, P. E. and Raats, J. G. (2009), 'Relationship between pre-slaughter stress responsiveness and beef quality in three cattle breeds', *Meat Sci.*, 81, 653–7.

Nakyinsige, K., Che Man, Y. B., Aghwan, Z. A., Zulkifli, I., Goh, Y. M., Abu Bakar, F., Al-Kahtani, H. A. and Sazili, A. Q. (2013), 'Stunning and animal welfare from Islamic and scientific perspectives', *Meat Sci.*, 95(2), 352–61.

Nanni Costa, L., Lo Fiego, D. P., Cassanelli, M. G., Tassone, F. and Russo, V. (2003), 'Effect of journey time and environmental condition on bull behaviour and beef quality during road transport in Northern Italy', *Dtsch. Tierarztl. Wochenschr.*, 110(3), 107–10.

Neves, J. E. G., Paranhos da Costa, M. J. R., Roca, R. O., Faucitano, L. and Gregory, N. G. (2016), 'A note comparing the welfare of Zebu cattle following three stunning-slaughter methods.', *Meat Sci.*, 117, 41–3.

Nicholson, J. D. W., Nicholson, K. L., Frenzel, L. L., Maddock, R. J., Delmore, R. J., J., Lawrence, T. E., Henning, W. R., Pringle, T. D., Johnson, D. D., Paschal, J. C., Gill, R. J., Cleere, J. J., Carpenter, B. B., Machen, R. V., Banta, J. P., Hale, D. S., Griffin, D. B. and Savell, J. W. (2013), 'Survey of transportation procedures, management practices, and health assessment related to quality, quantity, and value for market beef and dairy cows and bulls', *J. Anim. Sci.*, 91(10), 5026–36.

Novelli, S., Sechi, P., Mattei, S., Iulietto, M. F. and Cenci, G. B. T. (2016), 'Report on religious slaughter practices in Italy', *Vet. Ital.*, 52(1), 5–11.

Önenç, A. and Kaya, A. (2004), 'The effects of electrical stunning and percussive captive bolt stunning on meat quality of cattle processed by Turkish slaughter procedures', *Meat Sci.*, 66(4), 809–15.

Parker, A. J., Hamlin, G. P., Coleman, C. J. and Fitzpatrick, L. A. (2003), 'Quantitative analysis of acid-base balance in Bos indicus steers subjected to transportation of long duration', *J. Anim. Sci.*, 81(6), 1434–9.

Peña, F., Avilés, C., Domenech, V., González, A., Martínez, A. and Molina, A. (2014), 'Effects of stress by unfamiliar sounds on carcass and meat traits in bulls from three continental beef cattle breeds at different ageing times', *Meat Sci.*, 98(4), 718–25.

Petherick, J. C. (2007), 'Spatial requirements of animals: Allometry and beyond', *J. Vet. Behav.*, 2(6), 197–204.

Petherick, J. C. and Phillips, C. J. C. (2009), 'Space allowances for confined livestock and their determination from allometric principles', *Appl. Anim. Behav. Sci.*, 117(1–2), 1–12.

Pettiford, S. G., Ferguson, D. M., Lea, J. M., Lee, C., Paull, D. R., Reed, M. T., Hinch, G. N. and Fisher, A. D. (2008) 'Effect of loading practices and 6-hour road transport on the physiological responses of yearling cattle' *Aust. J. Exp. Agric.*, 48, 1028–33.

Pointon, A., Kiermeier, A. and Fegan, N. (2012), 'Review of the impact of pre-slaughter feed curfews of cattle, sheep and goats on food safety and carcase hygiene in Australia', *Food Control*, 26(2), 313–21.

Prendergast, D. M., Sheridan, J. J., Daly, D. J., McDowell, D. A. and Blair, I. S. (2003), 'Dissemination of central nervous system tissue from the brain and spinal cord of cattle after captive bolt stunning and carcass splitting', *Meat Sci.*, 65(4), 1201–9.

Probst, J. K., Neff, A. S., Leiber, F., Kreuzer, M. and Hillmann, E. (2012), 'Gentle touching in early life reduces avoidance distance and slaughter stress in beef cattle', *Appl. Anim. Behav. Sci.*, 139(1–2), 42–9.

Probst, J. K., Hillmann, E., Leiber, F., Kreuzer, M. and Neff, A. S. (2013), 'Influence of gentle touching applied few weeks before slaughter on avoidance distance and slaughter stress in finishing cattle', *Appl. Anim. Behav. Sci.*, 144(1), 14–21.

Probst, J. K., Neff, A. S., Hillmann, E., Kreuzer, M., Koch-Mathis, M. and Leiber, F. (2014), 'Relationship between stress-related exsanguination blood variables, vocalisation, and stressors imposed on cattle between lairage and stunning box under conventional abattoir conditions', *Livest. Sci.*, 164, 154–8.

Ramsay, W. R., Meischke, H. R. and Anderson, B. (1976), 'The effect of tipping of horns and interruption of journey on bruising in cattle', *Aust. Vet. J.*, 52(6), 285–6.

Randall, J. M. (1993), 'Environmental parameters necessary to define comfort for pigs, cattle and sheep in livestock transporters', *Anim. Prod.*, 57, 299–307.

Randall, J. M. and Patel, R. (1994), 'Thermally induced ventilation of livestock transporters', *J. Agric. Eng. Res.*, 57, 99–107.

Rezac, D. J., Thomson, D. U., Siemens, M. G., Prouty, F. L., Reinhardt, C. D. and Bartle, S. J. (2014), 'A survey of gross pathologic conditions in cull cows at slaughter in the Great Lakes region of the United States', *J. Dairy Sci.*, 97, 4227–35.

Romero, M. H., Uribe-Velasquez, L. F., Sanchez, J. A. and Miranda-de la Lama, G. C. (2013), 'Risk factors influencing bruising and high muscle pH in Colombian cattle carcasses due to transport and pre-slaughter operations', *Meat Sci.*, 95(2), 256–63.

Rosen, S. D. (2004), 'Physiological insights into Shechita', *Vet. Rec.*, 154(24), 759–65.

Ross, M., Widowski, T. M. and Haley, D. B. (2016), 'The effects of feeding space on the behavioural responses of cattle during rest periods offered as part of long-distance transportation', *Anim. Welfare*, 25(2), 217–25.

Rule, D. C., Beitz, D. C., de Boer, G., Lyle, R. R., Trenkle, A. H. and Young, J. W. (1985), 'Changes in hormone and metabolite concentrations in plasma of steers during a prolonged fast.', *J. Anim. Sci.*, 61(4), 868–75.

Sazili, A. Q., Norbaiyah, B., Zulkifli, I., Goh, Y. M., Lotfi, M. and Small, A. H. (2013), 'Quality assessment of longissimus and semitendinosus muscles from beef cattle subjected to non-penetrative and penetrative percussive stunning methods', *Asian-Australas. J. Anim. Sci.*, 26(5), 723–31.

Scanga, J. A., Belk, K. E., Tatum, J. D., Grandin, T. and Smith, G. C. (1998), 'Factors contributing to the incidence of dark cutting beef', *J. Anim. Sci.*, 76(8), 2040–7.

Schrama, J. W., Van Der Hel, W., Gorssen, J., Henken, A. M., Verstegen, M. W. A. and Noordhuizen, J. P. T. M. (1996), 'Required thermal thresholds during transport of animals', *Vet. Q.*, 18(3), 90–5.

Schwenk, B. K., Lechner, I., Ross, S. G., Gascho, D., Kneubuehl, B. P., Glardon, M. and Stoffel, M. H. (2016), 'Magnetic resonance imaging and computer tomography of brain lesions in water buffaloes and cattle stunned with handguns or captive bolts', *Meat Sci.*, 113, 35–40.

Shaw, F. D. and Tume, R. K. (1992), 'The assessment of pre-slaughter and slaughter treatments of livestock by measurement of plasma constituents-A review of recent work', *Meat. Sci.*, 32(3), 311–29.

Shaw, F. D., Baxter, R. I. and Ramsay, W. R. (1976), 'The contribution of horned cattle to carcase bruising', *Vet. Rec.*, 98(13), 255–7.

Shaw, N. A. (2002), 'The neurophysiology of concussion', *Prog. Neurobiol.*, 67, 281–344.

Silanikove, N. (1994), 'The struggle to maintain hydration and osmoregulation in animals experiencing severe dehydration and rapid rehydration: the story of ruminants', *Exp. Physiol.*, 79, 281–300.

Silanikove, N. (2000), 'Effects of heat stress on the welfare of extensively managed domestic ruminants', *Livest. Prod. Sci.*, 67(1–2), 1–18.

Smith, G. C., Tatum, J. D., Belk, K. E., Scanga, J. A., Grandin, T. and Sofos, J. N. (2005), 'Traceability from a US perspective', *Meat Sci.*, 71(1), 174–93.

Smith, R. A., Thomson, D. U. and Lee, T. L. (2015), 'Beef quality assurance in feedlots', *Vet. Clin. North. Am. Food Anim. Pract.* 31(2), 269–81.

Strappini, A. C., Metz, J. H. M., Gallo, C. B. and Kemp, B. (2009), 'Origin and assessment of bruises in beef cattle at slaughter', *Animal,* 3(5), 728–36.

Strappini, A. C., Frankena, K., Metz, J. H. M., Gallo, B. and Kemp, B. (2010), 'Prevalence and risk factors for bruises in Chilean bovine carcasses', *Meat Sci.*, 86(3), 859–64.

Strappini, A. C., Metz, J. H. M., Gallo, C., Frankena, K., Vargas, R., Freslon, I. and Kemp, B. (2013), 'Bruises in culled cows: when, where and how are they inflicted?', *Animal,* 7(3), 485–91.

Stull, C. L., Payne, M. A., Berry, S. L. and Reynolds, J. P. (2007), 'A review of the causes, prevention, and welfare of nonambulatory cattle', *J. Am. Vet. Med. Assoc.*, 231(2), 227–34.

Tarrant, P. V. (1990), 'Transportation of cattle by road', *Appl. Anim. Behav. Sci.*, 28(1–2), 153–70.

Tarrant, P. V., Kenny, F. J. and Harrington, D. (1988), 'The effect of stocking density during 4 hour transport to slaughter on behaviour, blood constituents and carcass bruising in Friesian steers', *Meat Sci.*, 24, 209–22.

Tarrant, P. V., Kenny, F. J., Harrington, D. and Murphy, M. (1992), 'Long distance transportation of steers to slaughter: effect of stocking density on physiology, behaviour and carcass quality', *Livest. Prod. Sci.*, 30(3), 223–38.

Teke, B., Akdag, F., Ekiz, B. and Ugurlu, M. (2014), 'Effects of different lairage times after long distance transportation on carcass and meat quality characteristics of Hungarian Simmental bulls', *Meat Sci.*, 96(1), 224–9.

Terlouw, E. M. C., Arnould, C., Auperin, B., Berri, C., Le Bihan-Duval, E., Deiss, V., Lefevre, F., Lensink, B. J. and Mounier, L. (2008), 'Pre-slaughter conditions, animal stress and welfare: current status and possible future research', *Animal,* 2(10), 1501–17.

Terlouw, E. M. C., Bourguet, C. and Deiss, V. (2012), 'Stress at slaughter in cattle: role of reactivity profile and environmental factors', *Anim. Welfare,* 21(Suppl. 2), 43–9.

Terlouw, E. M. C., Bourguet, C., Deiss, V. and Mallet, C. (2015), 'Origins of movements following stunning and during bleeding in cattle', *Meat Sci.*, 110, 135–44.

Thornber, P. M. and Adams, D. B. (2008), 'How scientific evidence is used in Australia to inform public policy on the long distance transportation of animals', *Vet. Ital.*, 44, 101–11.

Tucker, C. B., Coetzee, J. F., Stookey, J. M., Thomson, D. U., Grandin, T. and Schwartzkopf-Genswein, K. (2015), 'Beef cattle welfare in the USA: identification of priorities for future research', *Anim. Health Res. Rev.*, 16(2), 107–24.

Tume, R. K. and Shaw, F. D. (1992), 'Beta-endorphin and cortisol concentrations in plasma of blood samples collected during exsanguination of cattle', *Meat Sci.*, 31(2), 211–17.

Večerek, V., Malena Jr., M., Malena, M., Voslářová, E. and Bedáňová, I. (2006), 'Mortality in dairy cows transported to slaughter as affected by travel distance and seasonality', *Acta. Vet. Brno.*, 75(3), 449–54.

Veissier, I. and Boissy, A. (2007), 'Stress and welfare: two complementary concepts that are intrinsically related to the animal's point of view', *Physiol. Beh.*, 92, 429–33.

Velarde, A., Rodriguez, P., Dalmau, Fuentes, C., Llonch, P. von Holleben, K., Anil, M. H., Lambooij, J. B., Pleiter, H. Yesildere, T., Cenci-Goga, B. T. (2014), 'Religious slaughter: Evaluation of current practices in selected countries', *Meat Sci*, 96(9), 278–87.

Vial, F., Schaerrer, S. and Reist, M. (2015), 'Risk factors for whole carcass condemnations in the Swiss slaughter cattle population', *PLoS One*, 10(4), e0122717.

Villarroel, M., María, G. A., Sañudo, C., Olleta, J. L. and Gebresenbet, G. (2003), 'Effect of transport time on sensorial aspects of beef meat quality', *Meat Sci.*, 63, 353–7.

Voisinet, B. D., Grandin, T., O'Connor, S. F., Tatum, J. D. and Deesing, M. J. (1997), 'Bos indicus-cross feedlot cattle with excitable temperaments have tougher meat and a higher incidence of borderline dark cutters', *Meat Sci.*, 46(4), 367–77.

von Wenzlawowicz, M., von Holleben, K. and Eser, E. (2012), 'Identifying reasons for stun failures in slaughterhouses for cattle and pigs: a field study', *Anim. Welfare*, 21(Suppl. 2), 51–60.

Warner, R. D., Ferguson, D. M., Cottrell, J. J. and Knee, B. W. (2007), 'Acute stress induced by the preslaughter use of electric prodders causes tougher beef meat', *Aust. J. Exp. Agric.*, 47(7), 782–8.

Warner, R. D., Walker, P. J., Eldridge, G. A. and Barnett, J. L. (1998), 'Effects of marketing procedure and liveweight change prior to slaughter on beef carcass and meat quality', *Anim. Prod. Aust.* 22, 165–8.

Warren, L. A., Mandell, I. B. and Bateman, K. G. (2010a), 'An audit of transport conditions and arrival status of slaughter cattle shipped by road at an Ontario processor', *Can. J. Anim. Sci.*, 90(2), 159–67.

Warren, L. A., Mandell, I. B. and Bateman, K. G. (2010b), 'Road transport conditions of slaughter cattle: effects on the prevalence of dark, firm and dry beef', *Can. J. Anim. Sci.*, 90(4), 471–82.

Warriss, P. D. (1990), 'The handling of cattle pre-slaughter and its effects on carcass and meat quality', *Appl. Anim. Behav. Sci.*, 28(1–2), 171–186.

Warriss, P. D., Kestin, S. C., Brown, S. N. and Wilkins, L. J. (1984), 'The time required for recovery from mixing stress in young bulls and the prevention of dark cutting beef', *Meat Sci.*, 10(1), 53–68.

Warriss, P. D., Brown, S. N., Knowles, T. G., Kestin, S. C., Edwards, J. E., Dolan, S. K. and Phillips, A. J. (1995), 'Effects on cattle of transport by road for up to 15 hours', *Vet. Rec.*, 136(13), 319–23.

Chapter 8

# The effects of carcass chilling and electrical stimulation on visual beef quality and palatability

*Phillip E. Strydom, Agricultural Research Council and University of Stellenbosch, South Africa*



## 1 Introduction: beef quality and consumers

The marketing of any product, including beef, consists of a two-step process that first involves a 'customer' and then a 'consumer' (Smith, 2008). For beef, the 'customer' or person with the intent to purchase will be influenced by appearance characteristics such as fat and meat colour; amount of drip; fat, bone and meat proportions; the type of packaging and relation of all these to price (Grunert, 2004; Ngapo, 2006; Cho, 2007). At home, the quality of the meat will be tested by the 'consumer' for palatability characteristics. Most consumer surveys suggest that eating quality (defined by most consumers simply as 'taste') is the primary driver of purchase decisions (Shook, 2008). Bindon (2001) confirmed that variability in palatability, particularly tenderness, was the main reason for the decline in beef consumption in Australia during the early 1990s. Likewise, variation in palatability is also the primary driver for the quinquennial National Beef Quality Audit conducted in the United States since 1991 (Morgan, 1991; Howard, 2013). The palatability of meat is described as the simultaneous experience of tenderness, juiciness and meat flavour (Koohmaraie, 2002). The Australian MSA grading

http://dx.doi.org/10.19103/AS.2016.0009.08

system that grades various cuts of the beef carcass according to expected eating quality has indicated 'tenderness' as the most important attribute, closely followed by 'overall liking', while 'flavour' and 'juiciness' contributed less to the MQ4 score modelled for this system (Watson, 2008a,b).

Variation in 'experience' characteristics of beef depends on a complex of interactions between pre-harvest, slaughter, and post-slaughter factors and processes that have been thoroughly covered in the literature (Koohmaraie, 1996; Ferguson, 2001; Thompson, 2002; Watson, 2008b).

# 2 Chilling and rigor mortis

Normal commercial slaughter practices dictate that carcasses must be rapidly chilled as soon as possible after dressing so that microbiological shelf life is maintained and food safety risks are mitigated (Bowater, 1997). Chilling parameters are important for financial and product quality considerations. According to Ouellette (1980), chilling is the most expensive component of carcass processing. In addition, cooling rate and the cooling parameters will determine carcass weight loss, but these parameters will also have an impact on various biochemical processes that could influence visual appearance (Ledward, 1985; Rosenvold, 2011) and eating quality of beef (Marsh, 1966; Locker, 1963; Thompson, 2005). Attempts have been made over decades to adjust chilling parameters and other inputs during and after slaughter to improve on the economics of carcass chilling (cost and carcass weight loss), while maintaining or improving the quality of the final product (Locker, 1963; Watt, 1974; Rødbotten, 2001; Jacob, 2014).

Rigor mortis is the term used for individual muscle fibres becoming depleted of ATP (Bendal, 1969; Hwang, 2003). Jeacocke (1984) demonstrated that each fibre has its own course of going through the rigor mortis process, depending on initial glycogen. Due to irreversible cross bridges forming between actin and myosin, the bulk of the muscle will increase in stiffness as more fibres go into full rigor mortis. The stiffness becomes significant when pH in muscle reaches approximately 6.0, indicating that most of the muscle fibres have reached rigor mortis.

The effects of chilling rate on muscle tenderness have been studied since the 1960s (Locker, 1963; Marsh, 1966). If the bulk of muscle fibres have reached rigor mortis, it will prevent the remaining fibres to shorten excessively when exposed to low temperatures. However, in cases where low temperatures are reached before the bulk of muscle fibres has reached stiffness (high pH), extreme muscle shortening will occur. Therefore, Locker (1963) emphasized that the onset of the rigor mortis process should not occur at muscle temperatures of less than 14–19°C. Hannula (2007) narrowed the ratio down to a *longissimus dorsi* (LD) muscle pH below 5.7 before or when the temperature reaches 7°C and that the combination of temperature and pH at the onset of rigor mortis is the decisive factor of the degree of shortening that will take place. Taylor (1987) added that delayed ageing due to lower access of proteolytic enzymes to structural proteins (as a result of shortening) and their activity as influenced by slower pH decline and lower temperatures will also affect tenderization of muscle under low rigor temperatures, and preferred the term 'cold toughening' over that of 'cold shortening'.

# 3 Electrical stimulation, visual beef quality and palatability

## 3.1 Mechanisms involved in prevention of cold toughening and improvement of tenderness

Electrical stimulation (ES) was originally developed for commercial purposes in New Zealand in the late 1970s to prevent cold toughening in lambs that were exposed to an extreme chilling regimen directly after slaughter to be frozen for export purposes (Simmons, 2008). Therefore, the initial purpose of ES was to accelerate post-mortem glycolysis to prevent excessive muscle shortening when muscle entered rigor mortis (Swatland, 1981) and was mostly intended in slaughtering of sheep. However, Simmons (2008) reported that extreme chilling would be necessary to induce cold shortening in normal, well-finished, intact beef carcasses. Earlier (Cross, 1979) and recent reviews (Hwang, 2003) revealed that ES may affect tenderness in other ways as well. In addition, ES will also impact on other quality parameters such as colour and moisture characteristics as a result of its effect on glycolysis and the rigor mortis process.

Table 1 summarizes the reports that measured the effects of ES on different indicators of proteolysis (Hwang, 2003). While the results of the studies do not always agree, Hwang (2003) concluded that the increase in rate of glycolysis following ES that protects muscle fibres from cold shortening also initiates early activation of the calpain enzyme system through an increased release of calcium ions, which increases the rate of cytoskeletal protein degradation and contributes to tenderness. Hwang (2003) contributes the ambiguity among studies to the variation in parameters and study designs used by different authors. These may include the specific times during and/or after rigor that different indicators (enzyme activities) or outcomes (tenderness, colour and shelf life) were measured and different environments in which the muscles were studied, for example, excised muscles kept at constant temperatures or exposed to normal chilling pathways (intact carcass). Devine (1995) added that proteolytic enzymes may be active throughout the rigor process, although no significant tenderization is evident before rigor mortis is completed. Because individual fibres go into rigor at individual rates depending on their glycogen levels, ageing (tenderization) is delayed until all fibres reach rigor mortis. In addition, it is only when shortening of fibres during rigor is countered by the early proteolysis at elevated temperatures that peak force will decline (shear force tenderness).

When considering the mechanism of ES on tenderization (when cold toughening conditions do not prevail), one can assume that there will be no effects on tenderness due to ES, if the rate of pH decline is not enhanced, as was demonstrated by Hopkins (2001). Likewise, if pH decline is enhanced by ES and the process of rigor mortis is not entered at extreme temperatures (as discussed later), the rate of tenderization will always increase.

ES also contributes to tenderness by physical disruption. Hwang (2003) cited various studies describing the presence of contracture bands in stimulated muscle fibres that not only contribute to physical disruption (fractures in the fibres) but also release more $Ca^{2+}$ ions for proteolysis and increase substrate area for proteolytic actions.

## 3.2 Mechanisms influencing visual display of muscle

Initial appearance of meat is normally improved by ES but, depending on various interacting factors, this initial positive effect may deteriorate or remain. Earlier studies

**Table 1** Summary of reports that have measured the effect of electrical stimulation on different indicators of proteolysis

| Indicator of proteolysis | Type of stimulation/conditions | Outcomes[a] | Sources/species |
|---|---|---|---|
| Myofibrillar protein solubility | 280 V, 60 Hz for 12, 5 s pulses, 5 s apart and then 24 times at 420 V, within 45 min of death. | No effect of stimulation on protein solubility (LL) when measured at 44 h, irrespective of the chilling temperature. | Gilbert et al. (1984) – sheep |
| Degradation troponin-T and myofibrillar fragmentation index (MFI) changes | 300 V, 60 Hz for 80, 1.5 s pulses, 0.5 s apart giving a total of 120 s, within 30 min of death. Breed of cattle unspecified. | MFI was increased by stimulation (LL, SM, TB), but not significantly and there was no effect on the degradation rate of troponin-T. | Sonaiya et al. (1982) – beef |
| Degradation of α-actinin and troponin-T | 500 V, 60 Hz for 20, 2 s pulses, 1 s apart giving a total of 40 s, within 45 min of death. Breed of cattle unspecified. | Less α-actinin and troponin-T was found in stimulated muscle (LL), but there was an interaction with chilling temperature. | Salm et al. (1983) – beef |
| Myofibrillar protein solubility and degradation products of troponin-T | 550 V, 60 Hz for 80, 1 s pulses, 0.5 s apart for a total of 2 min, within 30 min of death. Breed of cattle unspecified. | Increased protein solubility (LL) when measured at 3 days post-mortem. Reported that 30 kDa and 32 kDa subunits appeared earlier in stimulated muscle (4-h post-mortem) | Ducastaing et al. (1985) – beef |
| Degradation troponin-T | 45 V, for 40, 2 s pulses, 2 s apart giving a total of 40 s, straight after death. Used Holstein veal calves. | Stimulation decreased the rate of degradation of the protein (LL). | Pommier et al. (1987) – beef |
| Myofibrillar protein solubility | 115 V, 60 Hz for 5, 12 s pulses, 3 s apart giving a total of 60 s, within stimulation 1 h post-mortem. Used Charolais crossbred cattle. | No effect of stimulation on protein solubility (LL) when measured at 7 days. | Bruce et al. (1990) – beef |
| Degradation troponin-T and myofibrillar fragmentation index (MFI) changes | 45 V, for 40, 2 s pulses, 2 s apart giving a total of 40 s, straight after death. Used Holstein steers. | No effect of stimulation on the rate of degradation of troponin-T, but a product of degradation did appear earlier in stimulated muscle (LL). MFI was higher for slow chilled, stimulated muscle than for control muscle slow chilled at 24 h. | Pommier (1992) – beef |
| Degradation of titin, nebulin, filamin and troponin-T | 600 V, 50 Hz for 120 s, within 1 h of death. Used Belgian white-red cattle. | After 1-day post-mortem, less troponin-T was present in stimulated muscle (LL) and by the 6th day there was less titin. | Uytterhaegen et al. (1992) – beef |

| | | | |
|---|---|---|---|
| Degradation troponin-T and myofibrillar fragmentation index (MFI) changes | 85 V, 14 Hz for 60 s within 30 min of death. Used Friesian/Holstein veal calves. | Stimulation had no effect on the appearance rate of the 30-kDa fragment in three different muscles (LL; SM; PS). MFI increased faster in stimulated muscle (LL), but had no effect on MFI in the SM. | Geesink et al. (1993) – beef |
| Degradation troponin-T and myofibrillar fragmentation index (MFI) changes | 85 V, 14 Hz for either 8 or 64 s within 5 min of death. Breed of cattle unspecified. | Stimulation for 64 s increased the appearance rate of the 30-kDa fragment in (LL), but by 21 days there was no difference in the amount of the fragment. Day 1 MFIs were higher in stimulated muscle (64 s), but there was no difference to stimulated muscle (8 s) or control at 7, 14 or 21 days. | Geesink et al. (1994) – beef |
| Degradation of tintin, nebulin, desmin and troponin-T | 200 V, 20 Hz for 15 s, 3 times with 30 s breaks within 1 h of death. Used Angus x Jersey cattle. | Titin, nebulin and troponin-T were degraded faster in *some* stimulated muscle (LL). | Ho et al. (1996) – beef |
| Degradation of titin, nebulin, desmin and troponin-T | 200 V, 20 Hz for 15 or 20 s, 3 times with 30 s breaks between within 1 h of death. Used Brahman x Simmental cattle. | Stimulation did not increase the degradation rate of the proteins (LL). | Ho et al. (1997) – beef |
| Myofibrillar protein solubility | 85 V, 14 Hz for 15 s straight after death. Used Friesian/Holstein cattle. | There was no effect of simulation on protein solubility in either the LL or SM muscle. | Den Hertog-Meischke et al. (1997) – beef |
| Degradation of titin and nebulin | 50 V, 60 Hz for 20 s within 3 min of death. Used Hanwoo bulls. | Degradation rate increased with stimulation and conditioning at higher temperatures (LL). | Rhee et al. (200) – beef |
| Degradation product of troponin-T and of a protein designated M1, changes in MFI and myofibrillar protein solubility | 45 V, with 36 pulses per s, a pulse width of 25 ms for 40 s. Used crossbred lambs. | Stimulation increased the degradation rate of protein M1, but had no effect on the appearance rate of the 30-kDA subunit. MFI and protein solubility increased faster in stimulated muscle (LL). | Hopkins and Thompson (2001a,b) – sheep |
| Degradation products of titin, desmin and troponin-T | 1130 V, 14.3 Hz for 90 s, within 30 min of death. Note also that spinal discharge was applied to the carcasses. Used Coopworth lambs. | Stimulation did not increase the degradation rate of the proteins (LL). | Geesink et al. (2000) – sheep |

[a] Indicates the muscle used where LL = longissimus, ST = semitendinosus, PS = psoas major, SM = semimembranosus, TB = triceps brachii.
Source: From Wang (2003).

on effects of ES on muscle appearance during USDA grading were summarized by Smith (1985). Significant improvement of marbling, lean firmness, lean colour and lean maturity scores were reported, especially when grading was done within the first 24 hours post-mortem. According to Renerre (1987), lower pH at a higher temperature due to ES increases the oxygen consumption rate (OCR) causing a thin layer of oxymyoglobin that leads to a more youthful colour at earlier stages post-mortem. Marbling was improved due to the early improvement in texture (finer) and colour (brighter) which advanced the 'setting up' of fat in the rib-eye. Interestingly, Davis (1981) recorded an improved marbling score for grain-fed steers but not for grass-fed steers. Roeber (2000) did not find any advantage for ES carcasses with regard to marbling assessed 36 h post-mortem, but confirmed improved colour scores with instrumental colour measurements. Roeber (2000) and earlier studies also emphasized that the positive effects of ES on grading parameters may be negated when grading is delayed, for example, for carcasses stored over weekends.

Scores for 'heat ring' were also reduced in ES carcasses (Smith, 1985). 'Heat ring' is the term for two colour tones in the same muscle developing as a result of rapid chilling. The differential chilling rates between the deeper and outer rib-eye muscle, for example, delay rigor mortis and colour development in the outer parts, while the deeper parts develop normal colour. ES overcomes these differences by advancing the time to reach rigor mortis.

## 3.3 The effect of ES on bleeding out of the carcass

Smith (1985) noted that ES improved bleeding out of carcasses on the slaughter and dressing floor, which reduced blood drip in chillers and also blood staining on the inside of the neck area and, therefore, led to less trimming and bacterial contamination. Effective bleeding will also improve the visual appeal of meat. According to Meat and Livestock Australia (MLA), low-voltage ES (LVES) of cattle (120–400 mA, 15 pulses per s) increases bleeding out rate (Devine, 2014), while the medium volt ES (MVES) system adopted by MLA for sheep (2 A, 15 pulses, 1 ms pulse width) increases the amount of blood collected during exsanguination.

# 4 Factors influencing the effectiveness of ES

Effectiveness of ES is determined by the rate and magnitude of pH decline in the muscle. Various factors inherent to the equipment but also to the carcass and environmental conditions will contribute in an integrated way to regulate pH decline and therefore, meat quality.

## 4.1 Specifications of an ES system

The specifications of an ES system are normally described in terms of electric potential (volts), current flow (m Amps) through the animal or carcass, pulse waveform and oscillation frequency. All of these parameters may influence the rate of pH fall and, therefore, quality (Devine, 2014). The parameters generally used must consider the combinations of electric potential, the appropriate waveform and pulse frequency, duration of ES and the species of animal. The rate of pH fall will further be influenced by muscle type, initial muscle glycogen level, muscle temperature and post-mortem delay of ES application. Wang (2003) noted that one of the reasons for variable effects of ES on quality parameters among the studies listed in Table 1 is variation in ES parameters.

There is no specific boundary that distinguishes between high-voltage ES (HVES) or LVES, but Devine (2014) refers to systems using electric potentials varying between 40V (beef) and as high as 1130V (sheep). Smith (1985) summarized earlier work on LVES and HVES. Both applications improved chiller-assessed lean colour score, USDA quality grade and 'heat ring' score, but only HVES improved marbling score. Furthermore, HVES improved tenderness more than LVES, although the effect of LVES was also significant. Purge was not affected by ES in these studies. Chrystall (1984) reported similar rates of pH fall with HVES and LVES, but the extent of pH fall was lower with LVES. Lower shear force values were recorded when peak current flow through the carcass exceeded 100mA, as opposed to 50 mA, but differences were small. However, the rate in pH decline follows an asymptotic curve with increased current flow and, therefore, it makes no sense to use extreme output currents (or voltages) (Devine, 2014). Chrystall (1984) found that pulse shape had no effect on pH decline and quality, although recent work showed that a rapid-rise-time, rectangular pulse seems to have a better stimulation effect than similar (V and A) sinusoidal pulses (Devine, 2014). In addition, 9–16 pulses per second presented the fastest fall in pH due to greater physical response and prolonged contraction of muscles. This is opposed to most early ES systems operating at 50 Hz (Table 1). Lower frequencies and narrow-pulse systems combined with LVES are also regarded as safer for ES operators. As an example, most beef and sheep abattoirs in Australia have been converted to MVES (e.g. 350 V 2 A, 15 Hz and 1 ms pulse widths for sheep) for best outcomes in terms of quality and the additional benefit of low power levels and narrower pulse width that comply with safety regulations (Pearce, 2009).

One of the mechanisms involved in accelerated ageing of muscle of ES carcasses is elevated 'free' $Ca^{2+}$. According to Geesink (2001), and Hopkins (2002), accelerated pH decline is mirrored by an increase in free $Ca^{2+}$ that will lead to higher rates of proteolysis at high muscle temperatures. Tornberg (2000) showed that the higher levels of free $Ca^{2+}$ were due to leakage from the sarcoplasmic reticulum as a result of cell membrane disruption during fast pH decline. HVES provided higher levels of free $Ca^{2+}$ and greater levels of calpain activation (Hwang, 2001a). Likewise, Wang (2003) noted that phosphate in muscle increased with increased wave frequencies of ES. Because higher levels of phosphate can reduce levels of free $Ca^{2+}$ due to precipitation, the increased phosphate could have restricted the activation of calpains. However, there are many other parameters that may also play a part in this complicated process such as interactions with muscle types.

## 4.2 ES and the effect of muscle fibre types

Different muscles in the carcass respond differently to ES (Olsson, 1994) due to variation in muscle fibre types and fibre-type ratios because different fibre types do not have the same glycolytic behaviours (O'Halloran, 1997). Interestingly, muscles with predominantly red muscle fibre types do not respond to ES through an increase in rate of pH fall but show evidence of super contracture and are prone to cold shortening. In contrast, muscle containing predominantly white muscle fibre types show no super contracture and increased rates of pH decline when stimulated but do not cold-shorten (Devine, 1984). Muscles with mixed fibre types (e.g. loin) have least desirable characteristics, that is, their rates of glycolysis are low and they cold-shorten. To these muscles, ES confers a great advantage. It should also be considered that the three factors involved in muscle tenderness, namely, connective tissue level and properties, ageing potential and tendency to shorten during rigor mortis, differ among different muscles of the carcass (Rhee, 2004). Therefore, the positive effect that ES may have on the rate of post-mortem ageing or prevention of cold-shortening in a specific muscle may be masked by higher baseline toughness as a result of high levels of connective tissue, for example, the high-connective tissue cuts of the fore quarter and certain buttock muscles (Smith, 1985).

## 4.3 Timing of ES application

ES may be applied on the undressed animal directly after exsanguination (5 minutes after stunning) or on the dressed carcass just before entering the chiller (20–45 minutes after stunning). When ES is delayed, the magnitude of the pH decline will be less due to lower muscle temperature and starting pH, but the final pH will not be influenced (Devine, 2014). Delayed ES can be used to manage the balance between pH and muscle temperature, as discussed under Section 5. Wahlgren (1997) used a 30-minute ES-delay approach to reach rigor mortis at ideal temperatures (15°C) and reported tenderer meat, that improved further with prolonged ageing. However, both ES treatments (delayed or conventional ES after exsanguination) produced more tender meat than NES. The results of Li (2011) also suggested that delayed ES may be an option to limit the risks of unfavourable rigor mortis conditions. They reported an improved initial colour (brighter) and no effect of ES on colour stability and purge when ES was applied 30 minutes after stunning. However, the effect of delayed application of ES was not tested against early application of ES, but only against NES.

## 4.4 Optimum utilization of ES

Longer durations of ES will lead to a faster decline of pH (Chrystall, 1984; Strydom, 2014) and, depending on temperature decline, may have variable effects on quality (as will be discussed later). Simmons (2006) introduced a new approach to the ES process where responses of key muscles in the carcass during stimulation are measured to derive information about initial and subsequent pH decline. The design is based on load cells that will measure muscle and carcass responses to initial test pulses, while decisions on duration and termination of ES are then based on real-time feedback processed through an algorithm.

In agreement with Simmons (2006), Geesink (2001) and Strydom (2014) showed that conservative use of ES may result in positive quality outcomes, as opposed to extensive durations of ES that were used in many earlier studies. Geesink (2001) recorded improved tenderness results when electrical inputs were limited to the immobilization of animals after stunning, as opposed to no electrical inputs. The opposite was experienced when multiple electrical inputs were used. Likewise, Strydom (2014) recorded improved tenderness, colour stability and water-holding capacity (WHC) when ES was limited to 15 sec, as opposed to 45 or 90 sec. The authors advise that the conservative treatment (15 sec) should not be regarded as an absolute recommendation, because Guransky (2010), for example, found no effect with shorter durations of ES. In the study of Strydom (2014), grain-fed animals exposed to minimum pre-slaughter stress were used, as opposed to grass-fed animals in the study of Guransky (2010). Grass- and grain-fed animals differ in weight, fatness and, more particularly, muscle fibre types and muscle glycogen (Pethick, 1999; Shibata, 2009), which will affect the muscle response to ES (Daly, 2006; Ferguson, 2008). The contrast between these two studies emphasizes the interaction between ES variables and biophysical carcass characteristics.

## 4.5 ES when tenderness is challenged

ES seems to be more effective to improve tenderness in cases where tenderizing mechanisms such as the calpain system are challenged (Smith, 1985). Hearsnhaw (1998) and Gursansky (2010) reported that the negative effect of *Bos indicus* genes on meat tenderness was overcome when ES was applied. The advantage of ES gained by *B. indicus* animals was higher than that of *Bos taurus* animals. Ferguson (2000) supported these findings by showing an increase in calpain activity in Brahman carcasses as a result of ES, whereas calpain activity of Hereford and Hereford crossbreds were unaffected. Likewise, Strydom (2011) demonstrated that loin muscles from beta agonist supplemented animals was tougher than non-treated animals, but showed larger improvement in tenderness after ES than in non-treated samples, even though the non-treated samples were still more tender in the end. Calpastatin level after 24 h was reduced by application of ES in both treated (beta agonist) and non-treated animals, but the effect of ES was more pronounced in the beta agonist treated group. In both *B. indicus* animals and beta agonist treated animals, high calpastatin activity impairs meat tenderization.

# 5 Meat quality defects related to ES

When the correct balance between pH and temperature decline is not maintained, it could impact on meat tenderness but also other quality characteristics such as WHC and colour (Marsh, 1981; Mancini, 2005; Savell, 2005; Thompson, 2006; Huff-Lonergan, 2007). Dransfield (1992) showed that a very fine balance exists between early activation of the calpain system (by early onset of the rigor mortis process) and the self-destruction of the enzymes due to autolyses prompted by high temperatures and low pH. Recent studies (Jacob, 2014; Warner, 2014a) suggest that this problem is becoming more common for beef due to increased electrical inputs during slaughter (apart from ES) and increased carcass weight and fatness that delay temperature decline. Table 2 summarizes the effects of rapid pH decline and/or higher than normal pre-rigor temperature on beef quality. These conditions are normally associated with the indiscriminate use of ES. Kim (2014) used the acronym high-temperature-low-pH (HTLP) that describes the condition of high-temperature, low pH (or high rigor mortis temperatures).

## 5.1 The effect of HTLP on meat tenderness

Tenderness is affected by HTLP through heat-induced sarcomere shortening (often termed 'heat shortening') and alterations of proteolytic enzyme activity (Hertzman, 1993; Devine, 1999; Rosenvold, 2008; Thomson, 2008; Jaime, 1992; Geesink, 2000; Devine, 2002).

**Table 2** Effects of rapid pH decline and/or higher than normal pre-rigor muscle holding temperature on meat quality attributes of beef muscles

| State (of muscle) | Muscle | Treatment | Effects observed on meat quality | Author |
|---|---|---|---|---|
| Excised | *Longissimus* | Pre-rigor excised muscle. Holding at 15, 20, 25, 30, 35°C until rigor, +/− wrapped | Tenderness: 20–35°C, tougher meat (WBSF and sensory) than 15°C, correlated with sarcomere length | Devine et al., 1999 |
| Excised | *Semitendinosus* | +/− ES and pre-rigor excised muscle holding at 0, 5, 10, 25, 35°C until rigor | Colour: 35°C higher *L** value, 25–35°C, higher *a** and *b** values; WHC: 0–10°C similar WHC, 25–35°C lower WHC;<br>ES: no effect on colour and WHC | Farouk and Swan, 1998 |
| Excised | *Semimembranosus* | +ES and pre-rigor excised muscle holding at fast and slow chilling for 24 h | Colour: higher *L** | Hector et al., 1992 |
| Carcass | *Biceps femoris, semimembranosus, semitendinosus* | +/− ES and 15 *v.* 37°C until rigor | Tenderness: 37°C tougher sensory at 15 days, not WBSF | Hertzman et al., 1993 |
| Carcass | *Longissimus* | +/− ES applied to a side within a carcass, achieved high rigor temperature of 41°C (+ES) and 33°C with (−ES) | Tenderness: no difference in WBSF or consumer sensory;<br>Colour: higher *a** value with ES;<br>WHC: increased purge after storage with ES, no effect on drip loss | Hopkins et al., 2014 |
| Excised | *Longissimus semitendinosus* | Pre-rigor excised muscle holding at 5, 15, 36°C for 24 h | Tenderness: 36°C, lowest WBSF followed by 15°C and 5°C, 36°C had greater extent of μ-calpain autolysis and troponin-T degradation | Hwang et al., 2004 |
| Excised | *Longissimus* | +/− ES, +/− wrapped and 15/38°C pre-rigor muscle holding until rigor mortis | Tenderness: higher WBSF (tougher), and no impact of either + ES or + wrapping on SF;<br>WHC: 38°C higher purge loss 38°C, higher purge loss | Kim et al., 2012 |
| Carcass | *Longissimus* | Carcasses from grass-fed and feedlot-fed steers; 0 *v.* 35°C for 3 h after slaughter | Tenderness: for the feedlot-fed animals, 35°C higher WBSF, lower sensory tenderness and shorter sarcomere length. No ageing effect on tenderness for 7 days storage | Lee and Ashmore, 1985 |
| Excised | *Sternomandibularis* | Pre-rigor excised muscle, +/− wrapped (or restrained) holding for 24h at 15°C or for 7 h at 37°C | Tenderness: 37°C, more tender meat than rigor at 15°C;<br>Colour: 37°C, paler colour;<br>Odour: sweaty smell observed | Locker and Daines, 1975 |

| | | | | |
|---|---|---|---|---|
| Excised | *Longissimus* | Pre-rigor muscle holding for 2 h at 27–40°C | Tenderness: most tender (sensory) if temp. at 2 h was 40°C | Lochner et al., 1980 |
| Carcass | *Longissimus* | Carcass +/− ES | Tenderness: ES lower;<br>WHC than non-ES counterpart;<br>Colour: ES higher *L**, *a** and *b** than non-ES counterpart | Roeber et al., 2000 |
| Excised | *Longissimus* | +/− ES, +/− wrapped and 15/35°C pre-rigor muscle holding until rigor mortis | Tenderness: 35°C, higher WBSF (tougher), but + ES and/or + wrapping offset negative impact of 35°C on meat tenderness;<br>WHC: no impact | Rosenvold et al., 2008 |
| Carcass | *Longissimus* | Carcass, +/− ES, 20°C Hz (no physical damage) and 37°C *v.* 2°C | Tenderness: +ES higher toughness (sensory) and reduced juiciness (sensory), 1 day ageing | Takahashi et al., 1984 |
| Excised | *Longissimus* | Post-rigor excised muscle holding at 15 *v.* 37°C for 3 h | Tenderness: 37°C, initial improvement in WBSF, but higher WBSF after 21 days ageing.<br>WHC: 37°C, higher cook loss | Thomson et al., 2008 |
| Carcass | *Longissimus* (L)<br>*gluteus medius* (GM) | Large dataset (*n* = 3865) containing meat aged from 1–35 days, +/− tender stretch, temperature at rigor from 2–42°C | Tenderness: GM decreased in eating quality (consumer sensory scores for tenderness, juiciness, flavour, overall liking) with increasing temperature at rigor. L at higher rigor temperature was initially better eating quality than ideal rigor temperature (15°C).<br>At 28 days, L had inferior eating quality from high rigor temperature carcasses. Tender stretch alleviated negative effects of high rigor temperature on eating quality<br>Colour: paler colour scores at grading in L with higher rigor temperature<br>Texture: higher number of L scores as coarse/soft, at high rigor temperature | Warner et al., 2014c |
| Carcass | *Longissimus* | Survey of occurrence of high rigor temperature (temp. >35°C when pH > 6.0) in 7 processing plants (*n* = 1512) | WHC: high rigor temperature muscles had more weep (exudate) on the meat surface (using filter paper method) | Warner et al., 2014a |

Note: This list does not contain all papers with data pertaining to the topic. It demonstrates that high temperatures and low pH at rigor influence quality traits and protein denaturation across many meat species. ES, electrical stimulation; SF, shear force, values measured by MIRINZ tenderometer; WBSF, Warner–Bratzler shear force; WHC, water-holding capacity.

Source: From Kim (2014).

Most of these studies showed that muscle entering rigor (pH = 6) at temperatures above 35°C will not age and tenderize to its full potential. Under these conditions, an early activation of μ-calpain by elevated free $Ca^{2+}$ occurs (Hwang, 2004), resulting in a faster degradation of cytoskeletal proteins and a positive effect on initial tenderness, as described earlier. However, these conditions will soon exhaust the proteolytic enzyme activity due to faster autolyses, leading to limited ageing potential (Whipple, 1990; Dransfield, 1994; Rhee, 2001; Thomson, 2008; Warner, 2014b,c).

## 5.2 The effect of HTLP on WHC

WHC describes the ability of meat to hold its own water (Pearce, 2011). Poor WHC will show in higher purge and even thaw loss, and will affect the yield and quality of fresh and processed meat. Excessive purge that accumulates in packaging could affect consumers' willingness to pay for a product (Grunert, 2004) and may also result in a negative perception of juiciness when cooked. Many factors may contribute to poor WHC, but the most important ones related to the condition of HTLP are denaturation of both sarcoplasmic and myofibrillar proteins (Tarrant, 1977; Kim, 2012), but specifically myosin (Offer, 1991).

Kim (2014) noted that most studies reporting on the effect of HTLP on WHC used ES to investigate accelerated pH decline on quality (Martin, 1983; Unruh, 1986). Warner (2014a) showed that the 'weep score' accessed during carcass grading increased from 0.3 to 0.6 as the loin muscle temperature at pH 6 rose from 25 to 40°C (Fig. 1). Other studies worked on pre-rigor excised muscles that were kept at high temperatures (>25°C), which in most cases resulted in poorer WHC.

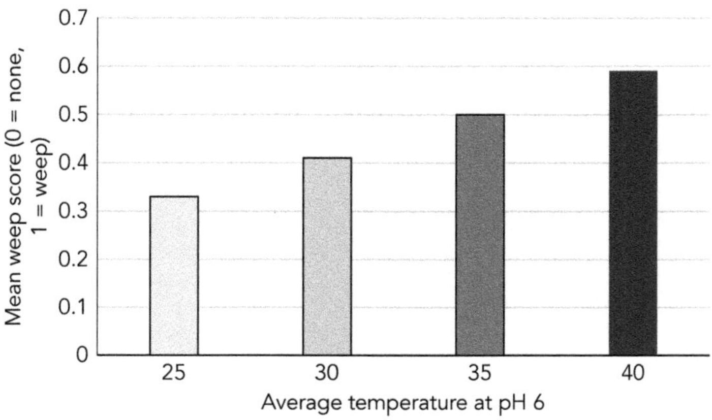


**Figure 1** Effect of temperature at pH 6 on the weep score at grading (0 = no weep; 1 = weep present). The weep score is assessed on the striploin at about the 12th rib (longissimus thoracis muscle), at grading, after a 2-h bloom. It is an assessment of the amount of fluid oozing out of the meat surface. Effect of temperature at pH = 6 on weep score, $P < 0.05$.

Source: From Warner (2014a).

## 5.3 The effect of HTLP on meat colour

Various studies have indicated positive colour development when ES is applied, especially during grading, as discussed in Section 3.2. However, according to the review study of Kim (2014), muscle exposed to high pre-rigor temperatures and rapid pH decline will lead to a paler meat colour and reduced colour shelf-life stability. Warner (2014a) showed that the proportion of pale scores (1b and 1c) during colour grading increased relative to darker colour scores as the loin muscle temperature at pH 6 rose from 25 to 40°C (Fig. 2). The paler colour was attributed to increased light scattering, while poorer colour stability was the result of myoglobin and/or myofibrillar protein denaturation. It is interesting that HTLP condition may also have a positive effect on initial colour properties. Various studies have recorded a brighter red colour upon initial display of PSE-like beef, but the colour deteriorated during display due to a faster rate of metmyoglobin accumulation (Ledward, 1985; Renerre, 1990; Young, 1999; Roeber, 2000; Sammel, 2002). The lack of colour stability results from the lower metmyoglobin-reducing ability due to denaturation of the metmyoglobin moiety at low pH (Zhu, 2003) and increased autoxidation caused by low pH and high temperatures (Ledward, 1985).

The PSE-like effect of HTLP conditions often results in different colour display within a single muscle, normally referred to as two-toning. The reason for this defect is the difference in temperature decline between the surface and deeper part of the same muscle and particularly in larger muscles (Sammel, 2002; Seyfert, 2004; Kim, 2010). Temperature differences as high as 15°C may occur in large muscles of the hindquarter. Normal colour development and stability will then occur in the outer parts, while the deeper parts that were exposed to HTLP will show defects as a result of lower metmyoglobin-reducing ability and oxygen consumption. This defect has a similar cause as 'heat ring' discussed earlier, except that with heat ring the outer part that chills quicker experiences a lower rate of glycolysis than the deeper muscle and develops colour at a slower rate. This is normally rectified by ES.

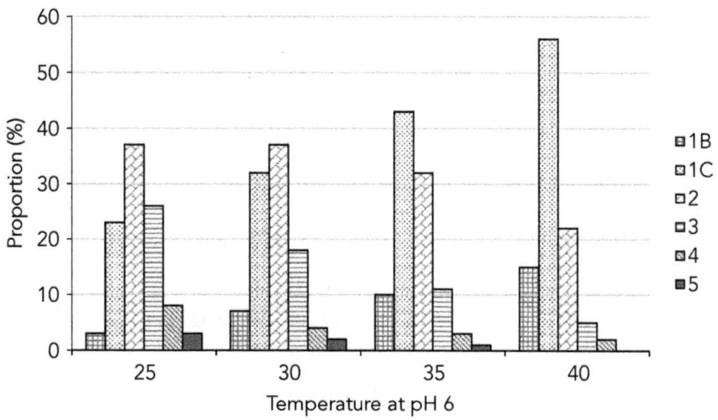


**Figure 2** Effect of temperature at pH 6 on the AUS-MEAT colour score (1B = palest, 5 = darkest; colour score >3 is unacceptably dark). The colour score is assessed on the striploin at about the 12th rib (longissimus thoracis muscle), at grading, after a 2-h bloom. Effect of temperature at pH 6 on AUS-MEAT colour score, $P < 0.05$.

Source: From Warner (2014a).

## 5.4 Controversies surrounding the use of ES and quality defects

The exact pathway of pH decline induced by ES and ultimate effect on quality depend on various factors such as the type and duration of ES application, the use of ES in combination with other electrical inputs, variation in response to different carcasses and on chilling conditions, to name but a few (Simmons, 2006, 2008; White, 2006a; Warner, 2014a). Moreover, it seems that many so-called defects associated with high temperatures during rigor mortis are the results of study design. Devine (2014) noted that without ES, muscle held at a constant elevated pre-rigor temperature will have reduced tenderization due to an impaired calpain system. However, with ES, pre-rigor exposure of enzymes and other proteins to low pH and high temperature is reduced due to individual fibres reaching rigor mortis rapidly while the rest follow soon. Likewise, Rosenvold (2008) found that ES loin samples exposed to rigor temperatures above 35°C did not toughen as would have been expected. They reasoned that ES accelerates the rate at which individual fibres complete rigor and once these fibres reach rigor mortis, other fibres will be protected from harmful effects of temperature. Both proteolytic enzymes and contractile proteins will then be protected from denaturation. The latter was supported by Offer (1991). Therefore, both heat-induced shortening and prevention of ageing by unfavourable pH/temperature conditions would have been prevented. Kim (2012) challenged their findings stating that only myosin (Offer, 1991) but not calpain will be protected under early rigor mortis conditions at high temperature and shortening will not take place, but tenderization will be impaired due to denaturation of u-calpain (Hwang, 2001b). These effects were independent of whether ES was applied or not and, therefore, the claim that ES protects remains questionable.

Rosenvold (2008) and Devine (2014) also reported an increase in drip loss with ES, not as a result of high temperatures at rigor mortis, but only in relation to amount of ageing. In other words, the accelerated ageing caused by ES is associated with more drip loss, because muscle tenderization (breaking up of muscle structure) is associated with production of free water. Interestingly, Rosenvold (2008) found that initial purge of muscle that went into rigor at 35°C was higher than those entering at 15°C. However, as ageing progressed, the differences were nullified as drip loss of the 15°C treatment increased in drip loss over that of muscles held at higher temperature.

When considering the reports on colour stability under HTLP, Strydom (2014) found that muscle was paler (higher $L^*$) and exhibited higher chroma values (more vivid) for the whole 7-day display period when HTLP conditions prevailed as a result of excessive ES compared to NES muscle.

# 6 Effective chilling of beef carcasses and the role of ES in quality management

## 6.1 Challenges of effective chilling

The heat transfer rate through any medium such as meat is a function of the thermal conductivity of the material, the heat gradient and the inverse of the distance between the surface and the deeper part of an object (muscle). The low thermal conductivity of meat, which is similar to that of wood, namely 0.45 W/(m.C°) (Fontana, 1999), combined with the relatively large distance between the carcass surface and deep muscle points are, therefore, key to chilling limitations of beef carcasses. This is exacerbated with large and fat carcasses and will have variable effects on different muscles depending on their size, shape and anatomical position in the carcass, and physiological type. Stolowski (2006) stated that effective cooling of larger and deeper muscles in beef carcasses, such as the *biceps femoris* (BF) and *gluteus medius* (GM) muscles, will always be a challenge as a result of their dimensions and poor thermal conductivity. Relatively small muscles and muscles with more surface exposure and better position, such as the *semitendinosus* (ST), *vastus lateralis* (VL) and *triceps brachhi* (TB) muscles, will show intermediate cooling rates, while the LD muscles, that is a flat muscle and with more surface exposed, will chill the quickest.

The solution to more effective chilling of carcasses or meat is two-fold. Firstly, the temperature gradient between the deeper muscles and carcass surface needs to be increased, while heat on the surface is removed as quickly as possible without freezing the surface with obvious limitations as explained earlier. The second method would be to decrease the muscle mass to be chilled so that the distance between the deep muscle part and the chilling environment is reduced. A third factor may be to manage factors contributing to initial carcass temperatures, but these interventions are beyond the scope of this chapter. Irrespective of the option, Thomson (2008) emphasizes that the final temperature is not the most important criteria for preventing temperature-related damage to muscle proteins, but rather the time the muscles spend under unfavourable conditions. They concluded that the first 5 hours spent above the MSA pH/temperature threshold is the most damaging to muscle physiology and, therefore, tenderness development, colour stability and moisture properties.

The methods applicable to the first approach include spray-chilling (Strydom, 1995; Savell, 2005), blast chilling (BC), immersion cooling and very fast chilling (VFC), while methods like hot boning belongs to the second approach (Stolowski, 2006; Seyfert, 2004; Van Moeseke, 2001). Some of these methods are often used in combination, such as hot boning and VFC, while ES is essential in managing the conditions of rigor mortis.

## 6.2 Hot boning and the role of ES

Hot boning is defined as the processing of carcasses or specific cuts before the completion of rigor mortis and was mainly invented to save energy required for chilling as well as chilling space (Jacob, 2014), but also due to improvements in certain functional properties and quality of processed meat products (Pisula, 1996). However, hot boning is also often cited as negative with regard to certain quality parameters, such as colour and tenderness, although Simmons (2006) ascribed this to poor management of pH and temperature decline. Colour (Brown, 1988; Seyfert, 2004) and tenderness defects (Pisula, 1996; White, 2006a) may occur due to increased mitochondrial respiratory rates, and propensity to cold shortening (Locker, 1960) as a result of delayed pH decline (Tarrant, 1977; Sammel, 2002) in deeper parts of muscles, respectively.

It is evident that hot boning is most applicable for effective chilling of large, bulky deep muscles such as the BF, GM and deeper parts of the *semimembranosus* (SM) muscles as indicated by Stolowski (2006), rather than flat-shaped superficial muscles like the LD and TB (in contrast to intact muscles of cold-boned carcasses). pH decline is normally delayed (Tarrant, 1977; Sammel, 2002) in deeper parts of muscles by hot-boning indicating a lower rate of glycolysis, although some studies did not record differences in pH decline due to hot-boning (Jeremiah, 1985; Syfert, 2004).

White (2006b) showed that ES improved tenderness of LD and SM hot-boned muscles irrespective of chilling temperature (2 and 10°C), while muscles of NES hot-boned carcasses were always tougher than cold-boned NES controls. Cold-boned muscles (ES and NES), in particular LD muscles, had shorter sarcomeres than hot-boned NES muscles even when carcasses were chilled at 10°C but ES limited shortening and thereby improved tenderness.

The faster chilling of hot-boned cuts tends to produce darker initial colour and darker colour during display in some muscles (BF, RF, VL). However, Seyfert (2004) and Sammel (2002) also noted that hot-boned buttock muscles (VL, RF and SM) exhibited better colour stability across the whole muscle (outer and deeper parts) than conventionally chilled cuts, probably due to less protein denaturation in the deeper parts resulting in a lower OCR than in deeper muscle parts of cold-boned carcasses. Partial hot boning did not always overcome two-toning. Seyfert (2004) applied ES and while this solved tenderness problems, it caused two-toning in large muscles like the SM of hot- and cold-boned carcasses, while less two-toning was found in the hot-boned BF muscle compared to cold-boned BF. Therefore, the effect of ES and hot boning on colour two-toning may be muscle type and muscle position dependent.

Less protein denaturation in hot-boned or partially hot-boned carcasses improved WHC of deeper muscle parts (Tarrant, 1977; Sammel, 2002). However, under conditions such as the use of ES, an unfavourable pH/temperature ratio may still occur and will be detrimental to the quality of certain muscles (Seyfert, 2004). Likewise, Taylor (1987) noted that hot boning will reduce drip loss and improve colour of deeper muscle parts or colour uniformity, but when ES was applied, the benefit of hot boning was diminished. However, it remains a fact that ES is needed to prevent shortening and toughening of excised muscle and, therefore, the correct management of ES and chilling is imperative to maximize the advantage of hot boning.

It is interesting to note that different muscles do not shorten to the same extent at low temperatures during hot boning. Olsson (1994) found that excised LD shortens much quicker and more than SM when exposed to low temperatures, such as with hot boning. LD has more intermediate and less white fibres than SM and, therefore, has less-developed sarcoplasmic reticulum and more mitochondria. Of further interest is the fact that excised muscles shorten to the same extent irrespective of ES, but the improvement in tenderness over time is higher for cold-stressed LD that were stimulated than for SM. Therefore, certain muscles benefit more from ES and most probably not due to prevention of shortening, but rather to improve ageing. Rhee (2004) indicated that SM has much higher connective tissue and much lower ageing capacity than LD that could have contributed to the difference in final quality between the two muscles.

## 6.3 VFC and the role of ES

Various studies have investigated the effect of VFC on quality parameters of meat with variable, and sometimes conflicting results. Most early studies were performed on intact carcasses, while hot boning was combined with VFC in later studies. ES is often not used as the aim is to reach rigor mortis below 0°C. The purpose of VFC is mainly to save on carcass weight loss by water condensing on carcasses when the surface is very cold (Hippe, 1991). However, depending on how VFC is defined, there are also other advantages to this method.

It is well known that pre-rigor muscle temperatures between 0°C and 5°C will induce cold-shortening (Locker, 1963), as opposed to muscle held between 10°C and 20°C that will develop into acceptable tenderness during ageing (Devine, 1999; Geesink, 2000). In contrast, it was found that muscle held between −1°C and −5°C (VFC) during rigor mortis produced meat with acceptable initial tenderness and good ageing ability (Joseph, 1996). Jaime (1992) proposed that VFC releases extra $Ca^{2+}$ ions from the sarcoplasmic reticulum in the microstructure of muscle that leads to increased action of proteolytic enzymes, which is further stimulated by high pH at rigor onset. In contrast, the release of high levels of $Ca^{2+}$ would also lead to excessive contraction. The increased action of proteases can be further enhanced when super contraction of muscle fibres under these conditions pulls sections apart in the muscle structure, thereby exposing more cytoskeletal proteins to enzyme activity. They further postulated that the breaks caused by proteolysis will prevent a continuous contraction of the length of muscle fibres. In this way, the enhanced proteolysis will overcome the negative effect of shortening.

Conflicting results recorded with VFC are probably due to the complexity of factors involved in reaching optimum muscle conditions at the right time and for the specific duration. These include carcass weight and fatness, chilling methods, pre-slaughter history of animals and other slaughter processes like ES. The general objective is to reach 0 to −1°C in 4 to 5 h (Dransfield, 1998; Joseph, 1996; Honikel, 1998). Different refrigeration methods can be used to achieve the correct cooling rate, but Jacob (2014) and Van Moeseke (2001) emphasized that effective chilling is probably possible only for meat cuts with a thickness of less than 80 mm removed during hot boning. However, in several studies carcasses were cold-boned after chilling at temperatures between −5 and −70°C and wind speeds of 1–5 $m.s^{-1}$ for 1–5 h, also referred to as blast chilling (Joseph, 1996; Janz, 2001; Li, 2006). The authors who studied hot-boned cuts used air temperatures of −20 to −25°C applied until the target muscles temperature was reached (Van Moeseke, 2001; Li, 2012), or liquid mediums such as NaCl brine (Van Moeseke, 2001) and propylene glycol (Jacob, 2012).

From the evidence available, it seems like the risk of severe cold shortening without the advantage of increased proteolysis, as well as poor colour, are the biggest risks when the correct temperature targets within a certain time are not reached with VFC. Van Moeseke (2001) recorded tougher meat with poorer colour when ST muscles were hot boned and chilled in a freezer at −20°C, or in brine solution at −2°C for 5 h. In agreement with other studies, they recorded sarcomere shortening in VFC samples, compared to conventional chilling, where ST muscles were left intact in carcasses. In addition, low pre-rigor temperatures also did not accelerate proteolysis as expected and these two conditions resulted in tougher meat for both VFC treatments. Aalhus (2002) achieved required temperatures for positive effects of VFC (<0°C in 5 h) only when carcasses were blast-chilled for 7 to 10 h at −35°C. These conditions improved tenderness, colour and marbling score and carcass shrink but increased purge, compared to carcasses chilled at 2°C. No ES was applied. From the available information in this work, it seems that muscle temperature may not have declined below 0°C in particular with the freezer treatment, which emphasizes the precariousness of the VFC method.

Li (2012) combined the principles of VFC and blast chilling by chilling hot-deboned loin samples at −21°C to 0°C core temperature and compared this scenario to more conventional chilling at 0, 7 and 14°C. In the absence of ES, carcasses chilled at 14°C fell within the ideal temperature/pH window for rigor mortis. However, the −21, 0 and 7°C treatments showed significant cold shortening as measured by sarcomere length directly in relation to chilling temperature. Surprisingly, the mean WBSF value of the VFC group (−21°C) was the same as that of the 14°C group at 1 day, improved the least of all treatments but was more tender than the 0°C group and similar to the 7°C group. The authors postulated that the improvement in tenderness under severely cold-shortened muscles of VFC conditions in contrast to cold-shortening conditions (0 and 7°C) was the result of myofibril stretching and breaking adjacent to areas of super contraction as a result of ice crystal formation that did not occur in 0 and 7°C samples. Purge loss was the highest for VFC (−21°C) and the 0°C group and slightly less for the 7°C group at 1 day post–mortem, but no differences were recorded at 21 days. Higher purge increased the *L** (lightness values) of the three groups, although the differences subsided at 21 days. Redness (*a**) and yellowness (*b**) were directly related to pre-rigor temperatures and could be explained by the effect of temperature on OCR and metmyoglobin-reducing ability (Ledward, 1985). However, for unaged samples, VFC samples exhibited the most stable colour during display and the 0°C treatment the poorest, but this effect was less prominent for 21-day aged samples. Because the 14°C treatment presented the best tenderness and also favourable colour and WHC results, it was proposed that VFC at −21°C be used in hot-boned muscle until a core temperature of 14°C is reached, which should then be maintained for 10 h before chilling at 2°C for hygiene purposes.

Aalhus (2001), Li (2006) and Zhu (2011) studied the effect of VFC on weight loss savings while using ES to prevent cold-induced toughening. The carcasses exposed to VFC treatments were chilled for 2 hours post dressing between −12 and −20°C. VFC decreased the rate of pH decline of ES carcasses compared to conventionally chilled ES carcasses. Zhu (2011) did not include any non-ES controls and, therefore, the effect of ES under these conditions could not be verified properly. However, ES probably lowered the risk of cold shortening (Locker, 1963) and both scenarios showed pH/temperature ratios in the optimum zone for tenderization (Thompson, 2002). No differences in tenderness were recorded between the treatments. Li (2006) showed that ES and conventional cooling will give the best tenderness results but the highest purge. In contrast, VFC and ES produced lower purge and more acceptable tenderness than VFC carcasses without ES. Aalhus (2001) recommended the combination of VFC (blast chilling) and ES to industry to reduce chilling times and shrink losses while producing beef with high eating quality. ES and VFC produced more tender LD than NES VFC sides or conventionally chilled NES sides and improved colour at grading, but not marbling score. Drip loss was lower with VFC in general.

# 7 Future research

The review study of Hwang (2003) and subsequent studies on ES highlight a number of topics that need further investigation. Although it is safe to say that ES improves meat tenderness through its effect on the calcium-dependent protease system (calpain), the exact condition in terms of pH and temperature pathways for optimum quality are still elusive after all of these studies and conflicting results are causing much confusion. In this regard, Hwang (2003) and Kim (2014) advise against extrapolating from *in vitro* conditions where excised muscles are studied under controlled and fixed environments, e.g. samples held at constant temperatures. While such studies have value with regard to better understanding of various biochemical and physical mechanisms, *in situ* studies where normal commercial conditions of steady pH and temperature declines prevail should be used to find answers for commercial conditions.

The study of Rosenvold (2008) was earlier referred to where excised muscles kept under high-temperature rigor conditions produced more tender meat when stimulated than those samples not stimulated. When considering that the NES muscles in this particular study design were exposed for longer times to high pre-rigor temperatures than ES muscles, it is understandable that ES muscles performed better. However, under commercial practices, such conditions are less likely. Kim (2014) suggested a better understanding of whether the conditions under which the different protein fractions in muscle denature differently among muscles and fibre types. Furthermore, a better understanding of the effects of duration of unfavourable conditions on denaturation will also improve the ability to manage these conditions.

Hwang (2004) also emphasizes that further investigation is needed to verify whether the disruption of muscle structure by ES increases access calpains to the muscle substrate. The work of Simmons (2006, 2008) should be continued to better understand variation in response of carcasses to ES and also that of individual muscles so that appropriate amounts of energy applied can be regulated by a self-response system.

Kim (2014) agrees that there is a limited understanding of underlying mechanisms by which proteolytic enzyme systems respond to high rigor temperatures, for example with ES. However, in response to the findings of Balan (2014) and Herrera-Mendez (2006) on the relationship between apoptosis and related actions of small heat-shock proteins (sHSP) on myofibrillar degradation, further investigation in this field would be valuable for a better understanding of the relationship among ES, high pre-rigor temperature and meat tenderness. Small HSPs play a role in maintaining cellular homeostasis by acting as a molecular chaperone to stabilize cell proteins during stress conditions and prevent irreversible aggregation of these proteins. Skeletal muscle undergoes apoptosis, or organized cell death, during conversion of muscle to meat, and these changes are associated with tenderization. Small HSP may play a role in this process by translocation from sarcoplasmic proteins to myofibrillar proteins and thereby affect their susceptibility to enzymatic degradation. The recent study of Contreras (2016) showed higher levels of sHSP in ES samples in later ageing time endpoints, but found no improvement of ES on tenderness. The use of a limited number of low final pH samples may have confounded the results, and further studies are encouraged.

With regard to colour and colour shelf life, Kim (2014) suggested that studies directed at a better understanding of the interaction of redox stability (exhaustion of reducing potential) and oxygen consumption (role of mitochondrial respiratory enzymes) under protein-denaturing conditions (HTLP) are better understood. This will assist in better management of ES parameters.

# 8 General summary

The conversion of muscle to meat during the rigor mortis process and the conditions under which this process occurs is most important for quality attributes of beef including colour, water-holding capacity, and tenderness. Electrical stimulation is used to prevent cold-induced toughening of beef, especially hot-boned beef, but is generally also applied because it is thought to enhances the ageing process and, therefore, improve tenderness of aged beef. However, indiscriminate use of electrical stimulation combined with improper chilling may reduce the muscle's capacity to age and give rise to pale colour, lack of colour stability and to increased drip loss. Electrical parameters (volts, amperes, wave form, wave frequency) of the ES system should be considered in combination with factors such as carcass chilling rate, animal specie, and time and duration of stimulation, as these may influence the critical balance between muscle pH and temperature. Maintaining this balance is becoming increasingly difficult with the global tendency towards higher carcass weight and fatness levels combined with the use of electrical inputs in addition to ES that could advance the rate of pH decline.

Hot boning and very fast chilling (VFC) are two methods used on their own or in combination to improve chilling efficiency, reduce carcass weight loss and save on chiller space. The correct parameters for VFC, namely to reach temperatures below −1°C within 4 to 5 h is important to benefit produce acceptable quality, while saving on weight loss. For hot-boning the management of pH decline and temperature is important to prevent cold-shrinkage and toughening of muscles that are not supported by attachment to the skeleton anymore.

# 9 Where to look for further information

The *Encyclopaedia of Meat Sciences* (2014). 2nd edition, in Chief Devine and Dikeman (Eds), Elsevier, Oxford has chapters on Electrical stimulation and Carcass chilling and boning.

The reviews of Kim, Y. H. B., Warner, R. D. and Rosenvold, K. (2014). Influence of high pre-rigor temperature and fast pH fall on muscle proteins and meat quality: a review. *Animal Production Science*, 54, 375–95 and Jacob, R. H. and Hopkins, D. L. (2014). Techniques to reduce the temperature of beef muscle early in the post-mortem period – a review. *Animal Production Science*, 54, 482–93 on effective chilling and the effects high rigor temperatures on beef quality.

Hwang, I. H., Devine, C. E. and Hopkins, D. L. (2003). The biochemical and physical effects of electrical stimulation on beef and sheep meat tenderness. *Meat Science*, 65, 677–91 provides an overview of effects and mechanisms of electrical stimulation on beef and sheep meat tenderness.

# 10 References

Aalhus, J. L., Robertson, W. M. and Dugan, M. E. R. (2002). Very fast chilling of beef carcasses. *Canadian Journal of Animal Science*, 82, 59–67.

Balan, P., Kim, Y. H. B. and Blijenburg, R. (2014). Small heat shock protein degradation could be an indicator of the extent of myofibrillar protein degradation. *Meat Science*, 97 (2), 220–2.

Bekhit, A. E. D., Farouk, M. M., Cassidy, L. and Gilbert, K. V. (2007). Effects of rigor temperature and electrical stimulation on venison quality. *Meat Science*, 75 (4), 564–74.

Bendall, J. R. (1969). *Muscles, Molecules, and Movement*. Heinemann, London.

Heinemann-Bindon, B. M. and Jones, N. M. (2001). Cattle supply, production systems and markets for Australian beef. *Australian Journal of Experimental Agriculture*, 41, 861–77.

Bowater, F. J. (1997). Economics of meat chilling and freezing. *Proceedings of the Institute of Refrigeration*, London, England. http://fjb.co.uk/wp-content/themes/fjb/publications/Economies%20of%20Meat%20Chilling%20&%20Freezing.pdf (accessed 20 February 2016).

Brown, T., Gigiel, A. J., Veronica, M., Swain, L. and Higgins, J. A. (1988). Immersion chilling of hot cut, vacuum packed pork primals. *Meat Science*, 22, 173–88.

Cho, S., Park, B., Ngapo, T., Kim, J., Kim, J., Dransfield, E., Hwang, I. and Lee, J. (2007). Effect of meat appearance on South Korean consumers choice of pork chops determined by image methodology. *Journal of Sensory Studies*, 22, 99–114.

Contreras-Castillo, C. J., Lomiwes, D., Wuc, G., Frost, D. and Farouk, M. M. (2016). The effect of electrical stimulation on post mortem myofibrillar protein degradation and small heat shock protein kinetics in bull beef. *Meat Science*, 113, 65–72.

Chrystall, B. B., Devine, C. E., Ellery, S. and Wade, L. (1984). Low voltage electrical stimulation of lamb: its effect on muscle pH and tenderness. *New Zealand Journal of Agricultural Research*, 27, 513–23.

Cross, H. R. (1979). Effects of electrical stimulation on meat tissue and muscle properties – a review. *Journal of Food Science*, 44, 509–14.

Daly, B. L., Gardner, G. E., Ferguson, D. M. and Thompson, J. M. (2006). The effect of time off feed prior to slaughter on muscle glycogen metabolism and rate of pH decline in three different muscles of stimulated and non-stimulated sheep carcasses. *Australian Journal of Agricultural Research*, 57, 1229–35.

Davis, G. W., Cole, A. B., Backus, W. R. and Melton, S. L. (1981). Effect of electrical stimulation on carcass quality and meat palatability of beef from forage- and grain-finished steers. *Journal of Animal Science*, 53, 651–7.

Devine, C. E., Ellery, S. and Averill, S. (1984). Responses of different types of ox muscle to electrical stimulation. *Meat Science*, 10, 35–51.

Devine, C. E. and Graafhuis, A. E. (1995). The basal tenderness of unaged lamb. *Meat Science*, 39, 285–91.

Devine, C. E., Wahlgren, N. M. and Tornberg, E. (1999). Effect of rigor temperature on muscle shortening and tenderisation of restrained and unrestrained beef M. longissimus thoracis et lumborum. *Meat Science*, 51 (1), 61–72.

Devine, C. E., Payne, S. R., Peachey, B. M., Lowe, T. E., Ingram, J. R. and Cook, C. J. (2002). High and low rigor temperature effects on sheep meat tenderness and ageing. *Meat Science*, 60 (2), 141–6.

Devine, C. E., Hopkins, D. L., Hwang, I. H., Ferguson, D. M. and Richards, I. (2014). Electrical stimulation. *Encyclopaedia of Meat Sciences*. Elsevier, Oxford.

Dransfield, E., Etherington, D. J. and Taylor, M. A. J. (1992). Modelling post-mortem tenderisation-II: enzyme changes during storage of electrically stimulated and non-stimulated beef. *Meat Science*, 31, 75–84.

Dransfield, E. (1994). Modelling post-mortem tenderisation – V: inactivation of calpains. *Meat Science*, 37, 391–409.

Dransfield, E. (1998). Very fast chilling in beef: 2. Muscle to meat, University of Bristol, Bristol, UK.

Ferguson, D. M., Jiang, S.-T., Hearnshaw, H., Rymill, S. R. and Thompson, J. M. (2000). Effects of electrical stimulation on protease activity and tenderness of M. longissimus from cattle with different proportions of Bos indicus content. *Meat Science*, 55, 265–72.

Ferguson, D. M., Bruce, H. L., Thompson, J. M., Egan, A. F., Perry, D. and Shorthose, W. R. (2001). Factors affecting beef palatability – farm gate to chilled carcass. *Australian Journal of Experimental Agriculture*, 41, 879–91.

Ferguson, D. M., Daly, B. L., Gardner, G. E. and Tume, R. K. (2008). Effect of glycogen concentration and form on the response to electrical stimulation and rate of post-mortem glycolysis in ovine muscle. *Meat Science*, 78, 202–10.

Fontana, A. J., Varith, J., Ikediala, J., Reyes, J. and Wacker, B. (1999). Thermal properties of selected foods using a dual needle heat pulse sensor. *Proceedings of American Society of Agricultural Engineers*, Paper No. 996063, Toronto, Canada.

Geesink, G. H., Bekhit, A. D. and Bickerstaffe, R. (2000). Rigor temperature and meat quality characteristics of lamb longissimus muscle. *Journal of Animal Science*, 78 (11), 2842–8.

Geesink, G. H.,Mareko, M. H. D., Morton, J. D. and Bicerstaffe, R. (2001). Electrical stimulation – When more is less. *Meat Science*, 57, 145–51.

Grunert, K. G., Bredahl, L. and Brunsø, K. (2004). Consumer perception of meat quality and implications for product development in the meat sector – A review. *Meat Science*, 66, 259–72.

Gursansky, B., O'Halloran, J. M., Egan, A. and Devine, C. E. (2010). Tenderness enhancement of beef from Bos indicus and Bos taurus cattle following electrical stimulation. *Meat Science*, 86, 635–41.

Hannula, T. and Puolanne, E. (2007). The effect of cooling rate on beef tenderness: The significance of pH at 7°C. *Meat Science*, 67, 403–8.

Hearnshaw, H., Gursansky, B. G., Gogel, B., Thompson, J. M., Fell, L. R., Stephenson, P. D., Arthur, P. F., Egan, A., Hoffman, W. and Perry, D. (1998). Meat quality in cattle of varying Brahman content: the effect of post-slaughter processing, growth rate and animal behaviour on tenderness. *Proceedings 44th International Congress of Meat Science and Technology*, Barcelona, Spain.

Herrera-Mendez, C. H., Becila, S., Boudjellal, A. and Ouali, A. (2006). Meat ageing: Reconsideration of the current concept. *Trends in Food Science and Technology*, 17, 394–405.

Hertzman, C., Olsson, U. and Tornberg, E. (1993). The influence of high temperature, type of muscle and electrical stimulation on the course of rigor, ageing and tenderness of beef muscles. *Meat Science*, 35 (1), 119–41.

Hippe, C. L., Field, R. A., Ray, B. and Russell, W. C. (1991). Effect of spray-chilling on quality of beef from lean and fatter carcasses. *Journal of Animal Science*, 69, 178–83.

Honikel, K. O. (1998). Very fast chilling in beef, 1. Peri-mortem and the chilling process, University of Bristol Press.

Hopkins, D. L. and Thompson, J. M. (2001). Inhibition of protease activity part 1. The effect on tenderness and indicators of proteolysis in ovine muscle. *Meat Science*, 59, 175–85.

Hopkins, D. L. and Thompson, J. M. (2002). The relationship between post-mortem calcium concentration or pH and indicators of proteolysis in ovine muscle. *Meat Science*, 61, 411–14.

Howard, S. T., Woerner, D. R., Scanga, J. A., Van Overbeke, D. L., Mafi, G. G., Igo, J. L., Salman, M. D., Tatum, J. D. and Belk, K. E. (2013). North American Beef Tenderness Survey 2011–12: Benchmarking tenderness and sample shipping procedures. *Journal of Animal Science*, 91, 5981–8.

Huff-Lonergan, E. and Lonergan, S. M. (2007). New frontiers in understanding drip loss in pork: recent insights on the role of postmortem muscle biochemistry. *Journal of Animal Breeding and Genetics*, 124, 19–26.

Hwang, I. H. and Thompson, J. M. (2001a). The effect of time and type of electrical stimulation on the calpain system and meat tenderness in beef longissimus dorsi muscle. *Meat Science*, 58, 135–44.

Hwang, I. H. and Thompson, J. M. (2001b). The interaction between pH and temperature decline early postmortem on the calpain system and objective tenderness in electrically stimulated beef longissimus dorsi muscle. *Meat Science*, 58 (2), 167–74.

Hwang, I. H., Devine, C. E. and Hopkins, D. L. (2003). The biochemical and physical effects of electrical stimulation on beef and sheep meat tenderness. *Meat Science*, 65, 677–91.

Hwang, I. H., Park, B. Y., Cho, S. H. and Lee, J. M. (2004). Effects of muscle shortening and proteolysis on Warner–Bratzler shear force in beef longissimus and semitendinosus. *Meat Science*, 68 (3), 497–505.

Jacob, R. H., Rosenvold, K., North, M., Kemp, R., Warner, R. D. and Geesink, G. H. (2012). Rapid tenderisation of lamb M. longissimus with very fast chilling depends on rapidly achieving sub-zero temperatures. *Meat Science*, 92, 16–23.

Jacob, R. H. and Hopkins, D. L. (2014). Techniques to reduce the temperature of beef muscle early in the post mortem period – a review. *Animal Production Science*, 54, 482–93.

Jaime, I., Beltran, A., Cena, P., Lopez-Lorenzo, P. and Roncales, P. (1992). Tenderisation of lamb meat: effect of rapid post mortem temperature drop on muscle conditioning and aging. *Meat Science*, 32, 357–66.

Janz, J. A. M., Aalhus, J. L. and Price, M. A. (2001). Blast chilling and low voltage electrical stimulation influences on bison (Bison bison bison) meat quality. *Meat Science*, 57, 403–11.

Jeacocke, R. E. (1984). The kinetics of rigor onset in beef muscle fibres. *Meat Science*, 11, 237–51.

Jeremiah, L. E., Martin, A. H. and Murray, A. C. (1985). The effects of various post mortem treatments on certain physical and sensory properties of three different bovine muscles. *Meat Science*, 12, 155–76.

Joseph, R. L. (1996). Very fast chilling of beef and tenderness – a report from an EU concerted action. *Meat Science*, 43, S217–S227.

Kim, Y. H, Lonergan, S. M. and Huff-Lonergan, E. (2010). Protein denaturing conditions in beef deep semimembranosus muscle results in limited m-calpain activation and protein degradation. *Meat Science*, 86 (3), 883–7.

Kim, Y. H. B., Stuart, A., Nygaard, G. and Rosenvold, K. (2012). High pre rigor temperature limits the ageing potential of beef that is not completely overcome by electrical stimulation and muscle restraining. *Meat Science*, 91 (1), 62–8.

Kim, Y. H. B., Warner, R. D. and Rosenvold, K. (2014). Influence of high pre-rigor temperature and fast pH fall on muscle proteins and meat quality: a review. *Animal Production Science*, 54, 375–95.

Koohmaraie, M. (1996). Biochemical factors regulating the toughening and tenderisation processes of meat. *Meat Science*, 43, 193–201.

Koohmaraie, M., Kent, M. P., Shackelford, S. D., Veiseth, E. and Wheeler, T. L. (2002). Meat tenderness and muscle growth: is there any relationship? *Meat Science*, 62, 345–52.

Lee, Y. B. and Ashmore, C. R. (1985). Effect of early postmortem temperature on beef tenderness. *Journal of Animal Science*, 60 (6), 1588–96.

Ledward, D. A. (1985). Post-slaughter influences on the formation of metmyoglobin in beef muscles. *Meat Science*, 15 (3), 149–71.

Li, C. B., Chen, Y. J., Xu, X. L., Huang, M., Hu, T. J. and Zhou, G. H. (2006). Effects of low-voltage electrical stimulation and rapid chilling on meat quality characteristics of Chinese yellow crossbred bulls. *Meat Science*, 72, 9–17.

Li, C., Li, J., Li, X., Hviid, M. and Lundstrom, K. (2011). Effect of low-voltage electrical stimulation after dressing on color stability and water holding capacity of bovine longissimus muscle. *Meat Science*, 88, 559–65.

Li, K., Zhang, Y., Mao, Y., Cornforth, D., Dong, P., Wang, R., Zhu, H. and Luo, X. (2012). Effect of very fast chilling and ageing time on ultra-structure and meat quality characteristics of Chinese yellow cattle M. longissimus lumborum. *Meat Science*, 92 (4), 795–804.

Locker, R. H. and Hagyard, C. J. (1963). A cold shortening effect in beef muscles. *Journal of the Science of Food and Agriculture*, 14, 787–93.

Locker, R. H. and Daines, G. J. (1975). Rigor mortis in beef sternomandibularis muscle at 37°C. *Journal of the Science of Food and Agriculture*, 26 (11), 1721–33.

Mancini, R. A. and Hunt, M. C. (2005). Current research in meat color. *Meat Science*, 71 (1), 100–21.

Marsh, B. B. (1954). Rigor mortis in beef. *Journal of the Science of Food and Agriculture*, 5 (2), 70–5.

Marsh, B. B. and Leet, N. G. (1966). Studies in meat tenderness. III. The effects of cold shortening on tenderness. *Journal of Food Science*, 31, 450–9.

Marsh, B. B., Lochner, J. V., Takahashi, G. and Kragness, D. D. (1981). Effects of early post-mortem pH and temperature on beef tenderness. *Meat Science*, 5 (6), 479–83.

Martin, A. H., Murray, A. C., Jeremiah, L. E. and Dutson, P. J. (1983). Electrical stimulation and carcass aging effects on beef carcasses in relation to post-mortem glycolytic rates. *Journal of Animal Science*, 57 (6), 1456–62.

Morgan, J. B., Savell, J. W., Hale, D. S., Miller, R. K., Griffin, D. B., Cross, H. R. and Shackleford, S. D. (1991). National beef tenderness survey. *Journal of Animal Science*, 69, 3274–83.

Ngapo, T. M. and Dransfield, E. (2006). British consumers preferred fatness levels in beef: Surveys from 1955, 1982 and 2002. *Food Quality and Preference*, 17, 412–17.

Offer, G. (1991). Modelling of the formation of Pale, Soft and Exudative meat effects of chilling regime and rate and extent of glycolysis. *Meat Science*, 30, 157–84.

O'Halloran, G. R., Troy, D. J., Buckley, D. J. and Revile, W. J. (1997). The role of the endogenous proteases in the tenderisation of fast glycolysing muscle. *Meat Science*, 47, 187–210.

Olsson, U., Hertzman, C. and Tornburg, E. (1994). The influence of low temperature, type of muscle and electrical stimulation on the course of rigor mortis, ageing and tenderness of beef muscles. *Meat Science*, 37, 115–31.
Ouellette, R. P., Lord, N. W. and Cheremisinoff, P. N. (1980). *Food Industry Energy Alternatives*, Food & Nutrition Press, Westport, CT.
Pearce, K. L., Hopkins, D. L., Williams, A., Jacob, R. H., Pethick, D. W. and Phillips, J. K. (2009). Alternating frequency to increase the response to stimulation from medium voltage electrical stimulation and the effect on objective meat quality. *Meat Science*, 81, 188–95.
Pearce, K. L., Rosenvold, K., Andersen, H. J. and Hopkins, D. L. (2011). Water distribution and mobility in meat during the conversion of muscle to meat and ageing and the impacts on fresh meat quality attributes – a review. *Meat Science*, 89, 111–24.
Pethick, D. W., Cummins, L., Gardner, G. E., Knee, B. W., McDowell, M., McIntyre, B. W., Tudor, G., Walker, P. J. and Warner, R. D. (1999). The regulation by nutrition of glycogen in the muscle of ruminants. *Recent Advances in Animal Nutrition in Australia*, 12, 145–51.
Pisula, A. and Tyburcy, A. (1996). Hot processing of meat. *Meat Science*, 43, S125–S134.
Renerre, M. (1990). Factors involved in the discoloration of beef meat. *International Journal of Food Science & Technology*, 25 (6), 613–30.
Rhee, M. S. and Kim, B. C. (2001). Effect of low voltage electrical stimulation and temperature conditioning on postmortem changes in glycolysis and calpains activities of Korean native cattle (Hanwoo). *Meat Science*, 58 (3), 231–7.
Rhee, M. S., Wheeler, T L., Shackelford, S D. and Khoomaraie, M. (2004). Variation in palatability and biochemical traits within and among eleven beef muscles. *Journal of Animal Science*, 82, 534–50.
Roeber, D. L., Cannell, R. C., Belk, K. E., Tatum, J. D. and Smith, G. C. (2000). Effects of a unique application of electrical stimulation on tenderness, color, and quality attributes of the beef longissimus muscle. *Journal of Animal Science*, 78, 1504–9.
Rosenvold, K., North, M., Devine, C., Micklander, E., Hansen, P., Dobbie, P. and Wells, R. (2008). The protective effect of electrical stimulation and wrapping on beef tenderness at high pre rigor temperatures. *Meat Science*, 79, 299–306.
Rødbotten, R., Lea, P. and Hildrum, K. (2001). Relative influence of low-voltage electrical stimulation, chilling rate and ageing time on tenderness of M. longissimus dorsi of Norwegian cattle. *Acta Agriculturae Scandinavica*, 51, 184–91.
Sammel, L. M., Hunt, M. C., Kropf, D. H., Hachmeister, K. A., Kastner, C. L. and Johnson, D. E. (2002). Influence of chemical characteristics of beef inside and outside semimembranosus on color traits. *Journal of Food Science*, 67, 1323–30.
Savell, J. W., Mueller, S. L. and Baird, B. E. (2005). The chilling of carcasses. *Meat Science*, 70, 449–59.
Seyfert, M., Hunt, M. C., Mancini, R. A., Hachmeister, K. A., Kropf, D. H. and Unruh, J. A. (2004). Accelerated chilling and modified atmosphere packaging affect colour and colour stability of injection-enhanced beef round muscles. *Meat Science*, 68, 209–19.
Shibata, M., Matsumoto, K., Oe, M., Ohnishi-Kameyama, M., Ojima, K., Nakajima, I., Muroya, S. and Chikuni, K. (2009). Differential expression of the skeletal muscle proteome in grazed 1300 cattle. *Journal of Animal Science*, 87, 2700–8.
Shook, J. N., Van Overbeke, D. L., Scanga, J. A., Belk, K. E., Savell, J. W., Lawrence, T. E., Morgan, J. B., Griffin, D. B., Hale, D. S. and Smith, G. C. (2008). The National Beef Quality Audit 2005. Phase I. Views of producers, packers and merchandisers on current quality characteristics of the beef industry. *Professional Animal Scientist*, 24, 189–97.
Simmons, N. J., Daly, C. C., Mudford, C. R., Richards, I., Jarvis, G. and Pleiter, H. (2006). Integrated technologies to enhance meat quality – an Australasian perspective. *Meat Science*, 74, 172–9.
Simmons, N. J., Daly, C. C., Cummings, T. L., Morgan, S. K., Johnson, N. V. and Lombard, A. (2008). Reassessing the principles of electrical stimulation. *Meat Science*, 80, 110–22.
Smith, G. C. (1985). Effects of electrical stimulation on meat quality, color, grade, heat ring, and palatability. *Advances in Meat Research: Electrical Stimulation*, Springer, Netherlands

Smith, G. C., Tatum, J. D. and Belk, K. E. (2008). International perspective: characterisation of United States Department of Agriculture and Meat Standards Australia systems for assessing beef quality. *Australian Journal of Experimental Agriculture*, 48, 1465–80.

Stolowski, G. D., Baird, B. E., Miller, R. K., Savell, J. W., Sams, A. R., Taylor, J. F., Sanders, J. O. and Smith, S. B. (2006). Factors influencing the variation in tenderness of seven major beef muscles from three Angus and Brahman breed crosses. *Meat Science*, 73, 475–83.

Strydom, P. E. and Buys, E. M. (1995). A pilot study on the effects of spray-chilling on carcass mass loss and surface bacteriology. *Meat Science*, 39, 265–76.

Strydom, P. E. and Frylinck, L. (2014). Minimal electrical stimulation is effective in low stressed and well fed cattle. *Meat Science*, 96, 790–8.

Strydom, P. E., Hope-Jones, M., Frylinck, L. and Webb, E. C. (2011). The effects of a beta-agonist treatment, Vitamin D3 supplementation and electrical stimulation on meat quality of feedlot steers. *Meat Science*, 89, 462–8.

Swatland, H. J. (1981). Cellular heterogeneity in the response of beef to electrical stimulation. *Meat Science*, 5, 451–5.

Tarrant, P. V. and Mothersill, C. (1977). Glycolysis and associated changes in beef carcasses. *Journal of the Science of Food and Agriculture*, 28, 739–49.

Taylor, A. A. (1987). *Accelerated Processing of Meat*, Elsevier Applied Science, London.

Thompson, J. M. (2002). Managing meat tenderness. *Meat Science*, 62, 295–308.

Thompson, J. M., Hopkins, D. L., D'Souza, D. N. Walker, P. J., Baud, S. R. and Pethick, D. W. (2005). Development of sensory protocol for testing palatability of sheep meats. *Australian Journal of Experimental Agriculture*, 45, 469–76.

Thompson, J. M., Perry, D. Daly, B., Gardner, G. E., Johnston, D. J. and Pethick, D. W. (2006). Genetic and environmental effects on the muscle structure response post-mortem. *Meat Science*, 74 (1), 59–65.

Thomson, K. L., Gardner, G. E., Simmons, N. and Thompson, J. M. (2008). Length of exposure to high post-rigor temperatures affects the tenderisation of the beef M. longissmus dorsi. *Australian Journal of Experimental Agriculture*, 48, 1442–50.

Tornberg, E., Wahlgren, M., Brondum, J. and Engelsen, S. B. (2000). Pre-rigor conditions in beef under varying temperature- and pH falls studied with rigometer, NMR and NIR. *Food Chemistry*, 69, 407–18.

Unruh, J. A., Kastner, C. L., Kropf, D. H., Dikeman, M. E. and Hunt, M. C. (1986). Effects of low-voltage electrical stimulation during exsanguination on meat quality and display colour stability. *Meat Science*, 18 (4), 281–93.

Van Moeseke, W., Smet, S. D., Claeys, E. and Demeyer, D. (2001). Very fast chilling of beef: effects on meat quality. *Meat Science*, 59, 31–7.

Wahlgren, N. M., Devine, C. E. and Tornberg, E. (1997). The influence of different pH-courses during rigor development on beef tenderness. In *Proceedings 43rd International Congress of Meat Science and Technology*, 622–3. Auckland, New Zealand.

Warner, R. D., Dunshea, F. R., Gutzke, D., Lau, J. and Kearney, G. (2014a). Factors influencing the incidence of high rigor temperature in beef carcasses in Australia. *Animal Production Science*, 54, 363–74.

Warner, R. D., Kerr, M., Kim, Y. H. B. and Geesink, G. (2014b). Pre-rigor carcass stretching counteracts the negative effects of high rigor temperature on tenderness and water-holding capacity – using lamb muscles as a model. *Animal Production Science*, 54, 494–503.

Warner, R. D., Thompson, J. M., Polkinghorne, R., Gutzke, D. and Kearney, G. A. (2014c). A consumer sensory study of the influence of rigor temperature on eating quality and ageing potential of beef striploin and rump. *Animal Production Science*, 54, 396–406.

Watson, R., Gee, A., Polkinghorne, R. and Porter, M. (2008a). Consumer assessment of eating quality – Development of protocols for Meat Standards Australia (MSA) testing. *Australian Journal of Experimental Agriculture*, 48, 1360–7.

Watson, R., Polkinghorne, R. and Thompson, J. M. (2008b). Development of the Meat Standards Australia (MSA) prediction model for beef palatability. *Australian Journal of Experimental Agriculture*, 48, 1368–79.

Watt, D. B. and Herring, H. K. (1974). Rapid chilling of beef carcasses utilizing ammonia and cryogenic systems: Effects on shrink and tenderness. *Journal of Animal Science*, 38, 928–34.

Whipple, G., Koohmaraie, M., Dikeman, M. E. and Crouse, J. D. (1990). Effects of high-temperature conditioning on enzymatic activity and tenderness of Bos indicus longissimus muscle. *Journal of Animal Science*, 68 (11), 3654–62.

White, A., O'Sullivan, A. Troy, D. J. and O'Neill, E. E. (2006a). Manipulation of the pre-rigor glycolytic behaviour of bovine M. longissimus dorsi in order to identify causes of inconsistencies in tenderness. *Meat Science*, 73, 151–6.

White, A., O'Sullivan, A. Troy, D. J. and O'Neill, E. E. (2006b). Effects of electrical stimulation, chilling temperature and hot-boning on the tenderness of bovine muscles. *Meat Science*, 73, 196–203.

Young, O. A., Priolo, A., Simmons, N. J. and West, J. (1999). Effects of rigor attainment temperature on meat blooming and colour on display. *Meat Science*, 52 (1), 47–56.

Zhu, L. G. and Brewer, M. S. (2003). Effects of urea denaturation and pH on the ability of porcine myoglobin to undergo reduction. *Meat Science*, 63 (4), 427–32.

Zhu, L. X., Gao, S. J. and Luo, X. (2011). Rapid chilling has no detrimental effect on the tenderness of low-voltage electrically stimulated M. longissimus in Chinese bulls. *Meat Science*, 88, 597–601.

Part 3

# Quality traits

Chapter 9

# Beef colour development and variation

*Ranjith Ramanathan, Oklahoma State University, USA and Richard A. Mancini, University of Connecticut, USA*



## 1 Introduction

Meat colour is a very important visual sensory attribute that influences purchasing decisions because consumers often associate bright-red colour with freshness and wholesomeness. It has been estimated that discolouration costs the US meat industry approximately $1 billion every year (Smith et al., 2000). Myoglobin is the protein that is primarily responsible for colour. More specifically, the oxidized form of myoglobin, metmyoglobin, causes discolouration. Van den Oord and Wesdorp (1971) reported that 50% metmyoglobin on the surface of beef results in rejection of meat by consumers. Therefore, maintenance of bright-red colour is critical to minimize the losses associated with discolouration.

Ante-mortem stress and post-mortem muscle pH influence meat colour through their effect on mitochondria. As a result, mitochondrial activity has received significant research focus to better understand processes that compete for oxygen and regenerate reducing equivalents in post-mortem muscle (Faustman and Cassens, 1990). More specifically, competition for oxygen between mitochondria and myoglobin is the limiting factor responsible for the development of bright cherry-red surface colour. Research also indicates that mitochondria can improve colour stability by regenerating reducing equivalents necessary for metmyoglobin reduction (Tang, 2005b). The practical implications of mitochondrial activity are evident in dark cutting beef (high pH) as well as the effects of ageing and injection-enhancement technology on colour.

http://dx.doi.org/10.19103/AS.2016.0009.09

Dark cutting beef is a condition where beef does not have a characteristic bright-red colour because ante-mortem stress depletes glycogen reserves prior to slaughter, which results in a high muscle pH. As a result, meat fails to bloom or brighten after the cut surface has been exposed to oxygen. Dark cutting beef is one of the most prominent meat quality issues worldwide. The US beef industry lost approximately $165–170 million in 2000 due to dark cutting carcasses (McKenna et al., 2002). An estimated $11 million loss per year to beef producers in Canada and a $35 million loss to the Australian beef industry have also been reported.

Although enzymes involved in mitochondrial function remain active post-mortem, beef ageing and injection-enhancement of whole muscles can influence colour development and stability because substrates required for mitochondrial oxygen consumption are depleted. As a result, ageing decreases mitochondrial activity, which can both improve red colour development and decrease colour stability. Injection-enhancement can increase the amount of available mitochondrial substrates and therefore, improve colour stability during display. This chapter will focus on the effects of post-mortem biochemistry, more specifically mitochondrial activity, on beef colour development and strategies to improve colour.

## 2 Myoglobin forms and beef colour

The sarcoplasmic protein primarily responsible for meat colour is myoglobin that consists of a colourless globin moiety comprising 153 amino acids arranged in 8 alpha helix segments wrapped around a haeme prosthetic group, which contributes to colour. The amino acids protect the hydrophobic haeme from the environment and also make myoglobin hydrophilic. The haeme has an iron in the centre. This central iron has six coordination sites – four are associated with the protoporphyrin ring; the fifth connects the haeme with the globin portion via the proximal histidine; and the sixth position is available to bind ligands such as oxygen, nitric oxide, carbon monoxide and water. Myoglobin can exist in two different valence states, that is, containing either ferrous or ferric iron. The valence state and the ligand present at the sixth position determine meat colour.

Myoglobin can exist in four forms: deoxymyoglobin, oxymyoglobin, metmyoglobin and carboxymyoglobin. Deoxymyoglobin is formed when no ligand occupies the sixth position and iron is in the ferrous state. This results in a purplish colour and occurs mainly in vacuum-packaged meat. Oxymyoglobin imparts a bright cherry-red colour to fresh beef when oxygen is the ligand bound to ferrous iron. The distal histidine can stabilize the ligand bound to myoglobin by linking the globin to the oxygen that is bound to iron. Oxymyoglobin can undergo oxidation, resulting in discolouration via the formation of brown metmyoglobin. In this case, water occupies the sixth position of ferric iron. When carbon monoxide binds with deoxymyoglobin, a characteristic bright-red colour is formed due to carboxymyoglobin. The human eye cannot distinguish oxymyoglobin from carboxymyoglobin. Myoglobin's haeme has a greater affinity for carbon monoxide than for oxygen, and the distal histidine can stabilize carbon monoxide more efficiently than oxygen, resulting in a stable red colour for carboxymyoglobin. The FDA has approved the use of carbon monoxide at a level of 0.4% in packaging to improve colour stability. To a lesser extent, haemoglobin and cytochrome can also contribute to colour. Warriss and Rhodes (1977) reported that in fresh meat cuts, residual haemoglobin contributes about

0.3% of total blood volume of live animal. Cytochrome is an electron transport protein present in mitochondria. Nevertheless, the contribution of cytochrome is minor.

# 3 Factors affecting meat colour

## 3.1 Biochemical factors affecting beef colour

Metmyoglobin reducing activity (MRA) and oxygen consumption are important biochemical parameters that can influence beef colour. MRA is the ability of muscle to convert metmyoglobin to deoxymyoglobin (Ledward, 1985). Although both enzymatic and non-enzymatic pathways can contribute to MRA, key requirements include an electron donor and an electron carrier. In enzymatic MRA, cytochrome b5 NADH-dependent reductase can act as an electron carrier, whereas in non-enzymatic MRA, methylene blue/cytochrome c can act as an electron carrier. The source of electrons can be from either NADH or mitochondrial activity. Bekhit and Faustman (2005) reviewed in detail the factors involved in MRA.

Mitochondria are important organelles involved in oxygen consumption. More specifically, mitochondria limit oxygen availability to myoglobin, maintaining a deoxygenated pigment state. This process also can contribute to electron transport-mediated MRA, minimizing discolouration because oxymyoglobin must first be converted to metmyoglobin before further reduction to deoxymyoglobin can occur (Hunt and Mancini, 2005). Therefore, when MRA and oxygen consumption are not optimum, colour stability may be limited. As a result, mitochondria influence both beef colour development (oxymyoglobin formation; Fig. 1) and colour deviations (discolouration due to metmyoglobin formation).

## 3.2 Myoglobin primary structure and oxidation properties

The primary amino acid sequence of myoglobin can also influence oxidation. Yin et al. (2011) compared myoglobin oxidation of seven livestock species and concluded that oxidation is species specific. More specifically, the number of histidine residues in the myoglobin of meat-producing livestock species can influence oxidation rate. Porcine myoglobin contains 9 histidine residues, whereas bovine myoglobin contains 13 histidine

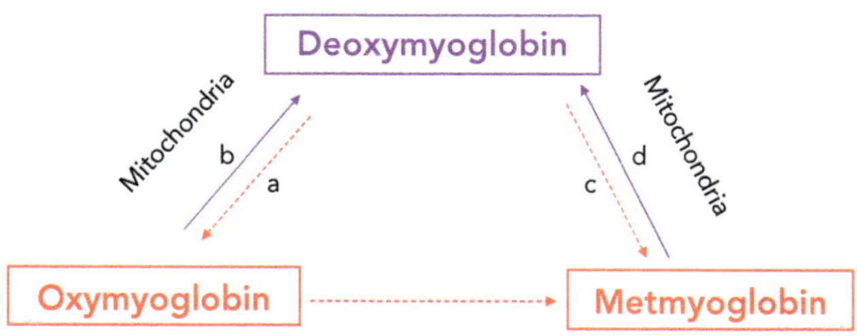


**Figure 1** Role of mitochondria in beef colour: (a) decreased mitochondrial oxygen consumption favours bloom and bright-red colour development, (b) increased mitochondrial oxygen consumption favours deoxymyoglobin and darker beef, (c) decreased mitochondrial activity favours metmyoglobin due to limited metmyoglobin reducing activity and oxygen partial pressure that promotes metmyoglobin formation and (d) increased mitochondrial activity favours metmyoglobin reduction and colour stability.

residues. Lipid oxidation can generate primary and secondary oxidation products such as aldehydes, ketones and peroxides. Primary products such as peroxide are short-lived and initiate further reactions. Secondary products such as aldehydes and ketones are long-lived, react with proteins and lipids, and freely diffuse in the sarcoplasm. *In vitro* studies have shown that incubating oxymyoglobin with aldehydes can increase myoglobin oxidation. 4-hydroxy-2-nonenal can preferentially bind with histidine residues of livestock species (Suman et al., 2006), which can destabilize myoglobin, as well as promote oxidation and haeme release. Furthermore, 4-hydroxy-2-nonenal binding makes myoglobin a poor substrate for enzymatic and non-enzymatic metmyoglobin reduction (Lynch and Faustman, 2000; Elroy et al., 2015). Application of proteomics has helped to study the covalent binding of lipid oxidation products to myoglobin.

Irrespective of primary structure, a decreased pH favours myoglobin oxidation by promoting protonation of oxygen within oxymyoglobin and destabilizing haeme-globin linkages (Livingston and Brown, 1982). A rapid decline in post-mortem pH can also promote myoglobin denaturation. This has much implication for the pork industry because the pale, soft and exudative condition is less common in beef.

## 3.3 Inherent muscle biochemistry

Anatomical location and the proportion of red and white fibres can influence beef colour. McKenna et al. (2005) determined the biochemical and physical factors affecting discolouration of 19 bovine muscles and classified beef muscles based on colour stability. These authors concluded that oxygen consumption and MRA are both the primary determinants of colour stability and are dependent on the proportion of red and white muscle fibres. For example, the psoas major has predominantly red fibres (type I; colour liable) and the longissimus lumborum has predominantly white fibres (type II; colour stable; Hunt and Hendrick, 1977). Type I fibres have greater myoglobin and mitochondrial concentration. A recent study suggests that the sarcoplasmic proteome differs significantly between the longissimus lumborum and psoas major (Joseph et al., 2012). More specifically, an overabundance and positive correlation of enzymes involved in colour stability such as aldose reductase, pyruvate dehydrogenase, β-enolase and triose phosphate isomerase were reported in longissimus lumborum muscle. Muscles abundant in red fibres (psoas major) will have more mitochondria and capillary density than muscles abundant in white fibres (longissimus lumborum). Hence tricarboxylic acid cycle and glycolytic substrates will be utilized at different rates (Kushmerick et al., 1992; Glancy and Balaban, 2011).

The semimembranosus muscle is an example of colour variation within a muscle. The outside portion of the semimembranosus is colour-stable whereas the interior portion is colour-labile (Sammel et al., 2002). This has been partly attributed to differences in temperature and pH decline during carcass chilling.

## 3.4 Effects of diet on beef colour

Pasture-fed cattle typically have darker meat than grain-fed cattle due to (1) less fat thickness and faster chilling, and (2) less muscle glycogen. More specifically, grain-fed cattle have more fat and glycogen, which promotes warmer post-mortem muscle temperature and increases glycolysis and lactic acid formation, especially in larger muscles. Although this will decrease pH and will lead to less water holding capacity and pale colour, increased lactic acid formation can also result in greater oxygen consumption by myoglobin than mitochondria.

### 3.5 Post-harvest practices to improve meat colour

Case-ready packaging has allowed beef purveyors to modify gas compositions within packages. Discolouration results when the thin layer of metmyoglobin formed between bright-red and purple portions of a steak migrates to the surface. This thin layer of metmyoglobin is due to decreases in oxygen partial pressure (Taylor, 1972). Although oxygen consumption and MRA limit the formation of metmyoglobin, these processes are decreased when meat cuts are aged. This has led to the use of greater oxygen concentrations in high-oxygen MAP, which allow oxygen to penetrate the interior of steaks to form deeper layers of oxymyoglobin and improved colour stability compared with PVC packaging. John et al. (2005) reported that sirloin steaks packaged in high-oxygen MAP had a stable red colour for 14 days. Oxygen scavengers have been used in meat packaging systems to scavenge residual oxygen because when beef does not have sufficient oxygen consumption to consume residual oxygen in the package, the decreased oxygen partial pressure can promote metmyoglobin formation. This also influences beef packaged in CO-MAP because carbon monoxide cannot bind to metmyoglobin. Uboldi et al. (2015) reported that the oxygen partial pressure decreased to 0.08 kP within 24 hours of storage with the inclusion of oxygen scavengers in the mother bag.

## 4 Case study 1: mitochondria effects on colour deviation

Post-mortem ageing increases palatability and tenderness (Dransfield, 1994). However, ageing also influences colour development and stability because mitochondrial activity can decrease myoglobin oxygenation and improve MRA. Although enzymes associated with mitochondrial function remain active post-mortem, substrates required for mitochondrial oxygen consumption are depleted (Andrews et al., 1951). As a result, the benefits of ageing on tenderness should be balanced with the effects of ageing on beef colour

1. How does ageing influence bloom?
2. What are the implications of ageing on colour stability?
3. How does ageing influence mitochondrial function, including mitochondria-mediated metmyoglobin reduction?
4. Can the addition of mitochondrial substrates in post-mortem muscle influence beef colour?

Competition for oxygen between myoglobin and mitochondria is the key determinant of beef colour development. As a result, post-mortem ageing can increase initial red colour development because decreased mitochondrial functionality will lead to less competition for oxygen. Mancini and Ramanathan (2014) reported that storage of beef will decrease mitochondrial activity and oxygen consumption because substrates are continuously depleted. Furthermore, Tang et al. (2005a) noted morphological changes in mitochondria during storage. Nevertheless, mitochondria remained active with compromised structure. However, the beneficial effects of limited mitochondrial activity on colour development may be negated by subsequent detrimental effects on colour stability. More specifically, the decrease in oxygen consumption associated with ageing also limits mitochondria-mediated

MRA. In addition, limited oxygen consumption can promote an oxygen partial pressure that is prone to myoglobin oxidation and discolouration. Hence, ageing can decrease colour stability.

Case-ready technology has allowed beef purveyors to enhance beef palatability traits with various ingredients and to modify the atmospheric conditions within packages (McMillin, 2008). Lactate is commonly added as an antimicrobial; however, this can result in darker beef. The mechanism of lactate-induced darkening suggests that the addition of lactate results in the formation of NADH by lactate dehydrogenase (Kim et al., 2006). This NADH can increase mitochondrial oxygen consumption, promoting a greater competition for oxygen, more deoxygenated myoglobin and a dark colour. However, increased NADH can also influence colour stability. More specifically, although lactate-enhanced loins are often darker than non-enhanced loins, this colour is more stable during display due to an increase in mitochondria-mediated MRA and NADH-dependent metmyoglobin reduction.

# 5 Case study 2: effects of ante-mortem stress on beef colour

Mitochondria are membrane-bound organelles primarily involved in energy production. Post-mortem muscle pH can significantly influence mitochondrial activity and, as a result, beef colour development and stability. Therefore, understanding the mechanism by which ante-mortem factors contribute to deviations in muscle pH will promote the development of post-mortem technologies that minimize darkening.

1. How does ante-mortem stress influence myoglobin oxygenation or bloom?
2. Can the post-mortem addition of mitochondrial inhibitors minimize darkening?

Typical post-mortem beef muscle pH ranges from 5.45 to 5.6. This is due to a shift from aerobic to anaerobic metabolism, which results in an accumulation of lactic acid and a decline in muscle pH from 7.2 to 5.6. Dark cutting beef results from the depletion of glycogen reserves prior to slaughter caused by both emotional and physical stress, and is often described as meat that fails to bloom after the cut surface has been exposed to oxygen (Hedrick et al., 1959). A limited amount of muscle glycogen at slaughter minimizes post-mortem lactic acid formation, resulting in a greater ultimate pH that influences both biochemical and physical muscle properties. More specifically, increased pH enhances mitochondria activity, which can decrease myoglobin oxygenation or red colour intensity (Lawrie, 1958; Ashmore et al., 1972; Ashmore, Parker and Doerr, 1972; Tang et al., 2005a). Furthermore, increased pH will slow mitochondrial degenerative reactions compared with normal muscle pH, allowing mitochondria to consume oxygen for a greater amount of time and making it less available for oxygenation (Fig. 2).

Dark cutting beef is a quality issue worldwide. Various studies have assessed post-harvest technologies for minimizing the incidence of dark cutting beef based on mitochondria function and biochemistry. For example, rotenone is a complex I inhibitor that blocks mitochondrial activity and significantly improves the redness of dark cutting muscle homogenates (Cornforth and Egbert, 1985; Ramanathan et al., 2009). From a more practical perspective, lactic acid enhancement has been used to decrease the pH of dark cutting beef and improve subsequent oxygenation (Sawyer et al., 2009).

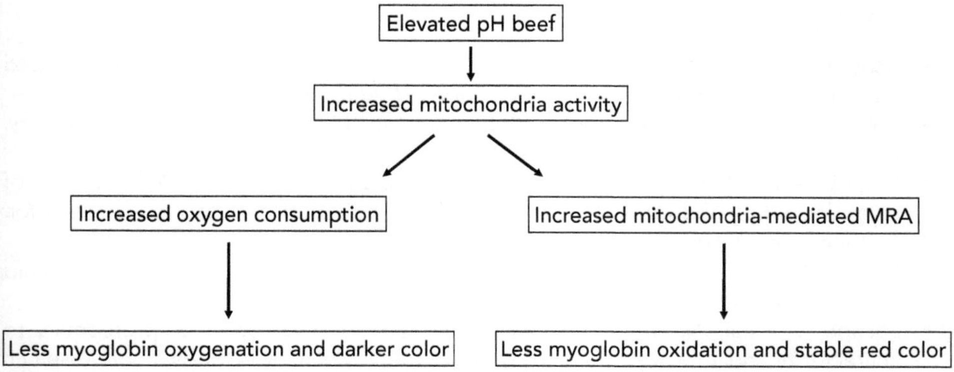


**Figure 2** Effects of elevated pH on beef colour.

# 6 Summary

Mitochondria remain active in post-mortem muscle and can influence myoglobin redox state via oxygen consumption and MRA. However, substrates are depleted continuously and oxidative conditions decrease mitochondrial activity. Hence, characterizing factors that influence mitochondrial function will help to develop pre- and post-harvest technologies that minimize the economic losses due to discolouration. Minimizing ante-mortem stress of cattle, utilizing optimum chilling and proper packaging can optimize beef colour development and stability.

# 7 Future trends

Meat colour will remain an important factor in purchasing decisions. Increasing knowledge of myoglobin chemistry and mitochondrial biochemistry are critical to develop technologies that improve colour. Understanding muscle-specific biochemistry will help to formulate packaging and injection-enhancement technologies that increase value. Developing techniques such as genomics, proteomics and metabolomics will increase our understanding of the biochemical processes that influence beef colour.

# 8 Where to look for further information

## Journals

- *Meat Science*
- *Journal of Agricultural and Food Chemistry*
- *Journal of Food Science*
- *Journal of Animal Science*
- *Food Chemistry*

## Recommended articles

- Ashmore, C. R., Parker, L. and Doerr, L. (1972). Respiration of mitochondria isolated from dark-cutting beef: Postmortem changes. *J. Anim. Sci.* 34, 46–8.
- Bekhit, A. E. D. and Faustman, C. (2005). Review. Metmyoglobin reducing activity. *Meat Sci.* 71, 407–39.
- English, A. R., Mafi, G. G., VanOverbeke, D. L. and Ramanathan, R. (2016). Effects of extended aging and modified atmospheric packaging on beef top loin steak color. *J. Animal Sci. Accepted*, In press.
- Faustman, C., Sun, Q., Mancini, R. A. and Suman, S. P. (2010). Myoglobin and lipid oxidation interactions: Mechanistic bases and control. *Meat Sci.* 86, 86–94.
- Joseph, P., Nair, N. M. and Suman, S. P. (2015). Application of proteomics to characterize and improve color and oxidative stability of muscle foods. *Food Research International*, 76, 938–45.
- John, L., Cornforth, D., Carpenter, C. E., Sørheim, O., Pettee, B. C. and Whittier, D. R. (2005). Color and thiobarbituric acid values of cooked top sirloin steaks packaged in modified atmospheres of 80% oxygen, or 0.4% carbon monoxide, or vacuum. *Meat Sci.* 69, 441–9.
- Mancini, R. A. and Ramanathan, R. (2014). Effects of postmortem storage time on color and mitochondria in beef. *Meat Sci.* 98, 65–70.
- Mancini, R. A. and Hunt, M. C. (2005). Current research in meat color. *Meat Sci.* 71,100–21.
- McKenna, D. R., Mies, P. D., Baird, B. E., Pfeiffer, K. D., Ellebracht, J. W. and Savell, J. W. (2005). Biochemical and physical factors affecting discoloration characteristics of 19 bovine muscles. *Meat Sci.* 70, 665–82.
- McMillin, K. W. (2008). Where is MAP going? A review and future potential of modified atmosphere packaging for meat. *Meat Sci.* 80, 43–65.
- Sammel, L. M., Hunt, M. C., Kropf, D. H., Hachmeister, K. A. and Johnson, D. E. (2002). Comparison of assays for metmyoglobin reducing ability in beef inside and outside semimembranosus muscle. *J. Food Sci.* 67, 978–84.
- Tang, J., Faustman, C., Hoagland, T. A., Mancini, R. A., Seyfert, M. and Hunt, M. C. (2005). Postmortem oxygen consumption by mitochondria and its effects on myoglobin form and stability. *J. Agric. Food Chem.* 53, 1223–30.

## Conferences

- International Congress of Meat Science and Technology
- Reciprocal Meat Conference, American Meat Science Association

## Institutes

- U.S. Meat Animal Research Center (MARC) http://www.ars.usda.gov/main/site_main.htm?modecode=30-40-05-00
- French National Institute for Agricultural Research – Inra, http://www.inra.fr/en/Scientists-Students
- Teagasc – Research and Innovation – Meat Technology and Quality, http://www.teagasc.ie/food/research/chemistry_technology/meat_technology/
- South Africa – http://www0.sun.ac.za/sarchi/

- Germany – http://www.mri.bund.de/en/de/institute/sicherheit-und-qualitaet-bei-fleisch.html
- AgResearch, New Zealand, http://www.agresearch.co.nz/

# 9 References


Andrews, M. M., Guthneck, B. T., McBride, B. H. and Schweigert, B. S. (1951). Stability of certain respiratory and glycolytic enzyme systems in animal tissues. *J. Biol. Chem.* 194, 715–19.

Ashmore, C. R., Parker, L. and Doerr, L. (1972). Respiration of mitochondria isolated from dark-cutting beef: Postmortem changes. *J. Anim. Sci.* 34, 46–8.

Bekhit, A. E. D. and Faustman, C. (2005). Review. Metmyoglobin reducing activity. *Meat Sci.* 71, 407–39.

Cornforth, D. P. and Egbert, W. R. (1985). Effect of rotenone and pH on the color of pre-rigor muscle. *J. Food Sci.* 50, 34–5, 44.

Dransfield, E. (1994). Optimization of tenderization, aging, and tenderness. *Meat Sci.* 36,105–21.

Elroy, N. N., Rogers, J., Mafi, G. G., VanOverbeke, D. L., Hartson, S. D. and Ramanathan, R. (2015). Species-specific effects on non-enzymatic metmyoglobin reduction in vitro. *Meat Sci.* 105, 108–13.

Faustman, C. and Cassens, R. G. (1990). The biochemical basis for fresh meat discoloration: A review. *J. Mus. Foods*, 1, 217–43.

Glancy, B. and Balaban, R. S. (2011). Protein composition and function of red and white skeletal muscle mitochondria. *American J. Physio. – Cell Physio.* 300, C1280–90.

Hedrick, H. B., Boillot, J. B., Brady, D. E. and Naumann, H. D. (1959). Etiology of dark-cutting beef. Columbia, MO: Missouri Agriculture Experiment Station Bull test, pp. 717–21.

Hunt, M. C. and Hedrick, H. B. (1977). Profile of fiber types and related properties of five bovine muscles. *J. Food Sci.* 42, 513–17.

John, L., Cornforth, D., Carpenter, C. E., Sørheim, O., Pettee, B. C. and Whittier, D. R. (2005). Color and thiobarbituric acid values of cooked top sirloin steaks packaged in modified atmospheres of 80% oxygen, or 0.4% carbon monoxide, or vacuum. *Meat Sci.* 69, 441–9.

Joseph, P., Suman, S. P., Rentfrow, G., Li, S. and Beach, C. M. (2012). Proteomics of muscle specific beef color stability. *J. Agric. Food Chem.* 60, 3196–203.

Kim, Y. H., Hunt, M. C., Mancini, R. A., Seyfert, M., Loughin, T. M., Kropf, D. H., et al. (2006). Mechanism for lactate-color stabilization in injection-enhanced beef. *J. Agric. Food Chem.* 54, 7856–62.

Kushmerick, M. J., Meyer, R. A. and Brown, T. R. (1992). Regulation of oxygen consumption in fast- and slow-twitch muscle. *Amer. J. Physio. – Cell Physio.* 263, C598–606.

Lawrie, R. A. (1958). Physiological stress in relation to dark-cutting beef. *J. Food Sci. Agri.* 11, 721.

Ledward, D. A. (1985). Post-slaughter influences on the formation of metmyoglobin in beef muscles. *Meat Sci.* 15,149–71.

Lynch, M. P. and Faustman, C. (2000). Effect of aldehyde lipid oxidation products on myoglobin. *J. Agric. Food Chem.* 48, 600–4.

Livingston, D. J. and Brown, W. D. (1982). The chemistry of myoglobin and its reactions. *Food Tech.* 35, 244–52.

Mancini, R. A. and Ramanathan, R. (2014). Effects of postmortem storage time on color and mitochondria in beef. *Meat Sci.* 98, 65–70.

Mancini, R. A. and Hunt, M. C. (2005). Current research in meat color. *Meat Sci.* 71, 100–21.

McKenna, D. R., Mies, P. D., Baird, B. E., Pfeiffer, K. D., Ellebracht, J. W. and Savell, J. W. (2005). Biochemical and physical factors affecting discoloration characteristics of 19 bovine muscles. *Meat Sci.* 70, 665–82.

McKenna, D. R., Roeber, D. L., Bates, P. K., Schmidt, T. B., Hale, D. S., Griffin, D. B., Savell, J. W., Brooks, J. C., Morgan, J. B., Montgomery, T. H., Belk, K. E. and Smith, G. C. (2002). National beef quality audit-2000: Survey of targeted cattle and carcass characteristics related to quality, quantity, and value of fed steers and heifers. *J. Anim. Sci.* 80,1212–22.

McMillin, K. W. (2008). Where is MAP going? A review and future potential of modified atmosphere packaging for meat. *Meat Sci.* 80, 43–65.

Ramanathan, R., Mancini, R. A. and Konda, M. K. R. (2009). Effects of lactate on beef heart mitochondrial oxygen consumption and muscle darkening. *J. Agric. Food Chem.* 57, 1550–5.

Sawyer, J. T., Apple, J. K., Johnson, Z. B., Baublits, R. T. and Yancey, J. W. S. (2009). Fresh and cooked color of dark-cutting beef can be altered by post-rigor enhancement with lactic acid. *Meat Sci.* 83, 263–70.

Sammel, L. M., Hunt, M. C., Kropf, D. H., Hachmeister, K. A. and Johnson, D. E. (2002). Comparison of assays for metmyoglobin reducing ability in beef inside and outside semimembranosus muscle. *J. Food Sci.* 67, 978–84.

Smith, G. C., Belk, K. E., Sofos, J. N., Tatum, J. D. and Williams, S. N. (2000). Economic implications of improved color stability in beef. In E. A. Decker, C. Faustman and C. J. Lopez-Bote (Eds), *Antioxidants in Muscle Foods: Nutritional Strategies to Improve Quality*, pp. 397–426. New York: Wiley Interscience.

Suman, S. P., Faustman, C., Stamer, S. L. and Liebler, D. C. (2006). Redox instability induced by 4-hydroxy-2-nonenal in porcine and bovine myoglobins at pH 5.6 and 4°C. *J. Agric. Food Chem.* 54, 3402–8.

Tang, J., Faustman, C., Hoagland, T. A., Mancini, R. A., Seyfert, M. and Hunt, M. C. (2005a). Postmortem oxygen consumption by mitochondria and its effects on myoglobin form and stability. *J. Agric. Food Chem.* 53, 1223–30.

Tang, J., Faustman, C., Mancini, R. A., Seyfert, M. and Hunt, M. C. (2005b). Mitochondrial reduction of metmyoglobin: Dependence on the electron transport chain. *Agric. Food Chem.* 53, 5449–55.

Taylor, A. A. (1972). Gases in meat packaging. *Meat World.* 5, 3–6.

Uboldi, E., Zanoletti, M., Franzetti, F. and Limbo, S. (2015). Master bag low-oxygen packaging system: Quality evolution of ground beef patties during storage, blooming and display presentation. *Food Pack. Shelf Life*, 5, 75–82.

Van den Oord, A. H. and Wesdorp, J. J. (1971). Analysis of pigments in intact beef samples. *J. Food Tech.* 6, 1–8.

Warriss, P. D. and Rhodes, D. N. (1977). Haemoglobin concentrations in beef. *J. Sci. Food Agri.* 28, 10, 931–4.

Yin, S., Faustman, C., Tatiyaborworntham, N., Ramanathan, R., Maheswarappa, N. B., Mancini, R. A., Joseph, P., Suman, S. P. and Sun, Q. (2011). Species-specific myoglobin oxidation. *J. Agric. Food Chem.* 59, 12198–203.

Chapter 10

# Beef carcass grading and classification

*Michael E. Dikeman, Kansas State University, USA*



## 1 Introduction

Global demand for beef has increased rather dramatically in recent years and is expected to continue to increase, primarily because of an emerging 'middle class' in several developing countries. There is clear evidence for beef as a choice of protein, if it meets consumers' expectations for palatability, and is convenient, consistent, safe, healthy and affordable. The goals for beef production, beef processing and retail marketing in all countries, therefore, need to include

1 appropriate integration of genetics and management systems that are profitable and sustainable;
2 cattle with optimum dressing percentages, meat yields and meat quality to meet beef processor demands and profitability;
3 meat that has the highest level of safety and wholesomeness;
4 meat that is optimum in nutritional value;
5 production and processing systems that meet societal expectations for animal welfare and are friendly to the environment and
6 meat that consistently meets consumers' expectations for palatability.

In general, there are three grading or classification systems in three continents that are used for a large proportion of beef production and processing in the world:

http://dx.doi.org/10.19103/AS.2016.0009.10

- The 'EUROP' system for EU countries.
- The U.S. Department of Agriculture (USDA) Beef Carcass Grading system in the United States and
- The Meat Standards Australia (MSA) system in Australia.

However, these three systems are quite different and reflect differing production systems. Cattle in beef production in the United States are predominantly finished on high-grain diets, whereas cattle in most other countries are predominately finished on grass and forage. Growth-promoting technologies are used in the United States and Australia, but not allowed/used in many other countries. Grading systems are less well established in some other countries. There are over 68 million cattle and yaks in China, but there are also umpteen breeds and production systems, which has resulted in the national standards of carcass grading not being enforced in that country (Meng, 2015). Brazil has a very large beef cattle population. However, because a majority of beef cattle in Brazil are of *Bos indicus* breeding finished on grass and known for relatively low marbling and less tender beef than that of *Bos taurus* breeding, Brazil does not have a national beef grading system. In Japan, a very specialized system based on extremely high marbling is used for 'Wagyu' cattle, but it does not work well for dairy and cross-bred cattle.

# 2 The purposes of classification and grading

The terms 'classification' and 'grading' often do not have the same connotation in all countries; however, they are sometimes used interchangeably. A widely accepted definition of carcass classification is the sorting of carcasses into classes based on specified criteria. Because biological variation is characterized by continuous variables, sorting carcasses into classes will inevitably lack precision, even though there will be less variation among carcasses or primal cuts within a class than among carcasses or primal cuts across different classes. In general, the objective of classifying or grading of carcasses is to reduce the amount of variation to a more manageable level. Grading, in contrast to classification, involves 'value judgements' to segment carcasses or meat according to their 'intrinsic value'. Both classification and grading schemes evolve over time in response to changes in the cattle population, changes in the market and as a response to new knowledge and developments in technology.

Traditional grading or classification systems in most countries have been based on 'carcass systems' in which carcasses are grouped according to similar appearance and/or composition and 'visual quality' traits, but most grading or classification systems are *not* based on actual consumer satisfaction for palatability. However, several countries are conducting research on developing systems with consumer satisfaction as the primary endpoint.

There are several beef-grading or classifying systems intended to predict or relate to meat appearance and palatability with various levels of accuracy, as well as systems to predict meat yield. 'Quality' sometimes is used in reference to carcass conformation or meat yield; however, it is more common and logical for 'quality' to refer to meat appearance and/or palatability, which is how 'quality' is used in this chapter. The focus of the *International Meeting on Beef and Lamb Carcass Grading to Underpin Consumer Satisfaction*, 20–21 August 2015, Paris, France was on eating quality as accepted by consumers.

The main purposes of beef classification and grading schemes include a target and reward system for cattle producers according to the composition and quality of the carcasses produced, and to facilitate trade in carcasses and meat by providing common and unambiguous language for describing commercially important attributes. Another purpose arises from the need for regulatory authorities to collect price information to enable market transparency, which requires a definition of standard carcasses on which comparisons may be based (Allen, 2014). Provided that the attributes have a real value in trade and that the classification system is operated with sufficient accuracy for all sectors of the industry to have confidence in it, composition and/or quality-based payments to producers are then linked to the classes or grades. This then becomes an effective means of sending signals about the type of carcasses desired by the market and, more importantly, consumer satisfaction.

When extra returns are available to producers who make adjustments by using different breeds or genetic selection emphasis and/or by making changes to the production system, these should have the effect over time for changing the population of harvested cattle in the direction of improved quality and/or composition as defined by market demands. This was most clearly demonstrated by the continuous reduction in fatness of pork carcasses over several decades in those countries that introduced effective grading or classification schemes. Due to the longer generation interval and the less clear signals being sent to producers in some countries, the effectiveness of beef carcass classification or grading schemes is less easily demonstrated. However, the principle that effective schemes aligned with meaningful price differentials will bring about change in the desired direction is nonetheless true.

## 3 Beef classification versus grading in Australia

In 1987, the AUS-MEAT 'classification' system included a set of descriptive terms describing features of carcasses that are useful to those involved in the trading of carcasses. The current AUS-MEAT classification system involves assessments of sex, dentition, carcass weight, P8 fat and butt shape on the slaughter floor; and marbling, meat colour and fat colour in the chiller. Those features deemed most important vary considerably among purchasers of carcasses and are not particularly effective in relating to eating quality. The AUS-MEAT classification system is in contrast to a 'grading' system that is defined as placing different values on carcasses for pricing purposes. Generally, this involves ranking carcasses in a 'hierarchy' for the traits of interest.

A fundamental change incorporated within the AUS-MEAT language was a shift from 'classifying' carcasses to the ultimate goal of grading various cuts for a cooked meal portion. Therefore, Meat Standards Australia (MSA) evolved beginning in 1995 to assign individual grades describing their predicted eating quality when cooked by a designated cooking method at a defined ageing period. Because only about 3 million cattle are graded by the MSA system, the country has both a 'classification' system and an individual muscle 'grading' system (MSA).

The MSA approach is distinctly different from other grading or classification systems that are in current use. Firstly, it is based on consumers' responses and, secondly, it independently grades 39 meat cuts rather than applying a common grade to the entire carcass. The system uses a total quality management (TQM) system approach to grading by including information about all the events that have occurred up until the point when steaks or roasts are cooked and eaten. This includes the genetics of cattle, backgrounding and finishing systems, pre-harvest handling of animals, and post-harvest treatment and processing of carcasses and meat (reviewed by Polkinghorne et al., 2008; Polkinghorne and Thompson, 2010).

It is of critical importance to the success of the MSA grading system that Australia has a national animal identification system and that Meat and Livestock Australia (MLA) is a producer-owned company. MLA works with the Australian Government and the wider red meat industry to deliver value to levy payers. Also, not nearly all cattle qualify to be graded by the MSA standards (about 3.2 million cattle qualify for grading every year).

Key grading factors include the following:

- ***Bos indicus*** content is assessed on a scale from 0% to 100%. It is widely known that as *Bos indicus* content increases, palatability decreases (primarily tenderness), although the rate of decrease varies with individual muscles. *Bos indicus* content is inputted from the vendor declaration or measured by using a combination of hump height and carcass weight.
- **Gender**. It is widely known that bulls generally have a lower eating quality score (this is based on age and other factors) than steers and heifers. There is no distinction between females and castrated males when corrected for other carcass grading parameters.
- **Hormonal growth promotants (HGPs)**. The MSA system uses the principle that muscles from animals implanted with steroidal growth promotants have lower MQ4 scores, with the effect ranging from 0 to 10 points, depending on the muscle. However, this system assumes that there is no difference between the use of one or two less aggressive implants that have only a subtle effect on meat palatability, and when three or more aggressive implants are used in the life of cattle, which can have a negative effect on palatability. The scientific literature points to β-agonists having a similar or greater negative effect on meat palatability as hormonal growth implants, yet Ractopamine™ has negligible effects on beef palatability, whereas Zilmax™ can have significant negative effects on palatability without prolonged ageing. However, the MSA system groups hormonal implants and β-agonists together, although work is currently underway to test for differences among hormonal implants (number and composition) and β-agonists.
- **Marbling and rib fat thickness**. Both marbling score and rib fat thickness are used to assess the impact of marbling on the palatability of individual muscles. An increase in marbling score from 250 to 550 results in an increase of eight palatability units for the striploin, with the magnitude of the adjustment varying with other muscles. Rib fat thickness is also used as a threshold effect in which carcasses must have more than 3 mm to minimize variation in chilling. However, the scientific literature shows that 3 mm is not adequate to prevent cold toughening, which results from chilling too rapidly.
- **Ossification score/carcass weight/weaning status**. The effect of ossification score on palatability interacts with carcass weight and varies according to muscle. The negative effect of increased ossification scores can be minimized by increased carcass weight (see example in Fig. 1). In addition, there is a milk-fed veal effect where muscles from young un-weaned calves are more palatable across all muscles.

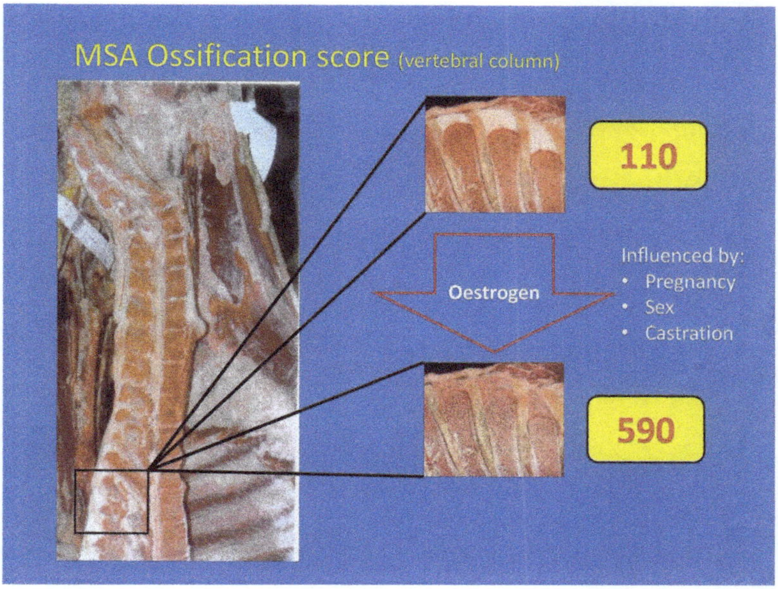


**Figure 1** Illustrations of differences in spinal column ossification and associated ossification scores in the Meat Standards Australia system.

- **Carcass hanging method**. Hanging carcasses by either the obturator foramen or the sacral ligament (often referred to as the tenderstretch method) results in increased palatability for the loin and hindlimb cuts. However, as ageing time exceeds five days, the benefit of the tenderstretch method decreases with longer ageing time.
- **Ultimate pH and meat colour scores of the loin**. Ultimate pH is used as a predictor up until 5.7, whereas at higher pH it becomes a threshold effect and those carcasses are not eligible for grading. High pH meat has reduced retail shelf life, a dark colour, an altered flavour and different cooking properties with respect to the degree of doneness.
- **Meat colour** is used as a threshold effect that excludes carcasses with meat colour scores greater than three.
- **Ageing**. The rate of ageing varies for individual muscles and is estimated within hanging options. The impact of ageing on palatability decreases with ageing time when carcasses are hung by the Achilles tendon. MSA-graded beef cannot be sold to consumers before five days of ageing.
- **Muscle.** The model predicts the palatability of 39 individual muscles, with up to a 30-point range in eating quality score.
- **Cooking method**. Palatability for individual muscles is predicted for specific cooking methods, such as grilling, roasting, stir frying and variations of thin slicing. A summary chart of the MSA system is shown in Fig. 2. The MSA-developed process is shown in Fig. 3. The MSA grading model for 39 meat cuts is shown in Fig. 4.

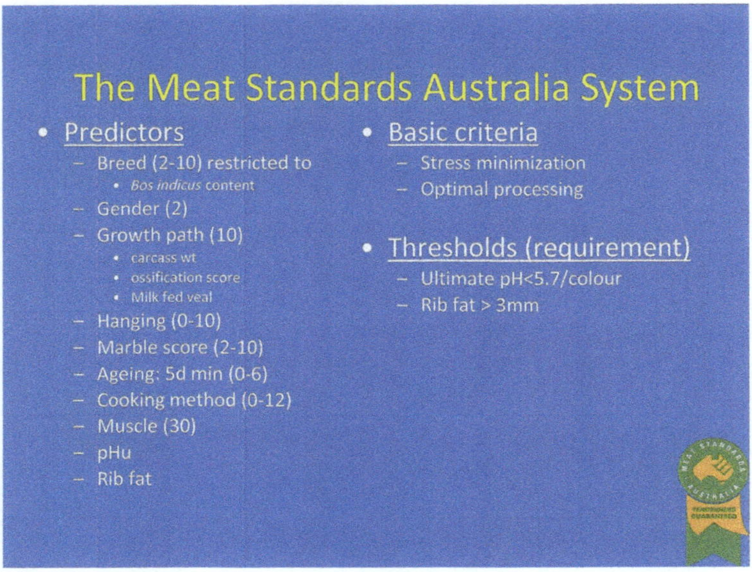


**Figure 2** Summary of factors evaluated in the Meat Standards Australia grading system.

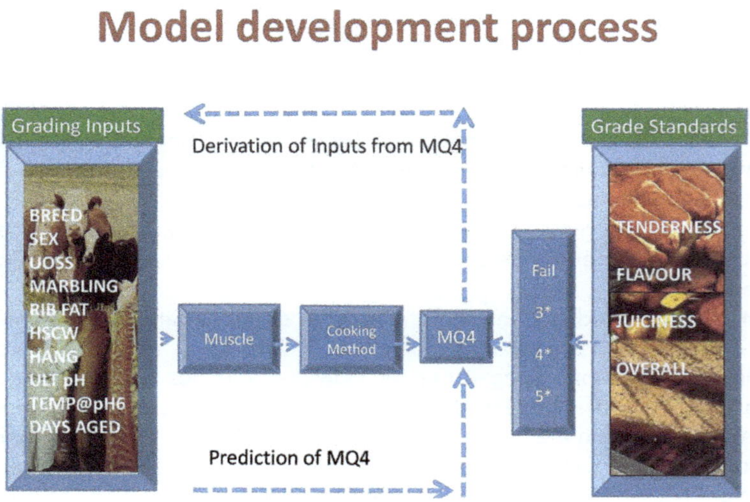


**Figure 3** Meat Standards Australia model development process.

Cattle within the MSA system must come from registered producers and undergo best practice management and stress minimization, such as not mixing different mobs of cattle. At harvest, all carcasses are graded once ultimate pH is obtained and then the grading data are used by the MSA model at the abattoir to predict the cut x cook outcome. MSA

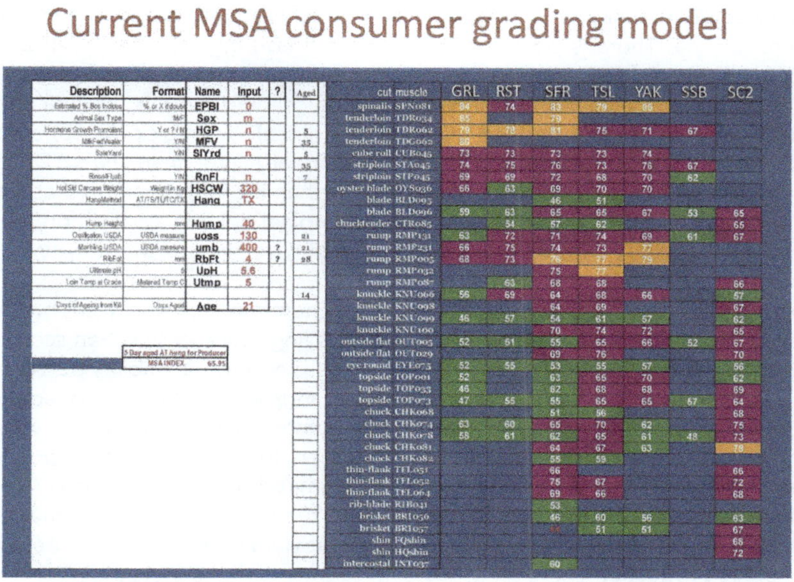


**Figure 4** Current Meat Standards Australia grading model for 39 different meat cuts.

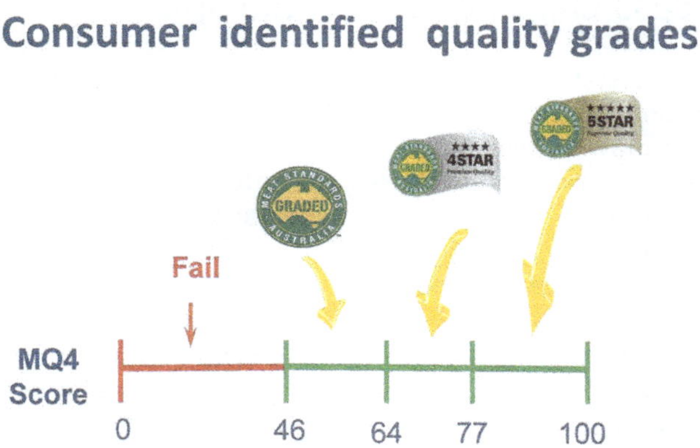


**Figure 5** Consumer-identified Meat Standards Australia quality grades.

currently grades in excess of 3.2 million carcasses in Australia. Compliance with threshold traits of rib fat thickness, pH and meat colour is relatively constant at 93%. Over a seven-year period, an annual survey of retail, wholesale and over the hooks prices for MSA and non-MSA product showed that the premiums for MSA were distributed across the industry with approximately 50% to producers, 30% to wholesalers and 20% to retailers (Polkinghorne, 2015). The grades are 'fail' (for grinding; score less than 48), '3-star' (meets 'every day' palatability; score of 49–63), '4-star' ('better than every day' quality; score of 64–79) and '5-star' ('Premium quality'; score of 80 or higher; Fig. 5). By just using MSA as a threshold grade for 3-star or above, the annual benefits are estimated to be of the order of $100 million pa. There is potential for this to increase if grades for 3-, 4- and 5-star products are marketed accordingly.

Several other European countries are developing grading models that are analogous to the MSA system (Poland, Ireland, N. Ireland, France). Wierzbicki (2015) stated that the MSA system is more attractive for the Polish beef sector than the USDA grading system and has more potential to be developed further. He further stated that the USDA system is based on rib-eye marbling, colour and spinal column ossification, but other factors affect palatability; rib-eye quality is not highly related to the quality of several other muscles.

It should be pointed out that the MSA grading system requires significant human input to assess *Bos indicus* content, rib fat thickness, marbling score, ossification score and pH and to monitor ageing time to segregate the 39 muscles into palatability outcomes. This system does not lend itself to instrument grading and/or to faster line speeds, such as in the United States. However, it should also be pointed out that the MSA system 'sets the standard' for predicting palatability for consumers that other countries should strive to achieve. Although the MSA system is to be admired for its emphasis on consumer satisfaction, it is economically very important that 'meat yield' differences must also be emphasized and that rewards and penalties for meat yield differences must be utilized in the beef cattle production and beef processing segments of the industry.

## 4 Beef carcass classification in Europe

In 1980, a common system of classifying beef carcasses was agreed upon in the European Union, and it was called the EUROP system (Fig. 6). The regulations state that beef carcasses must be classified according to their conformation and fat cover by trained classifiers. For conformation classes, the letters E, U, R, O and P are used, with E denoting carcasses with excellent conformation and P denoting poor conformation. There is an option to use an extra S (Superior) class for carcasses with extreme muscle development, such as double-muscled cattle. Fat cover is assessed on a five-point scale by using the numbers 1–5, with 1 denoting low fat covering and 5 denoting high fat covering. Some countries subdivide each of the categories for conformation and fat into three subclasses to give 15 classes for conformation and fat. In other countries, the most common fat class or classes are subdivided into two or three subclasses. The category of carcass (sex/age) must also be specified in the classification, and all carcasses must be dressed according to a detailed specification. Carcasses must be marked with the classification results and these results must be made available to producers who have provisions for making appeals.

National legislation governs the operation of the EUROP scheme in each member country. These include descriptions for conformation, fat cover and category classes and photographic standards. In most member countries, classifiers are employees of the abattoir but, in others, they are employed either by a public authority (Greece, Ireland and Portugal) or by a private company (Austria, Germany, the Netherlands, Great Britain and Northern Ireland) (Allen, 2014). In all cases, the classifiers are highly trained and must be regularly monitored and retrained if necessary. Standards throughout the European Union are maintained by an expert panel, which visits each country on a regular basis to check

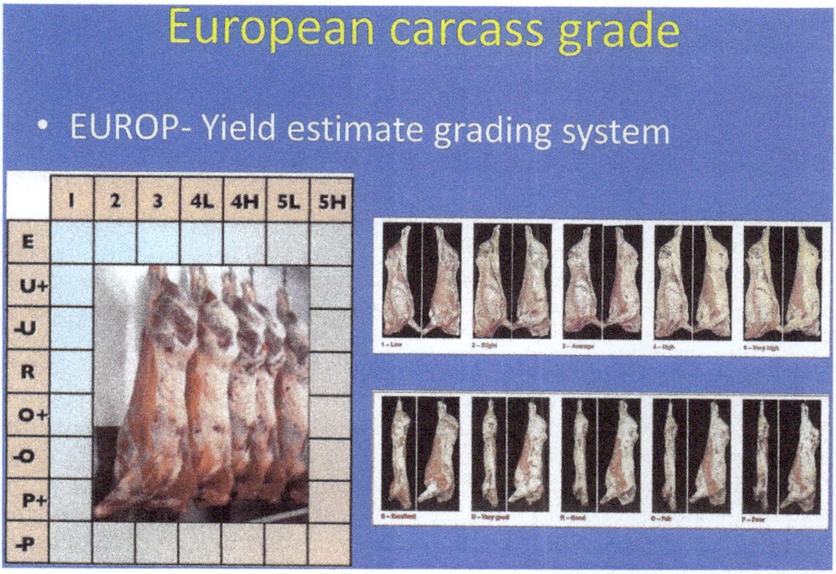


**Figure 6** The European system of 'EUROP' for classifying carcasses on conformation and meat yield.

whether the grading is in line with EU standards. The classification scheme is used by the European Union for price reporting and market intervention purposes, and by the industry for composition-based payments to producers and for carcass trading.

In May 2000, a decision was made to allow mechanical grading in the European Union by using video image analysis (VIA). This was followed in June 2003 by an agreement on the rules for conducting calibration trials for mechanical systems and criteria for determining their accuracy. Ireland was the first country to conduct a validation trial, which resulted in 25 VIA systems being installed in their main export factories in 2004. Calibration trials followed in some other countries and VIA systems have been installed in several countries. Information on classification is returned to the supplier by the harvest plant. Over 90% of carcasses are classified by machines by using VIA to carry out various measurements of the carcasses. As the determination of classification in this case is objective, no appeal is possible. In smaller plants, classification is carried out by factory employees who have been licensed by the Department of Agriculture, Fisheries and Food. In such classifications, the supplier can make an appeal against any decision of the classifier to the slaughter plant.

The EUROP system specifically relates to expected 'meat yield' and not to 'meat quality'. Farmer et al. (2015) stated that the EUROP system was never designed to measure 'meat quality' and that other measures must be integrated to predict or relate to beef palatability. Consequently, there is almost no focus on satisfying consumers' expectations for palatability. However, research is being conducted in some European countries to develop a prototype European eating quality prediction system to guarantee eating quality. Bonny et al. (2015) also stated that the EUROP system for carcass classification was never designed to measure eating quality and that there is almost no relationship between the EUROP classes and meat palatability. These authors further stated that research is underway to collect data from other collaborating countries to develop and test a prototype European eating quality prediction system to guarantee eating quality in the European market.

# 5 Beef grading in the United States

A federal system of beef carcass grading has evolved considerably since it was first initiated in 1927. The purposes of USDA grading are to aid quality-based marketing of cattle and to provide a common language for carcass trading and meat merchandising. Grading is voluntary and the service is provided by the USDA on a cost-recovery basis (USDA, 1989).

Beef carcasses can be graded for both quality and yield grades. The yield grades estimate the percentage of boneless, closely trimmed retail cuts from the high-value parts of the carcass – the round, loin, rib and chuck. The grades are numbered one to five, with YG1 having the highest expected meat yield and YG5 the lowest expected meat yield. The grades are calculated from a formula that includes the fat depth over the rib-eye; the percentage of kidney, pelvic and heart fat (KPH); carcass weight and rib-eye area. The rib-eye fat depth is measured at the 12th rib, three quarters of the length of the rib-eye from the chine bone, but skilled graders can make an adjustment to this measurement to reflect unusual amounts or distribution of fat in other parts of the carcass. In other words, they assess how representative this fat depth is of the total carcass fat. When carcasses are graded by instruments, fat thickness is measured at three locations over the rib-eye. The amount of KPH fat is evaluated subjectively and expressed as a percentage of carcass

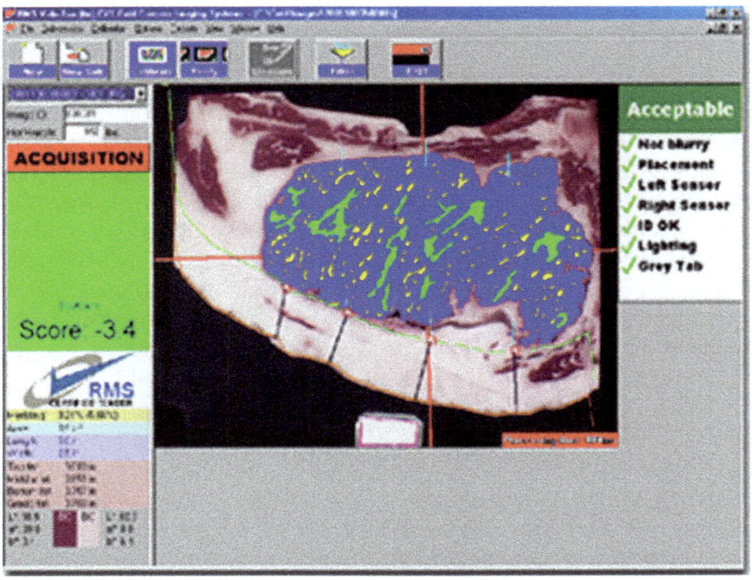


**Figure 7** Example of an image from a beef carcass graded by video image analysis in the United States.

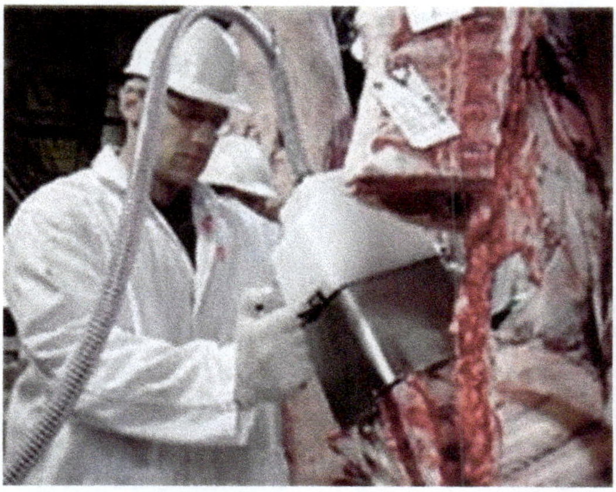

**Figure 8** Illustration of a video image analysis camera placed on a ribbed beef carcass.

weight, although one large processing company weighs carcasses before and after KPH fat removal so that the actual % KPH is determined.

The US industry developed and the USDA's Agricultural Marketing System recently approved the use of VIA for both yield grading and marbling assessment for quality grading (USDA, 2006). Most beef processors have implemented this technology because of greater accuracy and uniformity in application of the grading standards. A unique feature about the VIA system used in the United States is that it predicts the percentage of meat yield and yield grade and also accurately determines quality grade (Fig. 7 and 8).

The yield grades have descriptions in terms of the external and internal fat deposits, and a step-wise procedure is used to determine the yield grade. When graded by trained personnel, the preliminary yield grade (PYG) is first determined from the fat measurement over the rib-eye. This is then adjusted for rib-eye area relative to the carcass weight and then for the percentage KPH fat. Beef quality grades are designed to sort carcasses according to the meat's visual appearance and marbling, and expected meat palatability, which is a combination of tenderness, juiciness and flavour. Quality grading is based primarily on visual marbling, reflecting the amount of intramuscular fat, and on maturity as determined by ossification. Graders or the VIA system evaluate the amount of marbling fat in the rib-eyes after carcasses have been ribbed between the 12th and 13th ribs. Figure 9 shows the minimum marbling for the four grades of carcasses from young cattle.

The quality grades for young cattle are called Prime (most marbling), Choice, Select and Standard. Each quality grade is divided into one to three marbling score subclasses. For example, Prime is divided into Abundant, Moderately Abundant and Slightly Abundant. Each degree of marbling is divided into 100 subunits, but in practice, marbling scores are generally referred to in tenths within each marbling grade, for example, Slightly Abundant$^{90}$, Moderate$^{50}$, etc. Most large beef processing plants use the VIA instrument assessment of marbling. Extensive study and testing by the USDA has shown greater precision in assessment of marbling than when graded by trained graders.

Maturity is the second criterion of beef quality grading. Maturity refers to the physiological age of animals rather than the chronological age and is used mainly because chronological age is generally not available but also because physiological age (ossification) can be affected

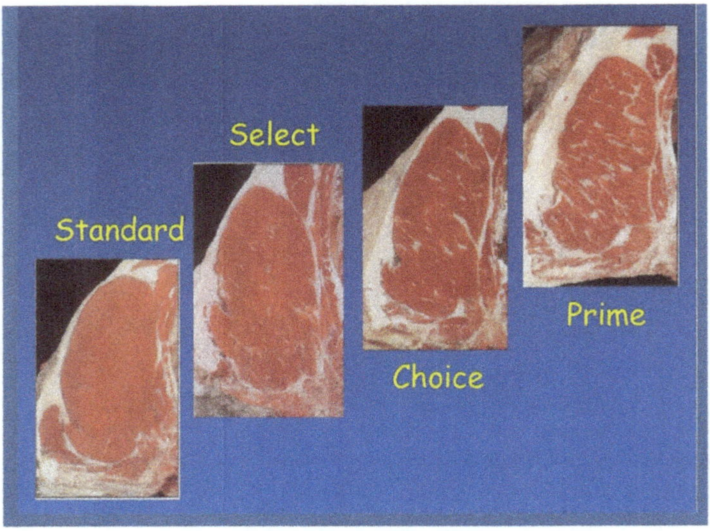


**Figure 9** Minimum degree of marbling for the four quality grades of young carcasses in the USDA beef grading system.

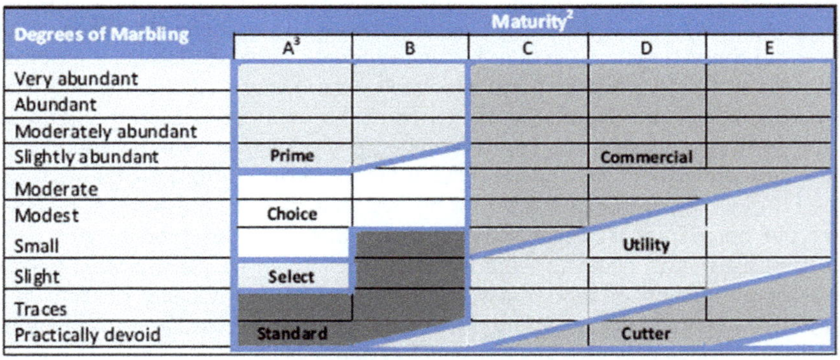


**Figure 10** U.S. Department of Agriculture Quality grading chart showing the relationship between marbling and carcass maturity (USDA, 1989).

by factors such as nutrition, gender and the biological type of cattle. The indicators of maturity are the bone characteristics (the degree of ossification of the cartilage of the sacral and lumbar vertebrae, and the spinous processes of the thoracic vertebrae). The MSA and USDA maturity systems are very similar. Greater ossification relates to older physiological age, and ossification is more highly related to meat palatability of young cattle than chronological age. Maturity is also based on colour of the rib-eye muscle (darkens with age) and texture (becomes more coarse with age). Carcass maturity grades are labelled A–E, A typically being 9–30 months and E typically being over 96 months. Grades of Commercial, Utility and Cutter typically are from cows that are more than 42 months of age (C, D or E maturity; Fig. 10). Carcasses are separately assessed for maturity on the basis of ossification and lean colour and texture and, when the two do not agree, they are balanced with more weight given to the bone ossification score. It should be emphasized that the VIA system cannot assess carcass maturity and, consequently, a trained USDA grader must be present to make this assessment or, in some cases, to override the marbling score determined by VIA when carcasses are not ribbed properly.

Marbling and per cent meat yield generally are negatively correlated, or they are antagonist traits. Figure 11 shows the relationship between marbling and per cent retail product (meat yield) obtained from data of 15 breeds of cattle (Dikeman, 2001). The antagonist relationship makes it challenging for seedstock producers to make genetic progress for both meat yield and marbling. However, there are sires that defy this antagonism, and the heritability of marbling, tenderness, rib-eye area and meat yield are high (Dikeman et al., 2005). Two Simmental bulls in two major AI studs from one moderate-sized breeding herd have the following expected progeny differences (EPDs): Bull #1 = top 1% marbling, rib-eye area and tenderness; Bull #2 = top 1% marbling, top 5% for rib-eye area and top 10% for tenderness. Both of these sires are above average for yearling weight, and in the top 2% for disposition, all-purpose index and terminal index. These genetic values are based on traditional EPD calculations without DNA-incorporated information. In the Angus breed, one Angus AI sire has EPDs of +1.31 for marbling score, +0.59 in$^2$ for rib-eye area and − 0.04 in. for fat thickness. Another Angus AI sire has EPDS of +0.81 for marbling, +0.71 for rib-eye area and −0.04 in. for fat thickness. These Angus examples show that both marbling and rib-eye area can be increased without increases in subcutaneous fat. Thus, there is great genetic potential to increase beef quality and meat yield regardless of whether grading systems stimulate such improvements.

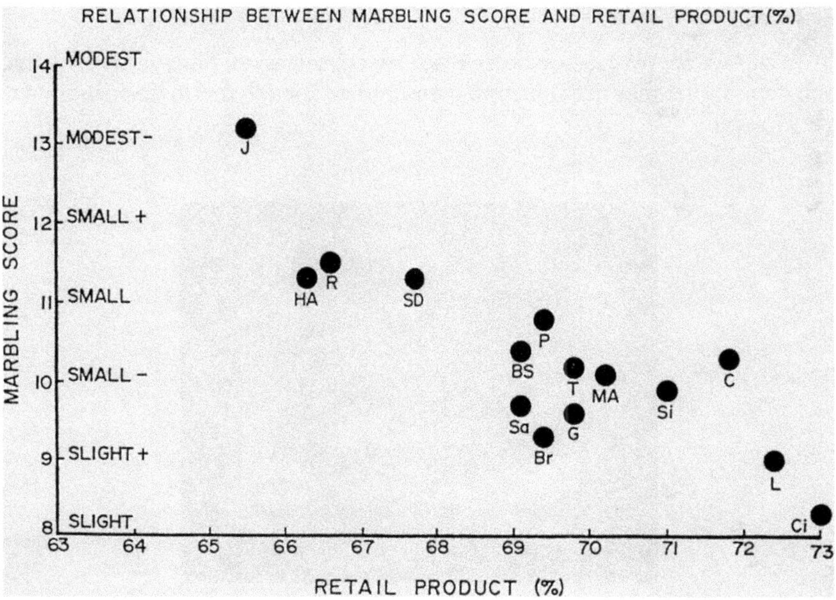


**Figure 11** Illustration of the negative relationship between marbling and meat yield (retail product) for 15 breeds of cattle.

The USDA recently approved (2012) a system by which processors, retailers or purveyors can label beef cuts as 'guaranteed tender' or 'guaranteed very tender' based on either Warner–Bratzler or slice shear force. It is anticipated that this system will be used in conjunction with grades to assure acceptable eating experiences for consumers most of the time. If these USDA labels are used extensively by the industry, then the importance of grades may diminish. It should be mentioned here that there are numerous 'branded beef' programmes in the United States; some of these are breed specific (Certified Angus or Certified Hereford Beef), some are production system specific (grass fed, organic) and some are combinations of the two and also of meat traits. Most of them have specific USDA beef quality grade minimums (such as marbling) or maximums (such as yield grade or maturity). Branded programmes are covered in Chapter 10 of this book.

The Canadian beef grading system is similar to the US system with grades of Prime, AAA, AA and A in contrast to the US grades of Prime, Choice, Select and Standard.

# 6 Beef grading in Japan

The beef carcass grading system in Japan is implemented according to the guidelines of the 'Beef Carcass Grading Standard' established by the Japan Meat Grading Association. This standard was established in 1988 and consists of the evaluation of 'yield' and 'meat quality'. The yield score is classified into three grades: A, B and C, determined by the estimated percentage yield. This percentage is calculated by a multiple regression equation as follows:

Yield estimated percentage (%) = 67.37 + (0.130 × Rib-eye area, $cm^2$) + (0.667 × Rib thickness, cm) – (0.025 × Cold left-side weight, kg) – (0.896 × Subcutaneous fat thickness, cm)

Note: Add 2.049% for Wagyu (Japanese Black breed) carcasses. Rib-eye area, rib thickness and subcutaneous fat thickness must be measured on the 6th to 7th rib section of the left side (see Fig. 12).

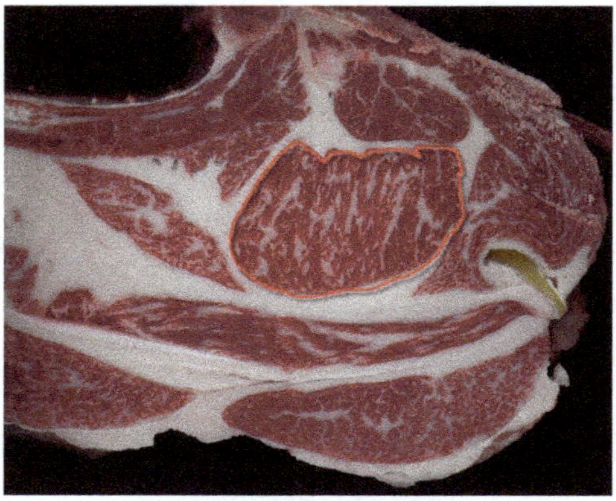

**Figure 12** Illustration of a beef carcass ribbed between the 6th and 7th vertebra in the Japanese grading system.

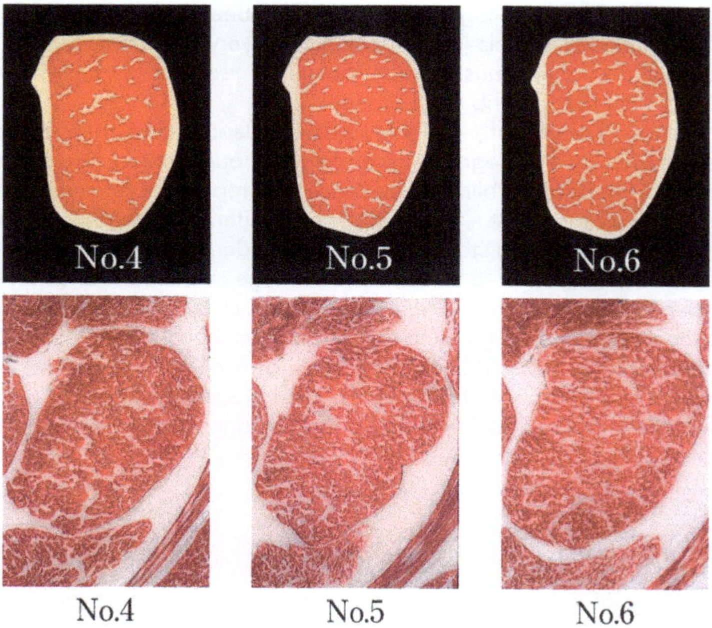


**Figure 13** Examples of marbling degrees in the Japanese beef grading system.

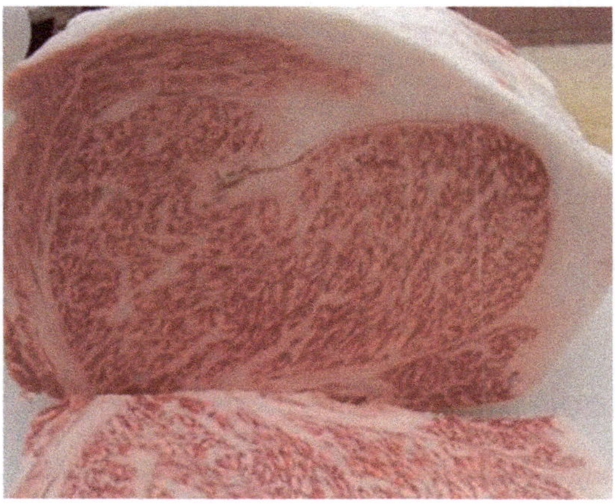

**Figure 14** Japanese beef that is more of a 'delicacy' than a protein source.

The meat quality score is classified into five grades, No.1 to No.5. The grades are determined in terms of (1) beef marbling, (2) meat colour and brightness, (3) firmness and texture of the meat and (4) colour, lustre and quality of the fat. Each of these four items is classified into five grades, and the meat quality score is graded down to the lowest scale among these items. (1) Beef marbling is evaluated by the Beef Marbling Standard (B. M. S.) prepared as 12 continuous models (No.1: poor, No.12: very abundant, see Fig. 13). The evaluation sites are the *longissimus thoracis*, dorsal *semispinalis dorsi* and *semispinalis capitis* muscles on the 6th to 7th rib cross section.

Fatty beef, such as 'Kobe beef', is highly valued in Japan. Thus, beef carcass prices are mainly determined by the degree of marbling, although breed also has an effect. Vitamin A is known to affect marbling as evidenced by reports of retinal suppression of adipogenesis and by a report that a low level of serum vitamin A results in a high degree of marbling. Vitamin A intake is, therefore, restricted during the late–early to middle fattening stages, which is about 15 to 23 months of age in the case of Japanese Black steers. The Japanese beef grading system places so much emphasis on extremely high marbling that the system is useful only in that country, and only upper-class citizens can afford the extremely highly marbled beef. The meat shown in Fig. 14 is more of a 'delicacy' than a protein source. In contrast to the Japanese beef grading system, the U.S. and MSA grading systems are important and very useful in exporting meat from those respective countries (Strong, 2004).

# 7 Summary

The primary purpose of producing beef cattle is to convert grass, forages and various by-products, plus some human-edible protein and energy, into highly nutritious and tasty beef. To accomplish this,

1 cattle enterprises must be profitable;
2 carcasses should yield a high percentage of meat;
3 beef should be safe, affordable, attractive, nutritious and highly palatable;
4 both production and processing systems need to be sustainable and socially responsible.

The above conditions must be factored-in when revising the existing beef grading standards or developing new ones. Few national grading systems place emphasis on all four of these conditions and, consequently, are not as effective as they should or could be in supporting the beef cattle industry and in meeting consumer expectations for beef palatability.

# 8 Where to look for further information

The website of MSA is http://www.mla.com.au/Marketing-beef-and-lamb/Meat-Standards-Australia.

Information about the Canadian beef grading system can be found at https://www.google.com/search?q=Canadian+Beef+Grading+System&newwindow=1&espv=2&tbm=isch&tbo=u&source=univ&sa=X&ved=0ahUKEwi7iofNqcbKAhVsl4MKHfjvCEcQsAQIQQ&biw=1280&bih=622#imgrc=G0NVFgdSFBxcGM%3A.

Recent research on measuring beef tenderness can be found at www.agmrc.org/media/cms/tenderbeefcard_4D130B39C5390.pdf.

# 9 References

Agricultural Marketing Service, U.S. Department of Agriculture (2013). Guaranteed Tender Beef: Opportunities and Challenges for a Differentiated Agricultural Product. Working Paper 04-WP 371.

Allen, P. (2014). Beef carcass classification and grading. In M. E. Dikeman and C. Devine (eds), *Encyclopedia of Meat Sciences*, 2nd ed., Elsevier, Academic Press.

Bonny, S., Pethick, D. W., Legrand, I., Wierxbicki, J., Allen, P., Farmer, L., Polkinghorne, R. J., Hocquette, J. F. and Garner, G. E. (2015). *International Meeting on Beef and Lamb Carcass Grading to Underpin Consumer Satisfaction*, 20–21 August, Paris, France.

Dikeman, M. E. (2001). Bull selection and meat quality. *American Association of Bovine Practitioners* 34: 74–82.

Dikeman, M. E. (2012). Animal breeding, management, and technology to efficiently increase productivity, meat yield, and meat quality of cattle. *Proceedings of Reciprocal Meat Conference*. North Dakota State University.

Dikeman, M. E., Pollak, E. J., Zhang, Z., Moser, D. W., Gill, C. A. and Dressler, E. A. (2005). Phenotypic ranges and relationships among carcass and meat palatability traits for fourteen cattle breeds, and heritabilities and expected progeny differences for Warner-Bratzler shear force in three beef cattle breeds. *Journal of Animal Science* 83: 2461–7.

European commission (2010). La politique agricole commune après 2013 – Débat public. http://ec.europa.eu/agriculture/cap-post-2013/debate/report/summary-report_fr.pdf.

Farmer, L. J., Devlin, D. J., Gault, N. F. S., Gordon, A. W., Moss, B. W., Polkinghorne, R. J., Thompson, J. M., Tolland, E. L. C. and Tollerton, I. J. (2009). Prediction of eating quality using the Meat Standards Australia system for Northern Ireland beef and consumers. In: *International Congress on Meat Science and Technology*, Copenhagen.

Farmer, L. J., Devlin, D. J., Gault, N. F. S., Gordon, A. W., Moss, B. W., Polkinghorne, R. J., Thompson, J. M., Tolland, E. L. C., Tollerton, I. J. and Watson, R. (2010). Adaptation of meat standards Australia quality system for Northern Irish Beef. *Advances in Animal Biosciences* 1: 127.

Farmer, L. J., Hagan, T. D. J., Devlin, Y., Gordon, A. W. and Oltra, O. R. (2012). Relationships between beef aroma compounds and other quality attributes. Paper presented at the International Congress on Meat Science and Technology Montreal.

Japan Meat Grading Association (1988). *New Beef Carcass Grading Standards*. Tokyo, Japan, Japan Meat Grading Association.

Legrand, I., Hocquette, J. F., Polkinghorne, R. J. and Pethick, D. W. (2013). Prediction of beef eating quality in France using the Meat Standards Australia system. *Animal* 7: 524–9.

Meng, Q.-X. (2015). *International Meeting on Beef and Lamb Carcass Grading to Underpin Consumer Satisfaction*, 20–21 August, Paris, France.

Polkinghorne, R. J., Philpott, J., Gee, A., Doljanin, A. and Innes, J. (2008a). Development of a commercial system to apply the Meat Standards Australia grading model to optimize the return on eating quality in a beef supply chain. *Australian Journal of Experimental Agriculture* 48: 1451–8.

Polkinghorne, R., Thompson, J. M., Watson, R., Gee, A. and Porter, M. (2008b). Evolution of the Meat Standards Australia (MSA) beef grading system. *Journal of Animal Science* 80: 3211–21.

Polkinghorne, R. J. and Thompson, J. M. (2010). Meat standards and grading: a world view. *Meat Science* 86: 227–35.

Polkinghorne, R. J. (2015). *International Meeting on Beef and Lamb Carcass Grading to Underpin Consumer Satisfaction*, 20–21 August, Paris, France.

Strong, J. (2004). Differences in carcass grading schemes used in the USA, Japan, and Australia. *Australian Journal of Experimental Agriculture* 44: 675–80.

Tonsor, G. (2012). USDA Certification of Beef Tenderness: An Overview of Economic Considerations. K-State Dept. of Agricultural Economics (Publication: AM-GTT-2012.4).

United Nations Economic Commission for Europe (2015). Specialized section on standardization of meat. Eating quality – proposals by the working group.

USDA (1989). United States standards for grades of carcass beef. Agricultural Marketing Service USDA, Washington, DC.

USDA (2006). Performance requirements for Instrument Marbling Evaluation (PRIME II). Implementation and verification of operational procedures. Livestock, Poultry, and Seed Program, Agricultural Marketing Service, USDA, Washington, DC.

Watson, R., Polkinghorne, R. J. and Thompson, J. M. (2008b). Development of the Meat Standards Australia (MSA) prediction model for beef palatability. *Australian Journal of Experimental Agriculture* 48: 1368–79.

Wierzbicki, J. (2015). *International Meeting on Beef and Lamb Carcass Grading to Underpin Consumer Satisfaction*, 20–21 August, Paris, France.

Chapter 11

# Branded beef programmes

B. N. Harsh and D. D. Boler, University of Illinois, USA



## 1 Introduction

Shrinking beef demand and heavy competition from poultry and pork in the 1970s led to the development of the first U.S. branded beef programmes. By differentiating beef that is likely to have desirable eating characteristics from traditional, unbranded beef, the industry was able to offer an assurance of consumer satisfaction for the first time. Since branded beef was designed to be consumer-driven and in-tune with consumer needs and desires, demand for branded beef continued to grow, eventually even outpacing demand for Choice beef.

In the two decades since the establishment of the first branded beef programme in 1978, interest or belief in these programmes was relatively low with only another 10 programmes certified by the USDA during this period (Speer, 2013). However, tremendous sales growth and demand for branded beef resulted in a rapid increase in the number of branded beef programmes by the late 1990s. As of 2016, there were 99 different certified beef programmes, each with different programme requirements and branding strategies (USDA, 2016a).

As the percentage of disposable income spent on food in developed countries continues to decrease worldwide, today's consumers have the means to make food-purchasing decisions based on lifestyle factors and personal values (Simmons, 2011). Growing consumer interest in food production practices, animal welfare and environmental concerns has created an even greater opportunity for branded beef programmes to capitalize on product differentiation.

http://dx.doi.org/10.19103/AS.2016.0009.11

Although consumer desires have changed substantially over the years, the basic principles of effective branded beef programmes are largely unchanged. By providing consumers a measure of dependability and consistency in eating experience, brands have the potential to directly influence consumers' perceptions of and willingness-to-pay for, branded beef.

# 2 The characteristics, history and types of branded beef programmes

Developing a singular, cohesive definition of a brand can be considerably more challenging than it initially appears. The earliest definitions date back to the days of hot-branding livestock and property with markings or symbols to designate ownership. And while a brand's symbol or logo is still used to differentiate the goods or services of one company from another, it has come to represent much more. In 1998, de Chernatony and Riley set out to establish the definition of a brand by summarizing the interviews of leading brand consultants. Ultimately, they found no two brand experts used the same definition. However, when distilled down to the basic components, the simplest definition of a brand is a multifaceted construct with different meanings to different people.

## 2.1 Branded beef programme characteristics

In the context of branded beef programmes, a brand is first and foremost a guarantee of reliability and consistency. The biological variability associated with beef palatability has proven one of the greatest challenges to ensuring consumer satisfaction. To this effect, Lusk and Schroeder (2006) reported the likelihood of beef to consistently deliver a positive eating experience is one of the most important factors affecting consumer preference and repeat purchasing. The ability to reduce consumer risk (in the form of wasted money, time or unsatisfactory eating experience) is an important aspect of brand reliability. In a historical study documenting consumer ratings of risk reduction methods, Roselius (1971) explained the top consumer responses were purchasing from well-established, reputable brands and relying on brand loyalty or previous experience.

Because consumers' purchasing decisions are influenced by personal and cultural values, a brand may also represent consumers' value systems (de Chernatony and Riley, 1998). Traditionally, a brand's values were those characteristics that added value for consumers such as consistency and responsibility. In many regards, the core values of a brand are the foundation for brand personality (Urde, 2003). With the growing interest in the beef production process, today's consumers increasingly want to know the beef that they consume comes from cattle raised and harvested in a way matching their personal values. For this reason, effectively communicating brand values may be more important than ever to improving the position of branded beef programmes.

## 2.2 History of branded beef programmes

In 1977, the U.S. Senate's Committee on Nutrition and Human Needs released *Dietary Goals for the American People*. The detailed document on best nutritional choices most notably included a recommendation for Americans to decrease red meat consumption.

Courtesy: Certified Angus Beef LLC

**Figure 1** Although Certified Angus Beef became the first USDA-certified beef programme in 1979, the logo was officially registered by the U.S. Patent & Trademark Office in 1983 (Minnick, 2010). Although the well-recognized 'shield' logo has been updated over the years, the founding brand principles remain largely the same. Other trademarks include the slogans, 'Clearly a Matter of Taste' and 'Angus Beef at Its Best'.

Around the same period, American epidemiologist Dr Ancel Keys's research regarding the role of saturated fats in cardiovascular disease was gaining national attention. As red meat was demonized by the media, beef demand saw substantial decline by the late 1970s (USDA, 2016b). In response, the beef industry worked to improve carcass yield and market leaner beef cuts. By in all regards, the industry was successful in this campaign, reducing external fat on retail cuts by 81% (McNeill, 2012).

However, an unintended consequence of the 'War on Fat' was the effect this focus would have on beef palatability. Renewed interest in beef quality prompted the concept for the first branded beef programmes. Designed to differentiate inherently, highly marbled Angus cattle from the uncertainty associated with commodity beef, the Certified Angus Beef (CAB) brand was developed in 1978. Recognizing the need for an impartial third-party evaluation, brand founders enlisted the help of the United States Department of Agriculture (USDA) to provide needed credibility (Minnick, 2010). These programme evaluation standards set forth in 'Schedule G1' would form the basis for the USDA first certified beef programme (Minnick, 2010). This practice is still in place today with all USDA-certified beef programmes having a schedule 'G' number.

Although each programme has approached creating competitive advantage or brand recognition a little differently, the creation of science-based programme specifications, or animal and carcass requirements, were critical to the development of all branded beef programmes. Based on the belief that superior grading Angus cattle could provide consumers a highly satisfactory eating experience, CAB's fundamental specifications were based on the selection of predominantly black-hided cattle producing A-maturity (physiologically young) carcasses with a Modest degree (approximately 6% extractable lipid; Savell et al., 1986) of marbling or greater (Figure 1).

In the early 1980s, Smith et al. (1980; 1982) documented the relationship between USDA quality grade and overall palatability adding an additional layer of scientific certainty to the notion of quality grade as a predictor of eating satisfaction. It is for this reason that all of today's USDA-certified branded beef programmes are required have some form of quality grade specification. Of the USDA-certified branded beef programmes in 2016, almost half have a quality grade specification of USDA Low Choice or better.

## 2.3 Types of branded beef programmes

All branded beef programmes have some programme specifications and marketing strategies in common. However, key differences among them have led to the creation of brand types or categories. Given the overlap in brand specifications, many branded beef programmes are actually a combination of two brand types (e.g. both store and company-specific branded beef programmes may have breed-specific brands).

### Breed association and breed-specific brands

Although breed claims were a common marketing tool well before the creation of certified beef programmes, prior to this point such marketing claims conveyed little in the way of quality cues. For this reason, Angus producers created CAB as both a guarantee and a differentiator. After a relatively short period, CAB's success as a fresh beef brand set the path for development of other breed-specific brands.

In 1996, the USDA set today's standardized specifications for eligibility in Angus-influenced certified branded programmes. This specification is commonly referred to as 'Schedule GLA'. In subsequent years, USDA specifications for Hereford, Texas Longhorn, and Wagyu-influence have followed (USDA, 2016c). Interestingly, USDA breed influence specifications are not equal. For example, cattle eligible for certified programmes claiming Hereford-influence may meet either the phenotypic or genotypic specifications, whereas to be eligible for Wagyu-influenced certified programmes, with no phenotypic specification in place, Wagyu cattle are required to meet genotypic specification for eligibility. Additionally, of the breeds with both phenotypic and genotypic specifications, USDA-certified programmes do not have to require both (requiring eligible cattle to have either phenotypic or genotypic specifications).

Unique to this type of branded beef programme, the only thing many breed-specific brands own is the brand itself. Not owning, marketing or selling cattle or beef products, breed-specific branded programmes, such as CAB, instead use demand for the brand to leverage marketing opportunities and increase profitability for its licensed packers, distributors, retailers and restaurants (CAB, 2011).

Breed-specific brands can have major influences on the breeds they represent. For example, the mission of CAB has always been to increase the demand for registered Angus cattle. From data collected by Superior Livestock Video Market sales over the past 10 years, Zimmerman et al. (2012) demonstrated feeder calves described as 'documented primarily Black Angus' received an average $5.53/cwt in premiums compared with other feeder calves of similar weights and backgrounds.

### Company-specific brands

Similar to other brands, company-specific brands may have a wide variety of quality and differentiation specifications such as breed, USDA quality grade, carcass and portion size, production methods and many more. However, where these brands differ from their breed specific counterparts is ownership. More common of company brands, most of the beef products sold under the brand are owned by the company, with the company owning a portion of the cattle as well. Where this strategy may incur some additional risk, it also has the reward of greater supply consistency.

### Store-specific brands

While breed- and company-specific beef brands may be regional or national in size, these brands carried by a number of different retailers are also known as 'national brands'. Looking to capitalize on the success of these national brands, retail chains soon developed their own branded beef programmes, referred to as 'private label' or 'store' brands. Initially positioned as lower-priced alternatives to national brands, private label brands were originally not a vehicle for acquiring brand premiums, instead enabling retailers greater opportunity for product differentiation and increased control over retail shelf space. The introduction of private label brands offered retailers the opportunity to increase overall profits in the meat case, earn greater gross margins compared with national brands, differentiate their chains from competitors, and attract value-conscious shoppers (Hyman et al., 2008). Similarly, price-sensitive shoppers may be more likely to switch from national brands to private label ones during times of economic challenge.

Although historically positioned as lower-priced options, private label brands have realized the potential to boost store loyalty and image through the creation of smart-value, high-quality national label alternatives. One example of a niche retailer is Whole Foods who have utilized this strategy of private label branding to increase the number of natural and organic options available, thereby building greater store loyalty. This perceived trustworthiness of specialty retail channels is reflected in natural and organic meat shoppers having above-average private label brand loyalty (FMI and NAMI, 2016).

Interestingly, the image of a store and its' private label products are interconnected. While store image may be improved or worsened by private label brands, consumer preference for those private labels may also be affected by store image as studies have shown attractive store layouts and ambience encourage more positive consumer evaluations of private label brands (Hyman et al., 2008).

# 3 The growth and economics of branded beef programmes

Despite the general decline in beef demand in the United States, demand for high-quality beef has continually risen as evidenced by a 17.7% unit increase in the number of carcasses grading USDA Prime and Choice from 2000 to 2015 (USDA, 2016d). Even so, just hitting the commodity Choice target has proven too little to meet consumer demand.

## 3.1 The growth of branded beef in the United States

Demand for branded Choice beef has outpaced commodity Choice beef with branded boxed beef sales taking a greater percentage of the wholesale beef total and has doubled since 2003 (Speer, 2013). In 2013, Zimmerman and Schroeder quantified this demand for branded beef by developing an index specific to CAB. They discovered CAB demand not only held steady during the 2008 recession, but had actually doubled since 2002, with demand for CAB increasing 108% compared with 51% growth for Choice-and-higher beef (Zimmerman and Schroeder, 2013).

Likely a result of decreasing demand for commodity Choice beef, approximately one-third of United States. Choice boxed beef was marketed under a brand designation in 2015 (Tatum, 2015). Zimmerman and Schroeder (2013) hypothesized that this response may be partially explained by the tendency of commodity beef products to reflect negative macroeconomic factors, such as the economic recession following the 2008 collapse of the U.S. housing market.

Today's quality focus has encouraged retailers to re-position store brands as having both price and quality equal to that of national brands (Hyman et al., 2008). Accordingly, consumer interest in private label brands has grown significantly in recent years. In 2010, store brands made up 36% of the beef items in the retail case; by 2015 the percentage increased to 51% (FMI and NAMI, 2015). And while the percentage of national brands in the meat case has remained largely unchanged in the past 10 years, brand preference as a whole has risen. These findings demonstrate the potential for national and private label brands to serve the full range of consumers desires ensuring mutual growth (FMI and NAMI, 2016).

## 3.2 Brand loyalty and consumer demographics

As a result of the quality and consistency associated with branded beef programmes, brand loyalty for both fresh and processed meats has steadily increased since the recession. In the 2016 Power of Meat study, the Food Marketing Institute and North American Meat Institute reported 46% of U.S. shoppers expressed brand preferences within the fresh meat category and 58% within the processed meats category (FMI and NAMI, 2016). Historically, U.S. shoppers have greater brand loyalty in the processed meats category compared with fresh meats, likely a result of greater product differentiation. Within each of these categories, a greater percentage of consumers preferred national brands to private label or store brands (FMI and NAMI, 2016).

Nonetheless, as both national and store brands are doing more to differentiate fresh meat options, developing above-average brand preference in the fresh meat category is highly dependent on demographics. Among the groups with the greatest brand loyalty for nationally branded fresh meat are men, high income (>$100k), and millennial shoppers (FMI and NAMI, 2016). Consumers with the greatest loyalty to store-branded fresh meat include natural and organic store shoppers as well as shoppers over 65 years of age. Interestingly, shoppers over 65 are also the demographic with the highest brand neutrality, or lack of brand preference, likely a result of greater meat purchasing and preparation knowledge (FMI and NAMI, 2016).

Shopper brand preferences are further validated by meat case trends with retailers reporting a 45% increase in the number of branded meat packages (FMI and NAMI, 2016). In 2015, the share of total meat packages with no brand information was down to just 4% nationwide (FMI and NAMI, 2016).

## 3.3 Barriers to future growth

Despite the tremendous demand for high-quality branded beef, there are still obstacles to ensuring long-term brand growth. One of the primary challenges of branded beef programmes is preserving brand differentiation. As the extrinsic quality cues initially used to differentiate brands become increasingly more ubiquitous, they may become less effective. For example, usage of the word 'Angus' as a quality indicator increased rapidly in the years since CAB's introduction in 1978. As of 2016, there were 68 USDA-certified beef programmes with Angus in the title and among these product specifications included the full spectrum of quality grades. As Angus marketing increases in popularity from high-end steakhouses to quick-service, fast-food chains, effective differentiation will need to go beyond breed claims to ensure brand effectiveness.

Another obstacle to growth, perceived quality of a brand may not directly translate into willingness-to-pay. Early willingness-to-pay studies for beef were primarily conducted in a hypothetical purchasing scenario. However, Lusk and Schroeder (2004) demonstrated that these kinds of hypothetical choice experiments tend to overestimate (by as much as 30% in some cases) consumers' actual willingness-to-pay. For this reason, non-hypothetical experimental auctions have become the preferred method for evaluating true willingness-to-purchase.

One such study (Froehlich et al., 2009) showed that although Canadian consumers were actually willing to pay a premium of approximately $1.31 per steak for branded steaks, as consumer confidence in selecting beef increased, their willingness to-pay decreased. Furthermore, the 'natural' brand and Angus brand evaluated in this study did not elicit different premiums, even though they represent distinctly different products that traditionally appeal to different market segments (Froehlich et al., 2009). Findings such as these may indicate that, with relatively few unbranded beef products remaining in the retail case, national or private label brands themselves may no longer be enough for consumers to differentiate quality, highlighting the increasing importance of brand reputation and advertising.

Certainly not unique to branded beef programmes, poor economic conditions will always be an impediment to brand growth. Consumer spending in the United States, and accordingly national brand preference, decreased dramatically in 2009 during the height of the recession and maintained a depressed plateau for the following four years (FMI and NAMI, 2016). By 2014, brand preference levels showed signs of a return to pre-recession levels and a willingness to ease up on money-saving measures adopted during the recession.

# 4 The brand effect

At their most basic level, brands convey information to consumers. As mentioned previously, for customers and shoppers, a brand may guarantee a certain quality level, reduce the risk of purchasing errors, and garner trust through conveyed company values. Thus, it comes to no surprise that branded beef products have the potential to affect consumer expectation and perception. The extent to which brands may affect perception and purchasing decisions, however, is dependent on a number of factors.

## 4.1 Brand lift

While blind testing is still the gold standard for evaluating differences in palatability attributes, consumers use quality cues, personal experience, packaging and brand information to make their beef selection and purchasing decisions. In a historical experiment documenting the effect branding can have on perception, Allison and Uhl (1964) reported no differences in overall liking of six beers when consumers were blinded to brand. However, when consumers were asked to rate the same six beers with all branding information presented, overall liking scores of all beers improved.

Although brand effect has been documented in other food products, Wilfong (2016) was the first to document the effect of branding on consumer palatability ratings of beef. Wilfong (2016) selected strip loins to represent five brands – CAB, USDA Prime, Choice, Select, and a fictional Select brand created for the study (Angus Select). For the first round of testing, consumers were asked to evaluate the five samples while blinded to treatment. To test the effects of brand influence in the second round, consumers were informed of the brand name and logo prior to evaluation. Wilfong (2016) reported no differences in palatability traits of CAB and Choice during blind testing; however, CAB was rated better during informed testing. Similarly, Angus Select, Prime and CAB all experienced a boost in flavour liking scores when brand information was disclosed. As for perceived quality, 18% more Prime samples were rated as 'premium quality' and 15% more CAB samples were rated as 'better than everyday quality' during informed testing compared with blind testing (Wilfong, 2016).

While results like these suggest that brands have the potential to boost consumer perceptions of a product or provide a 'brand lift' as described by Wilfong (2016), in some instances brands may have the opposite effect. In another study evaluating blind versus informed testing, Della Lucia et al. (2014) asked consumers to evaluate eight different brands of beer. While four of the brands received a boost in overall acceptability when brand information was provided to consumers, three brands actually decreased in acceptability compared to blind testing (Della Lucia et al., 2014). This is likely the effect of consumers brand familiarity, brand experience and perceived brand quality. Another example of this tendency, Wilfong (2016) showed that 83% of consumer panellists recognized the CAB brand, and associated the brand with a numerically greater perceived quality level than other Angus brands.

## 4.2 Product experience and familiarity

Consumers use quality cues to develop preference and purchasing decisions. In a survey of supermarket beef shoppers, Bredahl (2003) showed differences in quality cues used between high- and low-familiarity shoppers, where shoppers with more beef purchasing experience used intrinsic quality cues (e.g. colour, marbling, purge) to develop quality expectations and less experienced shoppers relied more heavily on brand information.

However, another study showed the ability of high-familiarity beef purchasers to form quality expectations predictive of actual quality experience was no better than low-familiarity purchasers (Banović et al., 2012). Such findings are important for beef given its inherent variability and the limited prediction capabilities of intrinsic quality cues. Zeithaml (1988) reported that even experienced shoppers may rely on extrinsic cues when product quality is perceived to be difficult to evaluate, making even clearer the potential for branding information to serve both as a quality cue and as a recognizable image for the basis of developing product familiarity.

## 4.3 Product framing

As consumers have become more familiar with brands, they have also grown to rely on them as a method to differentiate between products that may otherwise appear the same. Levin and Gaeth (1988) established the importance of product framing in consumer perception by asking consumers to evaluate quality attributes of ground beef labelled either '75% lean' or '25% fat'. Not surprisingly, consumer evaluations were more favourable toward ground beef exhibiting positive labelling ('75% lean') than negative labelling ('25% fat'). More interesting, however, is when order of events were manipulated (consumers were asked to sample ground beef before sharing labelling information), the magnitude of difference between positive and negative labelled ground beef decreased (Levin and Gaeth, 1988). Matching the results of Bredahl (2003), findings suggest increased product knowledge or familiarity may dilute the effect of brand framing.

In contrast, Wilfong (2016) showed in a population of consumer panellists where almost half reported beef consumption of 4 to 7 times per week, 25% of consumers stated they were familiar with the fictional 'Angus Select' brand created for the study. Moreover, consumer panellists rated Angus Select as representing similar quality to CAB.

This raises the question of whether the brand or the attributes a brand represents are responsible for consumer preference. Wilfong (2016) demonstrated further evidence of the perceived quality of Angus. While Angus Select received almost a 10% boost in the number of samples rated as 'premium' quality from informed versus blind testing, USDA Select had no perceived quality change.

# 5 Natural and organic branding

The natural and organic beef segment has shown notable growth in recent years with 40% of U.S. shoppers reporting a natural or organic meat purchase in the past three months (FMI and NAMI, 2016). A trend that shows little sign of slowing, 38% of these shoppers say they actually plan to increase their natural and organic purchases (FMI and NAMI, 2016).

In 2016, the most common reason for purchasing natural and organic meat was that it was 'free of substances I want to avoid' (FMI and NAMI, 2016). The 'free-from' trend or desire to avoid certain substances is certainly not new and has had far-reaching effects on meat labelling and branded beef programmes. The primary targets of 'free-from' labelling in beef include antibiotic-free and hormone-free. Similarly, long-term positive health benefit was listed as another reason for purchasing natural and organic meat (FMI and NAMI, 2016). To this effect, many natural and organic purchasers may be classified as a sub-set of the health-conscious consumer segment. The Laura's Lean Beef brand is one example of these brands targeting healthfulness more than tenderness or marbling. A brand traditionally recognized for marketing 92% lean ground beef and lean whole muscle cuts, Laura's Lean Beef expanded to include natural and organic product lines to capitalize on growing consumer demand.

Although there is very little evidence that natural and organic beef is healthier than conventionally raised beef, this consumer misconception is pervasive. With the average consumer being three to four generations removed from production agriculture, food production literacy is at an all-time low. To demonstrate, in an Oklahoma State Food Demand Survey, consumers stated 'added salt' was a natural ingredient while 'sodium chloride', the chemical name of table salt, was not (Lusk and Murray, 2013).

Regardless of the science, emotional appeal of the natural and organic meat segments present an enormous opportunity for branded beef programmes. Natural and organic shoppers have greater brand preference than conventional shoppers in both the fresh and processed meat categories (FMI and AMI, 2014). Since branded beef programmes were positioned as a differentiated, value-added segment of the retail meat case long before natural and organic interests peaked, they are uniquely positioned to expand existing product lines and capitalize on these trends.

## 5.1 Natural

As of 2016, 19 of the USDA-certified beef programmes included the word 'natural' in the brand name (USDA, 2016a). The USDA defines 'natural' as minimally processed products, containing no preservatives, artificial ingredients or colouring (UDSA, 2016e). By this definition, almost all fresh beef qualifies as 'natural'. However, the average consumer may associate the term 'natural' with many more attributes.

Recognizing the inconsistency between USDA and consumer definitions, many branded beef programmes have placed more rigorous specifications on their branded natural products. Examples of these more stringent specifications include on-farm practices such as no antibiotics, hormones, beta-agonists, or not fed animal by-products. While for many programmes these requirements for cattle eligibility are specific to the finishing period, others may be more general with requirements for the animal's entire lifetime. These more severe production requirements are frequently referred to as 'never-ever' programmes.

Today's consumers increasingly want to know 'how and where cattle were raised' (Igo et al., 2013). Many branded programmes have looked to third-party auditing services to authenticate animal welfare practices, operating procedures as well as cattle age and source. In fact, the terms 'no antibiotics' and 'no added hormones' cannot be used on meat product labels without sufficient documentation, usually in the form of auditing service affidavits (USDA, 2016e). An opportunity to demonstrate even more rigorous documentation, the USDA Process Verified Program enabled branded beef programmes, ranches, and packers meeting set criteria to market their programmes requirements as 'USDA Process Verified' (USDA, 2016f).

Although impressive demand exists for natural beef, producer incentive must be an important consideration to ensuring supply. In a two-year study of the effects of technology in beef production systems on carcass value and net returns, Maxwell (2014) reported naturally finished cattle needed significant slaughter premiums to equal net returns of conventionally fed cattle. Although feed production technologies (monensin, tylosin, and zilpaterol hydrochloride) increased the total feed costs of conventional cattle, conventionally fed cattle had considerably greater carcass value than natural cattle, a result of an additional 60 kg live weight and few quality grade differences. Consequently, natural cattle would need to command a premium of approximately $150/animal to have net returns equal to those of conventionally fed cattle (Maxwell, 2014). These findings demonstrate the importance of market climate evaluation to contracting slaughter premiums high enough to offset performance costs.

## 5.2 Organic

Unlike the term 'natural', use of the term 'organic' is highly regulated with specific requirements. Managed by the National Organic Program of the USDA, only products certified as meeting the USDA's organic requirements may use the USDA Organic seal. Although there may be some similarities between natural and organic requirements, the requirements for organic production are considerably more involved. For example, organically produced cattle must be raised separately from conventionally raised livestock, fed 100% organically grown feeds, and may not be given antibiotics at any point throughout the lifetime of the animal. For this reason, organic producers are required to treat sick animals and remove them from the programme if antibiotics are deemed necessary.

It should come as no surprise, given the required time, documentation and production costs, that organic beef carries a considerably higher price per unit than conventionally raised beef. Nonetheless, although the organic segment makes up a relatively small portion of the total meat case (3.8%), sales and volume continue to grow with a 31.5% increase in dollar growth and 29.3% increase in volume between 2014 and 2015 (FMI and NAMI, 2016).

## 5.3 Barriers to future growth

The natural and organic sector will likely continue to see growth in the coming years considering increased consumer spending, a shrinking price difference between natural and conventional, as well as greater availability. All said, price still remains the greatest barrier to market entry. In a scenario evaluating purchasing intention, the 2016 Power of Meat showed, at equal prices, 45% of consumers would purchase a conventional product, with 30% purchasing antibiotic-free and 25% purchasing organic (FMI and NAMI, 2016). However, when realistic market prices were applied to the three categories, the percentage of consumers purchasing antibiotic-free fell 10% units and the percentage purchasing organic fell 16% units (FMI and NAMI, 2016). The change was even more dramatic among shoppers who classified themselves as 'not current organic buyers'. This price barrier can be seen in economic trends as well, with natural and organic purchases decreasing slightly before plateauing through the height of the recession as consumers scaled back on unessential purchases.

# 6 International branded beef programmes

The development of branded beef programmes is certainly not unique to the United States. In fact, a number of international beef branding trends have had major impacts on the direction of U.S. brand programmes. Although clearly not an exhaustive list, below are a few examples of countries where branding of high-quality beef is common practice.

## 6.1 Japan

In Japan, the term 'Wagyu' is used to describe all Japanese beef cattle. Although the country's modern Wagyu are the result of cross-breeding with imported breeds during the late nineteenth century, the four officially recognized breeds include Japanese Black, Japanese Brown, Japanese Shorthorn and Japanese Polled. Of these four recognized breeds, Japanese Black make up approximately 90% of the Wagyu beef sold in Japan. The result of almost 100 years of intense selection, Wagyu beef is known for its extreme high degree of marbling and exceptional palatability.

Within Japan, some of the most famous Wagyu brands are differentiated by genetic lines within a breed as well as geographic region. The most recognizable Japanese beef brand in the world, Kobe, can only come from Japanese Black cattle of the Tajima bloodline and must be raised in the Hyogo prefecture where the city of Kobe is located (Gaskins, 2016).

Cattle fitting these criteria must then meet quality- and yield-related carcass specifications to qualify for the brand. Other examples of Japanese Wagyu branded beef with regional specifications are the Matsusaka and Ōmi brands.

Consumer demand for Wagyu has certainly not gone unnoticed, with numerous other countries having sought out Wagyu-type genetics to capitalize on this demand. Imports of Wagyu breeding stock during the 1980s set the foundation for Australia and U.S. national Wagyu breed associations. And although interest in Wagyu-type beef began in Japan, Wagyu-branding in the United States is at an all-time high. As of 2016, there were five Wagyu-type USDA-certified beef programmes (USDA, 2016a).

However, consumer desire for Wagyu-type beef does not necessarily translate into consumer knowledge of the segment. To this effect, Wagyu-type beef branding has created some consumer confusion regarding the difference between Wagyu and Kobe beef. A Google Trends search showed one of the top five search queries related to the term 'Wagyu' was 'Wagyu vs. Kobe' (Google, 2016). Although the production and labelling of Kobe beef is very tightly regulated in Japan, patent law differences between Japan and the United States have resulted in less restricted use of the term 'Kobe' in the United States. However, consumer pushback over what some considered misrepresentative labelling led to the development of marketing domestic beef with quality similar to Kobe beef as 'American Kobe-style beef'.

## 6.2 South Korea

For hundreds of years, native Korean cattle were used primarily as draft animals. During the mid-1960s, economic expansion as well as changing political and religious climates gave way to the first nationwide beef market (Jo et al., 2012). Although there are four breeds native to the Korean peninsula (Hanwoo, Jeju Black, Heugun, and Chikso), the majority of Korean beef is produced from cattle of the Hanwoo breed (FAO, 2016). Similar to Wagyu, however, the term 'Hanwoo' is used to describe all beef cattle raised domestically in South Korea.

As consumers' disposable income increased over the years, per capita beef consumption increased from 3 kg in 1985 to 12 kg by 2010 (Jo et al., 2012). The Korean cattle industry responded by focusing on increasing cattle numbers and improving yield attributes to meet the growing consumer demand. However, the liberalization of beef importation in South Korea in 2001 put substantial competitive pressure on Korean beef producers (Kim et al., 2004).

As domestic beef production declined and price increased, some companies adopted branded beef strategies to differentiate Hanwoo from imported beef and improve consumer perception (Ban and Wixom, 2016). Primary targets of differentiation focused on areas where Hanwoo was likely to have the competitive advantage over imported beef, such as beef quality grade and product traceability. Marketing domestic production has also had an important role in improving consumer demand and willingness-to-pay for Hanwoo.

# 7 Technology and branded beef programmes

The use of emerging technologies for the development or improvement of branded beef programmes may be an area of focus for the future.

As the demand for high-quality beef continues to grow, the beef industry needed the help of technology to help meet this demand. Advances in early genetic technologies such as Estimated Progeny Differences (EPDs) had a tremendous role in improving the genetic capability of U.S. feeder cattle. However, the arrival of genomic prediction through high-density genotyping may have the potential to revolutionize the rate at which genetic improvements such as marbling, tenderness or ribeye area can be made. While multiple years and progeny are needed to develop highly accurate EPDs, genomic testing can generate EPDs of similar accuracy with less time and fewer needed progeny. Although primarily used for the improvement of purebred cattle, these tools will be of increasing importance to commercial production in meeting the high-quality beef demand efficiently and with fewer inputs.

Another such technology with the potential to improve marketing of branded beef is the development of non-invasive primary tenderness indicators. Through visible and near-infrared spectroscopic (VISNIR) evaluation of the exposed loin muscle of ribbed beef carcasses, Shackelford et al. (2004) demonstrated the potential for VISNIR to identify carcasses possessing exceptional meat tenderness at normal production speeds.

Tools of this kind may present real opportunity to capitalize on consumers' valuation of tenderness. In a valuation of steak tenderness, Lusk et al. (2001) showed after an in-store taste test, consumers were willing to pay a premium of $2.71/kg for tender steaks compared with tough ones. Furthermore, when consumers were provided information about a steak's tenderness in addition to taste testing they were willing to pay a $0.61 greater premium than taste tests alone (Lusk et al., 2001). Findings of this kind make clear the potential for branded beef programmes to differentiate and market products using tenderness claims.

One strategic use of this technology is to identify carcasses with above-average palatability in traditionally under-valued populations that would not otherwise be expected to deliver a high-quality eating experience. To test this concept, Shackelford et al. (2001) used slice shear force (SSF) to identify exceptionally tender USDA Select steaks and evaluated consumer desire for a brand of this kind. Both consumer and trained taste panels identified USDA Select steaks with low SSF as more desirable for all palatability traits than high SSF steaks (Shackelford et al., 2001). Additionally, when put into a fictional branded product scenario 'Tender Select brand', 89% of consumers reported they would definitely or probably buy such a product (Shackelford et al., 2001).

In real-world application of this strategy, Nolan Ryan's All Natural Tender Aged Beef certified beef programme uses VISNIR technology to identify carcasses with above-average tenderness from heat-tolerant cattle. Because cattle with *Bos indicus* genetics would conventionally be associated with a tougher eating experience, the combination of VISNIR identification and traditional ageing techniques has enabled the brand to add value to carcasses that may have otherwise been overlooked by other branded beef programmes.

Although not a brand specifically, another example of technology that may be used by branded beef programmes is the USDA 'Guaranteed tender' and 'Guaranteed very tender' programme. Designed to provide consumers more information when making purchasing decisions, the Guaranteed Tender programme also gives producers and brand programmes an additional marketing tool for promoting high-quality beef. In 2011, standard specifications for marketing beef as 'Certified Tender' were developed by USDA and ASTM International, a worldwide standards organization. The standard to claim Certified Tender requires qualifying beef cuts to meet a minimum tenderness threshold value (MTTV) of a maximum of 4.4 kg for Warner-Bratzler shear force and 20.0 kg SSF (differences in MTTV are due to differences in shear force methods; ASTM, 2011).

# 8 Summary and future considerations

As branded beef becomes even more common in the retail case, it will be increasingly more important for branded beef programmes to focus marketing efforts on differentiating products according to consumer needs and desires. Although appropriate portion sizes, unique flavours, convenience, healthfulness and price are important considerations for the majority of consumers, the number of consumers using ethical considerations in cattle production to make purchasing decisions is growing. Branded beef programmes will need to continue balancing the use of industry advancements and technologies with authenticity, transparency and emotional connection to be successful with today's consumer.

# 9 Where to look for further information

Minnick, F. (2010), *The Brand that Changed Beef*, The Wooster Book Company. Wooster, OH.

Wilfong, A. K., McKillip, K. V., Gonzalez, J. M., Houser, T. A., Unruh, J. A., Boyle, E. A. E. and O'Quinn, T. G. (2016), 'Determination of the effect of brand and product identification on consumer palatability ratings of ground beef patties', *J. Anim. Sci.*, doi: 10.2527/jas.2016-089.

# 10 References


Allison, R. I. and Uhl, K. P. (1964), 'Influence of beer brand identification on taste perception', *J. Mark. Res.* 1, 36–9. doi:10.2307/3150054.

ASTM International. (2011), 'Standard specification for tenderness marketing claims associated with meat cuts derived from beef', https://www.ams.usda.gov/services/auditing/tender (accessed 10 September 2016).

Ban, Y. K. and Wixom, S. L. (2016), 'Korea- Republic of: Livestock and products semi-annual: Semi-annual livestock and products report', *USDA FAS Global Agricultural Information Network,* Livestock and Products Semi-annual_Seoul_Korea – Republic of_3-15-2016 (accessed 1 August 2016).

Banovic, M., Fontes, M. A., Barreira, M. M. and Grunert, K. G. (2012), 'Impact of product familiarity on beef quality perception', *Agribusiness,* 28(2), 157–75. doi: 10.1002/agr.21290.

Bredahl, L. (2003), 'Cue utilisation and quality perception with regard to branded beef', *Food Qual. Prefer,* 15, 65–75. doi: 10.1016/S0950-3293(03)00024-7.

Certified Angus Beef. (2011), 'The Cattlemen's Pocket Guide to the Certified Angus Beef Brand', http://www.cabpartners.com/educators/ (accessed 15 July 2016).

de Chernaony, L. and Riley, F. D. (1998), 'Defining a "Brand": Beyond the literature with experts' interpretations', *J. Mark. Manag.* 14, 417–43. doi: 10.1362/026725798784867798.

Della Lucia, S. M., Minim, V. P. R., Silva, C. H. O., Minim, L. A. and Cipriano, P. A. (2014), 'Use of relative risk test to evaluate the influence of the brand on beer acceptability', *Semina: Ciências Agrárias,* 35(1), 267–76. doi: 10.5433/1679-0359.2014v35n1p267.

FAO, 'Number of breeds by species and country', http://dad.fao.org/ (accessed 14 August 2016).

Food Marketing Institute and American Meat Institute. (2014), 'The Power of Meat: An in-depth look at meat and poultry through the shoppers' eyes' (accessed 15 July 2016).

Food Marketing Institute and North American Meat Institute. (2015), 'The Power of Meat: An in-depth look at meat and poultry through the shoppers' eyes' (accessed 15 July 2016).

Food Marketing Institute and North American Meat Institute. (2016), 'The Power of Meat: An in-depth look at meat and poultry through the shoppers' eyes', http://www.meatconference.com/PowerofMeat2016.pdf (accessed 15 July 2016).

Froelich, E. J., Carlberg, J. G. and Ward, C. E. (2009), 'Willingness-to-pay for fresh brand name beef', *Can. J. Agric. Econ.* 57(1), 119–37. doi: 10.1111/j.1744-7976.2008.01141.x.

Gaskins (2016), 'WSU Genetic Lines of Wagyu in the U.S.' *American Wagyu Association*, http://wagyu.org/breed-info/wagyu-origins (accessed 20 July 2016).

Google Trends, https://www.google.com/trends/explore?date=all&q=wagyu (accessed 14 August 2016).

Hyman, M. R., Kopf, D. A. and Lee, D. (2010), 'Review of literature – future research suggestions: Private label brands: Benefits, success factors and future research', *J. Brand Manage.* 17, 368–89. doi: 10.1057/bm.2009.33.

Igo, J. L., VanOverbeke, D. L., Woerner, D. R., Tatum, J. D., Pendell, D. L., Vedral, L. L., Mafi, G. G., Moore, M. C., McKeith, R. O., Gray, G. D., Griffin, D. B., Hale, D. S., Savell, J. W. and Belk, K. E. (2012), 'Phase I of The National Beef Quality Audit-2011: Quantifying willingness-to-pay, best-worst scaling, and current status of quality characteristics in different beef industry marketing sectors', *J. Anim. Sci.* 91, 1907–19. doi: 10.2527/jas2012-5815.

Jo, C., Cho, S. H., Chang, J. and Nam, K. C. (2012), 'Keys to production and processing of Hanwoo beef: A perspective of tradition and science', *Anim. Front.* 2(4), 32–8. doi: 10.2527/af.2012-0060.

Kim, R. and Milton, B. (2004), 'Identification of niche market for Hanwoo beef: Understanding Korean consumer preference for beef using market segment analysis', *Int. Food Agric. Manage. Rev.* 7(3), 46–64.

Levin, I. P. and Gaeth, G. J. (1988), 'How consumers are affected by the framing of attribute information before and after consuming the product', *J. Consum. Res.* 15, 374–8. doi: http://dx.doi.org/10.1086/20917.

Lusk, J. L., Fox, J. A., Schroeder, T. C., Mintert, J. and Koohmaraie, M. (2001), 'In-store valuation of steak tenderness', *Am. J. Agric. Econ.* 83(3), 539–50. doi: 10.1111/0002-9092.0017.

Lusk, J. L. and Murray, S. (2013). Food Demand Survey. 1:2 http://agecon.okstate.edu/faculty/publications/4559.pdf (accessed 1 August 2016).

Lusk, J. L. and Schroeder, T. C. (2006), 'Auction bids and shopping choices', *BE J. Econ. Anal. Poli.* 6(1), Article 4. doi: 10.2202/1538-0637.153.

Lusk, J. L. and Schroeder, T. C. (2004), 'Are choice experiments incentive compatible? A test with quality differentiated beef steaks', *Am. J. Agric. Econ.* 86(2), 467–82. doi: 10.1111/j.0092-5853.2004.00592.

Maxwell, C. L. (2014), 'Comparison of technologies in beef production systems' (thesis), https://shareok.org/bitstream/handle/11244/14992/Maxwell_okstate_0664D_13229.pdf?sequence=1 (accessed 1 August 2016).

McNeill, S. H., Harris, K. B., Field, T. G. and Van Elswyk, M. E. (2012), 'The evolution of lean beef: Identifying lean beef in today's U.S. marketplace', *Meat Sci.* 90, 1–8. doi:10.1016/j.meatsci.2011.05.02.

Minnick, F. (2010), *The Brand that Changed Beef,* The Wooster Book Company. Wooster, OH.

Roselius, T. (1971), 'Consumer rankings of risk reduction methods', *J. Mark.* 35(1), 56–61.

Shackelford, S. D., Wheeler, T. L., Meade, M. K., Reagan, J. O., Byrnes, B. L. and Koohmaraie, M. (2001), 'Consumer impressions of tender select beef', *J. Anim. Sci.* 79, 2605–14. doi: 10.2527/2001.79102605.

Simmons, J. (2011), 'Making safe, affordable and abundant food a global reality', *Range Beef Cow Symposium,* Paper 300. http://digitalcommons.unl.edu/rangebeefcowsymp/300 (accessed 8 July 2016).

Smith G. C., Carpenter, Z. L., Cross, H. R., Murphey, C. E., Abraham, H. C., Savell, J. W., Davis, G. W., Berry, B. W. and Parrish, Jr., F. C. (1984), 'Relationship of USDA marbling groups to palatability of cooked beef', *J. Food Qual.* 7, 289–308.

Smith, G. C., Murphey, C. E., Carpenter, Z. L., Savell, J. W. and Cross, H. R. (1980), 'An evaluation of USDA beef carcass quality grade standards', Final report to AMS, U.S. Dept. of Agric., Washington, D.C.

Speer, N. C. (2013), 'Consumers, business and breeding systems: Charting the beef industry's path' (white paper), Certified Angus Beef, http://www.cabpartners.com/news/research.php (accessed 7 July 2016).

Tatum, J. D. (2015), 'Recent trends: Beef quality, value, and price' (white paper), Certified Angus Beef, http://www.cabpartners.com/news/research.php (accessed 8 August 2016).

Urde, M. (2003), 'Core value-based corporate brand building', *Eur. J. Mark.* 37(7), 1017–40. doi: 10.1108/0309056031047764.

USDA (2016a), 'Comparison of certified beef programs', *USDA Agricultural Marketing Service,* https://www.ams.usda.gov/services/auditing/certified-beef-programs (accessed 15 August 2016).

USDA (2016b), 'Beef: Supply and disappearance (carcass weight, million pounds) and per capita disappearance', http://www.ers.usda.gov/data-products/livestock-meat-domestic-data.aspx#26063 (accessed 15 July 2016).

USDA (2016c), 'Schedule GLA – November 1996 USDA specification for characteristics of cattle eligible for approved beef programs claiming Angus influence', *USDA Agricultural Marketing Service,* http://www.ams.usda.gov/sites/default/files/media/LS-SCH-GLA.pdf (accessed 15 July 2016).

USDA (2016d), 'Beef grading historical records', *USDA Agricultural Marketing Service,* https://www.ams.usda.gov/sites/default/files/media/historicalbeefgradingvolumes.pdf (accessed 8 July 2016).

USDA (2016e), 'Meat and poultry labeling terms', *USDA Agricultural Marketing Service,* http://www.fsis.usda.gov/wps/portal/fsis/topics/food-safety-education/get-answers/food-safety-fact-sheets/food-labeling/meat-and-poultry-labeling-terms (accessed 15 July 2016).

USDA (2016f), 'Process verified program', *USDA Agricultural Marketing Service,* https://www.ams.usda.gov/services/auditing/process-verified-programs (accessed 15 July 2016).

Wilfong, A. K., McKillip, K. V., Gonzalez, J. M., Houser, T. A., Unruh, J. A., Boyle, E. A. E. and O'Quinn, T. G. (2016), 'Determination of the effect of brand and product identification on consumer palatability ratings of ground beef patties', *J. Anim. Sci.,* doi: 10.2527/jas.2016-089.

Zeithaml, V. A. (1988), 'Consumer perceptions of price, quality, and value: a means-end model and synthesis of evidence', *J. Mark.* 52, 2–22. doi: 10.2307/125144.

Zimmerman, L. C., Schroeder, T. C., Dhuyvetter, K. C., Olson, K. C., Stokka, G. L., Seeger, J. T. and Grotelueschen, D. M. (2012), 'The effect of value-added management on calf prices at Superior Livestock Auction video markets', *J. Agric. Resour. Econ.* 37(1), 128–43.

Chapter 12

# Ageing, physical and chemical methods for improving tenderness and palatability of beef

*D. L. Hopkins, NSW Department of Primary Industries, Centre for Red Meat and Sheep Development, Australia*



## 1 Introduction

The eating quality of meat is determined by tenderness, juiciness and flavour (Thompson, 2002), and the composite of these traits is called palatability, which can be described by the term 'overall liking'. Of these traits, tenderness is affected by both production and processing factors (Young, Hopkins and Pethick, 2005) and is the easiest of the three traits to measure objectively.

A measure of tenderness can be achieved by subjective consumers or by trained panellists in which a high score is desirable, and connective tissue and the amount of intramuscular fat can influence the score given. An objective measure of tenderness is the force required to shear a standardized piece of cooked meat, with low shear values being desirable (Hopkins et al., 2006). The objective measurement of tenderness of cooked meat can be undertaken with a variety of devices, as outlined by Purchas (2014).

There are three major factors that can impact on the tenderness of meat; these include collagen content and solubility (background toughness), muscle shortening (toughening) and ageing (tenderization) (Hopkins and Geesink, 2009). Two antagonistic processes (toughening and tenderization) take place during the post-mortem storage period (Hopkins and Thompson, 2001a), and the toughening process can be minimized by limiting the extent of muscle shortening during rigor development (Hopkins and Thompson, 2001a).

http://dx.doi.org/10.19103/AS.2016.0009.12

These factors are themselves affected by production traits such as genotype, age, gender and handling, and can be manipulated to some extent by processing post-slaughter; the cooking method also impacts on final tenderness.

Tenderization is the generalized term for the process that occurs during ageing, which leads to improvement in tenderness and, in reality, can only be measured post-rigor. Tenderization reflects proteolysis, which is the degradation of proteins (myofibrillar, cytoskeletal and costameric) into smaller sub-units that occurs with ageing, but proteolysis also leads to the turnover of living muscle (Goll et al., 1998). Factors that impact eating quality traits with a specific focus on the tenderness of beef will be outlined in this chapter, which will include an overview of methods that can be employed to improve eating quality at both the carcass and the primal level.

# 2 Ageing of beef products

The degradation of muscle proteins (proteolysis) during the post-mortem period has been studied by using a number of techniques, including fragmentation of myofibrils, determination of free amino acids, measurement of protein solubility, measurement of non-protein nitrogen and gel electrophoresis. It is this degradation during ageing that leads to improvements in tenderness, for example, tenderization. Recent developments in the proteomics field have provided more detailed information on the degradation of proteins post-mortem (Bendixen, 2005), but the challenge still remains to interpret which proteins actually impact directly on tenderization and which are non-influential.

Beef can be held to age either vacuum-packed (wet) or dry-aged without any protective covering, with the latter approach requiring a significant amount of time and space, and which also increases the loss of moisture. The surface of dry-aged beef develops a crust that must be trimmed prior to sale leading to additional weight loss. Warren and Kastner (1992) reported in a small study that dry-aged striploins had a more beefy and brown/roasted flavour compared to wet-aged beef, which produced a more bloody/serumy and metallic flavour. In a much larger, more recent study with three different cuts, there was no indication of a difference in consumer acceptability of meat from either ageing method (Laster et al., 2008), but there were small differences in shear force depending on the cut with no overall advantage to either method of storage. Having said this, when ribeye was aged for 44 days, there was an indication that dry-aged beef had a higher overall liking, and there was an increase in scores with longer ageing periods, whereas wet-aged product showed much less improvement.

Although the palatability outcomes are equivocal, a study by Kim et al. (2016) has provided evidence that the flavour profile of dry-aged beef is preferred by consumers and they went further to identify the metabolites contributing to the change in flavour. Using nuclear magnetic resonance (NMR), seven amino acids; tryptophan, phenylalanine, valine, tyrosine, glutamate, isoleucine and leucine were identified to have increased during dry ageing.

## 2.1 Process of proteolysis

Proteolysis of the myofibrillar proteins can precede (i.e. before rigor) tenderization (Hopkins and Thompson, 2001b), and the degree of disruption of the proteins (such as titin and nebulin) will influence the final tenderness of beef. More recently, changes in proteins due to apoptosis, which is the 'ordered' cell death that occurs under the anoxic conditions that immediately follow animal death, have been studied on the basis that this is the first step in a cascade of events leading to tenderization (Ouali et al., 2006).

The mechanism that drives tenderization post-rigor has demanded significant attention, and several explanations for this process have been proposed, but given the requirement for remodelling of protein in living muscle (Goll et al., 1998), it is apparent that such enzymatic pathways must also be central to post-mortem proteolysis. There are a number of reviews that summarize the biochemical changes that occur in meat during the post-mortem period (Hopkins and Thompson, 2002; Sentandreu et al., 2002; Ouali et al., 2006). Although there is still debate about their relative contribution to proteolysis of muscle proteins, cathepsins, calpains, proteasomes and, more recently caspases, have all been implicated, of which the calpains have the most dramatic effect (Hopkins and Geesink, 2009). Recent reviews deal with the latest evidence about which the enzymatic system is the most important in terms of improving meat tenderness (Kemp and Parr, 2012; Ouali et al., 2013), and this area will not be further covered here.

It should be noted that the disassembly of the sarcomere (the smallest contractile unit of muscle fibres) will release specific proteins from the myofibrillar matrix and yet remain undegraded (e.g. alpha-actinin), and it is also unrealistic to expect that the disappearance of a single myofibrillar protein could be used to explain tenderization in totality. The evidence indicates that a number of proteins associated with the thin and thick filaments and others found at the periphery of the Z-disk are degraded under normal post-mortem chiller conditions. Contrary to earlier suggestions that Z-disk degradation was significant during post-mortem storage of muscle, the evidence from the work of Taylor et al. (1995) shows that degradation of proteins near the Z-disk is what actually occurs. This conclusion can now extend at least for titin to other regions of the sarcomere (Greaser et al., 2000). In fact, several ultrastructural studies from the 1970s clearly show that the myofibre breaks in the I-band region and that the Z-disk is stable (e.g. Gann and Merkel, 1978).

## 2.2 Influential factors

There can be considerable differences in tenderness caused by post-slaughter chilling conditions as these can impact on the rate of post-mortem glycolysis. After death, muscle filaments are in a continuous state of contraction and relaxation, and as glycolysis proceeds, glycogen levels drop, adenosine triphosphate (ATP) is depleted and the filaments enter rigor and a contracted state.

### 2.2.1 Development of *rigor mortis*

The degree of shortening, or of tension development, during the onset of *rigor mortis* in muscle which is free to shorten, is a direct function of temperature down to about 15°C (Locker and Hagyard, 1963). If such isolated muscle is exposed to temperatures lower than about 12°C at this time, there is again an increasing tendency to shorten – it being as great at 2°C as at 40°C – and this is associated with decreased tenderness on cooking. As the temperature falls below 12°C, a pre-rigor contracture (cold shortening) takes place until *rigor mortis* is completed. This arises due to an increased level of cellular calcium from the sarcoplasmic reticulum as a result of the falling temperature, which in turn activates actomyosin ATPase. The rise of calcium in the cytoplasm under these conditions is due to the failure of the sarcoplasmic reticulum to sequester cytoplasmic calcium (Jaime et al., 1992). The release of calcium from mitochondria, when ATP concentration is not limiting (>3.5 μM), due to anoxic and cold conditions (below 15°C), causes inactivation of the ATP-driven calcium pump (Honikel and Hamm, 1978). The 'cold-shortening' phenomenon is less likely in beef carcasses due to the weight and fat levels, which slow temperature decline, compared to smaller species like sheep.

Converse to the cold-shortening effect is what can be called heat toughening/shortening due to low pH and high temperature during *rigor mortis* development. This effect has been reported in cattle (Warner et al., 2014), where it is defined as a pH below 6, with the temperature at 35°C or greater. Indeed, the speed of glycolysis under these conditions may lead to an exudative condition in beef, somewhat comparable to Pale, Soft, Exudative (PSE) in pigs, particularly in the deep hindquarter muscles. This is due to denaturation of myosin, which causes shortening of the myosin head and reduction in filament spacing. This induces water to be expelled from the muscle cells into the extracellular space (Offer, 1991), and such increased speed of *rigor mortis* onset can also limit ageing potential (recently reviewed by Kim et al., 2014). Alternative methods for manipulating the rate of glycolysis or increasing the rate of temperature decline have recently been reviewed by Jacob and Hopkins (2014) with a view to addressing this heat toughening issue in beef. One of these approaches is to target steps in the glycolytic pathway so as to manipulate the rate of pH decline. Jerez et al. (2003) reported that pre-rigor injection of either sodium fluoride or sodium citrate increased the level of muscle glycogen and thus reduced the absolute decline in pH. These compounds did result in a higher ultimate pH, and a reduction in sarcomere length, but in the case of sodium citrate, a reduction in shear force. Unfortunately, control samples were not excised from the carcass as were injected samples, so the treatment impact on sarcomere length and thus tenderness was confounded. Nevertheless, it was demonstrated that there are compounds that will slow the rate of pH decline, although no mode of action was proposed. For the immediate post-mortem period when the citric acid cycle is still operative, citrate is produced (Davies, 2004), which has an inhibitory effect on phosphofructokinase (Berg et al., 2003). As this enzyme is an important regulatory enzyme in glycolysis, it may partially explain the apparent reduction in glycolysis reported by Jerez et al. (2003) from the injection of sodium citrate.

The extent of post-mortem glycolysis, apart from its rate, also has an effect on the tenderness of beef (Purchas et al., 1999) in which tenderness is minimal at ultimate pH values between 5.8 and 6.2. As the ultimate pH increases from 5.5 to 6.0, tenderness appears to decrease; at ultimate pH levels above 6.2, however, tenderness increases once again. Watanabe and Devine (1996) subsequently showed that the breakdown of titin and nebulin is minimal at an ultimate pH in this region and concluded that one effect of ultimate pH on tenderness is exerted through its action on proteolytic enzymes. However, the study of Purchas et al. (1999) clearly showed that a combination of sarcomere length and proteolysis did not fully explain the curvilinear relationship between pH and shear force. This conclusion was based on examining the relationship in unaged and aged beef.

## 2.2.2 Impact of chilling temperature

It has long been clear that muscles contain proteolytic enzymes, which operate much more readily at 37°C than at 5°C (Sharp, 1963), and that, in general, higher holding temperatures post-rigor cause tenderization to occur in a faster time than at lower temperatures. Nevertheless, the rate of tenderization decreases gradually as holding temperatures rise from 40°C to 60°C. It then decreases sharply, ceasing altogether at 75°C (Davey and Gilbert, 1976). Proteolysis can continue even in shortened muscle without the meat-becoming tender (Locker and Wild, 1984), illustrating the interaction with sarcomere length and demonstrating that ultimate tenderness levels are not determined by only one mechanism (Starkey et al., 2016). Thus, the temperature applied during ageing directly impacts on the rate of tenderization, but for food safety reasons, storage temperatures must be between −1°C and 2°C to achieve maximum shelf life. Further, the time allowed for meat to age also impacts on the final tenderness presented to consumers. For example, as shown in Fig. 1, the greatest improvement in tenderness is early in the first 14 days of the ageing period. If meat is sold after very short ageing periods (e.g. within 2 days of death), this will increase consumer dissatisfaction, and for this reason, minimum ageing periods are employed across the beef processing and retailing sectors.

## 2.2.3 Impact of variation between and within muscles

The improvement in tenderness with ageing is also influenced by the muscle under consideration, as shown by Stolowski et al. (2006). In some muscles, there is minimal reduction in shear force with ageing (e.g. biceps femoris: Stolowski et al., 2006), but in the longissimus, reductions of at least 35% in shear force have been reported with ageing (Fig. 1). Monin and Ouali (1991) have extensively reviewed the reasons for differentiation between muscles in the rates and extents of ageing that they undergo post-mortem. In general, connective tissue is stable post-mortem (McCormick, 1994), although after extended periods of storage, intramuscular connective tissue (IMCT) does show signs of structural changes (Nishimura et al., 1995). In fact, Nishimura et al. (1998) described a method to examine the mechanical strength of IMCT and showed that after 10 days of ageing, there was a decrease in strength of this tissue in beef.

In addition to differences between muscles, there is also intramuscular variation. For example, in a study of the biceps femoris, semitendinosus, semimembranosus and

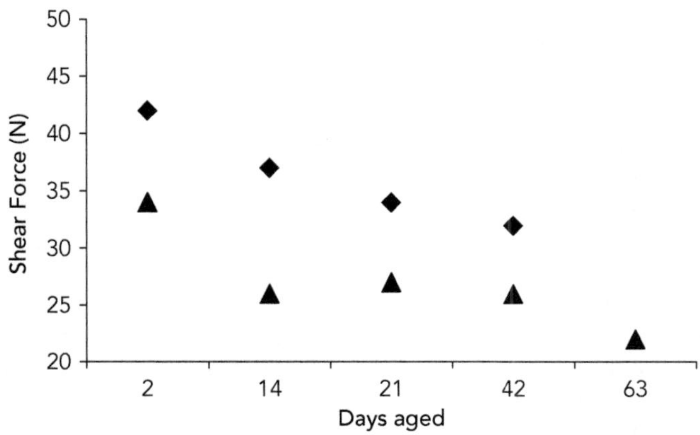


**Figure 1** Relationship between shear force and days aged of bovine *longissimus et lumborum* muscles aged for up to 42 days (Adapted from Stolowski et al. 2006) or 63 days (Adapted from Colle et al. 2015).

adductor muscles, it was shown that the intramuscular variation was greater than the variation among muscles (Reuter et al., 2002), with the biceps femoris exhibiting the greatest intramuscular variation.

It has been suggested that systematic and significant differences in tenderness between locations within muscles could be exploited, with known regions of toughness being removed for manufactured products (Denoyelle and Lebihan, 2003). This known variation, particularly among muscles, has been applied in the development of the Meat Standards Australia (MSA) grading system where the prediction model for the eating quality of beef (Thompson, 2002) is based on statistical models, where anatomical cut descriptions are replaced with prediction of the eating quality of muscles. However, the same muscle in different carcasses can exhibit detectable differences in traits like tenderness, and the model with a range of inputs provides predicted tenderness values. In a related way, the within-cut variation as a function of different muscles was fundamental to the development in the United States of muscle profiling (Belew et al., 2004) and the application of seam boning (Jones et al., 2005). This area will be dealt with in greater detail in Chapter 10.

Another factor that influences tenderness levels includes the effect of *Bos indicus* genes, which have been clearly shown to decrease the tenderness and increase the shear force of muscles such as the longissimus (Wheeler et al., 1996: Chapter 1); the age of the animal (Shorthose and Harris, 1990), with increasing age correlating with decreasing tenderness; and gender, with bulls more likely to produce less tender meat (Rodriguez et al., 2014).

# 3 Physical methods to improve eating quality: carcass methods

Most methods for improving eating quality that are classified as physically based operate on muscle structure and thus impact on tenderness. These methods can be applied at the carcass level or to primals. This section will focus on carcass methods, and then Section 4 will cover primal methods.

## 3.1 Electrical stimulation

Under conditions likely to lead to 'cold-induced' shortening (see Section 2.2.1), electrical stimulation (ES) has a role to minimize shortening by accelerating the onset of *rigor mortis*. ES involves passing an electric current through the body or carcass of freshly slaughtered animals. This electric current causes the muscles to contract, which increases the rate of glycolysis resulting in an immediate fall in pH (ΔpH that ranges from 0.6 pH units at 35°C to 0.018 units at 15°C) (Hwang et al., 2003). In recent years, there has been development of new ES technology in Australia that has enabled retrofitting of units in established abattoirs. This technology applies electricity with short square pulses and operates at medium voltages overcoming the safety concerns of high-voltage systems (Devine et al., 2014) and is most applicable to lightweight, lean carcasses that will cool quickly. In fact, as outlined previously (see Section 2.2.1), many beef carcasses do not require stimulation, because they exhibit rapid rates of glycolysis without stimulation (Hopkins et al., 2014). As such, ES operates on the metabolic processes after death, but there is also some evidence that super-contractions and physical disruption have been observed in muscle electrically stimulated (Hwang et al., 2003), which could lead to an improvement in tenderness; however, this cannot be considered the main mode of action from the use of ES to reduce toughness. The super-contractions appear to be caused through localized excessive release of calcium ions from the sarcoplasmic reticulum.

## 3.2 Tenderstretching

After the interaction between sarcomere shortening and tenderness was established (Locker, 1960), many studies have subsequently investigated possible techniques to increase sarcomere length and hence improve meat tenderness. Of the methods, tenderstretch is the oldest researched method and is achieved by suspending carcasses from the aitchbone (*obturator foramen*) in beef carcasses as they come off the slaughter chain. This results in the back legs falling into the walking position (Fig. 2), and this places increased tension on major leg and loin muscles, before they pass through *rigor mortis* (Hopkins, 2014), and prevents the shortening of sarcomeres and thus reduces the overlap between actin and myosin (Hopkins and Thompson, 2001a).

This technique has a dramatic effect on muscles such as the longissimus and hind leg muscles such as vastus lateralis and semimembranosus (Bouton et al., 1973: Fig. 3), but decreases the tenderness of the *psoas major*. The response to tenderstretching could also be partly attributed to disruption of the I-band proteins (Hopkins et al., 2000), and the level of improvement in tenderness is influenced by the cooking method used (Eikelenboomn et al., 1998) and the age of the animals, with data presented by Ahnström et al. (2012),

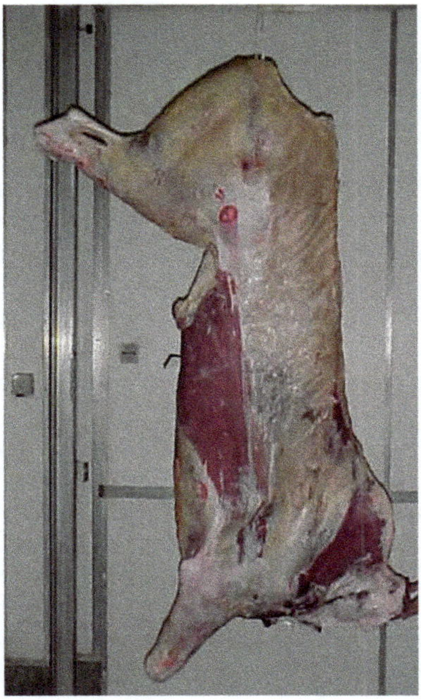

**Figure 2** The carcass is suspended by the aitchbone so that the back leg drops and the backbone straightens and the maximum tension is placed on these muscles. (Photograph courtesy of Dr J. M. Thompson)

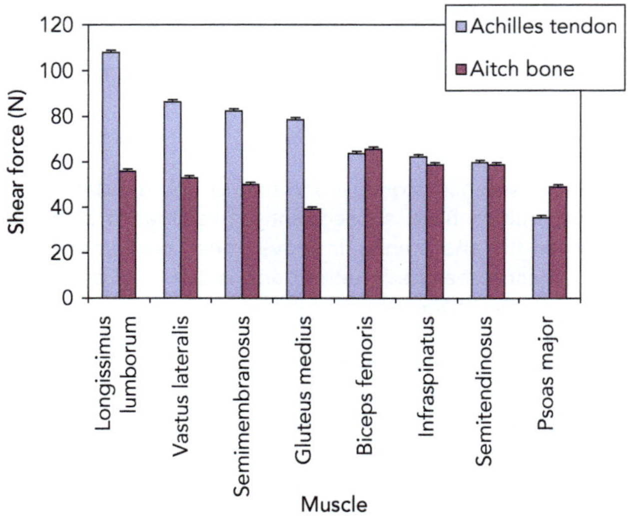


**Figure 3** Warner-Bratzler shear values (N) for muscles measured at 2–3 days post-mortem obtained from sides of beef hung by either the Achilles tendon or aitchbone.

Source: Adapted from Bouton et al., 1973.

suggesting that the response to tenderstretching in older animals is minimal, despite an increase in sarcomere length. This could reflect an increased cross-linking of collagen fibrils that counteracted the increase in sarcomere length, but this cannot be confirmed from the data of Ahnström et al. (2012).

The improvement in tenderness is so dramatic that in a number of muscles, the need for prolonged ageing is virtually eliminated; in addition, the variation in tenderness along the longissimus muscle is reduced. Commercial adoption of this technique has seen resurgence as processors (e.g. there is one processor in Australia who uses tenderstretching exclusively and the method has been adopted in several European countries; Ahnström et al., 2012) have developed ways to handle and store tenderstretched carcasses, such as adopting methods to re-hang carcasses from the Achilles tendon after attainment of *rigor mortis* and streamlining the processing of carcass movement and boning.

## 3.3 Tendercut

A new technique called 'tendercut' has been developed in the United States (Claus et al., 1997). This process involves sawing the vertebral column at the 12th/13th rib junction and/or the ischium at the rump/butt junction (Fig. 4). In addition to breaking the vertebrae at the 12th/13th rib junction, all tissues surrounding the loin are cut, such that only a dorsal component is holding the forequarter to the hindquarter. This technique allows the weight of the carcass to stretch selected muscles prior to the onset of rigor (Wang et al., 1994), whilst the carcass is still suspended by the Achilles tendon (Fig 4). It only affects tenderness in the loin region and thus has reduced applicability. Compared to tenderstretching, the tendercut process overcomes the need for additional chiller space and avoids the need to train boners on new cutting lines, but is much more difficult to do on a processing line.

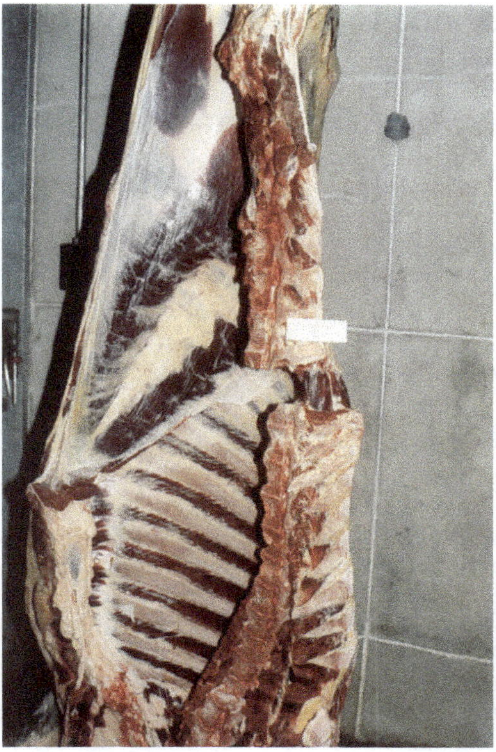

**Figure 4** A photograph showing a severed vertebral column used as part of the tendercut method. (Photograph courtesy of Dr J. R. Claus)

Cargill in the United States uses this technique routinely in their plants. A variation of this approach was examined by Aalhus et al. (1999), which they called the modified, online, altered suspension (MOLAS) method.

## 3.4 Foreshank manipulation

Based on the understanding that restraint of pre-rigor muscle will lead to longer sarcomeres and thus more tender meat, recent work has focussed on positioning the foreshank pre-rigor to prevent shortening in forequarter muscles (Grayson and Lawrence, 2013). Using simple metal hooks, a range of positions of the foreshank was examined. Of the six muscles in the forequarter subsequently tested for shear force, there was no effect of foreshank position for four of the muscles. For the *serratus ventralis*, there was a reduction (4.2 N compared with control) in the shear force when the foreshank was held cranially perpendicular (CP) to the chiller floor, compared to non-manipulated carcasses, but an increase (2.6 N compared with control) when the foreshank was caudally held at 30° in relation to the shoulder joint. This latter effect was similar in the *triceps brachii*, and, interestingly, the CP method also increased the shear force in the *triceps brachii*, despite a significant increase in sarcomere length. Based on these results, the technique offers little benefit to industry.

# 4 Physical methods to improve eating quality: primal methods

It has been demonstrated that if pre-rigor beef muscles can be excised from the carcass and held to prevent shortening (e.g. by wrapping; Devine et al., 1999) that this can provide a potentially economical way to speed up processing and, at the same time, minimize toughening. In this study, for example, wrapped beef longissimus muscle not aged had lower shear force values (mean 114 N) compared to unrestrained muscle (156 N). This principle has been applied in the development of two devices in recent years applicable to hot-boned beef.

## 4.1 Pi-Vac Elasto-Pack System®

The Pi-Vac Elasto-Pack System® is a method of tightly wrapping hot-boned muscles in an elastic wrapping material pre-rigor to prevent shortening and toughening of the meat. The system uses a highly flexible packaging sleeve, which is expanded by using a partial vacuum to allow the meat to be inserted. Once the vacuum is turned off, the flexible packaging retracts to its normal dimensions. This exerts longitudinal forces on the meat, preventing the contraction of the muscle. Almost all of the oxygen is also forced out of the packaging. The subsequent bound meat product has been labelled TenderBound. This technology can come in three different sizes, and some commercialization of this approach has occurred in Europe.

There is limited published data on the performance of the Pi-Vac Elasto-Pack System®, but one study reported on the results of subjecting hot-boned beef (*m. longissimus*) excised 90 minutes post-mortem to the Pi-Vac Elasto-Pack System® (O'Sullivan et al., 2003). This showed more than a 60% reduction in shear force for meat aged for 7 days, with the benefit lessening with another 7 days of ageing. Sarcomere length was increased from a very short 1.34 μm for control meat to 1.63 μm for meat packaged using the Pi-Vac Elasto-Pack System®. This was also reflected in an improvement in sensory tenderness, and the variability in tenderness was also reduced to provide a more consistent tenderness.

## 4.2 SmartStretch™

A more recent development is the SmartStretch™ technology that was patented by Meat and Livestock Australia, which can be used to stretch and hold hot-boned beef. The machine's operation is based on an externally ribbed flexible sleeve surrounded by inflatable bladders that are housed within an airtight chamber into which air can be pumped in or out of. Air is pumped out of the chamber to create negative pressure, which causes the sleeve to expand, allowing the meat to be inserted. It is then pumped into the inflatable bladders causing the meat to be compressed by force perpendicular to the direction of the muscle fibres. This also applies peristaltic action, moving the meat towards the same end of the sleeve that it was inserted into. Positive pressure is then applied to the exterior of the sleeve by pumping air into the chamber, forcing the meat upwards and into packaging, as shown in Fig. 5, where it is held during *rigor mortis* development. Further detail of the operation of the machine is provided by Taylor and Hopkins (2011).

The results from the application of SmartStretch™ have been varied for hot-boned beef. Increasing the length of hot-boned *m.* semimembranosus by 40% with an increase in

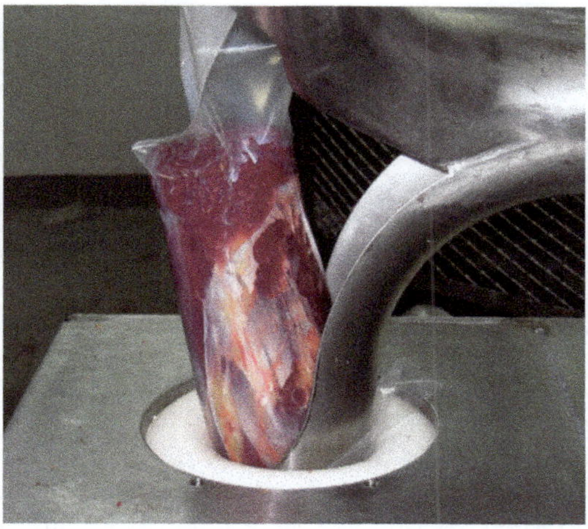

**Figure 5** Stretched beef being ejected from the flexible sleeve of the SmartStretch™ into the packaging. (Photograph courtesy of Dr. D. L. Hopkins)

sarcomere length from 1.78 to 2.68 μm had no impact on shear force (Toohey et al., 2012), whereas for the longissimus, a 17% increase in length did provide a significant reduction in shear force (13 N using a MIRINZ tenderometer) at 0 days ageing, but by 14 days of ageing, the benefit had disappeared (Taylor et al., 2013). Using younger cattle that are not normally hot boned, Taylor et al. (2012) showed significant improvements in sarcomere length and tenderness of hot-boned *m. gluteus medius* at 0 days of ageing, but the effect was nullified with ageing for 8 days. Younger cattle (with a dentition score of 2 or less) were sampled because the evidence suggested that the muscle of older cattle was less affected by the application of SmartStretch™. An interaction with collagen was postulated as a reason that SmartStretch™ did not improve tenderness in the semimembranosus from cull cows (Toohey et al., 2012), and this has some support based on the results of Ahnström et al. (2012) (see Section 3.1.2).

## 4.3 Pulse electric field

A potential stand-alone technology that can be used pre-rigor or post-rigor to enhance tenderization (by cell permeation/electroporation) is pulse electric field (PEF). PEF is based on the subjection of samples to electric current in a chamber (Fig. 6). The first study on the use of PEF for improving the quality of fresh beef was undertaken by O'Dowd et al. (2013). Although the study had a number of limitations (small sample size and numbers), it was reported that PEF resulted in structural changes as indicated by fragmentation of myofibrils, suggesting a mechanism for improving tenderness. Much more detailed studies have recently been completed (Bekhit et al., 2014; Suwandy et al., 2015a). The effect of PEF on the quality of cold-boned beef longissimus (1-day post-mortem) and semimembranosus (1 and 3 days post-mortem) was investigated by using a range of treatment intensities (Bekhit et al., 2014). The tenderness of the longissimus was found to benefit from PEF treatment (19.5% reduction in the shear force) regardless of the electrical input, whereas the level of tenderness of the semimembranosus was increased by increasing the treatment

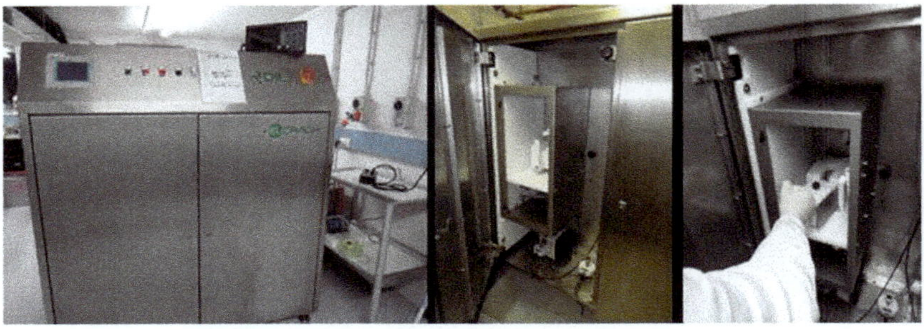

**Figure 6** The treatment chamber housed between the electrodes of a PEF system (Elcrack-HPV5, DIL, Quakenbrück, Germany). (Photograph courtesy of Dr. A. E. D Bekhit)

frequency (4.1, 10.4 and 19.1% reduction in the shear force at 20, 50 and 90 Hz, respectively). Higher purge loss and lower cooking loss were found in PEF-treated samples, but the total losses were similar (Bekhit et al., 2014). Since the tenderness improvement of the semimembranosus was not dependent on when PEF was applied post-mortem, extending at least up to 3 days post-mortem, this provides flexibility in application. Detailed examination by Suwandy et al. (2015a) showed increased proteolysis in PEF-treated longissimus, both early post-mortem and during subsequent post-mortem storage reflected by increased degradation of troponin-T and desmin.

When PEF was applied to hot-boned semimembranosus, there was a 21.6% reduction in the shear force regardless of the electrical inputs, whereas the level of shear force of the longissimus tended to increase with increasing the treatment frequency (Suwandy et al., 2015b). If the number of PEF treatments was increased from 1 to 3, then a reduction in shear force (2.5 N per repeat) was found for cold-boned longissimus, but no effect was observed for the semimembranosus (Suwandy et al., 2015c), whereas for hot-boned muscles, there was an increase in longissimus shear force if the PEF was repeated 3 times; for the semimembranosus, it was decreased, but the benefit disappeared with ageing (Bekhit et al., 2016). The results of these studies demonstrate the real potential of PEF to improve the tenderness of the semimembranosus, a muscle that has the lowest eating quality in the hindquarter of beef. Commercial application requires a batch processing system to be developed.

## 4.4 Other technologies

A number of other technologies have been investigated for improving beef tenderness of primals, and these include hydrodynamic pressure, hydrostatic pressure, blade tenderization, ultrasound and freezing and thawing, and these have been summarized by Hopkins (2014) and so will only be briefly covered here.

The latest developments in the use of hydrodynamic pressure are outlined by Bolumar et al. (2013), and to date, the evidence has suggested that this technology impacts on the structure of the muscle, as opposed to increasing the activation or proteolytic enzymes (Bolumar et al., 2014). The data suggest that this technology reduces the shear force of beef between 10 and 70% when applied pre-rigor, depending on the muscle and study (Bolumar et al., 2013). With the development of a prototype device in Germany that overcomes some of the limitations of the original design (uses electrical discharges under water) produced in the United States , which that relied on the use of explosives (Solomon et al., 1997), there is likely to be further investigation of this technology, but commercial application is still problematic.

High hydrostatic pressure processing (HPP) is another technology that has had a renewed focus. In the early work, McFarlane (1973) subjected six different pre-rigor beef muscles to ~100 MPa for 2 minutes at 35°C. It was found that the myofibrillar structure was disrupted, and significant improvements in tenderness (reduced shear force) and higher

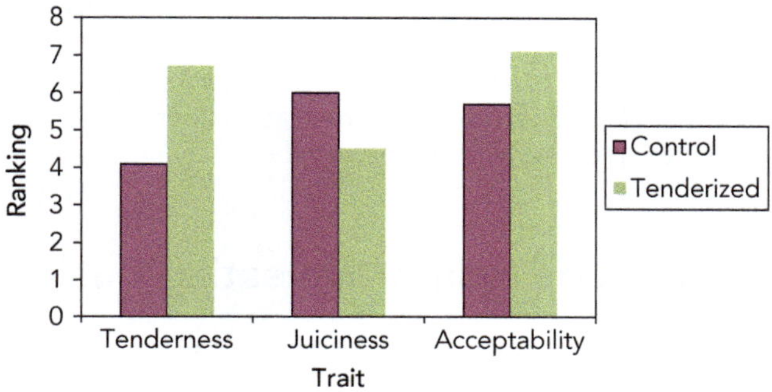


**Figure 7** Taste panel results for tenderness, juiciness and acceptability of pressurized and non-pressurized (control) longissimus each on a 1–9 scale with highest value equal to very tender, very juicy and most acceptable.

Source: Adapted from McFarlane 1973.

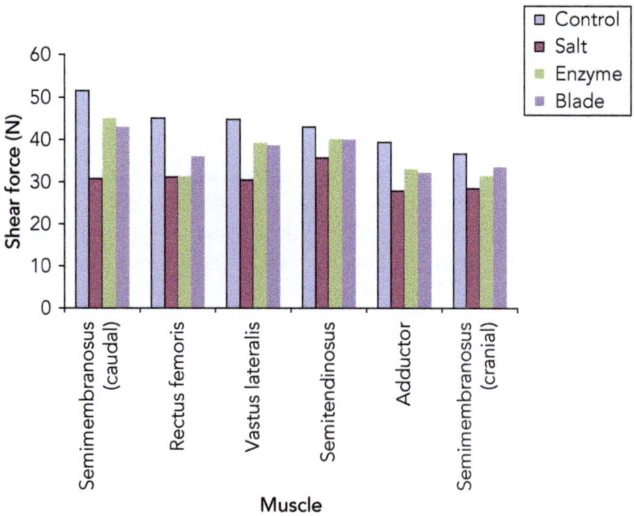


**Figure 8** Warner-Bratzler shear values (N) for beef muscles aged for 14–21 days post-mortem after injection with a sodium chloride (6%) and sodium polyphosphate (3.5%) solution (salt), or a solution containing bromelain (0.004%), or blade tenderized compared with controls.

Source: Adapted from Kolle et al., 2004.

sensory scores were observed compared to untreated pre-rigor excised muscle or post-rigor muscle untreated (Fig. 7). Subsequent work has demonstrated that the temperature and pressure applied to muscles impact on the response observed in terms of tenderness and also juiciness. Sikes et al. (2014) investigated a single-step pressure-heat process for post-rigor steaks 25 mm thick and suggested 76°C as being a suitable temperature for pressure treatment at 200 MPa to achieve optimal juiciness and tenderness with a 50% reduction in shear force and a reduction in water loss, compared to steaks cooked without pressure.

Blade tenderization is applied commercially to post-rigor meat, and often this approach is used to infuse solutions into meat, with the solutions often contributing to tenderization (e.g. Pietrasik and Shand, 2011). Such an approach either singly or in combination can improve meat tenderness (Fig. 8), but blade tenderization does impact on the appearance of the meat and requires good hygiene to ensure food safety.

# 5 Chemical methods to improve eating quality

There are a number of different 'chemical'-based approaches that have been studied as a means of improving tenderness and eating quality.

## 5.1 Application of ionic solutions

Pre-rigor injection of high ionic solutions creates an atypical environment within the muscle, and this may in itself result in premature membrane rupturing and the release of endogenous enzymes. Isolating the mode of action of ion injection on tenderization has been a major challenge because ions such as $Ca^{2+}$ can cause shortening when injected pre-rigor as well as activate endogenous enzymes (see Section 2.1). When meat is infused with calcium chloride solutions post-mortem, enhanced tenderness has resulted, and this has been attributed to stimulation of calpains during subsequent conditioning (Whipple et al., 1994), even when such infusion does not occur until 24 h post-mortem (Boleman et al., 1995). A major drawback of injection with $CaCl_2$ is that the meat tends to be regarded as having a bitter flavour and to be less acceptable to consumers (Rousset-Akrim et al., 1996) due to the fact that $Ca^{2+}$ can stimulate lipoxygenase activity and the mitochondrial respiration process leading to higher rate of oxidative processes. For this reason, commercial adoption of $CaCl_2$ infusion into meat has been limited.

Injection of sodium chloride (NaCl), combined with phosphates, will enhance the ability of muscle proteins to bind water and could reflect increased solubility of those proteins, and this is the principle applied for the brine treatment of pork. Such treatment has been shown to significantly reduce shear force (Kolle et al., 2004; Fig. 8), particularly in the caudal section of the semimembranosus.

Sodium chloride and polyphosphates accelerate the degradation rates of titin and troponin-T as well as the appearance of 95 and 30 kDa degradation products. This leads to higher tenderization rates, and these effects have been attributed to an increased pH due to the high buffering capacity of polyphosphates, with a summary of the results of infusing meat with ions provided by Hopkins and Bekhit (2014). The other advantage is that juiciness can be improved. Streiter et al. (2012) found that a solution of calcium ascorbate injected into longissimus from cull cows after one day of ageing significantly increased juiciness scores in a 14-day aged product, but when combined with the MOLAS method, there was no improvement. Interestingly, there was a benefit from combining the injection treatment with MOLAS for tenderness. If the pork industry is followed, there is likely to be an increased use of infusion in the beef industry, but this remains to be seen.

A novel technique to administer solutions to whole carcasses was reported by Farouk and Price (1994), and they reported that infusion of a solution containing maltose, glycerine, dextrose and sodium and potassium tripolyphosphates lowered the temperature by ~2°C compared to non-infused carcasses during the early post-mortem period (3 h), and also increased the rate of pH decline. Glycolysis was completed in 6 h compared to 12–24 h for the control. Data presented by Dikeman et al. (2003) in cattle confirm this response and also showed no consistent tenderness benefit from the infusion of a similar solution, and no appreciable benefits for eating quality have been reported using this approach, which has been commercialized as Rinse and Chill™ (MSPC, Inc., Eden Prairie, MN, USA). The technique involves flushing solutions into the body via the left carotid artery, immediately after exsanguination and before evisceration. This process effectively removes residual blood from the carcass, replacing it with a water-based solution to enhance the removal of blood, lower pH and reduce temperature earlier and more rapidly, and does increase carcass weight. This system is used in several abattoirs in a number of countries.

## 5.2 Application of acidic solutions

A number of organic acids such as acetic, citric and lactic have been used to tenderize meat. Administration is often by marination, but this requires extended storage periods for penetration into the meat. A number of studies have investigated the efficacy of these acids after injection into meat. These acids cause a significant decrease in pH, creating an environment that is optimal for cathepsin activity, which is confirmed by the degradation of proteins such as myosin, a protein that is not degraded under normal (control) post-mortem conditions. Associated with the drop in pH is swelling of the meat. Another effect of this treatment is a weakening/solubilization of perimysial collagen and a lowering of the temperatures required for denaturation of connective tissue (Aktas and Kaya, 2001). Overall, organic acids lead to an increase in tenderness (Table 1).

Because the effect of acid injection is largely dependent on the buffering capacity and overall pH reduction in meat, the type of acid used is very important to achieve desired quality attributes. Lactic acid lowers the pH compared with citric and acetic acids, and elicits more dramatic changes in the meat. Despite the benefits for tenderness of acid injection, there are reports that this results in a more pronounced sour chemical taste (Eilers et al., 1994), likely to make the meat less palatable.

**Table 1** Shear force (N; connective component) for beef *M. pectoralis profundus* treated with 0.5 mol $1^{-1}$ lactic acid at either 1 or 24 h post-mortem and aged for 2 or 14 days

| Lactic acid (1 h) | Lactic acid (24 h) | Control | Ageing period (days) |
|---|---|---|---|
| 29.1*a* | 27.5*a* | 44.2*b* | 2 |
| 34.2*a* | 27.5*a* | 44.2*b* | 14 |

Source: Adapted from Berge et al. (2001).
Means followed by a different letter (*a*, *b*) within a row are significantly different ($P < 0.05$).

## 5.3 Application of enzymes

Recognition that certain plants, fungi and bacteria produced nontoxic proteolytic enzymes (e.g. papain: Hwang and Ivy, 1951) was followed by their incorporation into commercial meat tenderizers. These were first used as dips. As such, they were somewhat unsatisfactory, because they over-tenderized the surface, producing a mushy texture (and sometimes unusual flavour), and because they were unable to penetrate within the meat, left the interior unaffected.

Different methods have been used to disperse solutions throughout meat, and the use of needling machines has been shown effective (Toohey et al., 2011) and the most practical. In species such as sheep, pre-slaughter and pre-rigor methods have been studied for introducing proteolytic enzymes into the muscle, but this does not appear to have been attempted in cattle.

There are a number of plant-derived enzymes that have been shown to degrade meat protein; these include papain (from paw paw), bromelain (from pineapple), ficin (from fig) and actinidain (from kiwi fruit), and then there are several enzymes derived from bacteria or fungi. The action of these has recently been extensively reviewed (Bekhit et al., 2014), and thus, limited detail will be provided here. The challenge with the utilization of these enzymes is to control the extent of their action because enzymes such as papain can easily result in meat having a mushy texture as they can indiscriminately degrade collagen and myofibrillar proteins. To this effect, recent work has focussed on characterizing the temperature and pH conditions required for these enzymes to exhibit optimal activity and identifying the individual proteins that are degraded (Ha et al., 2012, 2013). For example, Ha et al. (2013) reported that bacterial protease G and the fungal 31K protease preparations were similar in their ability to hydrolyse collagen, whereas a lower specific activity was obtained for a fungal 60K protease preparation; however, at the same time, the latter protease had a high specificity for collagen type 1. In contrast, protease G hydrolysed most myofibrillar proteins in a non-specific way and papain degraded proteins like titin, nebulin, actin and isomers of myosin, and this is the reason that this enzyme has such a dramatic effect on meat structure and thus tenderness (Ashie et al., 2002). Actinidain was found to be more effective at hydrolysing beef myofibril proteins than zingibain (from ginger), which was more effective at hydrolysing connective tissue proteins (Ha et al., 2012).

As shown in Fig. 8, treatment of meat with bromelain can decrease shear force in a range of muscles, and application of actinidain (from kiwi fruit) has also been shown under commercial conditions to improve tenderness (Toohey et al., 2011). There has also been some work to examine combining treatments of proteolytic enzymes and salt/phosphate-based solutions, which have shown additive effects in terms of reducing shear force (Pietrasik and Shand, 2011), providing another option for improving beef eating quality. Importantly, in this work, there was no reduction in overall palatability from the use of enzyme injection, which has not always been the case. There are now also commercial plant-derived enzyme products available for industrial use that are based on actinidain, and there is potential for combining proteases for targeting specific tenderizing applications.

The impact of the infused compounds on meat quality will be greatly dependent on the post-mortem time of the treatment (reflecting the pH and the temperature of the meat), the concentration of the infused compounds (level of activation or modification) and the method of infusion (the distribution of the compounds in the meat).

# 6 Conclusions

The tenderness of beef from several hindquarter muscles can be improved dramatically from the use of tenderstretching, and there is a case for wider adoption of the technology across the beef industry. At the primal level, there is scope to improve the tenderness of beef by the adoption of stretching technologies, and, for high connective tissue cuts, the application of ionic-based solutions in combination with selected proteolytic enzymes offers real opportunity to add value to such cuts. There is a need for further development of selected technologies so that they can be used in a batch mode (e.g. SmartStretch™ and PEF) again with a target on specific cuts. Even though there is potential to increase the rate of ageing, industry must still insist on minimum periods of ageing as consumer studies indicate the added benefit from flavour development that ageing can contribute to palatability.

# 7 Where to look for further information

http://www.sciencedirect.com/science/referenceworks/9780123847348
http://www.sciencedirect.com/science/journal/03091740?oldURL=y
http://www.ampc.com.au/resources/report
http://www.mla.com.au/Research-and-development/Search-RD-reports/RD-reports-search-results?keyWord=smartstretch
http://www.mla.com.au/Marketing-beef-and-lamb/Meat-Standards-Australia

# 8 References


Aalhus, J. L., Best, D. R., Costello, F. and Jeremiah, L. E. (1999), 'A simple, on-line processing method for improving beef tenderness', *Canadian Journal of Animal Science*, 79, 27–34.

Ahnström, M. L., Hunt, M. C. and Lundström, K. (2012), 'Effects of pelvic suspension of beef carcasses on quality and physical traits of five muscles from four gender–age groups', *Meat Science*, 90, 528–35.

Aktas, N. and Kaya, M. (2001), 'Influence of weak organic acids and salts on the denaturation characteristics of intramuscular connective tissue. A differential scanning calorimetry study', *Meat Science*, 58, 413–19.

Ashie, I. N. A., Sorensen, T. L. and Nielsen, P. M. (2002), 'Effects of papain and a microbial enzyme on meat proteins and beef tenderness', *Journal of Food. Science*, 67, 2138–42.

Belew, J. B., Brooks, J. C., McKenna, D. R. and Savell, J. W. (2003), 'Warner–Bratzler shear evaluations of 40 bovine muscles', *Meat Science*, 64, 507–12.

Bekhit, A. A., Hopkins, D. L., Geesink, G., Bekhit, A. A. and Franks, P. (2014), 'Exogenous proteases for meat tenderization', *Critical Reviews in Food Science and Nutrition*, 54, 1012–31.

Bekhit, A. E. D., van de Ven, R. J., Suwandy, V., Fahri, F. T. and Hopkins, D. L. (2014), 'Effect of Pulsed Electric Field treatment on cold boned muscles of different potential tenderness', *Food and Bioprocess Technology*, 7, 3136–46.

Bekhit, A. E. D., Suwandy, V., Crane, A., van de Ven, R. J. and Hopkins, D. L. (2016), 'Effect of repeated pulsed electric field treatment on the quality of hot boned beef loins and topsides', *Meat Science*, 111, 139–46.

Bendixen, E. (2005), 'The use of proteomics in meat science', *Meat Science*, 71, 138–49.

Berg, J. M., Tymoczko, J. L. and Stryer, L. (2003), *Biochemistry*, 5th ed., New York, W.H. Freeman and Company, p. 445.

Berge, P., Ertbjerg, P., Larsen, L. M., Astruc, T., Vignon, X. and Møllle, A. J. (2001), 'Tenderization of beef by lactic acid injected at different times post mortem', *Meat Science*, 57, 347–57.

Boleman, S. J., Boleman, S. L., Bidner, T. D., McMillin, K. W. and Monlezun, C. J. (1995), 'Effects of postmortem time of calcium chloride injection on beef tenderness and drip, cooking and total loss', *Meat Science*, 39, 35.

Bolumar, T., Enneking, M., Toepfl, S. and Heinz, V. (2013), 'New developments in shockwave technology intended for meat tenderization: Opportunities and challenges. A review', *Meat Science*, 95, 931–9.

Bolumar, T., Bindrich, U., Toepfl, S., Toldrá, F. and Heinz, V. (2014), 'Effect of electrohydraulic shockwave treatment on tenderness, muscle cathepsin and peptidase activities and microstructure of beef loin steaks from Holstein young bulls', *Meat Science*, 98, 759–65.

Bouton, P. E., Fisher, A. L., Harris, P. V. and Baxter, R. I. (1973), 'A comparison of the effects of some post-slaughter treatments on the tenderness of beef', *Journal of Food Technology*, 8, 39–49.

Claus, J. R., Wang, H. J. and Marriott, N. G. (1997), 'Prerigor carcass muscle stretching effects on tenderness of grain-fed beef under commercial conditions', *Journal of Food Science*, 62, 1231–4.

Colle, M. J., Richard, R. P., Killinger, K. M., Bohlscheid, J. C., Gray, A. R., Loucks, W. I., Day, R. N., Cochran, A. S., Nasados, J. A. and Doumit, M. E. (2015), 'Influence of extended aging on beef quality characteristics and sensory perception of steaks from the gluteus medius and longissimus lumborum', *Meat Science*, 110, 32–9.

Davey, C. L. and Gilbert, K. V. (1976), 'The temperature coefficient of beef ageing', *Journal of the Science of Food and Agriculture*, 27, 244–50.

Davies, A. S. (2004), 'Muscle structure and contraction', In *Encyclopedia of Meat Sciences*, in W. Jensen, C. Devine and M. Dikeman (eds), 882–901, Elsevier, Oxford.

Denoyelle, C. and Lebihan, E. (2003), 'Intramuscular variation in beef tenderness', *Meat Science*, 66, 241–7.

Devine, C. E., Wahlgren, N. M. and Tornberg, E. (1999), 'Effect of rigor temperature on muscle shortening and tenderisation of restrained and unrestrained beef m. longissimus thoracis et lumborum', *Meat Science*, 51, 61–72.

Devine, C. E., Hopkins, D. L., Hwang, I. H., Ferguson, D. M. and Richards, I. (2014), *Encyclopedia of Meat Sciences*, vol. 1, in C. Devine and M. Dikeman (eds), 486, Elsevier, Oxford.

Dikeman, M. E., Hunt, M. C., Addis, P. B., Schoenbeck, J. H., Pullen, M., Katsanidis, E. and Yancey, E. J. (2003), 'Effects of postexsanguination vascular infusion of cattle with a solution of saccharides, sodium chloride, and phosphates or with calcium chloride on meat quality and sensory traits of steaks and ground beef', *Journal of Animal Science*, 81, 156–66.

Eikelenboom, G., Barnier, V. M. H., Hoving-Bolink, A. H., Smulders, F. J. M. and Culioli, J. (1998), 'Effect of pelvic suspension and cooking temperature on the tenderness of electrically stimulated and aged beef, assessed with shear and compression tests', *Meat Science*, 49, 89–99.

Eilers, J. D., Morgan, J. B., Martin, A. M., Miller, R. K., Hale, D. S., Acuff, G. R. and Savell, J. W. (1994), 'Evaluation of calcium chloride and lactic acid injection on chemical, microbiological and descriptive attributes of mature cow beef', *Meat Science*, 38, 443–51.

Farouk, M. M. and Price, J. F. (1994), 'The effect of post-exsanguination infusion on composition, exudation, color and *post mortem* metabolic changes in lamb', *Meat Science*, 38, 477–96.

Gann, G. L. and Merkel, R. A. (1978), 'Ultrastructural changes in bovine longissimus muscle during postmortem ageing', *Meat Science*, 2, 129–44.

Goll, D. E., Thompson, V. F., Taylor, R. G. and Ouali, A. (1998), 'The calpain system and skeletal muscle growth', *Canadian Journal of Animal Science*, 78, 503–12.

Greaser, M. L., Boyer-Berri, C., Kumazawa, Y., Szalata, M. and Pospiech, E. (2000), 'Titin and tenderness', In *Proceedings 46th International Congress of Meat Science and Technology*, Buenos Aires, Argentina, 420–3.

Grayson, A. L. and Lawrence, T. E. (2013), 'Alternative pre-rigor foreshank positioning can improve beef shoulder muscle tenderness', *Meat Science*, 95, 36–41.

Honikel, K. O. and Hamm, R. (1978), 'Influence of cooling and freezing of minced pre-rigor muscle on the breakdown of ATP and glycogen', *Meat Science*, 2, 181–8.

Hopkins, D. L. (2014), '*Encyclopedia of Meat Sciences*', vol 3, in C. Devine and M. Dikeman (eds), 443, Elsevier, Oxford.

Hopkins, D. L. and Bekhit, A. A. (2014), '*Encyclopedia of Meat Sciences*', vol. 3, in C. Devine and M. Dikeman (eds), 431, Elsevier, Oxford.

Hopkins, D. L., Garlick, P. R. and Thompson, J. M. (2000), 'The effect on the sarcomere structure of super tenderstretching', *Australasian Journal of Animal Science*, 13(Supplement), 233.

Hopkins, D. L. and Thompson, J. M. (2001a), 'The relationship between tenderness, proteolysis, muscle contraction and dissociation of actomyosin', *Meat Science*, 57, 1–12.

Hopkins, D. L. and Thompson, J. M. (2001b), 'Inhibition of protease activity. Part 2. Degradation of myofibrillar proteins, myofibril examination and determination of free calcium levels', *Meat Science*, 59, 199–209.

Hopkins, D. L. and Thompson, J. M. (2002), 'Factors contributing to proteolysis and disruption of myofibrillar proteins and the impact on tenderisation in beef and sheep meat', *Australian Journal of Agricultural Research*, 53, 149–66.

Hopkins, D. L., Hegarty, R., Walker, P. and Pethick, D. (2006), 'Relationship between animal age, intramuscular fat, cooking loss, pH, shear force and eating quality of aged meat from sheep', *Animal Production Science*, 46, 879–84.

Hopkins, D. L. and Geesink, G. (2009), *Applied Muscle Biology and Meat Science*, in. M. Du and R. J. McCormick (eds), 149, CRC Press, Taylor & Francis Group, *USA*.

Hopkins, D. L., Warner, R. D., Ponnampalam, E. N. and van de Ven, R. J. (2014), 'The effect of fast pH decline on the meat and eating quality of beef carcases', *Animal Production Science*, 54, 407–13.

Hwang, K. and Ivy, A. C. (1951), 'A review of the literature on the potential therapeutic significance of papain', *Annals New York Academy of Science*, 54, 161–207.

Hwang, I. H., Devine, C. E. and Hopkins, D. L. (2003), 'The biochemical and physical effects of electrical stimulation on beef and sheep meat tenderness – a review', *Meat Science*, 65, 677–91.

Jacob, R. H. and Hopkins, D. L. (2014), 'Techniques to reduce the temperature of beef muscle early in the post mortem period- a review', *Animal Production Science*, 54, 482–93.

Jaime, I., Beltrán, J. A., Ceña, P., López-Lorenzo, P. and Roncalés, P. (1992), 'Tenderisation of lamb meat: Effect of rapid postmortem temperature drop on muscle conditioning and aging', *Meat Science*, 32, 357–66.

Jerez, N. C., Calkins, C. R. and Velazco, J. (2003), 'Prerigor injection using glycolytic inhibitors in low-quality beef muscles', *Journal of Animal Science*, 81, 997–1003.

Jones, S. J., Calkins, C. R., Johnson, D. D. and Gwartney, B. L. (2005), *Bovine Mycology*, Lincoln, NE, University of Nebraska. http://bovine.unl.edu.

Kemp, C. M. and Parr, T. (2012), 'Advances in apoptotic mediated proteolysis in meat tenderisation', *Meat Science*, 92, 252–9.

Kolle, B. K., McKenna, D. R. and Savell, J. W. (2004), 'Methods to increase tenderness of individual muscles from beef rounds when cooked with dry or moist heat', *Meat Science*, 68, 145–54.

Kim, Y. H. B., Warner, R. D. and Rosenvold, K. (2014), 'Influence of high pre-rigor temperature and fast pH fall on muscle proteins and meat quality: a review', *Animal Production Science*, 54, 375–95.

Kim, Y. H. B., Kemp, R. and Samuelsson, L. M. (2016), 'Effects of dry-aging on meat quality attributes and metabolite profiles of beef loins', *Meat Science*, 111, 168–76.

Laster, M. A., Smith, R. D., Nicholson, K. L., Nicholson, J. D. W., Miller, R. K., Griffin, D. B., Harris, K. B. and Savell, J. W. (2008), 'Dry versus wet aging of beef: Retail cutting yields and consumer sensory attribute evaluations of steaks from ribeyes, strip loins, and top sirloins from two quality grade groups', *Meat Science*, 80, 795–804.

Locker, R. H. (1960), 'Degree of muscular contraction as a factor on the tenderness of beef', *Food Research*, 25, 304–7.

Locker, R. H. and Wild, D. J. C. (1984), 'Ageing of cold shortened meat depends on the criterion', *Meat Science*, 10, 235–8.

Locker, R. H. and Hagyard, C. J. (1963), 'A cold shortening effect in beef muscles', *Journal of the Science of Food and Agriculture*, 14, 787–93.

McCormick R. J. (1994), 'The flexibility of the collagen compartment of muscle', *Meat Science*, 36, 79–91.

McFarlane, J. J. (1973), 'Pre-rigor pressurization of muscle: Effects on pH, shear value and taste panel assessment', *Journal of Food Science*, 38, 294–8.

Minh Ha, Alaa El-Din Bekhit, Carne, A. and Hopkins, D. L. (2012), 'Characterization of commercial papain, bromelain, actinidin and zingibain protease preparations and their activities toward meat proteins', *Food Chemistry*, 134, 95–105.

Minh Ha, Alaa El-Din Bekhit, Carne, A. and Hopkins, D. L. (2013), 'Characterization of commercial bacterial and fungal proteases and their activities toward meat proteins', *Journal of Food Science*, 78, C170–C177.

Monin, G. and Ouali, A. (1991), *Developments in Meat Science–5*, in R. A. Lawrie (ed.), 89, Elsevier Applied Science, London.

Nishimura, T., Hattori, A. and Takahashi, K. (1995), 'Structural weakening of intramuscular connective tissue during conditioning beef', *Meat Science*, 39, 127–33.

Nishimura, T., Hattori, A. and Takahashi. K. (1998), 'Changes in mechanical strength of intramuscular connective tissue during postmortem aging of beef', *Journal of Animal Science*, 76, 528–32.

Offer, G. (1991), 'Modelling of the formation of pale soft and exudative meat: Effects of chilling regime and rate of extent of glycolysis', *Meat Science*, 30, 157–84.

Ouali, A., Gagaoua, M., Boudida, Y., Beclia, S., Boudjellal, A., Hernan-Mendez, C. H. and Sentandreu, M. A. (2013), 'Revisiting the conversion of muscle into meat and the underlying mechanisms', *Meat Science*, 74, 44–58.

Ouali, A., Hernan-Mendez, C. H., Coulis, G., Beclia, S., Boudjellal, A., Aubury, L. and Sentandreu, M. A. (2006), 'Biomarkers of meat tenderness: Present knowledge and perspectives in regards to our current understanding of the mechanisms involved', *Meat Science*, 95, 854–70.

O'Sullivan, A., Korzeniowska, M., White, A. and Troy, D. J. (2003), 'Using a novel intervention technique to reduce the variability and improve tenderness of beef longissimus dorsi', In *Proceedings 49th International Congress of Meat Science and Technology*, 513–14, Campinas, Brazil.

Pietrasik, Z. and Shand, P. J. (2011), 'Effects of moisture enhancement, enzyme treatment, and blade tenderization on the processing characteristics and tenderness of beef semimembranosus steaks', *Meat Science*, 88, 8–13.

Purchas, R. W. (2014), *Encyclopedia of Meat Sciences*, vol. 3, in C. Devine and M. Dikeman (eds), 452, Elsevier, Oxford.

Purchas, R. W., Yan, X. and Hartley, D. G. (1999), 'The influence of a period of ageing on the relationship between ultimate pH and shear values of beef *m. longissimus thoracis*', *Meat Science*, 51, 135–41.

Reuter, B. J., Wulf, D. M. and Maddock, R. J. (2002), 'Mapping intramuscular tenderness variation in four major muscles of the beef round', *Journal of Animal Science*, 80, 2594–9.

Rodriguez, J., Unruh, J., Villarreal, M., Murillo, O., Rojas, S., Camacho, J., Jaeger, J. and Reinhardt, C. (2014), 'Carcass and meat quality characteristics of Brahman cross bulls and steers finished on tropical pastures in Costa Rica', *Meat Science*, 96, 1340–4.

Rousset-Akrim, S., Got, F., Bayle, M. C. and Culioli, J. (1996), 'Influence of $CaCl_2$ and NaCl injections on the texture and flavour of beef', *International Journal of Food Science and Technology*, 31, 333–43.

Sentandreu, M. A., Coulis, G. and Ouali. A. (2002), 'Role of muscle endopeptidases and their inhibitors in meat tenderness', *Trends in Food Science and Technology*, 13, 398–419.

Sharp, J. G. (1963), 'Aseptic autolysis in rabbit and bovine muscle during storage at 37°', *Journal of the Science of Food and Agriculture*, 14, 468–79.

Shorthose, W. R. and Harris, P. V. (1990), 'Effect of animal age on the tenderness of selected beef muscles', *Journal of Food Science*, 55, 1–8, 14.

Sikes, A. L. and Tume, R. K. (2014), 'Effect of processing temperature on tenderness, colour and yield of beef steaks subjected to high-hydrostatic pressure', *Meat Science*, 97, 244–8.

Solomon, M. B., Long, J. B. and Eastridge, J. S. (1997), 'The Hydrodyne: A new process to improve beef tenderness', *Journal of Animal Science*, 75, 1534–7.

Starkey, C. P., Geesink, G. H., Collins, D., Oddy, V. H. and Hopkins, D. L. (2016), 'Do sarcomere length, collagen content, pH, intramuscular fat and indicators of proteolysis explain variation in the tenderness of three *ovine* muscles?', *Meat Science*, 113, 51–8.

Streiter, P. J., Campbell, C. P. and Mandell, I. B. (2012), 'The effects of skeletal separation and moisture enhancement for improving the eating quality of cull cow beef', *Meat Science*, 92, 400–8.

Stolowski, G. D., Baird, B. E., Miller, R. K., Savell, J. W., Sams, A. R., Taylor, J. F., Sanders, J. O. and Smith, S. B. (2006), 'Factors influencing the variation in tenderness of seven major beef muscles from three Angus and Brahman breed crosses', *Meat Science*, 73, 475–83.

Suwandy, V., Crane, A., van de Ven, R. J., Bekhit, A. E. D. and Hopkins, D. L. (2015a), 'Effect of pulsed electric field on the proteolysis of cold boned *M. Longissimus lumborum* and M. *Semimembranosus* beef muscles', *Meat Science*, 100, 222–6.

Suwandy, V., Crane, A., van de Ven, R. J., Bekhit, A. E. D. and Hopkins, D. L. (2015b), 'Effect of pulsed electric field treatment on hot-boned muscles of different potential tenderness', *Meat Science*, 105, 25–31.

Suwandy, V., Crane, A., van de Ven, R. J., Bekhit, A. E. D. and Hopkins, D. L. (2015), 'Effect of repeated pulsed electric field treatment on the quality of cold-boned beef loins and topsides', *Food and Bioprocess Technology*, 8, 1218–28.

Taylor, J. M. and Hopkins, D. L. (2011), 'Patents for stretching and shaping meats', *Recent Patents on Food, Nutrition & Agriculture*, 3, 91–101.

Taylor, J. M., Toohey, E. S., van de Ven, R. J. and Hopkins, D. L. (2012), 'SmartStretch™ technology. IV. The impact on the meat quality of hot-boned beef rostbiff (*m gluteus medius*)', *Meat Science*, 91, 527–32.

Taylor, J. M., Toohey, E. S., van de Ven, R. J. and Hopkins, D. L. (2013), 'SmartStretch™ technology. VI. The impact on the meat quality of hot-boned beef striploin (m. longissimus lumborum)', *Meat Science*, 93, 413–19.

Taylor, R. G., Geesink, G. H., Thompson, V. F., Koohmaraie, M. and Goll, D. E. (1995), 'Is Z-disk degradation responsible for postmortem tenderisation?', *Journal of Animal Science*, 73, 1351–67.

Thompson, J. (2002), 'Managing meat tenderness', *Meat Science*, 62, 295–308.

Toohey, E. S., van de Ven, R. J., Thompson, J. M., Geesink, G. H. and Hopkins, D. L. (2012), 'SmartStretch™ Technology. V. The impact of SmartStretch™ technology on beef topsides (*m. semimembranosus*) meat quality traits under commercial processing conditions', *Meat Science*, 92, 24–9.

Tornberg, E. (1996), 'Biophysical aspects of meat tenderness', *Meat Science*, 43, 175–91.

Wang, H., Claus, J. R. and Marriott, N. G. (1994), 'Selected skeletal alterations to improve tenderness of beef round muscles', *Journal of Muscle Foods*, 5,137–47.

Warner, R. D., Dunshea, F. R., Gutzke, D., Lau, J. and Kearney, G. (2014), 'Factors influencing the incidence of high rigor temperature in beef carcasses in Australia', *Animal Production Science*, 54, 363–74.

Warren, K. E. and Kastner, C. L. (1992), 'A comparison of dry-aged and vacuum-aged beef strip loins', *Journal of Muscle Foods*, 3, 151–7.

Watanabe, A. and Devine, C. E. (1996), 'Effect of meat ultimate pH on rate of titin and nebulin degradation', *Meat Science*, 42, 407–13.

Wheeler, T. L., Cundiff, L. V., Koch, R. M. and Crouse, J. D. (1996), 'Characterization of biological types of cattle (cycle IV): Carcass traits and longissimus palatability', *Journal of Animal Science*, 74, 1023–35.

Whipple, G., Koohmaraie, M. and Arbona, J. R. (1994), 'Calcium chloride *in vitro* effects on isolated myofibrillar proteins', *Meat Science*, 38, 133–9.

Young, O. A., Hopkins, D. L. and Pethick, D. W. (2005), 'Critical control points for meat quality in the Australian sheep meat supply chain', *Australian Journal of Experimental Agriculture*, 45, 593–601.

Chapter 13

# Factors affecting flavour development in beef

*Chris R. Kerth, Texas A&M University, USA*



## 1 Introduction to flavour

Mackintosh and Hall (1936) reported that, as early as 1750, animal breeders initially gave some thought to the eating qualities of their product, which gave a basis for the modern breeds of cattle to be developed. Even then, most of the emphasis was on the tenderness and juiciness with little thought given to the overarching term of palatability. Not until about 1885 was there any emphasis on measuring palatability, but these studies were soon dropped in favour of studies regarding the economies of production. In 1924, however, there was a concerted effort on the part of animal scientists to analyse the palatability of meat by the USDA and a number of state agricultural experiment stations. It was a result of these experiments over a span of nearly a decade that Mackintosh and Hall concluded that the degree of finish or fatness of beef appeared to be associated with palatability. They concluded that an increasing degree of finish resulted in tenderness, juiciness and flavour being more intense and that the evidence seemed to justify the old-time belief that fat improves the palatability of the meat.

Palatability is a very general term used to describe the taste of something and is comprised of tenderness/texture, juiciness and flavour. Of all of the different senses used to evaluate steaks, the sense of taste is by far the most complex and requires a basic understanding of the most basic cellular biology of meat, but can be impacted by factors as broad as animal genetics and nutrition. Numerous researchers have studied the flavour of beef and all of the factors that impact the development of flavour. Although it is widely

http://dx.doi.org/10.19103/AS.2016.0009.13

**Table 1** Definition and reference standards for meat descriptive flavour aromatics and basic taste sensory attributes and their intensities where 0 = none; 15 = extremely intense from Adhikari et al. (2011)

| Sensory attribute | Definition | Reference, standard flavour scale value unless otherwise defined |
|---|---|---|
| Animal hair | The aromatics perceived when raw wool is saturated with water. | Caproic acid (hexanoic acid) = 12.0 |
| Beef identity | Amount of beef flavour identity in the sample. | Swanson's beef broth = 5.0<br>80% lean ground beef = 7.0<br>Beef brisket = 11.0 |
| Bitter | The fundamental taste factor associated with a caffeine solution. | 0.01% caffeine solution = 2.0<br>0.02% caffeine solution = 3.5 |
| Bloody/serumy | The aromatics associated with blood on cooked meat products. Closely related to metallic aromatic. | USDA choice strip steak = 5.5<br>Beef brisket = 6.0 |
| Brown/roasted | A round, full aromatic generally associated with beef suet that has been broiled. | Beef suet = 8.0<br>80% lean ground beef = 10.0 |
| Burnt | The sharp/acrid flavour note associated with over-roasted beef muscle, something over-baked or excessively browned in oil. | Alf's red wheat puffs = 5.0 |
| Chemical | The aromatics associated with garden hose, hot Teflon pan, plastic packaging and petroleum-based product such as charcoal lighter fluid. | Ziploc sandwich bag = 13.0<br>Clorox in water = 6.5 |
| Cocoa | The aromatics associated with cocoa beans and powdered cocoa and chocolate bars. Brown, sweet, dusty, often bitter aromatics. | Hershey's cocoa powder in water = 3.0<br>Hershey's chocolate kiss = 8.5 (flavour) |
| Cooked milk | A combination of sweet, brown flavour notes and aromatics cheese = associated with heated milk. | Mini Babybel original Swiss = 2.5 |
| Dairy | The aromatics associated with products made from cow's milk, such as cream, milk, sour cream or butter milk. | Dillon's whole milk = 4.5<br>Dillon's reduced fat milk (2%) = 8.0 |
| Fat-like | The aromatics associated with cooked animal fat. | Hillshire Farms Lit'l Beef Smokies = 7.0<br>Beef suet = 12.0 |
| Green | Sharp, slightly pungent aromatics associated with green/plant/vegetable matters such as parsley, spinach, pea pod and fresh cut grass. | Hexanal in propylene glycol (5,000 ppm) = 6.5 (aroma) Fresh parsley water = 9.0 |

| | | |
|---|---|---|
| Green-hay | Brown/green dusty aromatics associated with dry grasses, hay, dry parsley and tea leaves. | Dry parsley in medium snifter = 5.0 (aroma)<br>Dry parsley in ~30-mL cup = 6.0 |
| Leather | Musty, old leather (like old book bindings). | 2,3,4-Trimethoxybenzaldehyde = 3.0 (aroma) |
| Liver-like | The aromatics associated with cooked organ meat/liver. | Beef liver = 7.5<br>Braunschweiger liver sausage = 10.0 (must taste and swallow) |
| Metallic | The impression of slightly oxidized metal, such as iron, copper and silver spoons. | 0.10% potassium chloride solution = 1.5<br>USDA choice strip steak = 4.0<br>Dole canned pineapple juice = 6.0 |
| Overall sweet | A combination of sweet taste and sweet aromatics. The aromatics associated with the impression of sweet. | Post-shredded wheat spoon size = 1.5<br>Hillshire farms Lit'l beef smokies = 3.0<br>SAFC ethyl maltol 99% = 4.5 (aroma) |
| Rancid | The aromatics commonly associated with oxidized fat and oils. These aromatics may include cardboard, painty, varnish and fishy. | Microwaved Wesson vegetable oil (3 min at high) = 7.0<br>Microwaved Wesson vegetable oil (5 min at high) = 9.0 |
| Salty | The fundamental taste factor of which sodium chloride is typical. | 0.15% sodium chloride solution = 1.5<br>0.25% sodium chloride solution = 3.5 |
| Sour aromatics | The aromatics associated with sour substances. | Dillon's buttermilk = 5.0 |
| Sour dairy | Sour, fermented aromatics associated with dairy products such as buttermilk and sour cream. | Laughing Cow light Swiss cheese = 7.0<br>Dillon's buttermilk = 9.0 |
| Sour | The fundamental taste factor associated with citric acid. | 0.015% citric acid solution = 1.5<br>0.050% citric acid solution = 3.5 |
| Spoiled | The presence of inappropriate aromatics and flavours that is commonly associated with the products. It is a foul taste and/or smell that indicates the product is starting to decay and putrefy. | Dimethyl disulphide in propylene glycol (10000 ppm) = 12.0 (aroma) |

reported and accepted that raw meat has very little flavour other than a serumy, metallic flavour, it is the basic biology and composition of raw meat that determines the overall flavour of cooked meat. Everything down to the individual amino acid side chains and fatty acids can impact the flavour of beef and can be managed at some level.

Flavour is not a single attribute, although it has been measured for many years as a single attribute, defined originally by the American Meat Science Association (AMSA, 2016). Although many studies have considered flavour attributes when studying beef products, the industry did not have a complete library of terms or a general reference that described the actual flavour components of beef. The beef flavour lexicon (Adhikari et al., 2011; Table 1) is a library of flavour descriptors used to describe flavour notes in beef. As sensory panels implement training procedures, this particular chapter is important as the reference for the lexicon. With the recent work focused on flavour, the industry needed to standardize a flavour lexicon that could be used for research projects. It is important to be able to systematically identify and quantify flavour attributes that drive consumer acceptance.

## 1.1 Basic tastes

Most people are familiar with the five basic tastes of sweet, sour, salty, bitter and umami. These are the sensations that we most immediately associate with the taste of food and are the cornerstone of taste and flavour. Research has reported that each of these tastes are sensed on specific parts of the tongue, but more recent research has shown that the entire tongue can, in fact, sense all five of these tastes, but transmit the sensation to the brain with different intensities. Each taste bud on the tongue has 50–150 taste receptor cells that project microvilli on the surface of the taste bud to form a taste pore. Each of these receptors functions as a dedicated sensor wired to give appropriate response for each taste.

A seven trans-membrane alpha helix initiates signalling cascades utilizing ion channels to depolarize membrane structures resulting in taste. Sodium channel transduction mechanisms result in a salty taste. Proton concentration is naturally highly correlated to the sour taste, as it is related to the pH of the product and an acidic food like lemons results in the sour taste. Bitter taste is detected by a transducing G protein called gustducin, and umami is stimulated by glutamate in the form of monosodium glutamate. A review of Chandrashekar et al. (2006) goes into the details of all of the receptors and their function in the development of taste.

## 1.2 Trigeminal senses

The trigeminal senses (Laska et al., 1997) are much less well known, but are still an integral part of the overall taste of beef. The trigeminal senses are detected by the large trigeminal or fifth cranial nerve comprising ophthalmic, maxillary and mandibular branches. This nerve is responsible for both motor and sensory functions of the face and mouth. As such, it is responsible for the action of chewing or mastication in addition to sensing mouthfeel of the food being eaten. Because these nerves travel to the mucous membranes of the mouth and throughout the sinus cavity, their sensory functions are integral to tastes and aromas. Examples of these senses are the cool feel from menthol, the pungent quality of acetic acid and the warm sensation from ethanol (Kerth and Miller, 2015).

## 1.3 Aromas

According to Mozell et al. (1969), the role of smell in flavour generates much confusion. They studied human subjects that had either naturally or artificially disabled nasal chemoreceptors. The subject of their study was to determine whether sweet and salty were the result of basic taste or of aroma. Because sweet and salt are classically described as gustatory senses and are not normally considered to stimulate nasal chemoreceptors, they were surprised to find a dramatic loss in the ability of their subjects to identify salt water and sugar water when their nasal receptors were rendered inoperable. They concluded that smell plays an important role in all flavours. Furthermore, Shepherd (2005) reported on olfactory processing and its role in determining flavour and aromas. It was noted that the olfactory senses are sensitive to a one-carbon difference between stimulating molecules. He concluded that motor manipulation along with odour images, combined with taste, somatosensation, even vision and hearing, provide the basis for the perception of flavour. Additionally, he reported it as a higher cognitive function uniquely developed in humans with the help of language.

# 2 Beef aromas

As previously noted, raw beef has very little, if only a slightly sweet aroma, and it is not until it is heated that it imparts the aromas normally associated with beef. The generation of these aromas is incredibly complex and involves hundreds of chemical reactions taking place simultaneously. Each of the individual aromas has a different intensity and all of the aromas folding together to form what is commonly recognized as a beef aroma. Numerous review papers reference the specific aromas and the chemical compounds associated with each one of them. It is also important to note that each of these volatile aroma chemical compounds becomes volatile under different conditions, and most importantly, each one has a different level of detection to the human nose. Some compounds like 2,4 decadienal have very low thresholds, meaning we can detect their presence at concentrations as low as 70 parts per trillion. Other compounds, like acetic acid (vinegar), are detected at 2.5 parts per million requiring more than a million times more for us to detect it (Kerth and Miller, 2015). Additionally, some of the aroma compounds have different aromas depending on their concentrations, as shown by the fact that a single compound often will have several aroma descriptors.

## 2.1 Thermal degradation of lipids

The thermal degradation of lipids is very similar to the oxidation reaction, especially as it relates to the location, at which the lipid is detached. It is very important to understand the basic structure of lipids. Tri-acyl glycerides, or triglycerides, are comprised of three fatty acids esterified to a glycerol backbone found mostly in adipose tissue. A variation of this structure includes phospholipids found in cell membranes. Intramuscular fat or marbling is often thought to be one of the most valuable contributors to beef flavour, and it is very important to the juiciness and mouthfeel of beefsteaks. However, it has been shown that it is the differences in the structure of the fatty acids in the phospholipids of the cell membrane that largely determine species flavour. In fact, it is differences in the fatty acids of the phospholipids or polar lipids that are a direct result of nutrition and genetics. The structure and composition of the polar lipids, and the cell membrane lipids in general, have fewer saturated fatty acids (SFAs) and that, coupled with the lack of a fatty acid on the third glycerol carbon, makes the polar lipids more susceptible to fatty acid oxidation and thermal degradation of lipids. Thermal degradation of lipids causes a breakdown of fatty acids at double bonds and at ester bonds with no preference for which glycerol carbon is degraded first. After the fatty acid is degraded, the terminal end of the resulting carbon chain is modified to form volatile acids, alcohols, aldehydes, ketones, and others, contributing to flavour.

## 2.2 Thiamine degradation

Degradation of vitamins during the cooking process is a significant contributor to the aromas of beef, especially with regard to the degradation of thiamine. van der Linde et al. (1979) reported numerous sulphur-containing degradation products of thiamine including 4-methyl-5-(2-hydroxyethyl) thiazole and 5-hydroxy-3-mercapto-pentan-2-one, which give rise to numerous sulphur compounds, furans and thiophenes. Guntert et al. (1990) have proposed an extensive and very comprehensive reaction scheme for the thermal degradation of thiamine, and Werkhoff et al. (1989, 1990) have identified volatile sulphur compounds giving meaty notes to a model meat system. When a combination of cysteine, thiamine, glutamate, ascorbic acid and water were heated to 120°C for 30 minutes, they were able to characterize 70 sulphur components, of which 19 possessed individual odours described as meaty. These sulphur-containing compounds are of particular significance because the low detection threshold (Kerth and Miller, 2015) makes them especially noticeable when present.

## 2.3 The Maillard reaction

When temperatures between 55 and 80°C are reached, most proteins are denatured. Browning begins around 90°C and is related to the concentration of sugars in the meat (Maillard, 1912). Although the Maillard reaction does not require the very high temperatures of sugar caramelization and protein pyrolysis, it does require conditions of low moisture and sufficient temperature, as seen in dry heat cookery. In addition to temperature, the Maillard reaction also requires dry conditions like those found at the surface of meat during cooking. The substrates, for the formation of Maillard reaction products, are a reducing sugar and an amino acid. The most common reducing sugar in meat is ribose, which is derived from the degradation of ribonucleotides and residual products of ATP. The Maillard reaction is initiated by ribose reacting with free amino acids. The most classic reaction is the mixture of cysteine with ribose with heat applied to result in a beefy aroma.

## 2.4 Lipid–Maillard interactions

In examining the impact of lipids on flavour development, the importance of the intramuscular fat and phospholipids was noted. Without going into the details of the chemical interactions, it is worth noting the interaction of the lipid component with the Maillard reaction as it relates to the importance of lipid content of muscle on the development of the overall flavour of beef (Mottram and Edwards, 1983). In their research, Mottram and Edwards found that sensory panels and consumer evaluations failed to show any relationship between the flavour of lean meat and the amount of fat on the carcass. Thus, reinforcing the idea that the meat flavour was associated with the water-soluble flavour precursors; whereas species characteristics were associated with the lipid portion. When triglycerides were removed from meat samples, the aromas could not be differentiated from those of the original sample. However, when the phospholipids were removed from lean meat, a marked difference in aromas was found; the meaty aroma was replaced with a roasted, biscuit aroma, typically found in Maillard reactions. It was therefore determined that the meaty aroma was found in the lipid portion preventing many of the Maillard products.

# 3 Factors affecting beef flavour: nutrition and genetics

In reviewing the factors that impact the flavour of beef, the most dominant factor inherent in the beef itself is the lipid component, because most of the water-soluble components of flavour are developed in the Maillard reaction. While the amino acid composition may play a small role in the development of diverse precursors to flavour, fatty acid composition is much more important. Beef with the most desirable flavour has lower percentages of SFA and polyunsaturated fatty acids (PUFA) and higher percentages of monounsaturated fatty acids (MUFA) in the carcass fat (Oka et al., 2002). Fatty acid composition of bovine tissues is affected to some extent by factors like sex (Waldman et al., 1968), breed (Oka et al., 2002), diet (Melton et al., 1990) and age (Rule et al., 1997).

## 3.1 Nutrition

Melton (1990) provided a comprehensive review of the impact of different feeds on the flavour of meat. In considering the impact of nutrition on the flavour of beef, the primary factor is in the fatty acid composition of the muscle itself. Monogastric animals deposit fatty acids very much as they are consumed in the diet. Therefore, little modification occurs to the fatty acids, and the lipid component of the muscle closely resembles that of the diet. For example, if fish oil is fed to pigs, the meat will have a 'fishy' aroma to it and the lipid will be highly unsaturated because of the fatty acid composition of the fish oil. Cattle, on the other hand, have a rumen that is filled with microbes that modify lipids consumed in the diet. Most common is bio-hydrogenization of lipids making unsaturated fatty acids more saturated before they travel to the lower stomachs to be absorbed into the body and deposited in tissues. As a result, it is much more difficult to change the fatty acid composition of ruminants compared to monogastric animals. Theunissen et al. (1979) fed steers diets containing 10 or 20% canola oil with high levels of MUFA. The steers failed to deposit much 18:1 (oleic acid) in the muscle tissue compared to the control diet.

The most commonly discussed factor of nutrition on fatty acid composition and sensory traits of beef is in the finishing system, in particular grass-finishing versus grain-finishing. In a study by Cox et al. (2006), 1250 consumers were surveyed to determine their preference for grain- or grass-finished beef. Of those surveyed, 34% preferred grass-finished and 66% preferred grain-finished beef in a blind intercept survey. Larick and Turner (1990) reported that when forage-fed steers were then placed in a feedlot and fed a high-concentrate diet, as the days in the feedlot increased, there was an increase in phospholipid concentration.

Significant changes in the fatty acid composition were detected of the total phospholipid and several of the individual phospholipid fractions. In that study, phospholipid content and fatty acid composition were significantly correlated with differences in the flavour traits of ground beef determined by a descriptive sensory panel. Numerous other studies have shown differences in fatty acid profiles of forage-finished and grain-finished beef (Daley et al., 2010; Larick and Turner, 1990).

Tansawat et al. (2013) further found that volatile aroma compounds associated with pasture-fed beef, described as barny, gamey and grassy, were positively correlated to volatile benzaldehyde, toluene, dimethyl sulfone, 3-heptanone, 2-ethyl-1-hexanol and hexadecanoic acid methyl ester. Grain-fed beef had higher levels of the volatile compounds hexanal, 1-octen-3-ol, 2,3-octanedione and 2,6-bis(1,1-dimethylethyl)-4-ethyl-phenol, which were associated with umami and juicy flavours. Maughan et al. (2012) developed a lexicon of beef flavours unique to grass- and grain-finished cattle. The flavour descriptors that rated higher for grass-fed cattle included barny, bitter, gamey and grassy.

An additional benefit of finishing on grass or forage is the meat from these cattle naturally has a much higher level of alpha-tocopherol or vitamin E compared with those finished on grain. In fact, the level of vitamin E in the muscle of grass-finished cattle is equivalent to that of cattle finished on grain, supplemented with 2500 IU vitamin E/head/day (Yang et al., 2002). Vitamin E is a lipid-soluble antioxidant that has been shown to be very effective in reducing the amount of oxidation in meat, which results in improved colour and lipid oxidative stability. Because it is a lipid-soluble vitamin, it is deposited into the lipid bi-layer of cell membranes, making it in close proximity to the phospholipids, which are most susceptible to lipid oxidation.

## 3.2 Genetics

The genetic control of fatty acid variability has been shown in numerous studies (Taniguchi et al., 2004; Shennink et al., 2008; Conte et al., 2010). Stearoyl-CoA desaturase (SCD) is one of the most promising candidate genes to explain fatty acid variability. SCD is a rate-limiting enzyme responsible for the biosynthesis of MUFA from SFA. Two different isoforms of SCD (SCD1 and SCD5) have been located on chromosomes 26 and 6, respectively, in cattle. SCD1 has been of particular interest as it has been highly correlated with the content of the MUFA 18:1, oleic acid in beef. Furthermore, Baublits et al. (2009) reported that the MUFA oleic acid is moderately correlated (0.44 and 0.37) with beef/brothy and beef fat flavours in cooked beef, respectively. Westerling and Hedrick (1979) also reported that oleic acid was positively ($r = 0.67$) correlated to sensory flavour.

According to Oka et al. (2002), Japanese Black Wagyu steers have higher percentages of MUFA than Holstein, Japanese Brown, Charolais or Angus steers. Additionally, the percentage of MUFA in beef from Japanese Black Wagyu steers was also reported to depend on the regions in Japan from which the samples were obtained (Sturdivant et al., 1992). However, it was not clear whether the regional differences were due to nutritional or genetic differences. Oka et al. (2002) did note that MUFA, and in particular, oleic acid, could vary by as much as 3% from one sire to the next when fed identically. Chen et al. (2015) reported the heritabilities of fatty acids in subcutaneous fat and longissimus lumborum tissue. They found that while most fatty acids had fairly low heritability ($h^2 < 0.033$) palmitoleic acid, oleic acid and MUFA were moderately heritable ($h^2$ = 0.64, 0.47 and 0.47, respectively), indicating that genetic improvement could provide advancement for increasing the amount of these fatty acids in beef.

# 4 Factors affecting beef flavour: beef composition and processing

Because the chemical composition of meat plays such an important role in the development of flavour, a basic understanding of how compositional differences are determined in the product itself is of interest. The general composition, storage, handling and processing of meat determines the flavour of the beef product in the end.

## 4.1 Marbling effects

It has been documented for nearly 100 years that fat plays an important part in the palatability of beef and, in particular, the flavour of beef. Over the years, the primary focus of the role of marbling has been on overall palatability and on tenderness as these have been the main focus of quality grade and its ability to predict consumer acceptability (Savell et al., 1987; Wheeler et al., 1994). Wheeler showed the degree of marbling from traces to moderate had a significant impact on shear force and sensory overall tenderness, and juiciness in both *Bos taurus* and *B. indicus* cattle, but marbling in this range had no statistical influence on beef flavour intensity. Savell et al. (1987) tested a wider range in marbling from traces to slightly abundant and found that overall consumer acceptability was directly impacted by the level of marbling. However, sensory panel flavour intensity scores differed by only 0.5 point between traces and slightly abundant.

One of the problems associated with testing tenderness, juiciness and flavour of intact muscles is the tendency to confound one another. Thompson (2004) first tried adjusting the factor of flavour for tenderness and found that Australian consumer flavour ratings linearly increased up to about 13% fat. This was also shown by O'Quinn et al. (2012; Table 2) also reported an increase in consumer flavour liking increased up to USDA Prime (13.6% fat), but decreased for Wagyu (21.6% fat) steak samples. These measurements are for a single flavour or flavour liking score, and is therefore a composite of all of the beef flavours contained within the sample. Nevertheless, other than an increase in the rating for fat-like

**Table 2** Least squares means (SEM) for consumer ($n$ = 120) sensory scores[1] for palatability traits of beef strip steaks varying by fat level

| Quality treatment | Tenderness | Juiciness | Flavour | Overall liking |
|---|---|---|---|---|
| Wagyu (21.61% fat) | 75.99[a] (±3.17) | 80.88[a] (±2.85) | 69.27[bc] (±2.93) | 69.15[bc] (±2.95) |
| Prime (13.56% fat) | 79.57[a] (±2.28) | 78.33[ab] (±2.03) | 76.95[a] (±2.09) | 77.76[a] (±2.13) |
| High choice (7.21% fat) | 74.16[a] (±2.28) | 73.59[b] (±2.03) | 71.93[ab] (±2.09) | 72.48[ab] (±2.15) |
| Low choice (4.70% fat) | 63.78[b] (±2.28) | 61.60[c] (±2.04) | 66.44[c] (±2.09) | 63.27[c] (±2.13) |
| Select (3.00% fat) | 52.34[c] (±2.28) | 61.74[c] (±2.03) | 60.20[d] (±2.09) | 56.86[d] (±2.14) |
| Standard (1.28% fat) | 63.10[b] (±2.28) | 57.99[c] (±2.03) | 55.60[d] (±2.09) | 57.48[d] (±2.14) |
| Australian (2.29% fat) | 45.04[d] (±3.17) | 47.76[d] (±2.81) | 46.55[e] (±2.88) | 42.82[e] (±2.92) |

[a–e] Least squares means in the same column without a common superscript differ ($P < 0.05$).
[1] Sensory scores: 0 = not tender/juicy, dislike flavour/overall extremely; 100 = very tender/juicy, like flavour/overall extremely.
Source: Adapted from O'Quinn et al. (2012).

flavour in ground beef with 30% fat (compared to 10 or 20% fat; Blackmon et al., 2015), no other beef lexicon flavour descriptors were affected by fat level. These data point out the importance of having a small amount of fat in beef products to maintain the minimum flavour and mouthfeel scores, but in general, fat content has very little impact on flavour overall, unless levels below about 5% or above about 14% are reached whether the muscle is intact or ground. This indicates that it is the type of fat that is much more important to flavour than just the amount, and of the fat present, it is the impact of the phospholipids and much less the impact of the neutral lipids on the flavour profile.

## 4.2 Muscle location effects

As one would expect, the composition of fat and lean is different in different parts of the carcass. In very general terms, the posterior part of the carcass is very lean, and the muscles are largely used for locomotion. Moving towards the anterior and ventral sides of the carcass, there is a marked increase in the total amount of fat versus lean in addition to the function of the muscles changing. As the function and location of the muscle changes in the carcass, the composition of lipid and fatty acid composition also changes. The form and function of the muscle and its location determines the muscle fibre type and fatty acid composition and, in particular, its phospholipid fatty acid composition. Turk and Smith (2009; Table 3) did an extensive study on the differences in subcutaneous fatty acid composition of eight different carcass cut locations. They reported that SFA tended to be the highest in the rib, loin and sirloin and lowest in the brisket. Furthermore, they reported that MUFA, and in particular 18:1 n-9 (oleic acid), was by far highest in the brisket and lowest in the flank. Similar findings for brisket and flank were reported by Blackmon et al. (2015) and Kerth et al. (2015). Oka et al. (2002) reported that MUFA varied by as much as 15% when comparing subcutaneous (59.66%), intermuscular (55.12%), intramuscular (51.73%) and perinephric fat (44.4%).

When the sensory and consumer acceptance of different cuts are evaluated, one of the clearest examples is in examining the sensory of ground beef made from different parts of the carcass. Many times hamburgers are advertised for a particular trait (e.g. ground chuck, ground sirloin, ground brisket) because of perceived advantages each of these have over 'generic' ground beef. Blackmon et al. (2015) reported that ground

**Table 3** Monounsaturated fatty acid concentrations (g/100 g total fatty acids) and MUFA:SFA ratio in eight adipose tissue depots

| Fatty acid | Brisket | Chuck | Flank | Loin | Plate | Rib | Round | Sirloin | SE | *P*-values |
|---|---|---|---|---|---|---|---|---|---|---|
| 14:1n-5 | 1.91[a] | 1.58[b] | 1.07[e] | 1.38[cd] | 1.40[cd] | 1.26[d] | 1.46[bc] | 1.49[bc] | 0.023 | 0.001 |
| 16:1n-7 | 7.22[a] | 5.57[b] | 3.19[e] | 4.46[d] | 4.49[d] | 4.21[d] | 5.14[bc] | 5.04[c] | 0.081 | 0.001 |
| 18:1n-9 | 43.1[a] | 40.2[bc] | 36.8[e] | 38.4[d] | 39.5[cd] | 38.4[d] | 41.0[b] | 39.8[c] | 0.181 | 0.001 |
| 18:1c11 | 2.33[a] | 1.55[b] | 1.09[e] | 1.22[cde] | 1.33[cd] | 1.17[de] | 1.38[c] | 1.34[cd] | 0.028 | 0.001 |
| 18:1*t*11 | 2.02[b] | 3.08[a] | 3.47[a] | 3.56[a] | 3.61[a] | 3.38[a] | 2.94[a] | 3.21[a] | 0.100 | 0.002 |
| MUFA:SFA[a] | 1.47[a] | 1.14[bc] | 0.88[g] | 0.98[ef] | 1.05[de] | 0.96[f] | 1.16[b] | 1.08[cd] | 0.331 | 0.001 |

[a–g] Means with common superscripts are not different. Data are means for 50 carcasses.

Source: Adapted from Turk and Smith (2009).

beef made from brisket lean and fat had a more favourable fat-like flavour compared to flank, reflecting differences in oleic acid content. In Kerth et al. (2015), however, all of the ground beef had the same lean source, and the only difference was in the carcass location of the subcutaneous fat used to make the ground beef. Interestingly, no sensory or consumer liking differences were found even though the ground brisket tended to have a higher content of oleic acid. This likely reflects that the source of the fatty acid was in the subcutaneous fat and not the phospholipid, where the impact of fatty acid on flavour is found (Legako et al., 2015). This highlights the importance of knowing the fatty acids of the phospholipid fraction and not just the non-polar or triglycerides, nor the total fatty acids for that matter.

## 4.3 Dry vs. wet ageing

Ageing of beef is a common practice in the United States, as it is very useful in the development of tenderness of cuts of beef that are low in connective tissue. The naturally occurring enzymes in the muscle are effective at degrading the structural proteins found in the muscle, making the meat more tender. Most of the emphasis in ageing beef is in the tenderizing effect, and much less is reported on the impact of ageing on flavour development. Perhaps this is primarily the result of the advent of vacuum packaging in the 1970s, when most of the ageing changed from dry ageing to wet ageing. Dry ageing

**Table 4** Least squares means for trained panel flavour description of steaks stratified by ageing treatment × muscle

| | Flavor descriptor[a] | | | | | |
|---|---|---|---|---|---|---|
| Interaction effects | Beef flavour | Brown roasted | Bloody/ serumy | Musty | Putrid | Warmed-over flavor |
| *Ageing treatment x muscle* | | | | | | |
| Dry-aged, *M. spinalis thoracis* | 5.3[c] | 2.2[d] | 2.3[b] | 3.0[a] | 3.6[a] | 0.01[cd] |
| Wet-aged, *M. spinalis thoracis* | 7.1[b] | 2.8[b] | 3.0[a] | 0.7[d] | 0.4[c] | 0.24[ab] |
| Dry-aged, *M. longissimus thoracis* | 6.5[b] | 2.7[bc] | 2.4[b] | 1.7[bc] | 1.6[b] | 0.26[a] |
| Wet-aged, *M. longissimus thoracis* | 7.9[a] | 3.4[a] | 2.5[b] | 0.4[d] | 0.1[c] | 0.07[bcd] |
| Dry-aged, *M. gluteobiceps* | 5.5[c] | 2.3[cd] | 2.7[ab] | 2.1[b] | 2.8[a] | 0.08[bcd] |
| Wet-aged, *M. gluteobiceps* | 6.8[b] | 2.9[b] | 2.5[b] | 0.7[d] | 0.2[c] | 0.00[d] |
| Dry-aged, *M. gluteus medius* | 6.7[b] | 2.9[b] | 2.4[b] | 1.3[c] | 1.3[b] | 0.04[cd] |
| Wet-aged, *M. gluteus medius* | 7.2[a] | 2.7[bc] | 3.0[a] | 0.3[d] | 0.1[c] | 0.17[abc] |
| $P > F$ | 0.037 | 0.036 | 0.031 | 0.005 | 0.001 | 0.004 |
| RMSE[b] | 0.59 | 0.37 | 0.32 | 0.49 | 0.95 | 0.05 |

[a–d]Means within the same column lacking a common letter differ ($P < 0.05$).
[a]0 = none; 15 = extremely intense.
[b]Root mean square error.
Source: Adapted from Smith et al. (2014).

is best described as unpackaged meat aged at controlled temperature and humidity. Wet ageing is extended storage of meat in vacuum-sealed packages at refrigeration temperatures. The advantages of wet ageing is very little weight loss and, more importantly, much less oxidation that occurs because most of the oxygen is removed from the bag. Because oxygen is the primary cause of fatty acid oxidation, the benefit is clear.

Smith et al. (2014; Table 4) found beef flavour and brown/roasted flavours were higher while musty, putrid and warmed-over flavours (WOFs) tended to be lower in wet-aged rib and sirloin muscles compared with dry-aged muscles. Additionally, consumers scored overall like and flavour like about one point higher for wet-aged cuts compared with dry-aged steaks. Likewise, Sitz et al. (2014) found that consumers rated Prime, wet-aged beef more acceptable compared with dry-aged steaks. In fact, about 10% more consumers preferred the wet-aged steaks and were willing to pay more for them. Because a higher amount of oxidation would be expected with ageing, it was no surprise that 2,3-octanedione, pentanal, 3-hydroxy-2-butanone, 2-pentyl furan, 1-octen-3-ol, butanoic acid and hexanoic acid, compounds often associated with lipid oxidation, were reported in steaks that were aged and enhanced (Stetzer et al., 2008).

### 4.4 Further processing

The enhancement of pork and poultry (Brewer et al., 2002) and beef (Vote et al., 2000) with solutions mostly consisting of water, salt and phosphate have been practised to ensure a consistently tender and juicy product reaches the consumer. The components used in these solutions could alter the volatile flavour compounds found in different muscles. Sodium salts of tripolyphosphate, hexametaphosphate and pyrophosphate increase meat flavour and decrease stale and rancid flavour during storage by serving as indirect antioxidants by chelating heavy metals that often initiate the oxidation process. They can, however, produce a soapy aftertaste (Vote et al., 2000). Consumer acceptance of enhanced beefsteaks and roasts were studied (Robbins et al., 2003), and it was determined that enhancement improved the consumer overall acceptability score as well as the beef flavour score for both steaks and roasts. In both cases, the saltiness score was also about one point higher for enhanced steaks and roasts.

## 5 Factors affecting beef flavour: cooking

Heating of beef muscle modifies the water content, the micronutrient content and the colour of the muscle (Kondjoyan et al., 2014). It has been established that raw meat has very little aroma with only a bloody/serumy and slightly salty flavour. It is the process of cooking that takes the precursors like amino acids, fatty acids and carbohydrates within the meat itself and modifies them to give the flavour typically associated with meat. We have also discussed the fact that the water-soluble portion is mostly responsible for the roasted, caramelized flavour, while the lipid fraction of meat is responsible for the meaty and species flavours (Hornstein and Crowe, 1960). It is important to note here that the methods and specifications used to cook meat have a dramatic impact on the final flavour outcome of the meat product.

## 5.1 Temperature

Degree of doneness is scientifically determined by the internal temperature of the steak or roast. According to the National Cattlemen's Beef Association (2016), the following degrees of doneness are representative of the appropriate temperatures: very rare (55°C), rare (60°C), medium rare (65°C), medium (71°C), well done (77°C) and very well done (82°C). Consumers found steaks cooked to medium rare or less to have higher overall like scores than those for other degrees of doneness (Lorenzen et al., 1999). Bowers et al. (1987) found that sensory panel scores for mouth-filling blend and browned were highest, whereas bloody/serumy, sour and metallic scores were lowest for steaks cooked to 85°C. The beefy flavour was higher in well-done steaks compared with very rare steaks, while blood and fatty flavours are higher at lower degrees of doneness (Lorenzen et al., 2005). They also indicated that steaks cooked to a medium degree of doneness were associated with 4-heptanolide, while those cooked to higher temperatures were associated with pyrazines and hexanol. Surprisingly, much less is known about the flavour profiles of steaks with differing degrees of doneness compared to many other traits.

Additionally, the time and temperature at which a steak is cooked impacts the flavour and volatile aroma compounds generated (Kerth, 2016; Table 5). Aroma compounds described as fatty, tallowy and oily are highly related to the identity of beef flavour. These compounds are produced in the highest quantity when steaks are cooked either at low cooking temperatures (not degree of doneness) or for short periods of time. Aroma compounds described as roasted, nutty or fruity are developed from browning the surface of the steak as a result of cooking at high cooking temperatures or for long periods of time as would be the case if a steak was cut thicker. This study indicated that the amount of specific aroma compounds can be predicted with $r^2$ values of up to 0.62 from measuring cooking times and temperatures.

## 5.2 Cooking method

High heat for long periods of time will result in more Maillard products being generated, whereas cooking methods that result in slow cooking at low temperatures will favour the generation of lipid-derived flavour compounds. McKenna et al. (2004) found that consumers tended to have higher flavour and overall satisfaction scores from grilling compared to simmer/stewing. Further, flavour and overall acceptability scores were less influenced by USDA quality grade when steaks were grilled outside compared with other cooking methods, especially simmer/stewing. Another phenomenon commonly found in cookery is the generation of what is often referred to as WOF. When uncured meat is stored for a relatively short period of time after cooking, it develops WOF. This flavour problem of cooked meat has assumed much greater significance in recent years due to the rapid increase in fast food service facilities using large quantities of precooked or partially cooked meats and convenience items that are available and intended to be re-heated (Sato and Hegarty, 1971). Many researchers have hypothesized that peroxidation of unsaturated fatty acids is the cause of WOF and have evaluated the volatile compounds from that perspective. Among many of the products identified, hexanal was recommended as an indicator for WOF. Nevertheless, WOF not only is the result of peroxidation of unsaturated lipids, but is also caused by loss of desirable odorants involved in the pleasant flavour of freshly cooked meat (Kerler and Grosch, 1996).

**Table 5** Cooking surface temperature and steak thickness impacts on volatile compounds with published aroma descriptors

| Aroma compounds expressed in total volatiles ion counts | Skillet temperature, °C | | | Steak thickness, cm | | | $P > F$ | | RMSE | Published aroma descriptor[1] |
|---|---|---|---|---|---|---|---|---|---|---|
| | 177 | 204 | 232 | 1.27 | 2.54 | 3.81 | Temp | Thickness | | |
| (*E*)-2-Decenal | 679 | 1997 | 906 | 3668$^{c}$ | 219$^{cd}$ | −305$^{d}$ | NS | 0.049 | 6153.1 | Tallow |
| 1-Hexanol | 460$^{b}$ | 326$^{b}$ | 3629$^{a}$ | 2461$^{c}$ | 2173$^{c}$ | −219$^{d}$ | 0.035 | 0.11 | 4791.8 | Woody, cut grass |
| 1-Octanol | 3722 | 2497 | 2542 | 5511$^{c}$ | 1582$^{d}$ | 1668$^{d}$ | NS | 0.033 | 5943.5 | Soapy, waxy |
| 2-Ethyl-3-methyl pyrazine | 1407 | 1689 | 3614 | 498$^{d}$ | 2084$^{cd}$ | 4128$^{c}$ | NS | 0.086 | 5767.0 | Roasted |
| 2-Ethyl-3,5-dimethyl pyrazine | 687 | 749 | 2154 | 283$^{d}$ | 2268$^{c}$ | 1030$^{cd}$ | NS | 0.084 | 2973.6 | Nutty, roasted |
| 2-Ethyl-5-methyl pyrazine | 198$^{b}$ | 2413$^{b}$ | 9465$^{a}$ | 889$^{d}$ | 3635$^{cd}$ | 7554$^{c}$ | 0.010 | 0.082 | 10500.5 | Nutty, roasted |
| 2-Ethyl-6-methyl pyrazine | 863 | 2120 | 5895 | 400$^{d}$ | 1533$^{cd}$ | 6945$^{c}$ | NS | 0.071 | 10459.9 | Fruity |
| 2-Heptanone | 1157 | 466 | −20 | 1638$^{c}$ | 49$^{cd}$ | −85$^{d}$ | NS | 0.086 | 3038.0 | Fruity, spicy |
| 2-Methyl butanal | 15373 | 41488 | 66826 | 2702$^{d}$ | 25373$^{cd}$ | 95611$^{c}$ | NS | 0.021 | 119560.7 | Burnt, sickly |
| 2-Pentyl furan | 5134 | 3468 | 39 | 6821$^{c}$ | 142$^{d}$ | 1678$^{cd}$ | NS | 0.052 | 9849. 2 | Greenbean, sweet |
| 2,3-Butanedione | 79434$^{a}$ | 38761$^{ab}$ | 23908$^{b}$ | 76811 | 24713 | 40579 | 0.054 | 0.072 | 79350.2 | Buttery |
| 2,5-Dimethyl pyrazine | 7730$^{b}$ | 11876$^{b}$ | 38074$^{a}$ | 5089$^{d}$ | 21488$^{cd}$ | 31104$^{c}$ | 0.016 | 0.049 | 37637.7 | Roasted |
| 3-Methyl butanal | 17871 | 27468 | 40522 | 4852$^{d}$ | 16453$^{d}$ | 64555$^{c}$ | NS | 0.022 | 78214.9 | Meaty, roasted |
| Decanal | 1855 | 4257 | 2970 | 5591$^{c}$ | 2327$^{d}$ | 1163$^{d}$ | NS | 0.004 | 4795.2 | Waxy, smokey |
| Dimethyl sulphide | 1236 | 1455 | 1492 | 1041$^{cd}$ | 549$^{d}$ | 2594$^{c}$ | NS | 0.076 | 3115.7 | Sickly |
| Ethyl benzene | 10593 | 3402 | 3663 | 12178$^{c}$ | 3278$^{cd}$ | 2202$^{d}$ | NS | 0.058 | 16066.8 | Fruity |
| Heptanal | 87927 | 55096 | 94405 | 115079$^{c}$ | 88638$^{cd}$ | 33712$^{d}$ | NS | 0.035 | 113823.8 | Oily, rancid |
| Methyl pyrazine | 4809 | 6873 | 15640 | 1537$^{d}$ | 8873$^{cd}$ | 16914$^{c}$ | NS | 0.030 | 20234.3 | Roasted |
| Nonanal | 80950 | 95146 | 104488 | 151611$^{c}$ | 77333$^{d}$ | 51641$^{d}$ | NS | 0.001 | 94521.3 | Green, mouldy |
| Octanal | 58039 | 52305 | 70606 | 101429$^{c}$ | 49975$^{d}$ | 29548$^{d}$ | NS | 0.004 | 76611.9 | Soapy, fatty |
| Phenyl acetaldehyde | 28$^{b}$ | 872$^{ab}$ | 2433$^{a}$ | 519 | 1759 | 1056 | 0.097 | NS | 3772.9 | Rosy, perfume |
| Styrene | 64622 | 55131 | 53426 | 91312$^{c}$ | 37145$^{d}$ | 44722$^{d}$ | NS | 0.042 | 80401.9 | Sweet |
| Toluene | 36640$^{a}$ | 3574$^{ab}$ | 1777$^{b}$ | 5348 | 33571 | 3071 | 0.081 | NS | 58905.6 | Paint thinner |
| Trimethyl pyrazine | 2314$^{ab}$ | 449$^{b}$ | 8695$^{a}$ | −505 | 5743 | 6219 | 0.098 | NS | 13474.7 | Nutty, roasted |

[1] Published aroma descriptors taken from Brewer (2006), Calkins and Hodgen (2007), Kerth and Miller (2015), Moon et al. (2006) and Shahidi et al. (1986).

$^{a,b}$ Means in a row within the temperature treatment with different superscripts differ ($P < 0.05$).

$^{c,d}$ Means in a row within the steak thickness treatment with different superscripts differ ($P < 0.05$).

# 6 Assessing the flavour of beef products

While there are multiple sensory tools available to measure beef, it is important to note, as we have discussed throughout this chapter, that flavour is incredibly complex and a single tool is difficult to use to assess flavour. The fact that beef flavour is not a single attribute, and multiple attributes makes the sensory evaluation of its components challenging (Miller, 2012). In general, the evaluation of beef flavour is done by three primary methods: trained sensory descriptive panel, consumer sensory panel and gas chromatography/mass spectrometry/olfactory (GC/MS/O). In any of the cases, the Research Guideline for Cookery, Sensory Evaluation and Instrumental Tenderness Measurements of Meat (AMSA, 2016) is the best place to consult in determining which testing method best suits the experiment to be carried out.

## 6.1 Trained sensory panels

A trained sensory descriptive panel, when well trained (see Cross et al., 1978, for a description of panel training), should be considered an objective measurement of flavour just as much as the GC/MS/O. Because the method of training consists of anchoring very specific flavour traits to reference standards, it is possible for a trained panel to be extremely accurate in determining even very small differences among samples. Flavour in beef is not a single attribute, although it has been measured as a single attribute for many years as defined originally in the first editions of AMSA (2016). Instead, beef flavour consists of multiple attributes, including general categories of flavour aromas, basic tastes, feeling factors and aftertastes (Adhikari et al., 2011).

## 6.2 Consumer sensory panels

While we have discussed at length factors that impact beef flavour, ultimately it is consumers who decide what is acceptable. Consumers are the last step in the production chain; therefore, evaluating their assessment of a beef product is essential to determining the product quality. Consumer expectations can be divided into three basic types: psychological (an individual factor), sensory (a product-specific factor) and marketing (an environmental factor) (Font-i-Furnols and Guerrero, 2014). These aspects are interrelated and depend on additional factors that affect consumers' decision-making. One of the most important factors to consider when conducting consumer sensory panels is the amount of natural variation that exists. This includes variation in the product, and especially the variation that we know exists in flavour, as well as the variation in the population of people that is to be surveyed. Demographic information, such as cultural background, household income, education level, age, sex, experience with the product and others can have a profound impact on the ability of consumers to discriminate and will certainly impact their likes and dislikes. It is also important to understand the limitations of a consumer sensory panel. It is not possible to determine the level of a sensory trait like flavour from a consumer panel. A consumer is only able to tell us what he or she likes and the degree to which they like or dislike a certain trait, such as flavour, juiciness or tenderness.

As a result, it is of utmost importance that a ballot for consumer evaluation is carefully constructed to elucidate only those traits consumers are able to actually determine. They can only tell us their opinion and we should not rely on consumers to give us factual data regarding a sensory trait. Because of the large amount of variability in the consumer population we intend to sample, it is also necessary to use a much higher number of consumer panellists compared with that of a trained sensory panel in order for statistical significance to be determined. As a general rule, a minimum of 80 consumer panellists is required for the most basic of sensory trials. As the complexity of the data required increases, the number of consumer panellists also increases, and most often a statistician or someone who is familiar with conducting a consumer panel is needed to ensure the design of an experiment is statistically sound.

### 6.3 Gas chromatography/mass spectrometry/olfactory (GC/MS/O)

Because the measurement of aromas is so important to the understanding of beef flavour, it is necessary to have a reliable and objective method for measuring these aromas. The complexity of aromas makes the use of humans as instruments, even with a trained sensory panel, very limiting. Instead, it is useful to have an instrument that is capable of separating each individual aroma into its components for objective evaluation in addition to human recognition/evaluation. As alluded to previously in this chapter, the use of GC/MS/O has become a popular method for measuring and evaluating flavour compounds in beef (Kerth and Miller, 2015). The evaluation process starts at the point at which a meat sample is cooked and is ready to be served to either a trained or consumer panel. A sample of meat equivalent to that to be served to the panellists is placed into a glass container with a Teflon lid. A solid-phase micro-extraction (SPME) fibre is then inserted into the headspace above the meat sample and allowed to incubate there for a prescribed amount of time. The PDMS-carboxen coating on the exposed SPME absorbs the volatile compounds that are coming from the cooked meat sample and the volatiles become 'stuck' to the SPME. After the SPME is removed from the sample container, it is injected into the GC in which the injection port is held at a very high temperature, which desorbs all of the volatile compounds. The GC is basically a very well-controlled oven that starts at about room temperature and very slowly increases in temperature over time. Because the temperature of vaporization is different for each volatile compound, each volatile exits the GC at a different time, thus allowing for separation of each individual compound. As the volatile compound exits the GC, it enters a splitter, where part of the compound goes to the MS, which identifies and quantifies the volatile, and part of the compound simultaneously goes to an olfactory or 'sniff' port, where a panellist smells each volatile compound. In this way, the volatile compounds that are present in the cooked meat sample are identified, quantified and associated with an aroma in the meat sample.

## 7 Conclusion

Flavour is considered by many to be the 'guard rails' that keep consumers on the road to consuming beef, as it is a trait that is unique to beef. It is also one of the most complex traits we measure and, therefore, requires a comprehensive understanding of meat composition, basic chemical reactions and testing procedures for accurate measurement. Not only is meat composition important and influenced by numerous factors in the production of live animals; it is also critical to understand meat processing and cookery methods and how they influence the development of flavour.

# 8 Where to look for further information

*Flavor of Meat, Meat Products, and Seafoods*. Edited by F. Shahidi. ISBN 0751404845
*Development of a Lexicon for Beef Flavor in Intact Muscle* (Akhikari et al., 2011)
*Beef Flavor: A Review from Chemistry to Consumer* (Kerth and Miller, 2015)
*A Fresh Look at Meat Flavor* (Calkins and Hodgen, 2007)
*Effects of Feeds on Flavor of Red Meat: A Review* (Melton, 1990)
American Meat Science Association (www.meatscience.org)
Society of Sensory Professionals (www.sensorysociety.org)
National Cattlemen's Beef Association (beef.org)

# 9 References

American Meat Science Association (AMSA) (2016). Research Guidelines for Cookery, Sensory Evaluation and Instrumental Tenderness Measurements of Meat. 2nd Ed. AMSA, Champaign, Ill, USA.

Adhikari, K., Chambers IV, E., Miller, R., Vazques-Araujo, L., Bhumiratana, N. and Philip, C. (2011). Development of a lexicon for beef flavor in intact muscle. *J. Sens. Stud.* 26:413–20.

Baublits, R. T., Pohlman, F. W., Brown Jr., A. H., Johnson, Z. B., Rule, D. C., Onks, D. O., Murrieta, C. M., Richards, C. J., Sandelin, B. A., Loveday, H. D. and Pugh, R. B. (2009). Correlations and prediction equations for fatty acids and sensory characteristics of beef longissimus rib steaks from forage-fed cattle and retail USDA Choice and Select rib steaks. *J. Mus. Food.* 20:1–17.

Blackmon, T., Miller, R. K., Kerth, C. R. and Smith, S. B. (2015). Ground beef patties prepared from brisket, flank and plate have unique fatty acid and sensory characteristics. *Meat Sci.* 103:46–53.

Bowers, J. A., Craig, J. A., Kropf, D. H. and Tucker, T. J. (1987). Flavor, color, and other characteristics of beef longissimus muscle heated to seven internal temperatures between 55 and 85C. *J. Food Sci.* 52(3):533–6.

Brewer, M. S. (2006). The Chemistry of Beef Flavor – Executive Summary. National Cattlemen's Beef Association.

Brewer, M. S., Jensen, J., Prestat, C., Zhu, L. G. and McKeith, F. K. (2002). Visual acceptability and consumer purchase intent of enhanced pork loin roasts. *J. Mus. Food.* 13(1):53–68.

Calkins, C. R. and Hodgen, J. M. (2007). A fresh look at meat flavor. *Meat Sci.* 77:63–80.

Chandrashekar, J., Hoon, M. A., Nicholas, J. P. R. and Zuker, C. S. (2006). The receptors and cells for mammalian taste. *Nature* 444:288–94.

Chen, L., Ekine-Dzivenu, C., Vinsky, M., Basarab, J., Aalhus, J., Dugan, M. E. R., Fitzsimmons, C., Stothard, P. and Li, C. (2015). Genome-wide association and genomic prediction of breeding values for fatty acid composition in subcutaneous adipose and longissimus lumborum muscle of beef cattle. *BMC Genetics* 16:135

Conte, G., Mele, M., Chessa, S., Castiglioni, B., Serra A., Pagnacco, G. and Secchiari, P. (2010). Diacylglycerol acyltransferase 1, stearioyl-CoA desaturase 1, and sterol regulatory element binding protein 1 gene polymorphisms and milk fatty acid composition in Italian Brown cattle. *J. Dairy Sci.* 93:753–63

Cox, R. B., Kerth, C. R., Gentry, J. G., Prevatt, J. W., Braden, K. W. and Jones, W. R. (2006). Determining acceptance of domestic forage- or grain-finished beef by consumers from three southeastern U. S. states. *J. Food Sci.* 71(7):S542–S546.

Cross, H. R., Moen, R. and Stanfield, M. S. (1978). Training and testing of judges for sensory analysis of meat quality. *Food Tech.* 32:48.

Daley, C. A., Abbott, A, Doyle, P. S., Nader, G. A. and Larson, S. (2010). A review of fatty acid profiles and antioxidant content in grass-fed and grain-fed beef. *Nutr. J.* 9:10.

Font-i-Furnols, M. and Guerrero, L. (2014). Consumer preference, behavior and perception about meat and meat products: An overview. *Meat Sci.* 98:361–71.

Guntert, M., Bruning, J., Emberger, R., Kopsel, M., Kuhn, W., Thielmann, T. and Werkhoff, P. (1990). Identification and formation of some selected sulfur-containing flavor compounds in various meat model systems. *J. Agric Food Chem.* 38:2027–41.

Hornstein, I. and Crowe, P. F. (1960). Flavor studies on beef and pork. *Ag. Food Chem.* 8(6):494–8.

Kerler, J. and Grosch, W. (1996). Odorants contributing to warmed-over flavor (WOF) of refrigerated cooked beef. *J. Food Sci.* 61(6):1271–5.

Kerth, C. R. (2016). Determination of volatile aroma compounds in beef using differences in steak thickness and cook surface temperature. *Meat Sci.* 117:27–35.

Kerth, C. R., Braden, K. W., Cox, R., Kerth, L. K. and Rankins Jr., D. L. (2002). Carcass, sensory, fat color and consumer acceptance characteristics of Angus-cross steers finished on ryegrass (Lolium multiflorum) forage or on a high-concentrate diet. *Meat Sci.* 75:324–31.

Kerth, C. R., Harbison, A. L., Smith, S. B. and Miller, R. K. (2015). Consumer sensory evaluation, fatty acid composition, and shelf-life of ground beef with subcutaneous fat trimmings from different carcass locations. *Meat Sci.* 104:30–6.

Kerth, C. R. and Miller, R. K. (2015). Beef flavor: a review from chemistry to consumer. *J. Sci. Food Agric.* doi: 10.1002/jsfa.7204.

Kondjoyan, A., Kohler, A., Realini, C. E., Portanguen, S., Kowalski, R., Clerjoh, S., Gatellier, P., Chevolleau, S., Bonny, J. and Debrauwer, L. (2014). Towards models for the prediction of beef meat quality during cooking. *Meat Sci.* 97:323–31.

Larick, D. K. and Turner, B. E. (1990). Flavor characteristics of forage- and grain-fed beef as influenced by phospholipid and fatty acid compositional differences. *J. Food Sci.* 55(2): 312–17.

Laska, M., Distel, H. and Hudson R. (1997). Trigeminal perception of odorant quality in congenitally anosmic subjects. *Chem. Sens.* 22:447–56.

Legako, J. F., Dinh, T. T. N., Miller, M. F. and Brooks, J. C. (2015). Effects of USDA beef quality grade and cooking on fatty acid composition of neutral and polar lipid fractions. *Meat Sci.* 100:246–55.

Lorenzen, C. L., Davuluri, V. K., Adhikari, K. and Grun, I. U. (2005). Effect of end-point temperature and degree of doneness on sensory and instrumental flavor profile of beefsteaks. *J. Food Sci.* 70(2):S113–S118.

Lorenzen, C. L., Neely, T. R., Miller, R. K., Tatum, J. D., Wise, J. W., Taylor, J. F., Buyck, M. J., Reagan, J. O. and Savell, J. W. (1999). Beef customer satisfaction: cooking method and degree of doneness effects on the top loin steak. *J. Anim. Sci.* 77:637–44.

Mackintosh, D. L. and Hall, J. L. (1936). Fat as a factor in palatability of beef. Transactions of the Kansas Academy of Science (1903–), Vol. 39 (2–4 April 1936), pp.53–8. Kansas Academy of Science. http://www.jstor.org/stable/3625144.

Macleod, G. (1998). The flavour of beef. In *Flavor of Meat, Meat Products and Seafoods*, edited by F. Shahidi, pp. 27–60. Thomson Science, London, UK.

Maillard, L. C. (1912). Action des acids amines sur les sucres: formation des melanoidines par vois methodique. *Compt. Rend.* 154:66.

Maughan, C., Tansawat, R., Cornforth, D., Ward, R. and Martini, S. (2012). Development of a beef flavor lexicon and its application to compare the flavor profile and consumer acceptance of rib steaks from grass- or grain-fed cattle. *Meat Sci.* 90:116–21.

McKenna, D. R., Lorenzen, C. L., Pollok, K. D., Morgan, W. W., Mies, W. L., Harris, J. J., Murphy, R., McAdams, M., Hale, D. S. and Savell, J. W. (2004). Interrelationships of breed type, USDA quality grade, cooking method, and degree of doneness on consumer evaluations of beef in Dallas and San Antonio, Texas, USA. *Meat Sci.* 66:399–406.

Melton, S. L. (1990). Effects of feeds on flavor of red meat: a review. *J. Anim. Sci.* 68:4421–35.

Melton, S. L. (1983). Effect of forage feeding on beef flavor. *Food Tech.* 37:239

Miller, R. K. (2012). Sensory evaluation of beef flavor. In *Handbook of Meat, Poultry and Seafood Quality*, edited by L. M.L. Nollet, pp. 175–91. John Wiley and Sons, Inc., Ames, IA, USA.

Moon, S. Y, Cliff, M. A. and Li-Chan, E. C. Y. (2006). Odour-active components of simulated beef flavor analysed by soled phase microextraction and gas chromatography-mass spectrometry and-olfactometry. *Food Research International* 39:294–308.

Mottram, D. S. (1998). The chemistry of meat flavour. In *Flavor of Meat, Meat Products and Seafoods*, edited by F. Shahidi, pp. 5–26. Thomson Science, London, UK.

Mottram, D. S. and Edwards, R. A. (1983). The role of triglycerides and phospholipids in the aroma of cooked beef. *J. Sci. Food Agric.* 34:517–22.

Mozell, M. M., Smith, B. P., Smith, P. E., Sullivan, R. L. and Swender, P. (1969). Nasal chemoreception in flavor identification. *Arch. Otolaryngol.* 90:367–73.

National Cattlemen's Beef Association (2016). Beef Steak Color Guide. http://beefambassador.com/wp-content/gems/2015/06/doneness-chart.jpg (accessed 1 April 2016).

Oka, A., Iwaki, F., Dohgo, T., Ohtagaki, S., Noda, M., Shiozaki, T., Endoh, O. and Ozaki, M. (2002). Genetic effects on fatty acid composition of carcass fat of Japanses Black Wagyu steers. *J. Anim. Sci.* 80:1005–11.

O'Quinn, T. G, Brooks, J. C., Polkinghorne, R. J., Garmyn, A. J., Johnson, B. J., Starkey, J. D., Rathmann, R. J. and Miller, M. F. (2012). Consumer assessment of beef strip loin steaks of varying fat levels. *J. Anim. Sci.* 90:626–34.

Rule, D. C., MacNeil, M. D. and Short, R. E. (1997). Influence of sire growth potential, time on feed and growin-finishing strategy on cholesterol and fatty acids of the ground carcass and longissimus muscle of beef steers. *J. Anim. Sci.* 75:1525–33.

Sato, K. and Hegarty, G. R. (1971). Warmed-over flavor in cooked meats. *J. Food Sci.* 36:1098–102.

Savell, J. W., Branson, R. E., Cross, H. R., Stiffler, D. M., Wise, J. W., Griffin, D. B. and Smith, G. C. (1987). National consumer retail beef study: palatability evaluation of beef loin steaks that differed in marbling. *J. Food Sci.* 52(3):517–19.

Shahidi, F., Rubin, L. J., D'Souza, L. A., Teranishi, R. and Buttery, R. G. (1986). Meat flavor volatiles: A review of the composition, techniques of analysis, and sensory evaluation. *Crit. Rev. Food Sci. Nutri.* 24(2):141–243.

Schenni, K. A., Hick, J. M., Bovenjuis, H., Visker, M. H. P. W., van Valenberg, H. J. F. and van Arendonk, J. A. M. (2008). Milk fatty acid unsaturation: genetic parameters and effects of stearoyl=CoA desaturase (SCD1) and acyl CoA: diacylglycerol acyltransferase 1 (DGAT1). *J. Dairy Sci.* 91:2135–43.

Shepherd, G. M. (2005). Outline of a theory of olfactory processing and its relevance to humans. *Chem. Sens.* 30(Suppl. 1):i3-i5.

Sitz, B. M., Calkins, C. R., Feuz, D. M., Umberger, W. J. and Eskridge, K. M. (2014). Consumer acceptance and value of wet-aged and dry-aged beef steaks. *J. Anim. Sci.* 84:1221–6.

Smith, A. M., Harris, K. B., Griffin, D. B., Miller, R. K., Kerth, C. R. and Savell, J. W. (2014). Retail yield and palatability evaluation of indifidual muscles from wet-aged and dry-aged beef ribeyes and top sirloin butts that were merchandised innovatively. *Meat Sci.* 97:21–6.

Stetzer, A. J., Cadwallader, K., Singh, T. K., McKeith, F. K. and Brewer, M. S. (2008). Effect of enhancement and ageing on flavor and volatile compounds in various beef muscles. *Meat Sci.* 79:13–19.

Sturdivant, C. A., Lunt, D. K., Smith, G. C. and Smith, S. B. (1992). Fatty acid composition of subcutaneous and intramuscular adipose tissues and M. longissimus dorsi of Wagyu cattle. *Meat Sci.* 32:449–58.

Taniguchi, M., Utsugi, T., Oyama, K., Mannen, H., Kobayashi, M., Tanabe, Y., Ogino, A. and Tsuji, S. (2004). Genotype of stearoyl-CoA desaturase is associated with fatty acid composition in Japanese Black cattle. *Mam. Gen.* 15:142–8.

Tansawat, R., Maughan, C. A. J., Ward, R. E., Martini, S. and Cornforth, D. P. (2013). Chemical characterization of pasture- and grain-fed beef related to meat quality and flavour attributes. *Int. J. Food Sci. Tech.* 48:484–95.

Theunissen, T. J. J. M., Kouwenhoven, T. and Blauw, Y. H. (1979). Consumer's response to food products with increased levels of polyunsaturated fatty acids. *J. Food Sci.* 44:1483.

Thompson, J. M. (2004). The effects of marbling on flavor and juiciness scores of cooked beef, after adjusting to a constant tenderness. *Aus. J. Exp. Ag.* 44:645–52.
Turk, S. N. and Smith, S. B. (2009). Carcass fatty acid mapping. *Meat Sci.* 81:658–33.
van der Linde, L. M., van Dort, J. M., deValois, P., Boelens, H. and de Rijke, D. (1979). Volatile components from thermally degraded thiamine. In *Progress in Flavour Research*, edited by D. G. Land and H. E. Nursten, pp. 219–24. Applied Science, London.
Vote, D. J., Platter, W. J., Tatum, J. D., Schmidt, G. R., Belk, K. E., Smith, G. C., et al. (2000). Injection of beef strip loins with solutions containing sodium tripolyphosphate, sodium lactage and sodium chloride to enhance palatability. *J. Anim. Sci.* 78:952–7.
Waldman, R. C., Suess, G. G. and Brungardt, V. H. (1968). Fatty acids of certain bovine tissue and their association with growth, carcass and palatability traits. *J. Anim. Sci.* 27:632–5.
Werkhoff, P., Emberger, R., Guntert, M. and Kopsel, M. (1989). Isolation and characterization of volatile sulfur-containing meat flavor components in model systerms. In *Thermal Generation of Aromas*, edited by T. H. Parliament, R. J. McGorrin and C. T. Ho, pp. 460–78. American Chemical Society, Washington, DC.
Werkhoff, P., Bruning, J., Emberger, R., Guntert, M., Kopsel, M., Kuhn, W. and Surburg, H. (1990). Isolation and characterization of volatile sulfur-containing meat flavor components in model systerms. *J. Agric. Food Chem.* 38:777–91.
Westerling, D. B. and Hedrick, H. B. (1979). Fatty acid composition of bovine lipids as influenced by diet, sex, and anatomical location and relationship to sensory characteristics. *J. Anim. Sci.* 48(6):1343–8.
Wheeler, T. L., Cundiff, L. V. and Koch, R. M. (1994). Effect of marbling degree on beef palatability in Bos Taurus and Bos indicus cattle. *J. Anim. Sci.* 72:3145–51.
Yang, A., Brewster, M. J., Beilken, S. L., Lanari, M. C., Taylor, D. G. and Tume, R. K. (2002). Warmed-over flavor and lipid stability of beef: effects of prior nutrition. *J. Food Sci.* 67(9):3309–13.

Chapter 14

# Packaging systems for beef retailers and their effects on visual quality and palatability

J. W. S. Yancey, University of Arkansas, USA



## 1 Introduction

Fresh beef is one of the most variable items in the grocery store, with consumers given the choice of several cuts of various USDA grades and production systems (grass fed, organic, etc.) and, within cuts, options of size, colour, marbling levels, fat thickness and even bone. Consumers expect to have the ability to peruse the meat counter to choose that 'perfect' beef cut to take home for their family. To give beef consumers this experience, meat retailers and processors must have the optimal package in which to market their beef products.

Packages have several functions within the beef industry, with the most obvious to protect beef cuts from the outside environment. Packaging plays a pivotal role in ensuring food safety of beef cuts and enhancing the shelf life of beef, reducing waste and saving money for producers and, eventually, consumers. Packages also provide retailers with a tool for labelling and informing consumers about the quality and healthfulness of their beef purchase. Packaging systems must respect the desires of consumers, beginning with the

http://dx.doi.org/10.19103/AS.2016.0009.14

visual appraisal of the beef and continuing into the safety, wholesomeness, cooking and eating experience of consumers. In some cases, consumers will even expect packaging to reflect their values in maintaining environmental stewardship by minimizing waste of packaging materials and food lost because it was not purchased in a timely manner and then discarded by retailers.

Meat packaging is a vast topic and entire texts could be easily devoted to it alone; so, this chapter is a review of packaging relevant to beef production. To completely understand packaging in fresh meat, one must first have an understanding of fresh meat colour and the chemistry involved in providing beef for consumers with the appearance they expect in the meat case. Traditional packaging systems of foam trays and plastic overwrap are still very popular in US retail stores; therefore, the benefits and challenges of this system are discussed. The merits of vacuum packaging and variations on vacuum packaging also are covered, as these methods gain popularity with specialty beef products such as USDA organic and grass-fed beef. Case-ready and modified atmosphere packaging have gained prominence in the past 20 years and, having resolved several issues associated with other methods, have also created new challenges for processors and retailers. Finally, new developments in packaging are described, some of which have been implemented in commercial settings and some still in various stages of development.

# 2 The colour of fresh and ground beef

Meat colour is the single most important factor driving consumer purchasing decisions for beef in the retail case (Mancini and Hunt, 2005). Consumers expect beef to have a bright cherry-red appearance, and discolouration (indicated by darker, brown colouration) is associated with a lack of wholesomeness and freshness. Packaging systems are key to maintaining the bright, cherry-red colour consumers require and in inhibiting the formation of brown discolouration. To understand the role of packaging, one must first understand basic meat colour chemistry. A complete explanation of the myriad of factors that affect beef colour will be addressed in the Beef Colour Development Chapter.

The myoglobin pigment in beef that is responsible for aerobic energy production in the live animal is crucial for the development of the bright, cherry-red colour consumers expect beef to have in retail stores. Myoglobin exists in several forms, including oxymyoglobin, deoxymyglobin and metmyoglobin, which are primarily responsible for beef colour (Fig. 1). Myoglobin is a water-soluble protein, but within a hydrophobic pocket exists a haeme ring. The iron atom within this ring is bound to four pyrrole nitrogens with the haeme ring and to a histidine residue in the protein. Beef colour is dictated by the oxidation state of the iron and the presence of a reversible ligand at the 6th coordination site (Mancini and Hunt, 2005).

## 2.1 Oxygenation of myoglobin

When no ligand is present on the 6th coordination site of the haeme iron and the iron is in the reduced or ferrous state ($Fe^{2+}$), the pigment is known as *deoxymyoglobin* (Fig. 1). This form of myoglobin results in a purplish-red beef colour, commonly associated with freshly cut beef prior to blooming. When deoxymyoglobin is exposed to air, the oxygen will bind to the 6th site on the ferrous iron, resulting in a shift in the protein structure and the bright

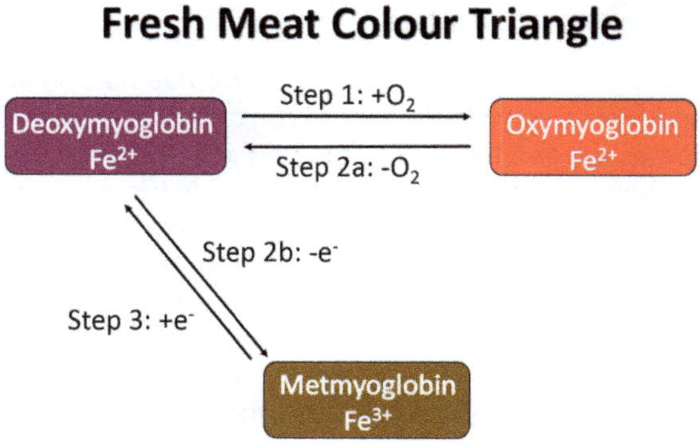


**Figure 1** Fresh meat colour triangle.
Source: Adapted from AMSA, 2012.

cherry-red beef colour associated with *oxymyoglobin* (Step 1, Fig. 1). The oxygenation of the myoglobin is commonly referred to as 'blooming' in the beef industry, and Lee et al. (2008) reported that 90% of the change in oxymyoglobin formation was completed within 60 minutes of cutting beef rib-eye steaks.

In freshly cut beef, a layer of oxymyoglobin will form at the cut surface. The depth of oxygenated pigment penetrating into the meat surface will be dependent on the partial pressure of oxygen in the atmosphere around the cut, temperature, muscle pH and the ability of the muscle to metabolize the oxygen. Respiratory enzymes within the muscle will be in competition with the myoglobin for oxygen, and oxygen's bond with iron may be lost.

## 2.2 Oxidation of myoglobin

Within the muscle, at the interface of the bright red, oxymyoglobin and the purplish-red deoxymyoglobin layers, a low partial pressure of oxygen exists (Mancini and Hunt, 2005). The oxygen concentration is low here because the oxygen is being consumed by the muscle for respiration and other metabolic activities. At low concentrations of oxygen, myoglobin is more susceptible to oxidation. Thus, when oxygen is removed from oxymyoglobin, forming deoxymyoglobin (Step 2a, Fig. 1), the haeme iron is quickly oxidized ($Fe^{2+}$ to $Fe^{3+}$) to form metmyoglobin (Step 2b, Fig. 1), a brown, discoloured pigment associated with spoiled beef (AMSA, 2012). In fresh beef, the muscle has an inherent capacity to reduce the metmyoglobin pigment back to the deoxymyoglobin form and return the beef to the purplish-red colour (Step 3, Fig. 1).

As bloomed beef cuts remain in oxygenated states, such as in a fresh beef case, a visible layer of metmyoglobin will form between the layers of oxy- and deoxymyoglobin (Fig. 2a and b). Eventually, the muscle's innate ability to reduce the brown, metmyoglobin pigment back to deoxymyoglobin will diminish, and the metmyoglobin layer will migrate closer and closer to the surface of the cut. Consumers will be able to distinguish the metmyoglobin discolouration through the oxymyoglobin layers and will begin to discriminate against the less-desirable cuts when visible metmyoglobin reaches about 20% of the surface pigment (MacDougal, 1982).

## 2.3 Other myoglobin pigments

When myoglobin is exposed to carbon monoxide, the haeme iron will interact with the carbon monoxide and form a very stable, *carboxymyoglobin* pigment (Fig. 3). The carbon monoxide out competes oxygen for the 6th binding site on the haeme iron and the beef will remain in bright cherry-red colour for several days. *Nitric oxide myoglobin* is formed when fresh meat is exposed to nitric oxide. This pigment may be heated to form the pink, nitrosylhaemochrome pigment associated with cured meats, but recently, scientists have found it may have a role to play in fresh beef colour as well.

Other forms of myoglobin also exist, but are rarely seen unless beef is nearing the end of its shelf life (Kropf et al., 2014). The presence of hydrogen peroxide, most likely from microbial contamination, may result in the formation of *choleglobin*. A reaction of hydrogen sulphide with myoglobin can result in *sulfmyoglobin*. Again, microbial

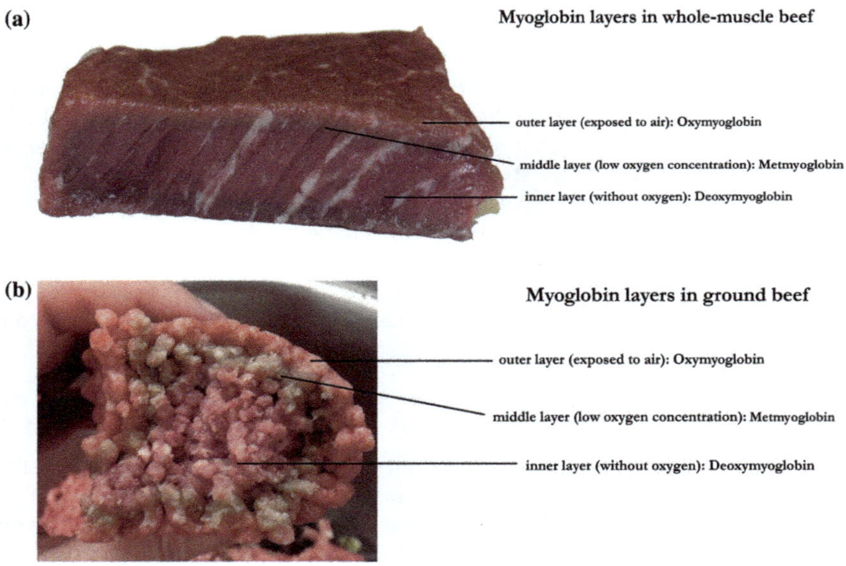


**Figure 2** (a) Myoglobin layers in whole-muscle beef and (b) myoglobin layers in ground beef.

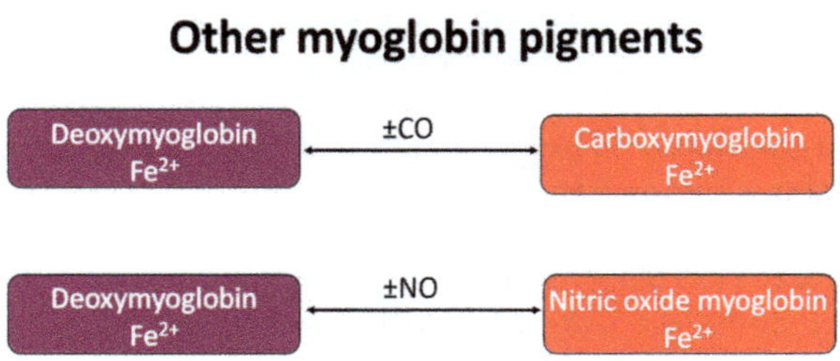


**Figure 3** Other myoglobin pigments important in fresh beef packaging.
Source: Adapted from AMSA (2012).

contamination may cause the presence of hydrogen sulphide, but it may also be the result of chemical residues used to treat carcasses or cuts. These forms of myoglobin are very undesirable and both result in a green colouration.

## 2.4 Effects of grinding

Ground beef is the most popular beef cut in retail sales, comprising over 50% of the pounds of beef sold at retail (NCBA, 2016). However, grinding of beef creates a new dynamic in packaging. During the grinding process, beef is subjected to cutting, mixing and grinding. These processes expose the entirety of the muscle to oxygen and begin the oxygenation of myoglobin throughout the cut. After grinding, the beef is repackaged and even when packaged aerobically, the centre of the ground beef package will lose access to oxygen. The respiratory enzymes will metabolize the oxygen and revert the myoglobin to metmyoglobin and eventually back to deoxymyoglobin (Kropf et al., 2014). The time taken for grinding and the extent to which the beef was exposed to oxygen creates an environment where oftentimes the entire mass of ground beef is oxygenated. This large oxygenation, and subsequent oxidation and reduction required for deoxygenation, depletes the ability of the beef to reduce metmyoglobin back to deoxymyoglobin over time. Ground beef will reach an undesirable, metmyoglobin condition much more quickly than intact muscle cuts like steaks and roasts (Kropf et al., 2014). Therefore, ground beef may require different, more protective packaging systems than intact muscle cuts.

# 3 Traditional aerobic packaging

After the development of plastic films in the 1920s, beef was commonly wrapped in clear plastic packaging for display in refrigerated cases in markets and grocery stores (Fig. 4). This packaging system was relatively inexpensive and allowed consumers to view beef cuts prior to purchasing. The plastic films were air-permeable, allowing oxygen to create

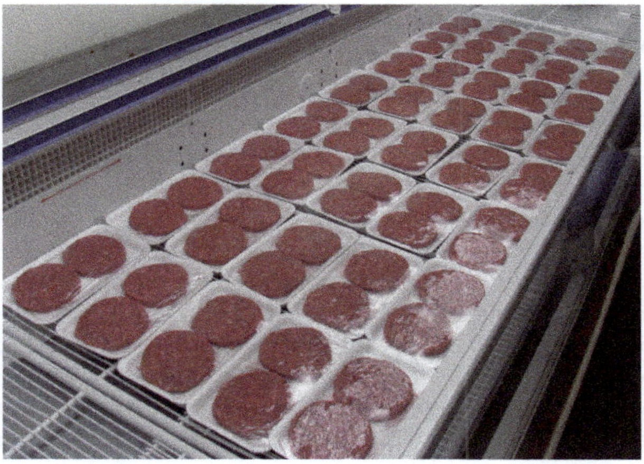

**Figure 4** Ground beef patties wrapped in overwrap packaging.

the bright red colour, and provided a moisture barrier, slowing the loss of moisture in the cuts (McMillin, 2008). Now that five generations of Americans have purchased beef in these types of packages, consumers expect beef to be displayed for easy viewing and to have a bright cherry-red colour they associate with freshness.

## 3.1 Trays

Overwrap packages must contain a sturdy, lightweight tray to allow for stacking and capture exudate from the beef cut. Although paper pulp trays were once used, they absorbed moisture from the beef cuts and had issues with leaking; therefore, they have been replaced by expanded polystyrene, oriented polypropylene or clear polystyrene (Kropf et al., 2014). Some companies have developed renewable foam trays from plants rather than petroleum products. As the technology to develop these plant-based, compostable packages advances, their use will increase in the industry, but today they are still cost-prohibitive in conventional markets.

## 3.2 Aerobic films

Films used in the aerobic packaging system should be oxygen permeable to allow the formation of oxymyoglobin, but should be impermeable to moisture and flexible to allow for contour to the varied shapes of beef cuts. They should be completely transparent and be resistant to condensation. Successful films are also sealed with low heat to allow for ease of in-store packaging.

Originally, cellophane coated with nitrocellulose was used in meat packaging, but it was replaced temporarily by pliofilm (Kropf et al., 2014). In the United States, the major overwrap film used is plasticized polyvinyl chloride (PVC). It stretches to about 0.017 mm in thickness and provides the oxygen permeability (5000–10 000 ml per $m^2$ per 24 h at 23°C) required to allow cuts to bloom without allowing moisture permeability. Some countries use low-density polyethylene (0.0254 mm thick) modified with vinyl acetate to improve oxygen permeability. This method lacks the clarity and gloss found with PVC and requires treatment to overcome condensation on the surface (Kropf et al., 2014).

## 3.3 Disadvantages of overwrap packaging

The goal of several packaging systems has been to maintain consumers expectation of a bright, cherry-red fresh beef colour; however, in the traditional, plastic overwrap, aerobic packaging system, fresh beef will only maintain the desired bright red colour for 5 to 7 days for intact cuts and 3 to 5 days for ground beef (McMillin, 2008). The activation of oxidation with aerobic packaging is detrimental not only to the visual aspects of the beef, but also beef flavour as the oxidation of the pigment will lead to oxidation of lipids in the cuts. Lipid oxidation will cause rancid flavour development and create an undesirable eating experience for consumers.

The aerobic packaging of PVC overwrap creates optimal conditions for the growth of aerobic spoilage bacteria such as *Pseudomonas*, *Acinetobacter* and *Moraxella* species (Montville and Matthews, 2005). Although these bacteria are not dangerous to humans, they metabolize glucose and amino acids within the meat creating putrid odours and green colouration associated with spoilage and unwholesomeness. Furthermore, overwrap packaging does not lend itself well to freezing without re-packaging.

Overwrapped packages are susceptible to freezer-burn because oxygen is readily available for oxidation.

# 4 Vacuum packaging

A simple method to increase the shelf life of fresh meats would be to use vacuum packaging. Removing the oxygen and the ensuing oxidation process would greatly slow the development of the brown colouration, and the anaerobic environment would inhibit the growth of spoilage bacteria. Although some specialty products, such as grass-fed or organic ground beef, are offered in vacuum packages, retailers have not embraced this packaging method for fresh beef because of the fear that consumers will reject the purplish-red colouring of vacuum-packaged beef (Fig. 5). Furthermore, purge is often visible within vacuum package beef cuts, which would be another detriment to its acceptance by consumers (Lagerstedt et al., 2011).

The purplish-red colouring of fresh beef packaged in vacuum packaging may be detrimental to its use in fresh retail, but the anoxic environment created with a vacuum bag may be ideal for other types of beef products. Whole-muscle cuts are commonly shipped, stored and aged in vacuum packages. Stores and brands are offering more meal-kit options that include beef, spices and vegetables packaged together for easy-to-prepare meals. These products typically have whole-muscle cuts packed in vacuum to lengthen the shelf life, and consumers are unable to see the meat through the kit packaging. Fully cooked, processed meats are commonly stored and marketed under vacuum. When colour is fixed as in cured meats or cooked products, oxygen only plays a detrimental role in storage, promoting oxidation and off-flavour development; so, thick, vacuum packages are warranted for optimal storage.

Chub packaging is commonplace for ground beef sold at retail in the United States and is essentially a form of vacuum packaging, although the vacuum is not perfect at the ends of the packages. Consumers are either unable to view the deoxymyoglobin pigment due

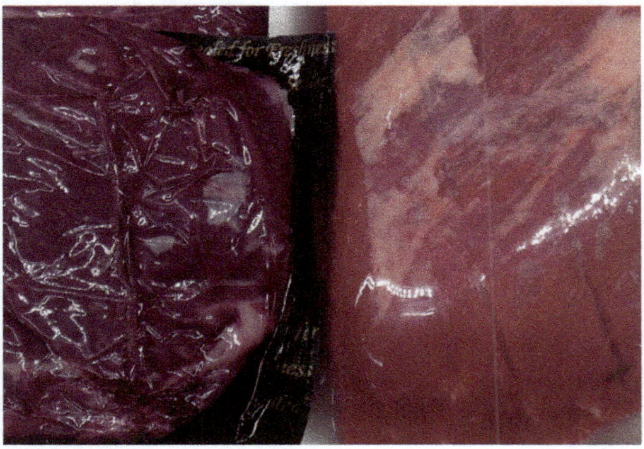

**Figure 5** Beef in vacuum packaging (left) and in overwrap packaging (right).

to the opaque bags or, in the case of transparent packaging, are not deterred by it for their ground beef. Nevertheless, many stores offer ground beef packaged in a variety of ways including vacuum, chub packages and overwrap.

## 4.1 Vacuum films

Unlike aerobic films, those used for vacuum packaging must be barriers to oxygen as well as moisture. A typical specification for vacuum bags for packaging large, sub-primal cuts in the United States is an oxygen permeability not exceeding 30 ml per 24 hour per $m^2$ (Lawrence and Kropf, 2014). Thicker, more dense films will be less permeable to oxygen, and oxygen barrier properties increase further with an increase in cross linkages and crystallinity of the polymers (Kropf et al., 2014b). Colour and shelf life will be most stable with greater barrier films, but with less permeability comes greater cost, and processors must balance maximum oxygen permeability with other factors that affect cost. Films must be flexible to contour around the shape of the product and costly, bone-guard material may be necessary to protect against puncture in bone-in cuts. Large companies that must consider packaging beef for national brands, and other factors affect the choice of films, including flowability, ease of fitting on machines, printability, stability at a wide temperature range, shrinkability and seal strength. Visual properties of gloss, clarity, transparency and absence of condensation are also important factors, especially when vacuum packaging retail cuts (Kropf et al., 2014b).

## 4.2 Concerns with vacuum packaging

Although not popular for retail beef, vacuum packaging is commonly used for beef sold into food service. Primals and sub-primals are sold in vacuum packaging to purveyors to be further processed into single-serving cuts, re-vacuum packaged and shipped to restaurants. During the cutting and packaging process, retail cuts may be allowed to bloom before being packaged under vacuum again (Lawrence and Kropf, 2014). After blooming and vacuum packaging, the low concentration of oxygen from the oxymyoglobin will cause the formation of metmyoglobin and will need to be reduced back to deoxymyoglobin in the anoxic environment of the vacuum package. In the case of long-aged cuts (aged over 21 days), like rib and loin steaks bound for high-end steak houses, the reducing capacity of the muscle may be diminished, resulting in a brown, metmyoglobin appearance on the surface of the steaks in the individual vacuum package. When these steaks are opened and viewed at the restaurant, they may be perceived as spoiled due to the brown colour and rejected. To prevent this phenomena from occurring, beef cuts aged past 21 days should be cut and packaged as quickly as possible, preventing the cuts from blooming initially and requiring the oxidation/reduction cycle from taking place (Lee et al., 2008).

## 4.3 Vacuum skin packaging

Vacuum skin packaging is a new method of packaging beef that is gaining popularity in Europe and somewhat in the United States. Beef cuts are placed on trays and a heated film is shrunk tightly around the cuts and adheres to the beef and the trays under vacuum (Lagerstedt et al., 2011). Although the beef will remain in the purplish-red, deoxymyoglobin state (Fig. 6), the visible purge is reduced and consumers have reported a lack of off-odour compared to vacuum packaging when the packages are

**Figure 6** Vacuum skin packaging.

opened (Lagerstedt et al., 2011). Even though the initial colour is not the expected bright red, the stability of colour was extended in vacuum skin packages compared to high-oxygen modified atmosphere and aerobic packages (covered in a subsequent section of this chapter) (Suman et al., 2014). McMillin (2008) reported that vacuum skin packages may have a shelf life as long as 90 days for whole-muscle beef and up to 60 days for ground beef. Purge loss was also reduced in vacuum skin packaged beef compared to vacuum and MAP (Lagerstedt et al., 2011). Research has also shown that cuts stored in vacuum skin packaging, then allowed to bloom were superior to steaks stored in vacuum and high-oxygen modified atmosphere packaging (Suman et al., 2014). Lagerstedt et al. (2011) observed that steaks packaged in vacuum skin were thinner than those packaged in vacuum or MAP after 14 days of storage.

# 5 Centralized packaging

## 5.1 The boxed-beef concept

In 1967, a beef packing company, Iowa Beef Packers, changed the dynamic of meat cutting and packaging by introducing their boxed-beef concept to retail stores. Rather than purchasing whole beef carcasses or quarters to be fabricated in-store, meat market managers could buy vacuum-packaged primal or sub-primal cuts and fabricate those into retail items that were popular with customers in their market (Cole, 1986). This drastically changed the way beef and other meat commodities were processed and marketed. By the mid-1980s, 84% of all wholesale beef cuts processed in the United States were sold vacuum packaged (Cole, 1986). Retail beef products were then cut and repackaged in overwrap at retail stores.

Even with the boxed-beef marketing system, fresh meat was still one of the few major food products that were processed and packaged at the retail level. Retail stores realized the benefits of purchasing only the cuts they needed to stock their cases. Deli meats and poultry were the first meats to move to centralized packaging systems, but the issues with maintaining the bright red colour as consumers expected halted the expansion of centralized packaging to fresh red meats. In the mid-1980s, some retail chains were cutting and packaging retail cuts in centralized facilities and shipping them relatively short distances to stores (Cole, 1986). These stores saw several benefits of this system, including better quality control through more regulation of sanitation and temperature, more uniform products, less waste through better utilization of trimmings, more efficient use of space, the use of automated equipment that may not be applicable at a store level and better inventory control (Cole, 1986). These benefits were driven by consumer demands for food safety and wholesomeness, as well as the greater demand for product to be in display for longer days as stores increased their business hours. Almost 30 years later, the removal of the local butcher and the move to producing case-ready cuts in a centralized 'factory' is still of concern to some consumers, and some stores advertise their in-house butchers. However, according to the 2010 National Meat Case Study, 71% of ground beef and 31% of beef was marketed to consumers in a case-ready package (NMCS, 2010). Ground beef has led the movement to case-ready packages of the red meat options because of concerns of its greater food safety risk than whole-muscle cuts (Eilert, 2005), but currently, poultry is greater than 94% case-ready in retail (NMCS, 2010).

## 5.2 Mother bag packaging

Although other packaging methods like vacuum packaging would improve the shelf life and ease-of-freezing, consumers still desire the overwrap package. Therefore, meat processors have developed a system in which beef is cut and packaged in a traditional overwrap package at a centralized facility, then several packages are placed together in a larger bag (mother bag) that is flushed with a gas mixture. The gas mixtures used in retail are proprietary, but research has shown that a mixture of carbon monoxide (0.4%), high levels of carbon dioxide (~20%) and a balance of nitrogen in the mother bag will create the bright, cherry-red beef colour and inhibit the growth of spoilage and pathogenic bacteria (Kropf et al., 2014a). These systems have used oxygen scavengers in combination with high levels of carbon dioxide to create an anoxic environment for storage, then allow the beef to bloom when opened at the store. Higher than ambient levels of carbon dioxide have been shown to inhibit the growth of spoilage organisms on meat products (Cole, 1986). Although the display life of packages in this system are similar to conventional overwrap (2–7 days) once the mother bag has been opened, the total storage life of beef cuts while in the mother bags is increased to 10–14 days for whole-muscle cuts and 7–10 days for ground beef compared to store-packaged overwrap (McMillin, 2008).

# 6 Modified atmosphere packaging

As several stores began to see the benefits of centralized packaging systems, but still struggled with the relatively short shelf life of aerobically packed fresh meats, new technology in packaging was being developed to control the microenvironment in meat packages. This new modified-atmosphere packaging (MAP) system utilizes coloured or opaque, thermoformed base trays with transparent-lidding films stretched over the opening. Meat cuts are placed fully within the trays and a gas is flushed in the headspace of the package. These gases may have high oxygen levels, or low oxygen with higher than ambient levels of carbon dioxide or carbon monoxide to optimize the fresh beef colour.

## 6.1 High-oxygen MAP

In typical high-oxygen MAP packages, the meat is placed in the package with a 1:1 headspace ratio with a gas mixture of 80% oxygen and 20% carbon dioxide. This level of oxygen created a very desirable bright-red colour and a complete oxygenation of myoglobin throughout the muscle (Seyfert et al., 2004a), rather than the surface oxymyoglobin layer of approximately 0.5 cm seen in beef exposed to ambient levels of oxygen. High levels of carbon dioxide also inhibit the growth of spoilage microorganisms, accentuating the improvement in shelf life (Kropf and Mancini, 2014). McMillin (2008) reported that cuts packaged in MAP had a storage life of 12–16 days for whole muscle and 10–12 days for ground or minced meats and a retail display life of 7–16 days.

Nonetheless, Belcher (2006) stated that once placed under the ultraviolet lights commonly found in retail display, the display life of high-oxygen MAP packages was limited to only 2–4 days. The combination of high oxygen levels and UV lights accelerated oxidation and the formation of metmyoglobin. The oxidation-induced rancid flavour development is also of concern for beef cuts packaged using high-oxygen MAP, and this packaging method commonly requires the addition of antioxidants to beef to prevent off-flavours from developing (McMillin, 2008).

Furthermore, bone-in beef cuts packaged in high-oxygen MAP packages had discolouration of the bone marrow that had the potential to be discriminated against by consumers (Mancini et al., 2005). The bone discolouration was especially noted in lumbar vertebrae that would be found in high-value T-bone steaks. The discolouration was not observed with low oxygen or carbon monoxide packaging and was attributed to the oxidation of haemoglobin in the marrow (Mancini et al., 2005).

As previously mentioned, steaks and patties packaged in high-oxygen MAP are completely oxygenated throughout the cut, and Seyfert et al. (2004a) found that the oxymyoglobin pigment of steaks packaged in 80% $O_2$ MAP packages denatured with cooking to a greater extent than controls, creating well-done, brown internal appearance when cooked to a medium degree of doneness. Furthermore, 100% of ground beef patties packaged in high-oxygen MAP packages browned prematurely when cooked to a medium degree of doneness (Seyfert et al., 2004b). Premature browning in whole-muscle cuts is not a food safety risk unless the cuts were blade tenderized or enhanced, translocating surface bacteria to the centre of the cuts (Seyfert et al., 2004b). In ground beef patties, however, premature browning is an especially great food safety risk because, in spite of years of education, consumers still use cooked colour as an indicator of doneness in ground beef, and patties browning prematurely could result in consuming ground beef that may not been properly cooked.

## 6.2 Low-oxygen MAP

Some MAP methods have been developed with no oxygen in the immediate package for meat storage (McMillin, 2008). These packaging systems may use a high level of carbon dioxide to inhibit microbial growth with a balance of nitrogen. Whole-muscle cuts may be stored in low-oxygen MAP for up to 60 days and ground beef up to 40 days (McMillin, 2008). Because metmyoglobin will form at low concentrations of oxygen, oxygen scavengers or scavenging technology embedded within the package may be required in low-oxygen MAP and would add to the cost of packaging (McMillin, 2008).

In some low-oxygen packaging systems, beef is packaged and stored in the purple, deoxymyoglobin state and a peelable lid or barrier film is removed to expose the product to ambient air and allow the beef to bloom (Belcher, 2006). This method increases the distribution life by 5–10 days over beef packaged aerobically, and the headspace required for each package is reduced by half (Belcher, 2006), which decreased space required in storage and transport. Low-oxygen MAP combines the ease of centralized packaging with the product quality of an aerobic packaging, but adds to the labour required at the store. Some research has shown that the post-packaging bloom in low-oxygen MAP is inconsistent (McMillin, 2008).

## 6.3 Carbon monoxide MAP

Several studies have examined the effects of exposing beef to carbon monoxide (CO) to form the very stable carboxymyoglobin for retail display (Kropf, 1980). The visual spectra of carboxymyoglobin is almost identical to oxymyoglobin and the carbon monoxide-induced pigment is much more stable. Some MAP systems have incorporated 0.4% carbon monoxide with 20% carbon dioxide and a balance of nitrogen (Kropf and Mancini, 2014). The whole-muscle shelf life of beef packaged under carbon monoxide MAP can be up to 35 days and for ground beef up to 28 days (McMillin, 2008). As with low oxygen MAP, the headspace requirement for carbon monoxide MAP is reduced compared to the greater headspace requirement for high-oxygen MAP. Very low oxygen levels are required for this packaging method to be successful; therefore, oxygen scavengers are required, increasing the packaging cost. Although it has not been shown to be of concern in beef cuts, lighter pigmented meats of pork and poultry have shown a pinking condition in cooked colour when packaged in carbon monoxide MAP (Mancini et al., 2005; McMillin, 2008).

Some researchers have expressed concern that beef packaged with carbon monoxide would remain red even under extended display or temperature abuse conditions and mask spoilage that may occur (Belcher, 2006). Furthermore, the use of a dangerous gas, regardless of the actual risk, is a perceived concern of consumer advocacy groups. These concerns have led some cities to take legislative action against carbon monoxide packaging.

# 7 Active packaging

'Active packaging' is a commonly used term in the industry and refers to adding certain components to a package with the aim of extending shelf life or maintaining food safety (Kerry et al., 2006). Active packaging provides more to the beef product than simply a barrier to the external environment, but helps maintain the colour and freshness that consumers desire. As the demand for fewer additives and preservatives in food increases, the options for ingredients to maintain product quality is decreased (Lee et al., 2010). Adding components to an active packaging system that inhibit oxidation and improve shelf life could be a useful alternative to ingredients that were removed from beef products to 'clean-up the label'. These packaging systems may incorporate oxygen scavenging or carbon dioxide emitting technology, films with antioxidant or antimicrobial properties or colour-enhancing nitrite embedded in the film.

## 7.1 Oxygen scavengers

It has already been stated that some low-oxygen packaging systems may require the use of oxygen scavengers. Even when oxygen is flushed out of a package or removed through vacuum, a small amount of oxygen may permeate back through films or remain trapped in the muscle itself (Kerry et al., 2006). The low-oxygen environment will facilitate the release of oxygen from the muscle back into the headspace of the package. The majority of oxygen scavengers are iron based and are activated by water (Lee et al., 2015). Iron oxygen scavenging sachets are commonly seen in food packages labelled DO NOT EAT. Glucose oxidase and alcohol oxidase are enzymatic oxygen scavengers, but are not widely used due to their cost. Oxygen scavengers may be incorporated into the packaging itself in the form of oxygen scavenging films, flat packets, labels or cards, although these technologies do not offer the speed and efficiency of the sachets (Vermeiren et al., 1999). Some oxygen-scavenging technologies embedded into multi-layer films are activated by UV light in the display case (Kerry et al., 2006). Although they are expensive, thinner films with oxygen scavenging technology may be less expensive or a better fit for certain products than a thicker film with a greater oxygen barrier (Vermeiren et al., 1999).

## 7.2 Carbon dioxide emitters

One challenge with anoxic packaging is the vacuum created when carbon dioxide is flushed into a package. Meat products will absorb the carbon dioxide and create a vacuum that may partially collapse the package (Kerry et al., 2006). This vacuum may lead to a loss of moisture and increased drip loss (Holck et al., 2014). Vermeiren et al. (1999) suggests that carbon dioxide emitting technology can be used to release carbon dioxide into the package to alleviate the collapse. Moisture absorbent, carbon dioxide-emitting pads in the package are comprised of ferrous carbonate or a sodium bicarbonate and ascorbic acid mixture and are activated by the moisture lost from the meat product (Kerry et al., 2006; Holck et al., 2014). Although studies of the use of this technology in beef products are limited, Holck et al. (2014) reported that chicken fillets packaged with pure carbon dioxide and a carbon dioxide emitter had increased shelf life without the package collapse and drip loss issues seen in packages without the carbon dioxide emitter. It is possible that carbon dioxide emitters could be used in beef packaging in a mother bag system where long-term storage in an anoxic environment is followed by retail display under aerobic conditions.

## 7.3 Antioxidant packaging

Antioxidants are commonly used in packaging materials to preserve the material itself, but scientists are incorporating antioxidants for the purpose of controlling oxidation in the packaged product. Inhibiting the formation of the oxidized metmyoglobin and the undesirable, brown beef colour that is associated with it is a main goal of this type of active packaging system. Furthermore, incorporating antioxidants into the packaging allows the active agents to be released over time to extend their action in the beef (Barbosa-Pereira et al., 2014).

Several studies have shown that synthetic antioxidants (BHA, BHT and propyl gallate) added to packaging are successful at inhibiting oxidation in beef (Barbosa-Pereira et al., 2014; Moore et al., 2003). Nevertheless, negative connotations associated with these synthetic antioxidants have led to the incorporation of naturally derived antioxidants from plant extracts and essential oils of rosemary, oregano, tea and other natural sources. Rosemary extract has been shown to be effective when coated on film (Barbosa-Pereira et al., 2014) and when immobilized within film (Nerín et al., 2006) in beef, and when coated on film in high-pressure-processed chicken patties (Bolumar et al., 2011). Camo et al. (2011) packaged beefsteaks in high-oxygen MAP packages with active film containing oregano extract, and found improved display life with at least 1% oregano. However, 4% oregano created an unacceptable aroma in the beef; thus, the optimum range was determined to be 1–2% oregano extract on the package. Eugenol, a phenolic compound found in clove essential oil, was incorporated into corn-zein-laminated linear low-density polyethylene bags that were used to vacuum-package ground beef patties. This active-packaging system inhibited lipid oxidation and improved colour stability compared to vacuum-packaged controls (Park et al., 2012). An extract containing natural antioxidants obtained from the cleaning process of brewing beer has been tested as a potential antioxidant in active packaging and was similar to propyl gallate in reducing lipid oxidation (Barbosa-Pereira et al., 2014).

## 7.4 Antimicrobial packaging

Some scientists and packaging companies have explored the option of incorporating antimicrobial agents directly in or on the packaging material. To increase the shelf life and safety of beef, antimicrobial packages increase the lag phase and decrease the growth phase of spoilage and pathogenic microorganisms in the product. Some of these antimicrobial agents are effective by migrating into the meat product, while others may not need to migrate into the food to be effective (Kerry et al., 2006).

Although it is unclear what antimicrobial technologies are being used in commercial applications in fresh beef packaging, scientists have examined several options. Silver ions incorporated into plastics for packaging of both fresh and processed meats have reduced growth of pathogens such as *Salmonella*, *Escherichia coli*, and *Campylobacter* (Realini and Marcos, 2014). Nisin, a compound produced by *Lactococcus lactis*, inhibits gram-positive bacteria, such as *Clostridium* spp., *Enterococcus*, *Listeria monocytogenes* and lactic acid bacteria (Lee et al., 2010). Natrajan and Sheldon (2000) found that coating packaging films with nisin reduced the growth of *Salmonella typhimurium* in chicken. *Listeria monocytogenes* growth was inhibited in inoculated fresh turkey, processed ham and fresh beef by bags coated with pediocin, a similar bacteriocin derived from *Pediococcus* (Ming et al., 1997).

As with antioxidants, some scientists have explored natural antimicrobial options to appease consumer demands. Ha et al. (2001) tested grapefruit seed extract incorporated into co-extruded multi-layered polyethylene film on ground beef and found reduced growth of aerobic and coliform bacteria with no effect on beef colour. Filter papers soaked in oregano essential oil were placed into MAP packaging but not touching the beef and extended the shelf life of beef samples by inhibiting the growth of spoilage organisms (Skandamis and Nychas, 2002).

## 7.5 Nitrite-embedded film

Vacuum packaging or vacuum skin packaging allows for the longest shelf life and minimizes lipid oxidation in beef, but the purplish-red colour of vacuum-packaged beef remains a looming problem in fresh beef because consumers discriminate against it (AMSA, 2012). When fresh beef is exposed to nitrite, the nitrite is reduced to nitric oxide and combines with the haeme iron in myoglobin to form nitric oxide myoglobin, which is a bright pinkish-red colour (Suman et al., 2014). Recent studies have explored the addition of nitrite to vacuum films to exploit this appealing colour formation, and utilize the shelf life longevity found with vacuum packaging. These films contain very low levels of nitrite (3–10 ppm for beef packaging vs. 150–200 ppm used for curing), and can be used in vacuum bags, vacuum skin packaging or in transparent ground beef chubs (Siegel, 2011). Claus and Du (2013) reported extended shelf life and improved redness in both fresh and frozen beef samples packaged with nitrite-embedded films, but also found very small, yet significant, levels of residual nitrite in the surface of the beef packaged in nitrite-embedded film. This technology combines the extended shelf life of vacuum packaging with the bright-red beef colour of aerobic packaging, but the cost effectiveness and consumer acceptability of residual nitrite in the final product may inhibit widespread use.

# 8 Intelligent packaging

Intelligent packaging or smart packaging senses the properties of the enclosed food or its environment and alerts manufacturers, retailers or consumers to the state of these properties. Intelligent packages will provide information about not only the current state of the packaged product, but may also indicate the history of the storage conditions to which the package has been exposed (Realini and Marcos, 2014). Many concepts of intelligent packaging include the use of sensors to communicate with retailers or consumers. Sensors are comprised of two basic functional units: the receptor transforms chemical or physical information into an energy form and the transducer transforms that energy into a readable signal (Kerry et al., 2006).

Time–temperature indicator labels may be adhered to packaging and activated by temperature changes, UV light or manually by processors or retailers (Realini and Marcos, 2014). These labels indicate freshness based on elapsed time and are accelerated at greater temperatures. Consumers may evaluate a change in colour on the label to determine freshness rather than merely a use-by date (Realini and Marcos, 2014). Some integrity indicators simply record and indicate how much time has passed since a product has been opened. Other temperature indicators reveal when a packaged product has been exposed to temperature abuse regardless of time. One marketed labelling system causes a change in a bar code assessed by retailers who may then sell, discount or reject incoming product based on the history of the package or the case.

In low-oxygen MAP packaging, oxygen indicators may be used in the package to monitor oxygen leakage into the package, but some are not really reliable because the indicators are reversible when oxygen levels have returned to the desired state. Oxygen leaked into a package may be absorbed and metabolized by the beef or metabolized by microbial populations, leading to a reversion of the low-oxygen state. Even more, intelligent packaging would monitor the actual freshness of the product as indicated by chemical changes in the meat, concentrations of metabolites or production of volatiles by microbial growth (Realini and Marcos, 2014), but these products are either still in commercial development or not yet cost-effective.

# 9 Additional issues: moisture control and reducing packaging waste

## 9.1 Moisture control

Cutting and packaging fresh beef ultimately results in moisture loss and accumulation of pigmented water in the package (purge), and consumers will discriminate against those packages with excess purge. Therefore, meat retailers have used absorbent pads in fresh beef packages since the inception of the packaged retail beef marketing system. The pads consist of a super-absorbent polymer between two layers of non-woven or micro-porous polymer and are commonplace in all fresh-meat packaging to absorb purge loss (Kerry et al., 2006). Although they were first used in aerobic packages with foam trays and plastic overwrap, MAP and vacuum skin packaging also contain absorbent pads. They may be loosely placed under the product or adhered to the tray with an approved glue.

## 9.2 Concerns with waste

In 2005, about two-thirds of the total packaging waste by volume was from food waste (McMillin, 2008), and concerns with waste and the environmental concerns associated with food are becoming a top priority with some consumers (see Environmental Impact Chapter). As indicated earlier, plant-based packaging materials are being developed for trays used for overwrap, vacuum skin and MAP systems, but cost limits their use beyond specialized, high-end beef products (Kropf et al., 2014).

Biodegradable polymers are another option for reducing the carbon footprint of beef packaging. The original plastics were actually derived from cellulose in 1862 (Marsh and Bugusu, 2007), but were later replaced by polymers produced from petroleum products. Biodegradable polymers may be made from starches and cellulose (cellophane) and are commonly produced from the action of bacteria on starches. Chitin is used to produce chitosan and is derived from crustacean and insect exoskeletons. Edible films, made from proteins sources from corn, milk, collagen or gelatin, may be future biopolymer options for beef packaging (Marsh and Bugusu, 2007).

# 10 Conclusions

Beef quality is dependent on many factors from calf conception to beef harvest and processing, but without proper packaging, all efforts of beef producers, feeders and processors are for naught. The dynamics of myoglobin chemistry in fresh beef and consumers requirement that beef have a bright red colour at purchase, make packaging of special importance in beef profitability and marketing. Successful packaging systems do much more than protect beef from the environment. They are essential in assuring the beef is exactly what consumers desire in regard to visual appraisal, food safety and wholesomeness, as well as ultimate palatability. Packages play a vital role in lengthening the shelf life of fresh and processed beef, thereby lowering cost by minimizing retail losses and food waste. The future of beef packaging is moving beyond mere product protection, into enhancement of shelf life and food safety with packaging technology that actively controls the beef product environment and can communicate with consumers based on the history of the package, further reducing the carbon foot print of retail beef by decreasing waste of both beef and packaging.

# 11 Where to look for further information

*The Encyclopedia of Meat Sciences, Volume 3*, Elsevier Ltd.
*Packaging Technology and Science*, John Wiley and Sons
*Meat Science*, Elsevier
www.meatscience.org, *American Meat Science Association*

# 12 References

AMSA (2012). *Meat Color Measurement Guidelines*. Champaign, IL, USA: American Meat Science Association.

Barbosa-Pereira, L., Aurrekeotxea, G., Angulo, I., Paseiro-Losada, P. and Cruz, J. M. (2014). Development of new active packaging films coated with natural phenolic compounds to improve the oxidative stability of beef. *Meat Science*. 97, 249–54.

Belcher, J. N. (2006). Industrial packaging developments for the global meat market. *Meat Science*. 74, 143–8.

Bolumar, T., Anderden, M. L. and Orlien, V. (2011). Antioxidant active packaging for chicken meat processed by high pressure treatment. *Food Chemistry*. 129, 1406–12.

Camo, J., Lorés, A., Djenane, D., Beltrán, J. A. and Roncalés, P. (2011). Display life of beef packaged with an antioxidant active film as a function of the concentration of oregano extract. *Meat Science*. 88, 174–8.

Claus, J. R. and Du, C. (2013). Nitrite-embedded packaging film effects on fresh and frozen beef color development and stability as influenced by meat age and muscle type. *Meat Science*. 95, 526–35.

Cole, A. B. (1986). Retail packaging systems for fresh red meat cuts. *Proceedings of the Reciprocal Meats Conference*. 39, 106–11.

Eilert, S. J. (2005). New packaging technologies for the 21st century. *Meat Science*. 71, 122–7.

Ha, J.-U., Kim, Y.-M. and Lee, D.-S. (2001). Multilayered antimicrobial polyethylene films applied to the packaging of ground beef. *Packaging Technology and Science*. 14(2), 55–62.

Holck, A. L., Pettersen, M. K., Moen, M. H. and Sørheim, O. (2014). Prolonged shelf life and reduced drip loss of chicken filets by the use of carbon dioxide emitters and modified atmosphere packaging. *Journal of Food Protection*. 77(7), 1133–41.

Kerry, J. P., O'Grady, M. N. and Hogan, S. A. (2006). Past, current and potential utilization of active and intelligent packaging systems for meat and muscle-based products: A review. *Meat Science*. 74, 113–30.

Kropf, D. H. (1980). Effects of retail display conditions on meat color. *Proceedings of the Reciprocal Meats Conference*. 33, 15–32.

Kropf, D. H. and Mancini, R. C. (2014). Packaging: Modified and Controlled Atmosphere. In *The Encyclopedia of Meat Sciences, Volume 3*. Elsevier Ltd.

Kropf, D. H., Yancey, J. W. S. and Yancey II, E. J. (2014a). Packaging: Overwrapping. In *The Encyclopedia of Meat Sciences, Volume 3*. Elsevier Ltd.
Kropf, D. H., Yancey, J. W. S. and Yancey II, E. J. (2014b). Packaging: Technology and Films. In *The Encyclopedia of Meat Sciences, Volume 3*. Elsevier Ltd.
Lagerstedt, Å., Ahnström, M. L. and Lundström, K. (2011). Vacuum skin pack of beef – A consumer friendly alternative. *Meat Science*. 88, 391–6.
Lawrence, T. E. and Kropf, D. K. (2014). Packaging: Vacuum. In *The Encyclopedia of Meat Sciences, Volume 3*. Elsevier Ltd.
Lee, K. T. (2010). Quality and safety aspects of meat products as affected by various physical manipulations of packaging materials. *Meat Science*. 86, 138–50.
Lee, M. S. (2008). The influence of wet-ageing and vacuum-package storage on bloom development of beef Gluteus Medius, Longissimus Thoracis, Semimembranosus, and Triceps Brachii. MS thesis. University of Arkansas Press.
Lee, M. S., Apple, J. K., Yancey, J. W. S., Sawyer, J. T. and Johnson, Z. B. (2008). Influence of wet-aging on bloom development in the *longissimus thoracis*. *Meat Science*. 80, 703–7.
Lee, S. Y., Lee, S. J., Choi, D. S. and Hur, S. J. (2015). Current topics in active and intelligent food packaging for preservation of fresh foods. *Journal of the Science of Food and Agriculture*. 95, 2799–810.
MacDougall, D. B. (1982). Changes in color and opacity of meat. *Food Chemistry*. 9, 75–88.
Mancini, R. A. and Hunt, M. C. (2005). Current research in meat color. *Meat Science*. 71, 100–21.
Mancini, R. A., Hunt, M. C., Hachmeister, K. A., Kropf, D. H. and Johnson, D. E. (2005). Exclusion of oxygen from modified atmosphere packages limits beef rib and lumbar vertebrae marrow discoloration during display and storage. *Meat Science*. 69, 493–500.
Marsh, K. and Bugusu, B. (2007). Food packaging – Roles, materials, and environmental issues. *Journal of Food Science*. 72(3), R39–55.
McMillin, K. W. (2008). Where is MAP going? A review and future potential of modified atmosphere packaging for meat. *Meat Science*. 80, 43–65.
Ming, X., Weber, G. H., Ayres, J. W. and Sandine, W. E. (1997). Bacteriocins applied to food packaging materials to inhibit *Listeria monocytogenes* on meats. *Journal of Food Science*. 62(2), 413–15.
Montville, T. J. and Matthews, K. R. (2008). *Food Microbiology: An Introduction, Volume 2*, American Society for Microbiology Press.
Moore, M. E., Han, I. Y., Acton, J. C., Ogale, A. A., Barmore, C. R. and Dawson, P. L. (2003). Effects of antioxidants in polyethylene film on fresh beef color. *Journal of Food Science*. 68, 99–104.
NCBA (2016). Retail marketing. *National Cattlemen's Beef Association*. Available at www.beefretail.org. Accessed on 13 January 2016.
Nerín, C., Tovar, L., Djenane, D., Camo, J., Salafranca, J., Beltrán, J. A. and Roncalés, P. (2006). Stabilization of beef meat by a new active packaging containing natural antioxidants. *Journal of Agricultural and Food Chemistry*. 54, 7840–6.
NMCS (2010). A snapshot of today's retail meat case. 2010 National meat case study executive summary. Available at http://www.beefretail.org/CMDocs/BeefRetail/research/2010NationalMeatCaseStudy.pdf. Accessed on 13 January 2016.
Park, H.-Y., Kim, S.-J., Kim, K. M., You, Y.-S. and Han J. (2012). Development of Antioxidant packaging material be applying corn-zein to LLDPE film in combination with phenolic compounds. *Journal of Food Science*. 77, E273–9.
Realini, C. E. and Marcos, B. (2014). Active and intelligent packaging systems for a modern society. *Meat Science*. 98, 404–19.
Seyfert, M., Hunt, M. C., Mancini, R. A., Kropf, D. A. and Stroda, S. L. (2004a). Internal premature browning in cooked steaks from enhances beef round muscles packaged in high-oxygen and ultra-low oxygen modified atmospheres. *Journal of Food Science*. 69(2), FCT142–6.
Seyfert, M., Mancini, R. A. and Hunt, M. C. (2004b). Internal premature browning in cooked ground beef patties from high-oxygen modified-atmosphere packaging. *Journal of Food Science*. 69(9), C721–5.

Siegel, D. (2011). An update on packaging fresh meat with nitrite containing film. *Proceedings of the Reciprocal Meats Conference.* Available from http://www.meatscience.org/publications-resources/rmc-proceedings/2011. Accessed on 12 January 2016.

Skandamis, P. N. and Nychas, G.-J. E. (2002). Preservation of fresh meat with active and modified atmosphere packaging conditions. *International Journal of Food Microbiology.* 79, 35–45.

Suman, S. P., Hunt, M. C., Nair, M. N. and Rentfrow, G. (2014). Improving beef color stability: Practical strategies and underlying mechanisms. *Meat Science.* 98, 490–504.

Vermeiren, L., Devlieghere, F., van Beest, M., de Kruif, N. and Debevere, J. (1999). Developments in the active packaging of foods. *Trends in Food Science and Technology.* 10, 77–86.

Chapter 15

# Measuring and assessing beef quality and sensory traits for retailers and consumers

*Derek A. Griffing and Christy L. Bratcher, Auburn University, USA*



## 1 Introduction

The 'experience qualities' of food products are those that can only be verified after purchase and consumption (Ford, 1988). Therefore, verifying experience qualities in products prior to consumption proves to be a challenging obstacle that the beef industry faces. The three primary factors associated with identifying the experience quality of beef are tenderness, juiciness and flavour (Neely, 1988), otherwise known collectively as 'palatability'. In the 2001 National Beef Quality Audit, five different sectors of the industry were asked to define eating satisfaction (Igo, 2013). The most common terms associated were as follows:

1. Tenderness (63.8% of government and allied industries, 44.1% of feeders, 65.4% of packers, 52.1% of food service operators and 66.7% of retailers);
2. Flavour (57.5% of government and allied industries, 53.8% of packers, 62.5% of food service operators and 70% of retailers).

Three of the five segments identified customer satisfaction as a key factor (34.0% of government and allied industries, 25.4% of feeders and 29.2% of food service operators). Identifying which products meet these expectations is often difficult, as most measures

http://dx.doi.org/10.19103/AS.2016.0009.15

of the quality of beef are destructive. The beef industry has often relied on indirect measurement for assessment of quality. One such measure of prediction is through the use of sensory panels. Measuring consumers' reactions to meat palatability is difficult because consumers' acceptance often is influenced by additional factors, such as price and nutrition (Savell, 1989, 1992; Chambers, 1993). Too many factors occur in consumers' households when preparing meat, such as cooking method, degree of doneness, seasonings added and person-to-person differences in preferences and thresholds for tenderness, juiciness and flavour. Therefore, it is difficult to predict from objective data, such as trained sensory panels or shear force, how consumers will rate meat at home (Lorenzen, 2003). It is important, however, to find the best way possible in assessing the quality of beef at retailer and purveyor levels as defined by an enjoyable eating experience for consumers.

## 2 Sensory evaluation

Properly structured sensory evaluation comprises a set of techniques for accurate measurement of human responses to foods and minimizes potentially biasing effects (Lawless, 1999). A definition that has been adopted by many is that sensory evaluation is a scientific discipline used to evoke, measure, analyse and interpret reactions to those characteristics of food and materials as they are perceived through the senses of sight, smell, taste, touch and hearing (Anonymous, 1975). This definition encompasses all of the senses, not just taste (Stone, 1993). Meilgaard et al. (2007) reported that attributes of a food item are typically perceived in the order of (1) appearance; (2) odour/aroma/fragrance; (3) consistency and texture; (4) flavour, including aromatics and chemical feeling; and (5) taste. Without training, sensory panellists will receive a jumble of near-simultaneous sensory impressions and will not be able to provide an independent evaluation of each (Meilgaard, 2007). Because instrumental assessments provide values that miss the important perceptual process of the interpretation of sensory experiences by the human brain prior to responding (Lawless, 1999), it is important to collect sensory data to assess the value of human responses. Sensory testing goes as far back as there were humans to evaluate the goodness or badness of anything that could be consumed (Meilgaard, 2007). Sensory testing can establish the worth or acceptability of a commodity as was seen by the rise of the trading industry when buyers would test a small sample of a load and sellers began to set their prices based on an assessment of the quality of the goods (Meilgaard, 2007).

There are various types of sensory evaluation. Most commonly, sensory panels consist of either a consumer panel or a trained panel to collect data. With consumer sensory panels, the number of necessary responses is pretty high; generally, a sample size of around 400 responses is required. The number of respondents has varied among different research protocols, from around 540 in the National Consumer Retail Beef Study (Savell, 1989) to as many as 2400 consumer responses in the Beef Consumer Satisfaction Surveys (Lorenzen, 1999, 2003; Neely, 1998, 1999; Savell, 1999). By utilizing a trained sensory panel, the number of responses needed is dramatically decreased to around eight responses; however, the amount of time and money it takes to train panellists increases rather dramatically. Panellists have to undergo a series of screening questions to identify if they are, in fact, appropriate individuals to serve on the panel and as many as four times the needed number of panellists may have to be screened before the panel is formed

(AMSA, 2015). During the pre-screening process, the ability to make judgements and follow directions should be determined because panellists who cannot do these tasks will not be successful (AMSA, 2015). Following pre-screening, a screening process will take place to determine the normal sensory acuity, interest, ability to discriminate and reproduce results, in addition to appropriate behaviour, cooperation, motivation and promptness (AMSA, 2015). After the members of the panel have been identified, training can take 3 weeks or longer. Once training is completed and tests have begun, performance must be monitored and the panel must be maintained, much like off-season workouts for sporting teams. For a trained panel, the goal is to have a panellist perform as accurately and precisely as a well-maintained piece of laboratory equipment. This takes time and money to assure that it is done properly.

In addition to the fact that samples must be destroyed in order for a sample to be tested, the difficulty in sensory evaluation arises from pitfalls in the variability of humans as test instruments. Any sensory test can be affected, depending on the extrinsic and intrinsic factors that surround a particular test instrument (panellist) on any given day. For this reason, panellists should be selected for sensitivity, trained and tested both before and during a sensory panel to maintain the acuity of each panellist. Because of the variability within and among every person, the panellist should not be interchanged during a given project. It is very important that the panellists understand the task at hand and the scales that have been assigned for use as a measurement tool. Correlation with others is an important task. Attitudes and illness as well as the basic human nature of perceptions will have an effect on perception. On any given day, if a panellist has had a negative encounter with a co-worker or a personal relationship preceding a scheduled sensory panel, they may have a different opinion of the samples presented than if they had just received praise from their boss or flowers from a loved one. In the same manner, if a panellist had an illness, such as a respiratory infection or a headache, they may not be fully focused on performing the task at hand or, more importantly, they may not be physically able to taste or smell the necessary components of a sample to fully be able to evaluate the necessary components.

Meilgaard et al. (2007) presented the 'Chain of Sensory Perception' as a three-step process: (1) the stimulus hits the sense organ and is converted to a nerve signal that travels to the brain; (2) with previous experiences in memory, the brain then interprets, organizes, and integrates the incoming sensations into perceptions and (3) a response is formulated based on the subject's perceptions. Therefore, a difference in response could be due to interference in any of these three steps. There could be an underlying cause, such as a respiratory infection that interferes with the nerve signal to the brain. There could be a lack of sufficient information in the memory of a particular odour or taste, or there could be an inability to convert the sensation into words or numbers. Training can be an attempt to shape the mental process, so that panellists move towards showing the same response to a given stimulus with a high repeatability (Meilgaard, 2007).

While many companies have had success with maintaining a reliable trained panel and have also used the occasional consumer panel, these techniques remain expensive and time-consuming. One such way that the beef industry has created a common language using predictor indices from beef carcasses is the US Department of Agriculture (USDA) beef grading system. The original examination of the quality grades published by Smith (1982, 1984, 1987) indicated that the grading system at that time effectively classified loin steaks (assessments of marbling are taken at the 13th rib interface) according to differences in overall palatability. As there have been slight changes in the grading system over the past 35 years, the newest research in the area from Colorado State University investigated the relationships between the USDA quality grades and sensory properties and included instrument-based quality grading (Acheson, 2014; Emerson, 2013).

# 3 Quality grading

The USDA beef quality grading system (USDA, 1997) uses the physiological maturity of beef carcasses determined by size, shape and ossification of bones and cartilage, in addition to the colour and texture of the lean. Maturity is combined with the marbling and firmness of the rib-eye muscle to assign a quality grade to carcasses. Quality grades for cattle less than 30 months of age (A maturity) can be classified as Prime, Choice, Select or Standard. Cattle that range from 30 to 42 months of age (B maturity) can be classified as Prime, Choice or Standard. Cattle above 42 months of age (C, D and E maturity) can be classified as Commercial, Utility or Cutter. Trained USDA graders use visual assessment to assign these scores and ultimately an overall quality grade. Traditionally, the classification of carcasses into quality grades has been used to indicate the perceived palatability of meat from these carcasses. However, research has indicated that the current grading system is ineffective at identifying meat tenderness (Savell et al., 1987).

Early studies have indicated that marbling is poorly correlated with tenderness and that only 5% of variation in beef tenderness can be attributed to marbling (Wheeler, 1994). Researchers have indicated that consumers are willing to pay more for guaranteed tender beef and that, in the current grading system, consumers are unsure if they are purchasing tender beef or not because quality grades are not a direct measure of tenderness (Lusk, 2001; Miller, 2001). Additionally, research by Boleman et al. (1997) showed that consumers can discern differences between 'categories' of tenderness and are much more likely to purchase tender steaks. According to Acheson et al. (2014), carcasses that were known to be from grain-fed animals less than 30 months of age were stratified using the current USDA grading system. The authors concluded that in the physiological categories of A or B–C maturity, there were no differences in shear force values or sensory panel ratings. By assessing the same data set using only degrees of marbling, the carcasses could be stratified according to shear force values and sensory properties (Acheson, 2014). Huffman et al. (1996) indicated that 98% customer satisfaction level can be achieved if Warner Bratzler shear values are less than 4.1 kg. Therefore, one could argue that by replacing the current quality grading system with a simple assigned tenderness rating system, fresh retail beef selection would be simplified for consumers.

# 4 Instrumental assessment of marbling measurements

Research has been focused on instrumental grading since the 1980s, and finally with the advancement of technology and persistent development, the USDA approved the use of instrument grading for marbling in 2006. Approvals for rib-eye area and for yield grade were sanctioned in 2003 and 2005, respectively. The two approved systems are Computer Vision System (CVS) from Research Management Systems, Inc. and VGB 2000 from E + V Technology GmbH. Researchers from Colorado State University compared the effectiveness of instrument grading for marbling to sensory attributes (Emerson, 2013). Results from this study indicate that instrument-grading of beef carcasses according to differences in marbling effectively identified differences in the sensory properties of strip loin steaks. For an overall sensory experience, 98–99% of the steaks with Moderately Abundant and Slightly Abundant and 80–90% of the steaks with Modest and Moderate marbling received positive ratings (Emerson, 2013). However, according to the 2011 National Beef Quality Audit (Moore et al., 2012), 2.3% of carcasses had Slightly Abundant or higher marbling, 5% Moderate and 17.3% Modest marbling. The impact of this technology has improved accuracy and precision of the beef grading system and reduced variation within plants and between shifts (Emerson, 2013). In addition, marbling assessments coupled with earlier approvals for rib-eye area determination, rib-eye area specifications can be more easily assessed to assist with certified or branded programmes. One additional advantage for using electronic assessments for grading purposes is that it can provide instantaneous feedback to producers. Research has also attempted to classify carcasses based on colour as an indicator of tenderness (Wheeler, 2002). Unfortunately, the indirect, non-invasive methods based on lean colour were not sufficient to be adopted by the industry.

# 5 Shear force measurements for tenderness

A direct and destructive way of measuring tenderness of beef is through shear force measurements. The two commonly used methods for assessing shear force are Warner-Bratzler Shear Force (WBSF) and Slice Shear Force (SSF). Both of these tests measure the force it takes to shear a cooked sample using standard protocols published by AMSA (2015). Researchers have compared each methodology within and among institutions (Wheeler, 1997, 2007), to each other, and to trained sensory panels (Shackelford et al., 1999). With the multitude of projects that have been conducted in this area of tenderness assessment, the National Cattlemen's Beef Association (NCBA) held a meeting to create a plan for the instrumental assessment of beef that was published in 2002, the National Beef Instrumental Assessment Plan II. The American Meat Science Association (AMSA) had previously published research guidelines in 1995 and more recently released the updated version of the Research Guidelines for Cookery, Sensory Evaluation and Instrumental Tenderness Measurements of Meat in 2015.

The WBSF method was widely used until the 1990s when the SSF methodology was developed. The WBSF measure can be obtained with either the original WBSF machine or with the WBSF blade attachments affixed to an automated testing machine such as the Instron or Texture Technology Texture Analyser (AMSA, 2015). The following protocol is summarized from AMSA (2015). The WBSF blade is a V-notched, 60° angle blade with the following specifications: (1) blade thickness of 1.1684 mm (0.046 inches); cutting edge bevelled to a half-round corner of a 2.363-mm-diameter circle; spacers providing gap for cutting blade to slide through a 2.0828-mm thickness. A minimum of six cores are taken from a 2.54-cm-thick steak, generally obtained from a handheld coring device with a diameter of 1.27 cm. Cores should be removed perpendicular to the longitudinal orientation of the muscle fibres. All cores obtained should be used to calculate a mean for each sample. One drawback of this method is that the steak needs to be cooked and then cooled prior to coring in order to facilitate ease of core removal, which adds a significant amount to time to the procedure. Generally, samples are cooked one day and chilled overnight, then cored and sheared the following day. The cores are sheared perpendicular to the muscle fibres. When using an automated testing machine, the crosshead speed of 200–250 mm/min should be used.

Consumers can differentiate among steaks varying in WBSF (Miller, 1995; Huffman, 1996). As WBSF decreased, tenderness scores increased, indicating that consumers could detect changes in tenderness similar to those found in instrumental measurement (Miller, 1998). Platter et al. (2003) revealed moderate to high correlations ($p < 0.05$) among mean marbling scores, WBSF and mean consumer panel palatability ratings. The correlation between consumer tenderness ratings and WBSF was moderately high ($r = 0.63$). However, marbling score correlations with WBSF, consumer tenderness ratings, consumer juiciness ratings and consumer flavour ratings were low ($r = -0.31$, $-0.27$, $-0.34$, $-0.22$, respectively). Of the three acceptance prediction models developed by Platter et al. (2003), the most accurate was the three variables (tenderness, juiciness and flavour) with $r^2 = 0.62$ to overall steak acceptance than the single consumer sensory rating models. The adjusted R-squared values for the marbling and the WBSF models were 0.053 and 0.225, respectively. When the Platter et al. (2003) models were tested against the models published by Wheeler et al. (2002), tenderness alone was the best predictor at 78.7% for determining if consumers would rate steaks as acceptable, with marbling accounting for 57.3% for determining if two-thirds of consumers would have rated steaks as acceptable.

SSF can be conducted immediately after cooking. The following protocol is again summarized from AMSA (2015). A 1-cm-thick, 3-cm-long slice is removed from the lateral half of the *longissimus* muscle from each sample generally parallel to the muscle fibres using a specially designed sample sizer with two parallel knife blades spaced 1 cm apart to make the cuts simultaneously through the length of the 5-cm-long sample at a 45° angle. The slice is sheared perpendicular to the muscle fibre using an automated testing machine equipped with a flat, blunt-end blade. The blade has the same thickness as the WBSF blade (1.1684 mm) and degree of bevel (half-round) on the shearing edge. The crosshead speed should be set to 500 mm/min. This measurement takes less overall time than WBSF because there is no a requirement for chilling of the samples prior to shearing and one sample can be removed more quickly than the six or more cores for WBSF. The NCBA's National Beef Assessment Plan II (2002) suggested that SSF be used as a baseline tenderness measurement and for the evaluation of raw material suppliers so that sourcing decisions could be made regarding tenderness and so that information could be passed back to producers to evaluate what production steps could be evaluated to facilitate selection of tenderness when possible. ASMA (2015) reported that at least one commercial company has implemented this suggestion and has been using this for several years. It should be noted, however, that two cattle breed associations in the United States publish 'expected progeny differences' (EPDs) based on WBSF, and that WBSF still is widely used in numerous countries.

# 6 Juiciness and flavour assessments

## 6.1 Juiciness assessment

Juiciness is both a sensory and chemical meat component that is important to overall beef palatability (Platter, 2003; Emerson, 2014). Primarily, the term 'juiciness' is associated with the moisture content of meat. Compositionally, water is the largest constituent of meat at 70–77%. However, research has revealed a direct link between juiciness and intramuscular fat content (Killinger, 2004; Emerson, 2014). Intramuscular fat can account for almost 50% of the variability in perceived juiciness (Emerson, 2014). Therefore, a combination of moisture and intramuscular fat are the principal components of overall juiciness. Still, it is necessary to mention the relationship between the degree of doneness or internal end-point temperature and juiciness. Research has shown a negative correlation between the degree of doneness and juiciness; as internal temperature increases, juiciness decreases due to moisture loss and intramuscular fat solubilization (Parrish, 1973; Aberle et al., 2001; Obuz, 2004; Yancey, 2011). Additionally, anatomical muscle location can influence juiciness because muscles of locomotion typically have a higher percentage of water content and a lower percentage of intramuscular fat content when compared to muscles of support (Carmack, 1995).

The assessment of juiciness can be separated into two major categories: objective assessment and subjective assessment. Common objective assessment tools/analysis(s) include drip loss analysis, cook loss analysis, proximate analysis for moisture and lipid content and a measurement of pH as a representative of water holding capacity. Arguably, human sensory evaluation is the most often used subjective tool to assess juiciness. However, an important distinction to address is the difference between trained human sensory panels and consumer sensory panels. Consumer-based sensory panels are hedonic in nature and focus on a combination of preference and acceptability, whereas a trained sensory panel is based on characteristic intensity and is more analytical in nature utilizing an eight-point scale or a linear scale (AMSA & NLMB, 1995). The 'halo-effect' must be minimized when human sensory evaluations are utilized to assess juiciness, meaning that one unfavourable or favourable characteristic should not influence the perceived assessment of other characteristics.

Performing juiciness assessments can be a challenge due to sample destruction, time and cost. The optimization of juiciness assessment can be an even greater challenge due to the differences among the abilities of beef purveyors, retailers and consumers to have facilities to test for juiciness subjectively, the same problem as for the assessment of tenderness.

## 6.2 Flavour assessment

Flavour is an intriguing and highly complex beef sensory attribute. Flavour can be described as the combination of taste and aroma leading to the possible sensory perceptions of sweet, sour, bitter, salty and umami. Beef flavour is primarily developed when heat is applied; however, the flavour intensity and occurrence of flavour-active volatiles are dependent upon both the water-soluble and lipid component precursor compounds present (Mottram, 1998; Brewer, 2007). Lipid-derived volatiles such as acids, aldehydes, ketones, alcohols or aromatic and heterocyclic compounds can have the greatest influence on beef flavour acceptability.

Most research has focused on tenderness as the most important beef sensory attribute due to historical consumer preference (Huffman, 1996; Platter, 2003); however, current research models are investigating the primary and secondary roles of flavour in overall beef palatability (Corbin, 2015; O'Quinn, 2015). Corbin et al. (2015) stated that with a large percentage of the US beef supply being classified as tender according to the National Beef Tenderness Survey (Guelker, 2013), the importance of flavour to overall beef eating satisfaction is magnified. In this study, steaks were selected on tenderness based on WBSF and analysed for fat percentage by near-infrared spectrophotometry. Steaks were subjected to both a consumer sensory panel and a trained panel for flavour descriptive analysis. The results revealed that fat content has a positive effect on beef flavour perception regardless of its effect on tenderness, and that consumer overall liking and flavour scores were closely related to the fat-like flavour, which increased in treatments with higher fat levels (Corbin, 2015). Along the same lines of research, O'Quinn et al. (2015) reported that differences in flavour characteristics among quality grade treatments could be attributed to increases in fat content. They found that increases in consumer eating satisfaction and flavour perception were related to a buttery/beef fat flavour with higher marbled steaks indicating that degree of marbling is important for consumers' satisfaction. Furthermore, when asked to describe beef flavour, consumers preferred descriptive terms such as beefy/brothy, brown/grilled, buttery/beef fat, nutty/roasted nut and sweet as opposed to bloody/metallic, grassy, gamey, livery, sour and bitter (O'Quinn, 2016).

# 7 Post-harvest optimization of palatability

## 7.1 Muscle profiling and merchandizing

By combining the science of muscle profiling and meat cookery, previously underutilized cuts can be appealing to both the industry and consumers. Obuz et al. (2003) showed that tenderness could be improved, given the proper cooking protocol for specific cuts. The use of moist heat cookery independently or in combination with dry heat cookery can produce a more palatable eating experience, especially for muscles with high connective content.

Post-mortem ageing affected the WBSF values for all muscles from the chuck and round evaluated by Bratcher et al. (2005) in a similar fashion. Therefore, consistent recommendations about post-mortem ageing can be given for these muscles to decrease the WBSF values; however, there are muscles from other subprimals that could be aged at different time periods to maximize tenderness. Bratcher et al. (2005) showed that the USDA grade had an effect on post-mortem ageing, in that WBSF for muscles from the upper two-thirds of the USDA Choice grade were not significantly improved after 7 days post-mortem ageing. Muscles from the USDA Select grade had significant improvements in WBSF at 14 and 28 days post-mortem. Location within a muscle was found to have an effect on the WBSF values in four of the nine muscles evaluated indicating that muscles need to be treated on an individual basis during fabrication and merchandizing of individual retail cuts or portions from muscles of the chuck and round. For some muscles, location within the cut can be ignored but for others location must be considered for tenderness enhancement or product utilization. Beef value cut research has been highly supported by the NCBA and merchandizing recommendations can be made according to current research.

## 7.2 Post-harvest palatability enhancement

A major component contributing to the toughness and quality characteristics of bovine muscles of locomotion is connective tissue, which is specifically related to background tenderness (Aberle et al., 2001). Connective tissue surrounds muscle at every level of organization. Mechanical tenderization and needle injection are processes utilized extensively in the United States to improve the palatability of whole-muscle meat. An estimated 555 processing facilities in the United States are responsible for the estimated 10.5% or 2.6 billion pounds of fresh beef product sold in the United States that is mechanically tenderized (Muth et al., 2012). Mechanical tenderization is commonly used on cuts from the round, sirloin, loin, rib and chuck. Quality grade and the specific muscle are used as determinants of passing rate and frequency (George-Evins et al., 2000). The majority of mechanically tenderized products are destined for food service as opposed to retail, 75.1% versus 13.3%, respectively (George-Evins et al., 2000).

Post-mortem ageing, electrical stimulation, post-mortem chilling conditions and mechanical tenderization are all important factors to consider when optimizing assessment. These topics are covered in depth in another chapter in this text.

# 8 Optimization of assessment and future research

Unfortunately, there is no silver bullet in the assessment or measurement of beef sensory properties. The beef industry has long been searching for the perfect tool for prediction of consumer acceptability and as a marker of economic incentive. Consumers have been able to at least detect differences in the tenderness of beef and have often indicated that they would be willing to pay more for a guaranteed tender product. The most reliable measures of tenderness to assure that every carcass is classified according to tenderness are destructive measures and this presents problems within the test itself. Usually, one muscle is tested and that one muscle will not necessarily be indicative of the tenderness of every muscle in that carcass. Shear force determination, both WBSF and SSF, and trained sensory panels provide moderately high correlations to consumer perceptions of tenderness; however, any of the three of these tools come with time constraints and a cost for the labour and destruction of a part of the saleable product. With technological advancements and advancements in research, it is clear that marbling plays a significant role in consumer perceptions of flavour, and, to a lesser extent, tenderness; research has indicated that the beef industry has done a good job of increasing the amount of carcasses that are considered tender. Producer pre-harvest tenderness strategies have been critical. Management of temperament, components of finishing diets and finishing implants have resulted in a more consistent product (Tatum et al., 2007). Additional tenderness research related to natural proteinase activity (u-calpain) and the inhibitor calpastatin as well as genetic markers has driven breed selection (Tatum et al., 2007; Riley et al., 2007). When compared to *Bos taurus* breeds, *Bos indicus*-influenced cattle have a higher calpastatin activity level leading to an ultimately tougher product (Chase et al., 2002). Dikeman et al. (2005) evaluated strip loins from nearly 7200 cattle from 14 different breeds and determined the heritability of tenderness is moderately high (0.4) according to WBSF and (0.37) by a trained sensory panel. Only three breeds of cattle have EPDs for shear force; therefore, breeders of other breeds do not have any information to use to improve tenderness other than indirectly via marbling, which has a heritability of 0.68 according to Dikeman et al. (2005). With the current research indicating that marbling plays a key role in flavour, it is quickly becoming and may increasingly be the new area of future research. The time-honoured assessment via a trained sensory panel is most likely the best approach to ensure that desired sensory characteristics are being met. The filtering of that information from the sector performing the evaluation back through the supply chain, all the way to the producer, is extremely important to make changes to breeding selection and management practices from the live animal standpoint. However, flavour is much more complex than tenderness or marbling, and has moderately low heritability. Of course, as tools are used to assess the sensory characteristics of beef, decisions can also be made as to the fate of the product itself. If a product is classified as less than tender, decisions can be made to further process those muscles or entire carcasses into processed meats or maybe into a tenderized end product through the use of marinades or needle tenderization. In the evaluation of current research, it appears that at the processor level, the use of the electronic assessment cameras is doing a superb job of classifying carcasses into appropriate marbling categories that are then correlating with consumer acceptance of both flavour and overall palatability.

Modern consumers want as much information related to and about their food as possible. The dissemination of applicable research is warranted. Future research is needed to simplify consumer education. Additionally, the incorporation of palatability assessment values could be included in purveyor and retailer product specifications. Future innovative research is needed to discover, modify and implement palatability assessments that do not sacrifice time or increase costs and yet meet the demands of beef purveyors, retailers and consumers.

# 9 Where to look for further information

To learn more about the beef grading system in the United States, please access the following website: http://www.beefresearch.org/CMDocs/BeefResearch/Beef%20Grading.pdf.

For more specific information on the background and specific research protocols for WBSF and SSF, please visit: http://www.meatscience.org/publications-resources/printed-publications/sensory-and-tenderness-evaluation-guidelines and http://www.ars.usda.gov/Main/docs.htm?docid=21720.

# 10 References


Aberle, E. D., Forrest, J. C., Gerrard, D. E., Hedrick, H. B., Judge, M. D., Merkel, R. A. and Mills, E. W. (2001), *Principles of Meat Science*, 4th ed. Kendal/Hunt Publishing Company.

Acheson, R. J., Woerner, D. R. and Tatum, J. D. (2014), 'Effects of USDA carcass maturity on sensory attributes of beef produced by grain-finished steers and heifers classified as less than 30 months old using dentition', *J. Anim. Sci.*, 92, 1792–9.

AMSA. (2015), *Research Guidelines for Cookery, Sensory Evaluation and Instrumental Tenderness Measurements of Fresh Meat*, American Meat Science Association, Chicago, IL.

Anon. (1975), 'Minutes of Division Business Meeting', Institute of Food Technologies, Sensory Evaluation Division, IFT, Chicago, IL.

Boleman, S. J., Boleman, S. L., Miller, R. K., Taylor, J. F., Cross, H. R., Wheeler, T. L., Koohmaraie, M., Shackelford, S. D., Miller, M. F., West, R. L., Johnson, D. D. and Savell, J. W. (1997), 'Consumer evaluation of beef of known categories of tenderness', *J. Anim. Sci.*, 75, 1521–4.

Bratcher, C. L., Johnson, D. D., Littell, R. C. and Gwartney, B. L. (2005), 'The effects of quality grade, aging, and location within muscle on Warner–Bratzler shear force in beef muscles of locomotion', *Meat Science*, 70, 279–84.

Brewer, S. (2006), *The Chemistry of Beef Flavor – Executive Summary*, Cattlemen's Beef Board.

Carmack, C. F., Kastner, C. L., Dikeman, M. E., Schwenke, J. R. and Garcia Zepeda, C. M. (1995), 'Sensory evaluation of beef-flavor-intensity, tenderness, and juiciness among major muscles', *Meat Science*, 39, 143–7.

Corbin, C. H., O'Quinn, T. G., Garmyn, A. J., Legako, J. F., Hunt, M. R., Dinh, T. T. N., Rathmann, R. J., Brooks, J. C. and Miller, M. F. (2015), 'Sensory evaluation of tender beef strip loin steaks of varying marbling levels and quality treatments', *Meat Science*, 100, 24–31.

Emerson, M. R., Woerner, D. R., Belk, K. E. and Tatum, J. D. (2014), 'Effectiveness of USDA instrument-based marbling measurements for categorizing beef carcasses according to differences in longissimus muscle sensory attributes', *J. Anim. Sci.*, 91, 1024–34.

Ford, G. T., Smith, D. B. and Swasy, J. L. (1988), 'An empirical test of the search, experience and credence attributes framework', *M.A. Adv. Consum. Res.*, 15, 239–44.

George-Evins, C. D., Unruh, J. A., Marsden, J. L. and Kastner, C. L. (2000), 'A survey on the use of blade tenderizers by beef fabrication plants', In *2000 Cattlemen's day (119–120)*, Manhattan, KS: Kansas State University.

Guelker, M. R., Haneklaus, A. N., Brooks, J. C., Carr, C. C., Delmore, R. J., Griffin, D. B., Hale, D. S., Harris, K. B., Mafi, G. G., Johnson, D. D., Lorenzen, C. L., Maddock, R. J., Martin, J. N., Miller, R. K., Raines, C. R., VanOverbeke, D. L., Vedral, L. L., Wasser, B. E. and Savell, J. W. (2013), 'National Beef Tenderness Survey—2010: Warner-Bratzler shear force values and sensory panel ratings for beef steaks from United States retail and food service establishments', *J. Anim. Sci.*, 91(2), 1005–14.

Huffman, K. L., Miller, M. F., Hoover, L. C., Wu, C. K., Brittin, H. C. and Ramsey, C. B. (1996), 'Effect of beef tenderness on consumer satisfaction with steaks consumed in the home and restaurant', *J. Anim. Sci.*, 74, 91–7.

Igo, J. L., VanOverbeke, D. L., Woerner, D. R., Tatum, J. D., Pendell, D. L., Vedral, L. L., Mafi, G. G., Moore, M. C., McKeith, R. O., Gray, G. D., Griffin, D. B., Hale, D. S., Savell, J. W. and Belk, K. E. (2013), 'Phase I of the national beef quality audit-2011: Quantifying willingness-to-pay, best-worst scaling, and current status of quality characteristics in different beef industry marketing sectors', *J. Anim. Sci.*, 91, 1907–19.

Killinger, K. M., Calkins, C. R., Umberger, W. J., Fuez, D. M. and Eskridge, K. M. (2004), 'Consumer sensory acceptance and value for beef steaks of similar tenderness, but differing in marbling level', *J. Anim. Sci.*, 82, 3294–301.

Lawless, H. T. and Heymann, H. (1999), *Sensory Evaluation of Food: Principles and Practices*, New York: Springer.

Lorenzen, C. L., Neely, T. R., Miller, R. K., Tatum, J. D., Wise, J. W., Taylor, J. F., Buyck, M. J., Reagan, J. O. and Savell, J. W. (1999), 'Beef customer satisfaction: cooking method and degree of doneness effects on the top loin steak', *J. Anim. Sci.*, 77(3), 637–44.

Lorenzen, C. L., Miller, R. K., Taylor, J. F., Neely, T. R., Tatum, J. D., Wise, J. W., Buyck, M. J., Reagan, J. O. and Savell, J. W. (2003), 'Beef customer satisfaction: Trained sensory panel ratings and Warner-Bratzler shear force values', *J. Anim. Sci.*, 81, 143–9.

Lusk, J. L., Fox, J. A., Schroeder, T. C., Mintert, J. and Koohmaraie, M. (2001), 'In-store valuation of steak tenderness', *Amer. J. Agr. Econ.*, 83, 539–50.

Meilgaard, M. C., Civille, G. V. and Carr, B. T. (2007), *'Sensory Evaluation Techniques'*, 4th ed., Boca Raton: CRC Press.

Miller, M. F., Carr, M. A., Ramsey, C. B., Crockett, K. L. and Hoover, L. C. (2001), 'Consumer thresholds for establishing the value of beef tenderness', *J. Anim. Sci.*, 79, 3062–8.

Miller, M. F., Hoover, L. C., Cook, K. D., Guerra, A. L., Huffman, K. L., Tinney, K. S., Ramsey, C. B., Brittin, H. C. and Huffman, L. M. (1995), 'Consumer acceptability of beef steak tenderness in the home and restaurant', *J. Food Sci.*, 60, 963.

Miller, M. F., Ramsey, C. B., Hoover, L. C., Carr, M. A. and Crockett, K. L. (1998), 'Consumer thresholds for establishing the value of beef tenderness', *Proc. Recip. Meat Conf.*, 51, 4.

Moore, M. C., Gray, G. D, Hale, D. S., Kerth, C. R., Griffin, D. B., Savell, J. W., Raines, C. R., Belk, K. E., Woerner, D. R., Tatum, J. D., Igo, J. L., VanOverbeke, D. L., Mafi, G. G., Lawrence,

T. E., Delmore, R. J., Christensen, L. M., Shackelford, S. D., King, A., Wheeler, T. L., Meadows, L. R. and O'Connor, M. E. (2012), 'National Beef Quality Audit-2011: In-plant survey of targeted carcass characteristics related to quality, quantity, value and marketing of fed steers and heifers', *J. Anim. Sci.*, 90, 5143–51.

Mottram, D. S. (1998), 'Flavour formation in meat and meat products: a review', *J. Food Chem.*, 62(4), 415–24.

Muth, M. K., Michaela, M. B. and Coglaiti, C. (2012), 'Expert elicitation on the market shares for raw meat and poultry products containing added solutions and mechanically tenderized raw meat and poultry products', *FSIS*, http://www.fsis.usda.gov/wps/wcm/connect/3a97f0b5-b523-4225-8387-c56a1eeee189/Market_Shares_MTB_0212.pdf?MOD=AJPERES (accessed 2016).

NCBA. (2002), 'National Beef Assessment Plan II: Focus on Tenderness', National Cattlemen's Beef Association and the Cattlemen's Beef Board, Beef Update.

Neely, T. R., Lorenzen, C. L., Miller, R. K., Tatum, J. D., Wise, J. W., Taylor, J. F., Buyck, M. J., Reagan, J. O. and Savell, J. W. (1998), 'Beef customer satisfaction: Role of cut, USDA quality grade, and city on in-home consumer ratings', *J. Anim. Sci.*, 76, 1027–33.

Neely, T. R., Lorenzen, C. L., Miller, R. K., Tatum, J. D., Wise, J. W., Taylor, J. F., Buyck, M. J., Reagan, J. O. and Savell, J. W. (1999), 'Beef customer satisfaction: cooking method and degree of doneness effects on the top round steak', *J. Anim. Sci.*, 77(3), 653–60.

Obuz, E., Dikeman, M. E. and Loughin, T. M. (2003), 'Effects of cooking method, reheating, holding time, and holding temperature on beef longissimus lumborum and biceps femoris tenderness', *Meat Science*, 65, 841–51.

Obuz, E., Dikeman, M. E., Grobbel, J. P., Stephens, J. W. and Loughin, T. M. (2004), 'Beef longissimus lumborum, biceps femoris, and deep pectoralis Warner-Bratzler shear force is affected differently by endpoint temperature, cooking method, and USDA quality grade', *Meat Science*, 68, 243–8.

O'Quinn, T. G., Woerner, D. R., Engle, T. E., Chapman, P. L., Legako, J. F., Brooks, J. C., Belk, K. E. and Tatum, J. D. (2016), 'Identifying consumer preferences for specific beef flavour characteristics in relation to cattle production and post-mortem processing parameters', *Meat Science*, 112, 90–102.

Platter, W. J., Tatum, J. D., Belk, K. E., Chapman, P. L., Scanga, J. A. and Smith, G. C. (2003), 'Relationships of consumer sensory ratings, marbling score, and shear force value to consumer acceptance of beef strip loin steaks', *J. Anim. Sci.*, 81(11), 2741–50.

Parrish Jr., F. C., Olson, D. G., Miner, B. E. and Rust, R. E. (1973), 'Effect of degree of marbling and internal temperature of doneness on beef rib steaks', *J. Anim. Sci.*, 37, 430–4.

Savell, J. W., Branson, R. E., Cross, H. R., Stiffler, D. M., Wise, J. W., Griffin, D. B. and Smith, G. C. (1987), 'National consumer retail beef study: palatability evaluations of beef loin steaks that differed in marbling', *J. Food Sci.*, 53, 512–17.

Savell, J. W., Lorenzen, C. L., Neely, T. R., Miller, R. K., Tatum, J. D., Wise, J. W., Taylor, J. F., Buyck, M. J. and Reagan, J. O. (1999), 'Beef customer satisfaction: cooking method and degree of doneness effects on the top sirloin steak', *J. Anim. Sci.*, 77(3), 645–52.

Savell, J. W. and Shackelford, S. D. (1992), 'Significance of tenderness to the meat industry', *Proc. Recip. Meat Conf.*, 45, 43.

Shackelford, S. D., Wheeler, T. L. and Koohmaraie, M. (1999), 'Evaluation of slice shear force as an objective method of assessing beef longissimus tenderness', *J. Anim. Sci.*, 77, 2693–9.

Smith, G. C., Cross, H. R., Carpenter, Z. L., Murphey, C. E., Savell, J. W., Abraham, H. C. and Davis, G. W. (1982), 'Relationship of USDA maturity groups to palatability of cooked beef', *J. Food. Sci.*, 47, 1100–7.

Smith, G. C., Carpenter, Z. L., Cross, H. R., Murphey, C. E., Abraham, H. C., Savell, J. W., Davis, G. W., Berry, B. W. and Parrish Jr., F. C. (1984), 'Relationship of USDA marbling groups to palatability of cooked beef', *J. Food Qual.*, 7, 289–308.

Smith, G. C., Savell, J. W., Cross, H. R., Carpenter, Z. L., Murphey, C. E., Davis, G. W., Abraham, H. C., Parrish Jr., F. C. and Berry, B. W. (1987), 'Relationship of USDA quality grades palatability of cooked beef', *J. Food Qual.*, 10, 269–86.

Stone, H. and Sidel, J. (1993), '*Sensory Evaluation Practices*', 2nd ed., San Diego: Academic.
Tatum, J. D. (2015), 'Recent trends: Beef quality, value and price', Fort Collins: Colorado State University.
USDA. (1997), 'Official United States Standards for Grades of Carcass Beef', United States Department of Agriculture, Agricultural Marketing Service, Washington, DC.
Wheeler, T. L., Cundiff, L. V. and Koch, R. M. (1994), 'Effect of marbling degree on beef palatability in Bos taurus and Bos indicus cattle', *J. Anim. Sci.*, 72, 3145.
Wheeler, T. L., Shackelford, S. D., Johnson, L. P., Miller, M. F., Miller, R. K. and Koohmaraie, M. (1997), 'A comparison of Warner-Bratzler shear force assessment within and among institutions', *J. Anim. Sci.*, 75, 2423–32.
Wheeler, T. L., Shackelford, S. D. and Koohmaraie, M. (2007), 'Beef longissimus slice shear force measurement among steak locations and institutions', *J Anim. Sci.*, 85, 2283–9.
Wheeler, T. L., Shackelford, S. D. and Koohmaraie, M. (1999), 'Tenderness classification of beef: IV. effect of USDA quality grade on the palatability of "tender" beef longissimus when cooked well done', *J. Anim. Sci.*, 77, 882–8.
Wheeler, T. L, Vote, D., Leheska, J. M., Shackelford, S. D., Belk, K. E., Wulf, D. M., Gwartney, B. L. and Koohmaraie, M. (2002), 'The efficacy of three objective systems for identifying guaranteed tender beef cuts', *J. Anim. Sci.*, 80(12), 3315–27.
Yancey, J. W. S., Wharton, M. D. and Apple, J. K. (2011), 'Cookery method and end-point temperature can affect the Warner–Bratzler shear force, cooking loss, and internal cooked color of beef longissimus steaks', *Meat Science*, 88, 1–7.

Chapter 16

# The role of beef in human nutrition and health

*Chunbao Li, Nanjing Agricultural University, China*



## 1 Introduction

Beef is an important source of both macronutrients and micronutrients. The macronutrients include water, high-value protein and fat comprising saturated and unsaturated fatty acids. The micronutrients in beef and edible offal include haeme iron; zinc; selenium; and vitamins D, B1 (thiamine), B2 (riboflavin), B3 (niacin), B5 (pantothenic acid), B6 (pyridoxin) and B12 (cobalamin). All these nutrients play a critical role in human physiological functions and deficiencies in any of them can cause disorders or diseases. Nevertheless, concerns have been raised regarding the effect of red and processed meat intake on human health. Indeed, based on epidemiological studies, a working group of the International Agency for Research on Cancer (IARC) has classified the consumption of processed meat as 'carcinogenic to humans' (Group 1) (on the basis of sufficient evidence of an association with colorectal cancer), and further classified the consumption of red meat as 'probably carcinogenic to humans' (Group 2A).

In this chapter, we review the nutrients available in beef and then assess the concerns raised about the health impacts of consuming red meat. We argue that there is no convincing evidence to fix the association between intake of red meat and the risk of cancer. It is suggested that eating a balanced diet is a far more significant factor in maintaining a healthy status.

http://dx.doi.org/10.19103/AS.2016.0009.16

# 2 Macronutrients in beef

Beef is an important source of some nutrients for humans, and is mainly composed of water, protein and fat. However, the proportions of these constituents vary greatly with breed, sex, age at slaughter, marbling, and cuts or muscles (Schönfeldt et al., 2010; Sexten et al., 2012).

## 2.1 Water

Water in beef exists in three forms, namely, free water, immobilized water and bound water, which show different physicochemical properties. The water content is inversely proportional to fat content. Normally, meat from veal, young bulls, heifers and steers contains more water but less fat than that from old animals. Marbled meat from Japanese Wagyu and Korean Hanwoo cattle contains much higher intramuscular fat but lower water than other breeds. Belgian blue cattle, a double-muscle breed, contain higher levels of protein and water than other breeds.

## 2.2 Protein

It is well known that meats are good sources of high-quality proteins. Lean meat contains relatively constant percentage of protein (20 to 24%) and most meat proteins have high biological values because of the enrichment of essential amino acids. In beef, about 40% of the amino acids are essential amino acids that cannot be synthesized in the body and have to be supplied by the diet.

The bioavailability of beef protein depends, to a certain extent, on its composition and process conditions. For example, leg muscles contain higher levels of connective tissue (collagen) than loin; collagen is composed of hydroxyproline and other amino acids and cross-links (Li et al., 2007), which are resistant to digestive enzymes. Protein oxidation occurs during storage under oxygen, long-term curing, high-temperature cooking, radiation, microwave and other practices. Protein oxidation is characterized by the formation of disulphide and carbonyl bonds. Such a change has been shown to substantially affect protein digestibility. Under the same conditions of cooking and *in vitro* digestion, beef showed lower digestibility than chicken and fish. The digestion products of beef were shown to differ greatly from pork, chicken and fish (Wen et al., 2015).

## 2.3 Fat

In beef, fat exists between the muscles (i.e. intermuscular fat), within the muscles (i.e. intramuscular fat or marbling) or under the skin (i.e. subcutaneous fat). Although fat is not preferred in some European countries, the abundance of intramuscular fat (marbling) is the primary determinant of beef-quality grades in the United States and some Asian countries (Park et al., 2002; USDA, 1997). However, it is widely concerned that fats in beef are mainly composed of saturated fatty acids, including palmitic acid (16:0) and stearic acid (18:0), and small amounts of myristic acid (14:0); oleic acid (18:1n-9) is the main monounsaturated fat (MUFA) present in beef. In addition, beef contains a small amount of trans-fatty acids and is the only source of conjugated linoleic acid. Generally, the amount of polyunsaturated fatty acids in beef is greater than that of MUFAs. Pasture feeding has been shown to increase the amount of unsaturated fatty acids in beef as compared to grain feeding (Nuernberg et al., 2005). The problem of pasture feeding is yellow colour of fat due to the accumulation of carotenoids (Zhou et al., 1993). In recent years, the feeding system has been changed greatly to more grain or concentrates in order to improve growth performance and profits.

# 3 Micronutrients in beef

## 3.1 Iron

Beef is a very good source of iron, varying with cuts from 2.1 mg per 100 g in loin to 10 mg per 100 g in liver. Thus, the intake of beef as well as other red meat is a good way to solve the iron deficiency problem. There are two types of iron in meat: haeme and non-haeme. Haeme iron is highly absorbable as compared to non-haeme iron and mainly exists in animal origin foods, which accounts for 50 to 87% of the total iron in meat. Haemoglobin and myoglobin are the main carriers of haeme in meat (Buzala et al., 2016). Meat source non-haeme iron could also be easily absorbed compared to plant source iron. The absorption of non-haeme iron is facilitated by ascorbic acid, citric acid and some amino acids. Citric acid enhances the absorption of non-haeme iron through chelation, and ascorbic acid and cysteine reduces iron to a more readily available form of Fe (II), thus facilitating its intestinal absorption (Buzala et al., 2016)

## 3.2 Zinc

Meat is rich in zinc and the percentage of zinc is the greatest in beef, amounting to 4 mg zinc per 100 g in lean beef. The zinc in meat is more bioavailable than other sources. Meat has been shown to enhance the absorption of zinc and 20 to 40% of zinc is absorbed from meat. However, its mechanism is still less known. Zinc is a co-factor of many enzymes, for example, carbonic anhydrase, RNA polymerase and Cu-Zn superoxide dismutase. Zinc plays an important role in physiological functions involving growth, cognitive development, sense of taste and reproduction. Zinc deficiency may cause dwarfism, dysfunction of vision, loss of appetite, gonadal dysgenesis, loss of memory and other problems.

## 3.3 Selenium

Beef is also relatively rich in selenium, containing about 7 μg selenium per 100 g meat. Organ meats contain higher levels of selenium, for example, the percentages of selenium amount to approximately 150 μg per 100 g in kidney and 42 μg per 100 g in liver. Selenium is a component of the enzyme glutathione peroxidase, which removes reactive oxygen species (ROS) by redox reactions, including hydrogen peroxide and hydroperoxides. Selenium has some other functions. It enhances immunity; delays senescence; and regulates the absorption of vitamins A, C and E, and protein synthesis as well.

## 3.4 Vitamins

### Vitamin D

Vitamin D is very important as it protects against rickets, a kind of disease characteristic of the softening of the bones (Dunnigan et al., 2005). Beef contains 0.5 μg vitamin D per 100 g meat. In the body, vitamin D can be synthesized under the sunshine by converting 7-dehydrocholesterol to cholecalciferol, which would be converted to 25-hydroxycholecalciferol in the liver and dihydroxycholecalciferol in the kidney. Dihydroxycholecalciferol is the most bioactive and stimulates calcium absorption. In meat, vitamin D exists in three forms, that is, cholecalciferol, 25-hydroxycholecalciferol and dihydroxycholecalciferol. The activity of 25-hydroxycholecalciferol is five times that of cholecalciferol and 25 times that of dihydroxycholecalciferol. Vitamin D aids the calcification of the bones and the development of the immune system, and protects against such diseases as tuberculosis, diabetes and coronary heart disease (Mulvihill, 2002).

## Vitamin B12 (cobalamin)

B vitamins are water soluble and a proportion of them could be lost during cooking. The amount of loss is dependent on cooking time and cooking temperature. However, meat is still a good source of B vitamins, including vitamin B12.

Beef and offal meats (e.g. liver and kidney) contain different amounts of vitamin B12. For example, there are 1 to 3 μg vitamin B12 per 100 g in meat, 23 to 110 μg per 100 g in liver and 15 to 54 μg per 100 g in kidney. Vitamin B12 plays a critical role in the formation of red blood cells and the development of nervous systems. It is known that vitamin B12 could be synthesized by gut bacteria and can be absorbed through epithelium entericum.

## Vitamin B6 (pyridoxine, pyridoxal and pyridoxamine)

Vitamin B6 has three forms, namely, pyridoxine, pyridoxal and pyridoxamine. Beef contains 0.3 to 0.5 mg vitamin B6 per 100 g. The liver contains slightly higher levels of vitamin B6. In the body, pyridoxine and pyridoxamine can be changed to pyridoxal 5′-phosphate, which is a co-factor for metabolic enzymes involving protein and amino acid metabolism, and conversion of fat to energy. In addition, vitamins B6 and B12 have the common function to decrease the level of homocysteine in blood because both vitamins are the co-factors for enzymes that are involved in the degradation of homocysteine. High level of homocysteine could be associated with coronary artery diseases. Vitamin B6 is the co-factor for cystathionine B synthase and cystathionase that convert homocysteine to cysteine, while vitamin B12 is the co-factor for methionine synthase that converts homocysteine to methionine. Vegetarians may have lower intakes of vitamins B6 and B12 and have higher levels of plasma homocysteine than those who eat meat and edible organs, indicating that meat gives a great importance to maintain low level of homocysteine (Mulvihill, 2002).

## Vitamin B3 (niacin)

Meat is the richest source of niacin and beef contains about 5 mg niacin per 100 g. Compared to plant sources, niacin from meat is easily absorbed by the body because the former is bound to glucose. Niacin mainly functions in the conversion of carbohydrates and fat into energy.

## Vitamin B1 (thiamine)

Beef contains 0.1 mg thiamine per 100 g, only one-tenth of that in pork (1 mg thiamine per 100 g), but it is still an important source of thiamine, which is also a co-factor for enzymes involving the conversion of fats and carbohydrates into energy. In addition, thiamine plays a role in appetite and nervous system function. A deficiency of thiamine may cause polyneuritis, beriberi, and inflammation and degeneration of peripheral nerves.

### Vitamin B2 (riboflavin)

Beef contains 0.2 mg riboflavin per 100 g. Riboflavin involves energy production by acting as a co-factor for enzymes of the Krebs cycle. Riboflavin also promotes healthy skin, eyes and vision, and may prevent cataracts. A deficiency of riboflavin may cause orogenital syndrome.

### Vitamin B5 (pantothenic acid)

Pantothenic acid exists widely in living bodies and thus both plant and animal foods contain enough quantity of this vitamin. In beef, pantothenic acid amounts to 0.75 mg per 100 g. Normally, pantothenic acid deficiency does not happen if we eat enough balanced food. Pantothenic acid plays a role in energy production and the formation of steroid hormones.

## 4 Beef intake and health concerns

In recent years, there has been wide concern about the association between the consumption of red meat and processed meat and human health problems, especially of colorectal cancers. In October 2015, a working group of the IARC classified consumption of processed meat as 'carcinogenic to humans' (Group 1) on the basis of 'sufficient evidence' for colorectal cancer, and classified consumption of red meat as 'probably carcinogenic to humans' (Group 2A) (Bouvard et al., 2015). In this report, red meat refers to 'unprocessed mammalian muscle meat, including beef, veal, pork, lamb, mutton, horse, and goat meat', whilst processed meat refers to 'meat that has been transformed through salting, curing, fermentation, smoking, or other processes to enhance flavour or improve preservation', mainly including pork or beef (Bouvard et al., 2015). However, such a viewpoint has been argued by many meat producers, processors or meat scientists (McAfee et al., 2010). Arguments are focused on two points: (1) components of diets are not fixed for a healthy person, and it is difficult to imagine that a person eats too much red meat or processed meat each day for a long time; (2) balanced diets are more important to protect against metabolic disorders and related cancers.

It takes more than ten years to develop from normal growing cells to malignant cancer cells, which comprises at least three phases: initiation, promotion and progression. The initiation phase is usually characteristic of gene mutations due to factors such as radiation, ultraviolet light and toxic chemicals. The promotion phase is normally induced by certain compounds including nitrosamines, polycyclic aromatic hydrocarbons (PAHs) and other substances. Cancer cells then progress and finally malignant properties appear. It has been reviewed that aberrant ROS burst and that dysregulation of those ROS-dependent cellular processes are strongly associated with many cancers, and ROS levels are elevated in cancer cells partially due to their higher metabolism rate (Ding et al., 2015). Environmental factors, especially of dietary factors (e.g. excessive intake of carbohydrates), might either induce ROS production or alter gut bacteria, and finally cause diseases. In a recent animal study, meat protein intake at recommended levels was found to maintain a more balanced composition of gut bacteria, thereby reducing the antigen load and inflammatory response in the host (Zhu et al., 2015).

Until now, there has been no direct scientific evidence to show the association between meat intake and cancer risks. Here, we give a brief introduction to the risks of several intrinsic and extrinsic carcinogenic factors. Intrinsic factors include fat and iron, while extrinsic factors mainly include heterocyclic amines (HA), PAHs and *N*-nitroso compounds (nitrosamines).

## 4.1 Fat

Beef contains higher amounts of saturated fatty acids than found in pork and much higher amounts than in chicken. It has been found that high intakes of red meat (including beef) may cause higher levels of total cholesterol and low-density lipoprotein cholesterol and triglycerides in plasma (Li et al., 1999). The major saturated fatty acids in beef are myristic acid C14:0, palmitic acid C16:0 and stearic acid C18:0; all of them have been shown to have an association with risk to cardiovascular disease (Hu et al., 1999). However, stearic acid has been considered neutral in regard to increasing serum cholesterol in some cases. Meanwhile, there is no convincing evidence to show the causal effect of high fat intake on cardiovascular disease because plant oil in diets also contain quite a high proportion of saturated fatty acids. In fact, the changes of fatty acids composition during meat cooking or processing might be more important for health concerns. For example, low dose irradiation was shown to increase total saturated fatty acids, monounsaturated fatty acids and atherogenicity index; however, neither did it change the ratios of monounsaturated fatty acids to saturated fatty acids in the total lipid fraction nor thrombogenicity index (Chen et al., 2007). Li et al. (2016) indicated that extended cooking may modify fatty acid profiles, decrease the derived cardiovascular indices and reduce cholesterol content. This indicates that long-term stewed meat, which is popular in Asian countries, could be healthier than other kinds of cooked meat because of the alteration of fatty acid composition.

## 4.2 Iron

Beef contains four times more iron and ten times more haeme iron than that found in chicken and other meats. A high intake of beef will increase cell proliferation in the mucosa, through lipoperoxidation and/or cytotoxicity of gut contents (Santarelli et al., 2008). A high intake of beef will also increase the haeme iron in the body cells, where haeme iron may induce ROS production and oxidative stress, and further DNA damage as a result of increasing risk of cancer development (Buzala et al., 2016). During this process, iron initiates lipid oxidation (i.e. the Fenton reaction) that produces ROS capable of abstracting a proton from unsaturated fatty acids. Free iron ions may bind to membrane phospholipids (e.g. phosphatidylcholine) and catalyse the degradation of lipid hydroperoxides to form peroxide radicals or alkoxyl radicals. The alkoxyl radicals may be further degraded to produce aldehydes, including malondialdehyde (Carlsen et al., 2005; Orino and Watanabe, 2008), which may induce DNA damage by reacting with deoxyguanosine, deoxyadenosine and deoxycytidine to form DNA adducts. The major DNA adduct is 1, N2-malondialdehyde-deoxyguanosine (M1dG) (Bastide et al., 2011). However, there is a complex antioxidant system, including SOD and glutathione peroxidase, which can remove ROS. Both the rate of ROS production and the rate of antioxidation determine the status of ROS, which is influenced by many factors other than iron. On the contrary, quite a lot of people around the world still suffer from iron deficiency due to low meat intake.

## 4.3 Heterocyclic amines

All meats contain creatinine, which can be converted into heterocyclic amines by Maillard reactions when cooked for longer periods at high temperature. Beef is more preferred for roasting, grilling, frying and barbecue. As cooking time and cooking temperature increases, the levels of heterocyclic amines in meat will increase greatly. Epidemiological studies indicate that there are some associations between the way meat is cooked and risk of cancer. Several forms of heterocyclic amines have been detected in some cooked meat, including grilled and roasted beef. Of these heterocyclic amines, 2-amino-1-methyl-6-phenylimidazo[4,5-b] pyridine (PhIP) and 2-amino-3,8-dimethylimidazo [4,5-f] quinoxaline (MeIQx) are relatively highly carcinogenic. In the liver, these heterocyclic amines are *N*-hydroxylated by cytochrome P450 or acetylated by *N*-acetyl transferases in the target organ. The *N*-hydroxylated and acetylated heterocyclic amines may bind to DNA and then disturb the replication, transcription and expression of target genes, for example, CYP1A1, CYP1A2, CYP2E1, NAT1, NAT2 and GSTM1 genes (Sugimura et al., 2004; Le Marchand et al., 1999). In animal models, 50% toxic dose (TD50) of PhIP is 2.2 mg/(kg·d) for rats and 64.6 mg/(kg·d) for mice, while MeIQx TD50 is 0.7 mg/(kg·d) for rats and 11 mg/(kg·d) for mice (Sugimura et al., 2004).

In grilled, fried or roasted meat, the concentrations of PhIP and MeIQx vary greatly, but they are below 1000 ng per 100 g meat (Mahfouz et al., 2014). It means that if humans have the similar sensitivity of heterocyclic amines to animals, tonnes of meat should be eaten a day to reach the carcinogenic level.

## 4.4 Polycyclic aromatic hydrocarbons

PAHs are environmental pollutants. Many PAH compounds, including benzo[*a*]pyrene, are carcinogenic and mutagenic. In grilled or smoked meat, there are relatively high quantities of PAH. Normally, grilled meat and smoked meat are safe to eat because the actual content is so low not as to cause any health problems. During grilling, meat fat is melted out and directly contacts fire or very hot surface to produce PAH compounds. During smoking, the PAH compounds are produced from the burning of sawdust and other wood materials. When the smoke containing PAH produced during grilling or smoking comes in contact with meat, some of PAH compounds are then deposited on the meat. Previous studies showed that there was a certain association between high intake of smoked meat and the risk of some cancers. Gunter et al. (2005) observed that a 6% increased risk of large (>1 cm) adenoma per 10 ng/day consumption of benzo[a]pyrene [OR = 1.06 (95% CI, 1.00–1.12)] and an incremental increase of 10 g of barbecued red meat per day was associated with a 29% increased risk of large adenoma [OR = 1.29 (95% CI, 1.02–1.63)]. However, no convincible scientific data verify this.

## 4.5 *N*-nitroso compounds (nitrosamines)

Nitrate and nitrite are commonly used for the purpose of curing meat products because they have functions such as colour enhancement and stability, anti-oxidation, preservation and taste enhancement. However, bacteria may convert nitrite and nitrate into nitrosamines during curing or during digestion in the gut. Nitrosamines have been shown to be both mutagenic and carcinogenic. In an *in vitro digestion and animal study, incubation* of faecal inoculum with overcooked nitrite-cured meat (90°C, 30 min) resulted in up to nine-fold higher 4-hydroxy-2-nonenal concentrations as compared to incubation with nitrite-cured raw and non-overcooked meat; intake of overcooked nitrite-cured meat also caused higher concentrations of the NOC-specific DNA adduct *O*-6-carboxy-methylguanine in colonic digesta (Van Hecke et al., 2015). However, nitrate/nitrite exists widely in some vegetables, which means that cured meat is not the sole source of nitrate/nitrite. In addition, nitrite/nitrate can be formed by gut bacteria from the amino acid arginine, indicating that undigested and unabsorbed dietary proteins could also have a certain contribution to the formation of nitrosamines.

Some earlier animal and epidemiological studies have shown a relationship between the exposure of nitrite, nitrate and *N*-nitroso compounds and the development of gastric cancer (Kuroiwa et al., 2008). The association between intake of meat, red meat or processed meat and risk of cancer were also extensively studied. However, those epidemiologic studies varied by study design (including ecologic, case-control and cohort studies), agent under consideration (nitrate, nitrite or *N*-nitroso compounds) and exposure assessment (estimates from meat intake based on food frequency questionnaires) (Brayn et al., 2012). Thus, the association between the intake of cured meat and the morbidity or mortality from cancer is doubtful.

# 5 Summary

Beef is a very good source of high-value protein, iron, zinc, selenium and B vitamins. During cooking or processing, several carcinogenic substances could be formed, including heterocyclic amines and PAHs. The risks of these substances causing cancers depend on a lot of factors, such as the amount, the chemical status and human physiology. Although some epidemiological studies have shown an association between higher intake of beef and the risk of cancer, there is no convincing evidence to prove this. Balanced diets could be the most important to avoid metabolic disorders and certain cancers.

# 6 Where to look for further information

Further information can be found in Dr. Chunbao Li co-authored peer-reviewed publications via website (https://www.researchgate.net/profile/Chunbao_Li/contributions).

# 7 References

Bastide, N. M., Pierre, F. H. F. and Corpet, D. E. (2011). Haeme iron from meat and risk of colorectal cancer: a meta-analysis and a review of the mechanisms involved. *Cancer Prevention Research*, 4: 177–84.

Bouvard, V., Loomis, D., Guyton, K. Z., Grosse, Y., Ghissassi, F. E., Benbrahim- Tallaa, L., Guha, N., Mattock, H. and Straif, K. (2015). Carcinogenicity of consumption of red and processed meat. *The Lancet Oncology*, http://dx.doi.org/10.1016/S1470-2045(15)00444-1.

Bryan, N. S., Alexander, D. D., Coughlin, J. R., Milkowski, A. L. and Boffetta, P. (2012). Ingested nitrate and nitrite and stomach cancer risk: an updated review. *Food and Chemical Toxicology*, 50: 3646–65.
Buzala, M., Słomka, A. and Janicki, B. (2016). Haeme iron in meat as the main source of iron in the human diet. *Journal of Elementology*, 21(1): 303–14.
Carlsen, C. U., Møller, J. K. S. and Skibsted, L. H. (2005). Haeme-iron in lipid oxidation. *Coordination Chemistry Reviews*, 249: 485–98.
Chen, Y. J., Zhou, G. H., Zhu, X. D., Xu, X. L., Tang, X. Y. and Gao, F. (2007). Effect of low dose gamma irradiation on beef quality and fatty acid composition of beef intramuscular lipid. *Meat Science*, 75: 423–31.
Ding, S., Li, C., Cheng, N., Cui, X., Xu, X. and Zhou, G. (2015). Redox regulation in cancer stem cells. *Oxidative Medicine and Cellular Longevity*, Article ID 750798.
Dunnigan, M. G., Henderson, J. B., Hole, D. J., Barbara, M. E. and Berry, J. L. (2005). Meat consumption reduces the risk of nutritional rickets and osteomalacia. *British Journal of Nutrition*, 94: 983–91.
Gunter, M. J., Probst-Hensch, N. M., Cortessis, V. K., et al. (2005). Meat intake, cooking-related mutagens and risk of colorectal adenoma in a sigmoidoscopy-based case-control study. *Carcinogenesis*, 26: 637–42.
Hu, F. B., Stampfer, M. J., Manson, J. E., Ascherio, A., Colditz, G. A., Speizer, F. E., et al. (1999). Dietary saturated fats and their food sources in relation to the risk of coronary heart disease in women. *American Journal of Clinical Nutrition*, 70: 1001–8.
Kuroiwa, Y., Yamada, M., Matsui, K., Okamura, T., Ishii, Y., Masumura, K., Tasaki, M., Umemura, T., Mitsumori, K., Nohmi, T., Hirose, M. and Nishikawa, A. (2008). Combined ascorbic acid and sodium nitrite treatment induces oxidative DNA damage-associated mutagenicity in vitro, but lacks initiation activity in rat forestomach epithelium. *Toxicological Sciences*, 104 : 274–82.
Le Marchand, L., Sivaraman, L., Franke, A. A., et al. (1999). Gene diet interactions and the high colorectal cancer risk of Japanese Americans[C]//UTSUNOMIYA J, MULVIHILL J J, WEBER W. Familial cancer and prevention: molecular epidemiology: a new strategy toward cancer, 329–35.
Li, C. B., Zhou, G. H. and Xu, X. L. (2007). Comparisons of meat quality characteristics and intramuscular connective tissue between beef longissimus dorsi and semitendinosus muscles from Chinese yellow bulls. *Journal of Muscle Foods*, 18: 143–61.
Li, D., Sinclair, A. J., Mann, N., Turner, A., Ball, M., Kelly, F., et al. (1999). The association of diet and thrombotic risk factors in healthy male vegetarians and meat-eaters. *European Journal of Clinical Nutrition*, 53: 612–19.
Li,Y., Li,C., Li,H., Lin,X., Deng, S. and Zhou, G. (2016). Physicochemical and fatty acid characteristics of stewed pork as affected by cooking method and time. *International Journal of Food Science and Technology*, 51: 359–69.
Mahfouz, E. M., Sadek, R. R., Abdel-Lateif, W. M., et al. (2014). The role of dietary and lifestyle factors in the development of colorectal cancer: case control study in Minia, Egypt. *Central European Journal of Public Health*, 22: 215–22.
McAfee, A. J., McSorley, E. M., Cuskelly, G. J., Moss, B. W., Wallace, J. M. W., Bonham, M. P. and Fearon, A. M. (2010). Red meat consumption: an overview of the risks and benefits. *Meat Science*, 84: 1–13.
Nuernberg, K., Dannenberger, D., Nuernberg, G., Ender, K., Voigt, J., Scollan, N. D., Wood, J. D., Nute, G. R. and Richardson, R. I. (2005). Effect of a grass-based and a concentrate feeding system on meat quality characteristics and fatty acid composition of longissimus muscle in different cattle breeds. *Livestock Production Science*, 94: 137–47.
Orino, K. and Watanabe, K. (2008). Molecular, physiological and clinical aspects of the iron storage protein ferritin. *Veterinary Journal*, 178: 191–201.
Park, G. B., Moon, S. S., Ko, Y. D., Ha, J. K., Lee, J. G., Chang, H. H. and Joo, S. T. (2002). Influence of slaughter weight and sex on yield and quality grades of Hanwoo (Korean native cattle) carcasses. *Journal of Animal Science*, 80: 129–36.
Santarelli, R. L., Pierre, F. and Corpet, D. E. (2008). Processed meat and colorectal cancer: A review of epidemiologic and experimental evidence. *Nutrition and Cancer*, 60: 131–44.

Schönfeldt, H. C., Naudé, R. T. and Boshoff, E. (2010). Effect of age and cut on the nutritional content of South African beef. *Meat Science*, 86: 674–83.

Sexten, A. K., Krehbiel, C. R., Dillwith, J. W., Madden, R. D., McMurphy, C. P., Lalman, D. L. and Mateescu, R. G. (2012). Effect of muscle type, sire breed, and time of weaning on fatty acid composition of finishing steers. *Journal of Animal Science*, 90: 616–25.

Sugimura, T., Wakabayashi, K., Nakagama, H. and Nagao, M. (2004). Heterocyclic amines: mutagens/carcinogens produced during cooking of meat and fish. *Cancer Science*, 95: 290–9.

USDA. (1997). United States standards for grades of carcass beef. Agriculture Marketing Service, USDA, Washington, DC.

Van Hecke, T., Vossen, E., Hemeryck, L. Y. et al. (2015). Increased oxidative and nitrosative reactions during digestion could contribute to the association between well-done red meat consumption and colorectal cancer. *Food Chemistry*, 187: 29–36.

Wen, S., Zhou, G., Song, S., Xu, X., Voglmeir, J., Liu, L., Zhao, F., Li, M., Li, L., Yu, X., Bai, Y. and Li, C. (2015). Discrimination of in vitro and in vivo digestion products of meat proteins from pork, beef, chicken, and fish. *Proteomics*, 15: 3688–98.

Zhou, G. H., Yang, A. and Tume, R. K. (1993). A relationship between bovine fat colour and fatty acids composition. *Meat Science*, 35: 205–12.

Zhu, Y., Lin, X., Zhao, F., Shi, X., Li, H., Li, Y., Zhu, W., Xu, X., Li, C. and Zhou, G. (2015). Meat, dairy and plant proteins alter bacterial composition of rat gut bacteria. *Scientific Reports*, 5:15220 | DOI: 10.1038/srep15220.

Part 4

# Emerging trends

Chapter 17

# The future of DNA technologies for improving beef quality: marbling, fatty acid composition and tenderness

*Elly Ana Navajas, Instituto Nacional de Investigación Agropecuaria, Uruguay*



## 1 Introduction

Agricultural production worldwide faces the challenge of increasing food production to meet the demands of the increasing human population. There is also a growing demand for food that meets key nutritional requirements, including beef, which is one of the principal sources of protein and other nutrients (Scollan et al., 2005). Biotechnology and, in particular, DNA technologies have created new opportunities in achieving sustainable intensification of production (FAO, 2011) by improving meat production and quality with minimum impacts on the environment.

In the last decade, significant developments of DNA technologies have increased the tools available to improve meat quality traits. Genomes of several livestock species have been sequenced, including cattle. In addition, technologies for measuring hundreds of thousands of polymorphisms across the entire genome are also available in different species (Meuwissen et al., 2016). Both sequencing and genotyping technologies are accessible at decreasing costs, making them easier to use at an academic and industry level. The availability of genomic information for a large number of animals allows the implementation of genome-wide association studies (GWAS), which provides a deeper and more comprehensive understanding of the genetic basis for traits of interest (Bolormaa

http://dx.doi.org/10.19103/AS.2016.0009.17

et al., 2014). In related fields, such as whole-genome expression studies and metabolomics, novel technologies provide valuable information about the biological processes from genotypes to phenotypes. By integrating these data, it will also be possible to improve the understanding of the genetic architecture of the traits of interest, which will help to improve current breeding strategies (Ibanez-Escriche and Simianer, 2016) and facilitate new ways of achieving the attributes desired by the consumer market.

Current breeding strategies of livestock include genomic selection (GS), which can accelerate the genetic progress already seen in current breeding programmes. It can even help to incorporate new traits that have not been included before, due to the difficulties and expense of measuring them. This is frequently seen when dealing with quality traits in the meat industry. When selection schemes are in place for livestock species, little emphasis has been given to attributes such as carcass composition and, in particular, the quality of the meat (Meuwissen et al., 2001, 2016). However, recently breeders and geneticists have given far more attention to the quality traits of meat. The need to satisfy increasing consumer expectations in the last few decades has increased efforts towards the genetic improvement of these traits (Simm et al., 2009).

Beef quality can be defined in many different ways and by several different traits. Palatability (which is key for consumer satisfaction) comprises tenderness, texture, juiciness, and flavour (Verbeke et al., 2010). The tenderness of meat is the primary driver, while the latter traits are taken into consideration once meat is suitably tender (Aaslyng, 2009). Nutritive value aspects include composition, protein and fat content, fatty acid profile, and mineral level (Leheska et al., 2008). Intramuscular fat content (often called marbling) is a very relevant trait from a nutritional perspective, given its influences on the risks of developing cardiovascular disease, obesity and various forms of cancer. This has led to some recommendations to limit consumption of some foods, including beef (Pan et al., 2012). The composition of fat, particularly the proportions of fatty acids, is very relevant because of the controversial associations between some fatty acids and consumer health. Interestingly, both fat content in the muscle and its fatty acid profile are traits associated with beef flavour and tenderness at genetic and phenotypic levels. Although environmental pre- and post-slaughter factors are important, these traits are strongly influenced by genetics (Dikeman, 1990). Reported estimates of heritability indicate there is potential for genetic improvement, but it has been limited by the difficulties experienced when recording the data (Simm et al., 2009). GS is therefore a revolutionary new technology that may well contribute to incorporating of quality traits in beef cattle breeding programmes.

In this chapter, the information provided by new DNA technologies and the use of genomics in both GWAS and GS are described in detail. Because the use of genomic information in beef breeding programmes shows its potential in improving the quality of beef, the current use of GS in beef cattle is discussed. The progress on the genomics applied to the genetic improvement of beef quality is presented here, as well as research findings for marbling, fatty acids profile and tenderness. The benefits of building large training populations and the alternatives to this, are also discussed.

# 2 Genotyping and sequencing technologies in the genomic era

Genomes of many livestock species have been sequenced in the last ten years including beef cattle. The Bovine Genome Sequencing and Analysis Consortium completed sequencing one of the largest genomes in 2009 (The Bovine Genome Sequencing and Analysis Consortium, 2009). Since then, significant developments of DNA technology and genome sequencing led to the detection of millions of single-nucleotide polymorphisms (SNP), making it possible to obtain a very dense coverage of the genome (Van Tassel et al., 2008).

The approaches that have been used for genome sequencing and genotyping have become faster and cheaper in the past decade. Dense, high-throughput genotyping assays have been developed for many livestock species, allowing the screening of tens to hundreds of thousands of SNPs simultaneously. The Illumina BovineSNP50 chip was developed by a consortium of animal scientists using SNP discovery populations in Holstein, Angus and other mixed breeds of beef cattle (Matukumalli et al., 2009; Van Tassell et al., 2008) and this provided a density of 50k SNPs per animal. In the following years, different SNP genotyping platforms have become commercially available, even new versions of the 50k. They differ in the number of SNPs included and in the potential role they play in GWAS and GS. Table 1 presents the SNP panels commercially available by private companies, such as Illumina, GeneSeek (Neogen) and Affymetrix.

This major innovation was rapidly adopted by the dairy industry, with the implementation of the first genetic evaluation with genomic data in 2007 (Wiggans et al., 2011). Many GWAS were carried out for different economically relevant traits in both beef and dairy cattle using this panel (dairy, Cole et al., 2011; Raven et al., 2014; beef, Saatchi et al., 2013a, Lu et al., 2013). The development by Illumina and Affymetrix of higher density panels with almost 800k SNPs that uniformly span the entire bovine genome (Table 1)

**Table 1** SNP panels available for cattle genotyping

| Product name | Acronym | Number of SNPs | Company |
|---|---|---|---|
| GoldenGate Bovine3K | 3k | 2,900 | Illumina |
| BovineLD BeadChip | 6k | 6,090 | Illumina |
| BovineLD v2.0 BeadChip | 8k | 7931 | Illumina |
| EuroGenomics 10K | EuroG10k | 10,000 | EuroGenomics |
| GeneSeek Genomic Profiler LD | GGP-LD | 8,762 | GeneSeek-Neogen |
| GeneSeek Genomic Profiler-LD chip | GGP-LD v4 | 30,125 | GeneSeek-Neogen |
| BovineSNP50K | 50kv | 54,001 | Illumina |
| BovineSNP50 v2 BeadChip | 50kv2 | 54,609 | Illumina |
| GeneSeek Genomic Profiler HD | GGP-HD | 76,992 | GeneSeek-Neogen |
| GeneSeek Genomic Profiler 150k | GGP-150K | 138,962 | GeneSeek-Neogen |
| GeneSeek Genomic Profiler Indicus HD | GGP-HD Indicus | 74,677 | GeneSeek-Neogen |
| Axiom Genome-Wide BOS 1 Array | BOS1 | 648,855 | Affymetrix |
| BovineHD BeadChip | 777k | 777,962 | Illumina |

provides precise resolution for mapping casual mutations, and increasing the power of GWAS (Rolf et al., 2010). The shorter distances between markers also give the high-density panels an important role to play in multi-breed analysis, which are very relevant for beef cattle GS (Berry et al., 2016; Meuwissen et al., 2016).

Following these significant developments, panels with a lower number of SNPs were made available with the objective of increasing the uptake of this technology by lowering genotyping costs. In 2010, the Bovine3K®, with 2900 SNPs, was released but was replaced in 2011 with the BovineLD® chip with 6909 SNPs (Boichard et al., 2012a). Custom SNP panels that include individual gene tests, as well as markers spread throughout the genome are also commercialised for cattle genomics by GeneSeek (8k SNP GGP-LD, 80k SNP GGP-HD and recently released GGP-150k) and a 10,000 SNP panel (EuroG10k) developed by a consortium of European partners based on the Illumina technology (http://www.fabretp.eu/eurogenomics.html).

The information of low- and medium-density panels can be used for inferring or deducing genotypes of any missing SNPs in the high-density panels. In addition, genomic data of un-genotyped animals can be imputed from genotyped relatives. Although there are many different imputation approaches (Sargolzae et al., 2014; Hickey et al., 2012; Browning et al. 2007), the accuracy tends to be high and the computational costs are reasonable. Studies using different cattle breeds and cross breeds reported high accuracies, with concordance rates above 0.90 when imputing from low-density panels to 50k and from 50k to 700k (Piccoli et al., 2014; Chud et al., 2015). Very similar values were also found when imputing un-genotyped parents from at least 3–4 genotyped progeny (Berry et al., 2014; Piccoli et al., 2014).

These methodologies are relevant for the implementation of GS because all animals should have genotypes of equivalent density, when databases currently include animals genotyped with different panels (Piccoli et al., 2014). In addition, they provide the opportunity of increasing the number of genotyped animals by inferring the genomic information of phenotyped animals, without causing the additional costs of re-genotyping.

Whole-genome sequencing on a large scale is now possible due to the advances in genome sequencing and the reduced cost. The 1000 bull genomes project (www.1000bullgenomes.com) is another international initiative that it is building a large database for imputation of genetic variants for GS and GWAS studies in all cattle breeds. The information provided by these sequences is very useful for the identification of casual variants for economically relevant traits (Daetwyler et al., 2014) and, as a by-product, many thousands of SNPs markers are identified (Meuwissen et al., 2016).

The innovation in genotyping and sequencing technologies as previously described, are accompanied by further developments in related fields such as transcriptomics (Ibanez-Escriche and Simianer, 2016). The current capacity of direct transcriptome sequencing (also known as RNA-seq, RNA sequencing) has increased the capacity to study gene expression, becoming the preferred approach for functional genomics studies. These developments may contribute in identifying casual mutations in the traits of interest.

## 3 Genomic selection and genome-wide association studies

GS is the use of genetic markers spread throughout the genome to assist in predicting individual genetic merit. This concept was introduced by Meuwissen et al. (2001), before

DNA technologies needed for its implementation were even available (VanEenennaam et al., 2014). With the later developments of DNA technologies and high-throughput genotyping platforms that are available, GS has become a reality in the livestock industry with varying levels of use among different species.

Considerable benefits in the genetic improvement of meat quality is expected by GS, due to the possibility of including economically relevant meat quality and enhancing the genetic progress of those that are already considered in conventional genetic evaluation. Genomics selection has the potential of accelerating the rate of genetic gain, by providing more accurate estimations of genetic values at early ages, which reduces the generation interval. The speed of genetic improvement) $\Delta G$/year depends on (Falconer and Mackay, 1996):

$$\Delta G / year = \frac{i r \sigma_a}{L}$$

- selection intensity (i), which is the proportion of animals in a population that are selected to become parents of the next generation,
- accuracy of selection (r), quantified as the correlation between the estimated breeding value (EBV) and the true breeding value,
- additive genetic standard deviation of the trait of interest ($\sigma_a$) that is related to the genetic variation in the population available for selection,
- generation interval (L), defined as the average age of parents when their offspring are born.

GS also provides the opportunity of including traits in the breeding programmes that have received less attention due to the difficulties or high cost of collecting the phenotypic information needed for the estimation of genetic merit (Garrick, 2011). This is often the case for meat quality, and therefore, it is one of the groups of economically relevant traits in which substantial benefits are expected to be seen by introducing genomics into breeding programmes. In addition, meat quality traits can be measured only post-mortem and, thus, require information to be collected on relatives or progeny of the selected individuals, resulting in limited accuracies of selection and a longer generational interval.

Conventional genetic evaluations rely on pedigree and phenotypic information to estimate the genetic merit of selection candidates using best linear unbiased prediction (BLUP) method (Kinghorn et al., 2015). Genetic merit is usually expressed as estimated breeding values (EBV) or expected progeny difference (EPD, one half of the EBV), which is the most common way of delivering the information of genetic merit in the beef industries in North and South America (Garrick and Fernando, 2015). By incorporating the genomic data, selection candidates are then compared based on genomic EBV (GEBV) or genomic EPD (GEPD).

Different ways of integrating the genomic information in genetic evaluations (blending) have been developed (reviewed by Swan et al., 2012). Initially, GEBVs were used as an additional trait in multi-trait BLUP evaluation (MacNeil et al., 2010) or combined using a selection-index approach (Hayes et al., 2009a). More recently, the single-step evaluation was proposed by Aguilar et al. (2010), in which phenotypic, genomic, and pedigree information are combined together in one model, in which information from genotyped and non-genotyped animals is used simultaneously.

The implementation of GS depends on building training or reference populations, which comprise of a large number of genotyped and phenotyped animals, providing the information to estimate the SNP effects (Fig. 1). In general terms, the genetic merit for any particular trait is the result of adding up the effects of all genes involved in the determination of that trait. For any specific animal, its breeding value is given by the summation of substitution effects of the alleles carried by the animal in all loci. Estimated SNP effects depend on the association between the segregating SNP genotypes and the segregating causal polymorphisms and the magnitude of this association reflects the extent of linkage disequilibrium (LD). With the high density of SNPs obtained across the genome with the current genotyping technologies, any QTL affecting a specific trait should be close to the genotyped markers in sufficient LD (Garrick, 2011).

In GS, SNP effects are estimated using training populations, which comprise information of production traits and genomic data. The estimates of SNPs effects are then applied to predict GEBV for the selection candidates based on their own genotyping information. Over time, genotypic and phenotypic information collected from selected animals are used to re-estimate SNP effects (recalibration). This is an important step because LD between SNPs and QTL may erode over time, decreasing the accuracy of previous estimates of SNP effects and, therefore, the accuracy of GEBV.

There are many aspects to take into account when considering training populations, given their associations with the accuracy of genomic predictions (Goddard, 2009; Hayes et al., 2009b). The accuracy of genomic predictions is influenced by several factors that depend on the specific breed and trait. These include effective population size, heritability and architecture of the trait (Goddard, 2009), which are relevant to consider but which cannot be improved. However, the size of the training population (Goddard, 2009; Hayes

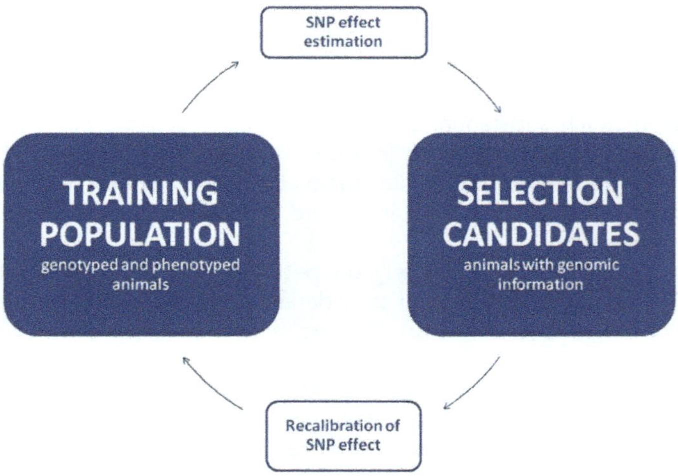


**Figure 1** Genomic selection in cattle requires a training (reference) population of animals with genomic and phenotypic information of the trait of interest, which is used for the estimation of the SNP effects. These estimations are used for the prediction of merit to selection candidates that do not have phenotypic records based on their individual genomic information obtained by genotyping DNA samples. Phenotypic information recorded later and the genomic information can be integrated to training populations and used to recalibrate the estimation of SNP effects.

et al., 2009b) the relationship among animals and the relationship between training and the candidate animal populations (Pszczola et al., 2012) are also relevant factors and these can be optimised when designing training populations and GS strategies.

The accuracy of estimated SNP effects and therefore the accuracy of GS increase with larger training populations (Meuwissen et al., 2001; Goddard, 2009). Accuracy improvements will be faster and maximum values higher for traits with high heritabilities (Goddard, 2009). Density of the SNP panel might affect the accuracy that can be achieved, with a denser panel resulting in higher accuracies, particularly when the effective population size is larger (Pszczola et al., 2012).

When phenotypes are difficult to measure and/or expensive, it is more difficult to build large training populations and genomic accuracies are expected to be low. Therefore, the design in the case of small reference populations may be more relevant to maximise the accuracies that can be achieved. Accuracies can be improved by a minimum relatedness among animals in the training population and a strong genetic relationship between the training population and the evaluated animals (Pszczola et al., 2012).

For the traits already under genetic evaluation, training populations are usually built using phenotypic information that was used in genetic evaluations and the resulting estimations of genetic merit. However, for most meat quality traits the estimations of SNP effects are usually based on existing experimental databases or research projects in which data are collected from reference animals in experimental herds (Garrick, 2011). Examples for beef quality traits are Beef Co-operative Research Centre in Australia (Bolormaa et al., 2013, 2014), the Qualvigène programme in France (Allais et al., 2014), the Carcass Merit Project (Dikeman et al., 2005; McClure et al., 2012), and the Meat Animal Research Center Germplasm Evaluation Program (Weber et al., 2012) in the United States.

Although GS does not rely on the actual identification of QTLs influencing the traits being evaluated, training populations are valuable databases to investigate the underlying genetics by GWAS. All markers with small or large effects are considered for genomic prediction. In the case of polygenic traits, Garrick and Fernando (2015) indicated that excluding some of the markers with small effects had no impact on predictive ability, though discarding many of them can result in a reduction of predictive ability. On the other hand, GWAS aim to identify the genes and corresponding alleles that contribute to the genetic variation in a particular trait, which is of interest from a biological perspective and provides relevant information to GS (Garrick and Fernando, 2015). One example of this contribution is the development of custom SNP panels that include markers linked to economically relevant traits. Furthermore, GWAS provide an insight into the architecture of the traits that is defined by the number of loci and the size of their effects, which is also relevant for GS.

The advances in molecular genetics in the 90´s were the basis for marker-assisted selection (MAS), which was expected to enhance the rate of genetic improvement by using DNA information. MAS required two steps. The first was the detection and mapping of QTLs related to the traits of interest. Secondly, the QTL information was used to predict the genetic merit of the animals (Meuwissen et al., 2016). Besides the important discovery of QTLs with large effects, genome-wide association studies with low-density markers failed to identify hundreds of QTLs with small effects that explained the majority of the genetic variation of quantitative traits.

Most of the economically relevant traits, including meat quality, are quantitative polygenic traits and recent research indicates, for example, that dairy traits are influenced by 2000–10 000 genes (Meuwissen et al., 2016). With the availability of high-density marker coverage of the genome and the possibility of genotyping large numbers of animals, QTL detection became more effective (Ducrocq and Wiggans, 2014).

GWAS, as well as GS, requires the estimation of SNP effects. Fitting every marker individually was one of the approaches used initially for GWAS in livestock, as previously applied in human studies. However, statistical methods in which all SNPs are fitted simultaneously but assuming different SNP effect distributions (i.e. normal distribution, t-distribution, etc.) have been proposed (for a review, Gianola 2013; de los Campos et al., 2013; Gianola et al., 2009). The mixed-linear model approach assumes that effects of all SNPs are normally distributed with equal variance. Because these GBLUP models perform well and have low computational requirements, they have become popular approaches for practical genomic prediction (Hayes et al., 2009b). Different approaches based on Bayesian regression models have also been proposed. The first approaches, described by Meuwissen et al. (2001), were Bayes A, which assumes a nonlinear distribution of the SNP effects with many SNPs having small effects but allow some of them to have a moderate to large effect, and Bayes B, which assumes a mixed distribution, with some fraction (p) of the SNPs with zero effect and the remainder (1p) having a t-distribution. Since then, several others have been also implemented, such as the Bayes Cπ and Bayes R methods (Erbe et al., 2012; Habier et al., 2011).

Genomic prediction accuracies obtained by different estimation approaches of SNP effects have been investigated by simulation and empirical studies (de los Campos et al., 2013). In general, simulation studies indicate that higher accuracies are obtained by those methods that assumed different SNP effects (Daetwyler et al., 2010; Meuwissen et al., 2001). Results obtained by field data analysis tend to confirm theoretical conclusions, but differences of accuracy estimates obtained by different methods are smaller. Nevertheless, higher accuracies are obtained when approaches such as Bayes B are used in traits influenced by large-effect QTLs. Bolormaa et al. (2013), who investigated the prediction accuracy of feed efficiency and carcass and meat quality traits, reported that Bayes R gave greater accuracy of GEBV than GBLUP and the advantage was greatest (up to 0.23) for traits with mutations of moderate effect and a high number of significant SNPs.

Denser SNP panels will increase the probability that any QTL has a SNP that is in perfect LD with it. With whole-sequence data, the causative SNP resulting in the QTL will be among the millions of polymorphisms that are genotyped. Therefore, genomic prediction will rely on the actual causative mutations with favourable effects on genomic prediction accuracy. Meuwissen and Goddard (2010), based on a simulation study, reported that whole-genome-sequence information improved the accuracy of GS relative to the accuracies available with dense SNP panels. Because the causative mutations for one trait are hidden among SNPs with no effect, maximum accuracies were achieved with Bayesian methods that assumed that many SNPs had no effect (Meuwissen and Goddard, 2010).

# 4 Genomic selection in beef cattle

The improvement of accuracy for many traits included in beef cattle breeding programmes by GS is less significant than in dairy cattle, which has slowed its adoption (Berry et al., 2016; Van Eenennaam et al., 2014). Accuracies of GBV for different breeds and traits are presented in Table 2. Accuracies in this case were assessed as the correlation between the GEBV and de-regressed estimated breeding values. De-regressed breeding values

**Table 2** Accuracy of genomic predictions[1] of growth, carcass traits and meat quality traits in beef cattle breeds

| Animals training population | Angus[3] 3570 | Simmental[4] 2703 | Limousin[4] 2239 | Hereford 1081[5] | Hereford 2980[6] |
|---|---|---|---|---|---|
| Traits[2] | | | | | |
| Birth weight | 0.56 | 0.65 | 0.58 | 0.45 | 0.68 |
| Weaning weight | 0.33 | 0.52 | 0.58 | 0.32 | 0.52 |
| Yearling weight | 0.36 | 0.45 | 0.74 | 0.32 | 0.60 |
| Rib-eye muscle area | 0.60 | 0.59 | 0.63 | 0.42 | 0.49 |
| Fat thickness | 0.60 | 0.29 | – | 0.45 | 0.48 |
| Marbling | 0.69 | 0.63 | 0.65 | 0.33 | 0.43 |
| Tenderness | | 0.53 | | | |

[1] Correlation between the genomic values and the de-regressed estimated breeding values
[2] Fat thickness, marbling and rib-eye muscle area are carcass traits evaluated based on actual carcass data and *in vivo* measurements by ultrasound scanning. Tenderness is measured by force required to shear a rib-eye steak.Based on [3] Saatchi et al. (2011), [4] Saatchi et al. (2012), [5] Saatchi et al. (2013b) and [6] Berry et al. (2016).

are the EBV provided by the genetic evaluations adjusted by the reliability of EBV and heterogeneity of the residual variance (Garrick and Fernando, 2015; Garrick et al., 2009).

Accuracy values for beef cattle traits reviewed by Garrick and Fernando (2015) and Van Eenennaam et al. (2014) are between 0.3 and 0.7 (see also Table 2). Although these average values are lower than those reported for economically relevant traits in dairy, GEBVs are more accurate than conventional parent average EBVs (Garrick and Fernando, 2015). Parental average EBVs are the estimations of genetic merit available in animals with no phenotypic and no offspring phenotypic records. The higher accuracy of GBV implies that including genomic information leads to better accuracies that may reflect a faster genetic improvement.

There are several reasons explaining the lower accuracies in beef cattle compared with dairy cattle. Larger training populations are needed to improve accuracies because of the larger effective population sizes of beef cattle breeds. In addition, the large number of breeds used in the beef industry complicates achieving high accuracies. Different strategies have been investigated. Applying in a breed genomic prediction equations estimated in another breed was not effective, as reported by Katchman et al. (2013), because SNP effects may differ among breeds (Purfield et al., 2015). Unfortunately, the use of prediction equations developed in one breed and in one country did not have satisfactory results when used for genomic prediction in the same breed from another country (Saatchi et al., 2013b). Similar results were also reported in dairy cattle (Hayes et al., 2009b). Nevertheless, using a multi-breed training population can improve the genomic prediction accuracy compared to a single-breed one, particularly for small breeds, given the reasonable representation in the training population and provided that the ancestors of the selection candidates are included (see Hozé et al., 2014). Similar findings were reported by De Haas et al. (2015) when pooling genotypes and dry matter intake phenotypes from 10 Holstein-Friesian research herds of Europe, North America and Australasia.

In general, phenotypes recorded on farms for the beef cattle evaluations are for traits that can be easily measured in selection candidates mainly related to growth, together with visual scores of muscularity and measurements or scores of height or even skeletal development (Kinghorn et al., 2015; Garrick, 2011). In addition, ultrasound assessment of carcass traits such as eye muscle area, subcutaneous fat depth and marbling may also be included (Bertrand, 2009; Nkrumah et al., 2007).

However, many of the key traits that influence profitability of beef production are not considered in genetic evaluations. Most of them are difficult to measure (e.g. feed efficiency and fertility), or occur late in life (e.g. stayability) (Van Eenennaam et al., 2014). Important traits such as tenderness and feed efficiency have had minimal prospects for selection by traditional means, as there are no easy and cost-effective phenotypic measures (Garrick, 2011).

Genetic improvement of carcass and meat quality traits by conventional selection have been limited by the difficulties and high cost of recording these traits, as well as the complexity of implementing progeny tests that require individual traceability from birth to slaughter and in the case that traits can be assessed on the carcass, or at the specialised facility for measuring meat quality.

There have been few projects in the world that have recorded a large volume of information on carcass and meat quality traits, but these few have provided very useful information on genetic variation within and between breeds and genetic associations between traits (i.e. Carcass Merit Project (Dikeman et al., 2005; McClure et al., 2012); Meat Animal Research Center Germplasm Evaluation Program (Weber et al., 2012)). In addition, there are limited examples of genetic evaluations that have included the measurements of meat quality traits such as tenderness (i.e. Limousin and Simmental in the United States) and in general, due to the difficulties mentioned, the data that was recorded have only allowed for the estimation of EBV of a relatively small number of bulls.

Both large projects and genetic evaluation databases are valuable resources for the implementation of GS with high accuracies. Nevertheless, there is a general absence of historical databases on traits that are more difficult to measure, which limits the implementation of GS in these cases. This has constrained the expected economic benefits from GS in beef cattle and, consequently, the associated genotyping costs are barrier to implementation.

Although implementation of GS in beef cattle has generally been slow, there are within-breed genomic evaluations already in place or close to official implementation status, and research on genomic evaluations has been carried out in several countries (Berry et al., 2016). American Angus was the first to implement genomic evaluation on a national scale with a training population of more than 50,000 genotyped animals (Lourenco et al., 2015). Later, Australian Angus introduced genomic prediction equations developed with accuracies between 0.20 and 0.45 (Johnston et al., 2012), values that are below those reported for American Angus, Limousin, Simmental and Hereford by Saatchi et al. (2011), (2012) and (2013b). In the United States, the American Hereford Association has also incorporated genomic data into the national genetic evaluation by developing their own training population and prediction equations (Berry et al., 2016). Ireland and the United Kingdom are very close to releasing genomic prediction for economically relevant traits. The Irish Cattle Breeding Federation (http://www.icbf.com) will release GEBV for all beef breeds based on a training population of more than 100,000 animals in 2016, while in the

UK EGENEs (www.sruc.ac.uk/info/120275/egenes) will undertake the genomic evaluation for carcass traits in the Limousin breed, based on training populations built on phenotypic information from carcass video image analysis.

# 5 Genomic selection and genome-wide association studies of beef quality traits

Conventional genetic improvement of beef carcass and meat quality traits has been based on pedigree and carcass data recorded on a sample of progeny through organised progeny tests (Bertrand, 2009). As selection candidates are usually relatives of the slaughtered cattle, EBVs for these traits will usually have low accuracies. Another factor is that carcass measurements are taken late in the animals life after it has been harvested, leading to a longer generation interval, which reduces selection response per year (Falconer and Mackay, 1996), compared with the growth traits that can be easily measured in the selection candidates. The low genetic progress by progeny testing, the complexity of data recording in abattoirs, and the relatively high cost per animal have limited the genetic improvement of carcass and beef quality traits.

One way of augmenting genetic progress of carcass and meat quality traits is by developing methods to measure or predict these traits to be used in breeding programmes (for a review, Simm et al., 2009). The availability of *in vivo* measurements by ultrasound scanners enables cost-effective records in yearling bulls and heifers as this can be carried out on farms. This increases the number of evaluated animals at a younger age compared to traditional progeny testing (Bertrand, 2009; Williams, 2002). Consequently, many beef cattle breeds have EPDs for carcass traits, as reported by Garrick (2011). Beyond the great impact of ultrasound measurements for carcass quality, there is a lack of precise *in vivo* methods to predict meat quality traits (Simm et al., 2009). GS is a possible solution to genetic improvement of these traits.

GEBV are delivered for a comprehensive list of economically relevant traits in the beef industry, which were already included in conventional genetic evaluations. Although the most important contribution of GS is in traits that are difficult to select, such as meat quality (Meuwissen et al., 2016), they have not been incorporated apart from marbling, which is already in the conventional genetic evaluations in many beef breeds (Garrick, 2011). Nevertheless, research on genomics of beef quality traits is ongoing. Results from investigations of GS and GWAS are presented for marbling, fatty acid profiles, and also tenderness.

## 5.1 Marbling

Marbling has been included in carcass traits since the 90´s for the genetic evaluation of beef cattle. Marbling scores are traditionally assessed in the quartered beef carcasses based on the visible intramuscular fat (IMF) in the *longissimus dorsi* muscle and by ultrasound based on the prediction of IMF. Because a premium is paid to carcasses with high marbling, there is a clear economic motivation to improve marbling in North American, Australian and East Asian markets. More recently, interest in marbling in several exporting countries has been fuelled by its popularity in the Japanese beef market (Kinghorn et al., 2015).

In addition, given the difficulties of measuring quality traits and the association between intramuscular fat and the eating quality traits, carcass marbling has been used in place of eating quality (Garrick, 2011). In North America, the premium paid for high marbling is based on its value as an indicator of good eating quality (Kinghorn et al., 2015).

Accuracies of genomic predictions of marbling are presented in Table 2. The magnitude of the estimates reported by Saatchi et al. (2011), (2012) and (2013b) are similar to other carcass and growth traits. These values estimated in single-breed training populations are greater than those reported by Bolorma et al. (2013) based on a multi-breed training population that also included cross-bred animals. The estimates for marbling were 0.25 and 0.27 based on Bayes R and GBLUP approaches, respectively. These results are slightly higher than estimates obtained by Weber et al. (2012). They investigated the predictive ability for growth and carcass traits of two multi-breed beef cattle training populations: 3,358 cross-bred cattle of the U.S. Meat Animal Research Center Germplasm Evaluation Program (USMARC_GPE) and 1,834 high accuracy bull sires of the 2,000 Bull Project (2000_BULL) representing influential breeds in the US beef cattle industry. The estimate reported for marbling was 0.23 when USMARC_GPE was used as the training population and 2000_BULL as a validation population. When 2000_BULL was used as a training population and then validated with the USMARC_GPE population the estimate was 0.19. Onogi et al. (2015) reported a realised prediction accuracy for marbling scores assessed in Japanese Black cattle of 0.44. This accuracy obtained by ssGBLUP was 44% greater than the accuracy of conventional EPD by BLUP. It is difficult to compare the accuracies from different studies because of differences in trait heritabilities for different populations, the size of training and validation populations, and the statistical methods to estimate marker effects. However, the results suggest multi-breed populations tend to provide less accurate genomic predictions.

## 5.2 Fatty acid profiles

Due to the perceived and controversial association between beef fat and coronary diseases and obesity, consumers have been encouraged to reduce their beef intake. Although beef has a high nutritional value as it is a rich source of essential amino acids, minerals, B vitamins and essential polyunsaturated fatty acids such as linoleic and linolenic acid, human nutritionists have criticised beef for its high proportion of saturated fatty acids (SFA) and low proportion of polyunsaturated fatty acids (PUFA). Yet, beef has nearly as much monounsaturated fat (mostly oleic acid) as saturated fat, and the main saturated fat is stearic, which has been shown to be neutral in increasing serum cholesterol. Beef has considerable amounts of n-3 PUFA, which is reflected in beneficial low ratio n-6/n3 fatty acids. Beef is also a major dietary source of conjugated linoleic acid (CLA), which has health-promoting biological properties (Mateescu, 2015; Scollan et al., 2005).

Quantifying the composition of IMF requires expensive and labour intensive chemical analysis, which has limited the inclusion of fatty acid profiles of IMF as a trait in beef breeding programmes, even though estimates of genetic heritability of fatty acids are low to moderate (Mateescu, 2015). Although these difficulties have also restricted understanding of the underlying genetic mechanisms, recent GWAS studies have provided more information on genomics and GS potential application (Hamill et al., 2012). Accuracies of genomic predictions for groups of fatty acids and some key individual acids are presented in Table 3.

**Table 3** Realised accuracy of genomic EBV for fatty acids of longissimus muscle

| | Chen et al. (2015) | | Saatchi et al. (2013a) | Onogi et al. (2015) |
|---|---|---|---|---|
| Trait | GBLUP | BayesCπ | BayesB | GBLUP |
| C14:0 | 0.45 | 0.73 | 0.57 | 0.66 |
| C14:1 | | | | 0.70 |
| C16:0 | 0.36 | 0.50 | 0.53 | 0.51 |
| C16:1 | | | | 0.72 |
| C18:0 | 0.35 | 0.38 | 0.27 | 0.58 |
| C18:1 | | | 0.35 | 0.65 |
| C18:2 | | | | 0.56 |
| Total saturated fatty acids | 0.37 | 0.43 | 0.45 | |
| Total polyunsaturated | 0.30 | 0.33 | 0.08 | |
| Total monounsaturated | 0.27 | 0.31 | 0.38 | 0.55 |
| PUFA/SFA | 0.38 | 0.39 | 0.12 | |
| Omega 3 | 0.40 | 0.40 | 0.01 | |
| Omega 6 | 0.29 | 0.32 | 0.07 | |
| Omega 6/omega 3 | 0.16 | 0.14 | 0.00 | |

Chen et al. (2015) carried out GWAS for 81 and 83 fatty acid traits that were measured in subcutaneous adipose and longissimus lumborum (LL) muscle based on a training population of 1366 purebred and cross-bred steers and heifers. All the animals were genotype with 50K SNP panel and genomic accuracies were estimated using GBLUP and Bayes Cπ approaches. The estimates obtained from both approaches are presented in Table 3. Reported average accuracies for all fatty acid traits from GBLUP and Bayes Cπ were 0.32, and 0.35, respectively, in subcutaneous adipose, and 0.39, and 0.46, respectively, in longissimus lumborum muscle. These estimates were greater than those obtained by 'conventional' BLUP (0.23 and 0.17), confirming the effectiveness of genomic prediction, although BLUP accuracies might be underestimated given that only one generation of pedigree information was available. Accuracies between 0.51 and 0.70 were calculated by Onogi et al. (2015) for seven individual fatty acids and MUFA (see Table 3).

Accuracies of genomic prediction for 24 fatty acid traits in Angus beef cattle were reported by Saatchi et al. (2013b) based on a training population of 2177 animals. In general, accuracies were lower than those reported by Chen et al. (2015). This is the case for most of the traits in Table 3, with the exception of SFA and MUFA. Estimates of 0.70 and 0.40 for PUFA and MUFA were presented by Ahlberg et al. (2014) in a training population of 227 cross-bred animals. These high estimates may be due to the fact that these are the accuracies of the training analysis, which tend to be greater than realised accuracies.

Predictive ability should not be determined directly from training analyses. The realised accuracy must be determined in separate validation analyses (Garrick and Fernando, 2015). However, it is often very difficult to have independent validation populations. An alternative approach for the validation in real beef cattle data is to partition the training population into groups, and undertake training analyses in all but one group, then use the final group for validity. In general, groups are made up of related animals, so validation animals will not have close relatives in training (Garrick, 2011; Saatchi et al., 2011). It is important to take this into consideration; otherwise, the expected responses of GS could be overestimated.

Marbling, intramuscular fat and fatty acids are polygenic traits, for which GWAS have determined several relevant genome regions. Previous studies, using low maker density, mapped several QTL across the genome. Findings were reported in most of the 29 chromosomes and candidate genes have also been identified (Hamill et al., 2012; Pannier et al., 2010). Current techniques that allow genotyping for very high marker density increase the power of detecting markers in linkage disequilibrium with causative SNP, which is the final aim of DNA association studies.

Chen et al. (2015) identified 360 significant SNPs associated with one or more fatty acid traits in the LL tissues, with proportions of genotypic variance explained by individual significant SNPs that varied from 0.005 to 24.28% in LL. A total of 87, 117 and 120 SNPs were found for PUFAs, SFAs and MUFAs, respectively. Many regions in the genome that harbour good candidate genes have been mapped by WGAS. Saatchi et al. (2013a) and Chen et al. (2015) found very significant associations of different fatty acids with SNPs at BTA 19, BTA 29 and BTA 26, in which fatty acid synthase (FASN), thyroid hormone responsive (THRSP) and sterol-CoA desaturase (SCD) are major candidate genes. These results are in agreement with GWAS of fatty acid profile of subcutaneous fat by Kelly et al. (2014).

In all studies, there are regions associated with different fatty acid traits but no information is available to identify candidate genes. In general, there is very poor information for regions that explain the low proportion of genetic variation. Interestingly, regions that harbour genes that have been associated in previous studies were not relevant in all studies, as pointed out by Saatchi et al. (2013a). Genes such as SREBP1, ACACA, ACSL1, FABP4, LEP, LXRA and PPARG were associated with marbling, intramuscular fat, and its composition (Saatchi et al., 2013a). Potential candidate genes GULP1 and ITGAV located on BTA2 in close proximity to C313Y were identified by Ahlberg et al. (2014), but more research is needed. In general, further investigation on genomics and gene expression will contribute to furthering the understanding of the genetic basis of fatty acid composition.

## 5.3 Tenderness

The eating quality attribute that is most relevant for consumers is the tenderness of the meat. Given its importance, several studies on genetic variation of tenderness have been carried out and candidate genes have been identified (Mateescu, 2015; Allais et al., 2014). One of the most relevant findings, which have been confirmed in many recent GWAS, is the identification of candidate genes that are associated with post-mortem proteolysis. The calpain enzyme family of calcium-dependent cysteine proteases, in particular the μ- and m-calpain enzymes break down the meat fibres during the post-mortem proteolysis increasing tenderness (Allais et al., 2014; Hamill et al., 2012), although studies showed that m-calpain plays a minor role in beef tenderisation (Camou et al., 2007). The calpastatin activity, regulated by the CAST gene, also plays an important role in proteolysis by inhibiting the calpains. High levels of calpastatin reduce the proteolysis activity of calpain, and thus decreases beef tenderness (Pringle et al., 1997).

The μ- and m-calpain are encoded by calpain 1 (CAPN1) and calpain 2 (CAPN2) genes, respectively. Significant associations with shear force (an objective measure of tenderness) were identified for three SNPs at the μ-calpain locus (White et al., 2005; Page et al., 2002). The 316 and 4751 polymorphisms in particular have been validated in a number of independent populations (Hamill et al., 2012). A SNP in the calpain 3 gene has also been identified and patented for its association with beef tenderness (Barendse et al., 2008). Four SNPs associated with tenderness have been located in the CAST gene region and 3 of them confirmed in validation studies (Van Eenennaam et al., 2007). Given the relevance of these findings and the significance of beef tenderness for the industry, markers for both calpain and calpastatin genes were included in the commercial DNA test GeneSTAR Tenderness and Igenity Tender-GENE (Mateescu, 2015). These tests were validated in several breeds by Van Eenennaam et al. (2007), who reported that markers in both panels have very significant associations with tenderness. Candidates genes associated with beef tenderness were reviewed by Allais et al. (2014).

Although these findings indicate that tenderness is regulated by major genes, GWAS studies show that the architecture of the trait is complex and many SNPs across the genome have a relatively small effect on tenderness. GWAS studies confirmed the relevance of CAST and CAPN1 in different breeds and populations (i.e. Allais et al., 2014; Bolormaa et al., 2014; Bolormaa et al., 2013; McClure et al., 2012; Bolormaa et al., 2011) but also identified many other QTL or genomic regions that can affect tenderness. McClure et al. (2012) found that even CAPN1 and CAST being the QTL most strongly associated with shear force, there were 78 regions associated with this trait in at least 3 of the 5 taurine breeds considered in the WGAS. Similarly, Bolormaa et al. (2013) located very significant SNPs in the CAPN1 and CAST regions, among several other significant SNPs for shear force.

Shear force is an instrumental indicator of tenderness that is commonly used, but it does not consider other sensory factors that influence consumer preference, which can only be evaluated by taste test panels (Mateescu, 2015). Trained tasting panels are made up of individuals with the ability to distinguish textures and flavours and they are trained in how to score different sensory characteristics (Simm et al., 2009). Allais et al. (2014) performed GWAS of tenderness assessed by shear force and a tenderness score given by a sensory panel in Charolais, Limousin and Blonde d'Aquitaine. Despite the strong genetic correlation between shear force and the tenderness score estimated in all the breeds (−0.86 to −0.91), the QTL for these traits were not the same. This indicates that results should be considered carefully, given that even correlating traits can have different QTL affecting them.

GWAS can provide important information in pleitropic effects as showed by Bolormaa et al. (2014). In the last decade, an improvement of beef tenderness has been reported in North America, which has been attributed to an increase of the level of intramuscular fat. The magnitude of the association between intramuscular fat and tenderness has been somewhat controversial. Moderate and positive genetic correlations between intramuscular fat and consumer panel tenderness were estimated in temperate breeds by Reverter et al. (2003) and Wheeler et al. (2010). In agreement with this, Bolormaa et al. (2014) reported that SNPs with an effect on tenderness will usually influence marbling as well.

GWAS provides information to help understand the basis of genetic correlations among economically relevant traits. In some cases, genetic correlations are not favourable to breeding objectives such as improving tenderness and keeping fat content constant due to consumer concerns regarding health. Another example is the unfavourable negative correlation between beef retail yield and intramuscular fat (range: −0.60 to +0.12, reviewed by Marshall, 1999). The characterisation of how QTL affect different traits may contribute to the optimisation of genetic gain in both traits by different selection pressures. In addition, QTL with favourable patterns for traits with undesirable correlations can be identified as good targets for selection. Bolormaa et al. (2014) reported a QTL on BTA 4 with an allele that increases retail beef yield and marbling but also decreases subcutaneous fat. This would be an interesting selection target for beef production systems aimed towards markets paying for high yields and marbling such as the United States and Australasian countries, without unnecessary increases of subcutaneous fat that generally enters the by-product stream.

The realised accuracy of 0.53 for genomic predictions of tenderness in Simmental was reported by Saatchi et al. (2012) (Table 2). Based on a multi-breed training population of 5358 genotyped animals, Bolormaa et al. (2013) estimated realised accuracies of 0.41 and 0.25, using BayesR and GBLUP, respectively. As expected, due to the effect of major QTL on tenderness, the estimates were larger using BayesR. In agreement with Saatchi et al. (2012), accuracies for tenderness were similar to those traits already included in genetic evaluation systems.

# 6 Future trends

GS provides the opportunity of including meat quality in breeding programmes enabling the selection for these traits with higher accuracies at a younger age. This is clearly the case of those characteristics that are already in genetic evaluation systems. There are examples of genetic evaluations that provide EBV for some carcass traits and marbling, in which adding genomic information to the prediction of genetic merits will increase accuracies of EBV for these traits. It also provides the opportunity of increasing selection intensity if more selection candidates are also evaluated based on their genomic information. Increasing accuracies, selection intensities and reducing generation intervals, are aspects that directly relate to faster genetic gains.

Genomics offer the possibility of integrating the very expensive and difficult to measure meat quality traits which have not been considered in breeding programmes, besides the relevance that they have for consumers and markets. In this chapter only two of these traits have been discussed. Beef tenderness is evaluated in very few breeds and countries in the world, while no EBVs are available for intramuscular fat composition. There are also other very relevant traits such as flavour, palatability, or mineral content that may be taken into consideration in the future. The final decision about the new traits to be included will depend on market demands and economic incentives, but GS is a novel tool to genetically improve them.

Current results indicate that accuracies of EBV for marbling, tenderness and fatty acids profile are moderate with the current sizes of training populations. These results are even more promising because theoretical and field studies indicate that accuracies of all traits improve if training population sizes increase. Several complementary strategies should be considered with the aim of increasing training populations and improving genomic accuracies. Strategies such as building multi-breed training populations and nonlinear analysis based on high-density SNPs or genome sequence data have been suggested. Combining training populations from different countries may improve statistical power for genome-wide association studies and genomic prediction accuracies. Large training populations will help with the identification of genomic regions influencing meat quality traits and other economically relevant traits, as well as the basis of the genetic correlations among them. Genome sequencing data will also greatly improve the understanding of the genetic basis of these traits, and in complementation with transcriptomic and other 'omics' sciences, identify genes and gene networks.

Minimising the cost of phenotyping is another important challenge of building large training populations of traits that are difficult and expensive to measure. Carcass and meat quality recorded by progeny testing is a valuable resource for training populations. A large volume of phenotypic data from commercial animals can be included in training populations, which can be particularly relevant for genomic prediction of carcass traits and marbling score. In addition, developments of fast and non-invasive methods to predict meat quality traits would be very valuable and would increase training populations. These may facilitate a fast screening of large numbers of carcasses with minimum additional phenotyping cost. This would contribute to a more cost-effective implementation of the training populations and help lead to a continuous expansion, which will improve accuracies and encourage the uptake within the industry.

# 7 Where to look for further information

Implementation of GS and GWAS are recent but have evolved rapidly. Information on methodology, simulations studies and results based on field data are published in many different journals (i.e. *Animal, Animal Genetics, Animal Production Science, BMC genomics, Genetics, Genetics Selection Evolution, Journal of Animal Science, Journal of Dairy Science, Livestock Science, Nature Genetics, PLoS Genetics*). In addition to this, many books on this topic have been published such as Gondro, C., van der Werf, J. and Hayes, B. (Eds), 'Genome-Wide Association Studies and Genomic Prediction', Berlin, Springer Series, 2013.

Several free software for the analysis of genomic data are available. Examples are fastStructure (https://rajanil.github.io/fastStructure/4), BLUPF90 family (http://nce.ads.uga.edu/wiki/doku.php) and GenSel (http://www.biomedcentral.com/content/supplementary/1471-2105-12-186-s1.pdf).

There are also several resources online. Information about marked genes and their functions is available in the Ensembl database (http://www.ensembl.org) for many livestock species, reported QTLs are in the AnimalQTL database (http://www.animalgenome.org/cgi-bin/QTLdb/index) and genomic pathway information is in http://www.genome.jp/kegg/pathway.html.

# 8 References

Aaslyng, M. D. (2009), 'Trends in meat consumption and the need for fresh meat and meat products of improved quality', In: *Improving the Sensory and Nutritional Quality of Fresh Meat*, Eds J. Kerry and D. A. Ledward, Cambridge, UK: Woodhead Publishing Ltd, 3–18.

Aguilar, I., Misztal, I., Johnson, D. L., Legarra, A., Tsuruta, S. and Lawlor, T. J. (2010), 'Hot topic: a unified approach to utilize phenotypic, full pedigree, and genomic information for genetic evaluation of Holstein final score', *Journal Dairy Science*, 93, 743–52.

Ahlberg, C. M., Schiermiester, L. N., Howard, J. T., Calkins, C. R. and Spangler, M. L. (2014), 'Genome wide association study of cholesterol and poly- and monounsaturated fatty acids, protein, and mineral content of beef from crossbred cattle', *Meat Science*, 98, 804–14.

Akanno, E. C., Plastow, G., Woodward, B. W., Bauck, S., Okut, H., Wu, X-L., Sun, C., Aalhus, J. L., Moore, S. S., Miller, S. P., Wang, Z. and Basarab, J. A. (2014), 'Reliability of molecular breeding values for Warner-Bratzler shear force and carcass traits of beef cattle – An independent validation study', *Journal of Animal Science*, 92, 2896–904.

Allais, S., Levéziel, H., Hocquette, J. F., Journaux, L., Rousset, S. and Denoyelle, C. (2014), 'Fine mapping of quantitative trait loci underlying sensory meat quality traits in three French beef cattle breeds', *Journal of Animal Science*, 92, 4329–41.

Barendse, W., Harrison, B. E., Bunch, R. J. and Thomas, M. B. (2008), 'Variation at the Calpain 3 gene is associated with meat tenderness in zebu and composite breeds of cattle', *BMC Genetics*, 9, 41.

Berry, D. P., Garcia, J. F. and Garrick, D. J. (2016), 'Development and implementation of genomic predictions in beef cattle', *Animal Frontiers*, 6, 32–7.

Berry, D. P., McParland, S., Kearney, J. F., Sargolzaei, M. and Mullen, M. P. (2014), 'Imputation of ungenotyped parental genotypes in dairy and beef cattle from progeny genotypes', *Animal*, 8, 895–903.

Bertrand, J. K. (2009), 'Using actual and ultrasound carcass information in beef genetic evaluation programs', *Revista Brasileira de Zootecnia*, 38, 58–63.

Boichard, D., Chung, H., Dassonneville, R., David, X., Eggen, A., Fritz, S., Gietzen, K. J., Hayes, B. J., Lawley, C. T., Sonstegard, T. S. et al. (2012), 'Design of a bovine low-density SNP array optimized for imputation', *PLoS ONE* 7, e34130.

The Bovine Genome Sequencing and Analysis Consortium (2009), 'The genome sequence of taurine cattle: a window to ruminant biology and evolution', *Science*, 324, 522–8.

Browning, S. R. and Browning, B. L. (2007), 'Rapid and accurate haplotype phasing and missing-data inference for whole-genome association studies by use of localized haplotype clustering', *American Journal of Human Genetics*, 81, 1084–97.

Bolormaa, S., Neto, L. R., Zhang, Y. D., Bunch, R. J., Harrison, B. E., Goddard, M. E. and Barendse, W. (2011), 'A genome- wide association study of meat and carcass traits in Australian cattle', *Journal of Animal Science*, 89, 2297–2309.

Bolormaa, S., Pryce, J. E., Kemper, K., Savin, K., Hayes, B. J., Barendse, W., Zhang, Y., Reich, C. M., Mason, B. A., Bunch, R. J. et al. (2013), 'Accuracy of prediction of genomic breeding values for residual feed intake and carcass and meat quality traits in Bos taurus, Bos indicus, and composite beef cattle', *Journal of Animal Science*, 91, 3088–3104.

Bolormaa, S., Pryce, J. E., Reverter, A., Zhang, Y., Barendse, W., Kemper, K., Tier, B., Savin, K., Hayes, B. J. and Goddard, M. E. (2014), 'A multi-trait, meta-analysis for detecting pleiotropic polymorphisms for stature, fatness and reproduction in beef cattle', *PLoS Genetics* 10: e1004198.

Camou, J. P., Marchello, J. A., Thompson, V. F., Mares, S. W. and Goll, D. E. (2007), 'Effect of postmortem storage on activity of μ-and m-calpain in five bovine muscles', *Journal of Animal Science*, 85, 2670–81.

Chen, L., Ekine-Dzivenu, C., Vinsky, M., Basarab, J., Aalhus, J., Dugan, M. E. R., Fitzsimmons, C., Stothard, P. and Li, C. (2015), 'Genome-wide association and genomic prediction of breeding values for fatty acid composition in subcutaneous adipose and longissimus lumborum muscle of beef cattle', *BMC Genetics*, 16, 135.

Chud, T. C. S., Ventura, R. V., Schenkel, F. S., Carvalheiro, R., Buzanskas, M. E., Rosa, J. O., Mudadu, M. A., Vinicius, M., Mokry, F. B., Marcondes, C. R., Regitano, L. C. A. and Munari, D. P. (2015), 'Strategies for genotype imputation in composite beef cattle', *BMC Genetics*, 16, 99.

Cole, J. B., Wiggans, G. R., Ma, L., Sonstegard, T. S., Lawlor, T. J., Jr, Crooker, B. A., Van Tassell, C. P., Yang, J., Wang, S., Matukumalli, L. K. and Da, Y. (2011), 'Genome-wide association analysis of thirty one production, health, reproduction and body conformation traits in contemporary U.S. Holstein cows', *BMC Genomics*, 12, 408.

Daetwyler, H. D., Capitan, A., Pausch, H., Stothard, P., van Binsbergen, R., Brøndum, R. F., Liao, X., Grohs, C., Jung, S., Baur, A., Esquerré, D., Gollnick, N., Legrain, S., Laurent, P., Djari, A., Rossignol, M. N., Klopp, C., Rocha, D., Fritz, S., Eggen, A., Bowman, P., Coote, D., Chamberlain, A., VanTassel, C. P., Hulsegge, I., Goddard, M., Guldbrandtsen, B., Lund, M. S., Veerkamp, R., Boichard, D., Fries, R. and Hayes, B. J. (2014), 'Whole-genome sequencing of 234 bulls facilitates mapping of monogenic and complex traits in cattle', *Nature Genetics*, 46, 858–65.

de Haas, Y., Pryce, J. E., Calus, M. P. L., Wall, E., Berry, D. P., Løvendahl, P., Krattenmacher, N., Miglior, F., Weigel, K., Spurlock, D., Macdonald, K. A., Hulsegge, B. and Veerkamp, R. F. (2015), 'Genomic prediction of dry matter intake in dairy cattle from an international data set consisting of research herds in Europe, North America and Australasia', *Journal of Dairy Science*, 98, 6522–34.

de los Campos, G., Hickey, J. M., Pong-Wong, R., Daetwyler, H. D. and Calus, M. P. L. (2013), 'Whole genome regression and prediction methods applied to plant and animal breeding', *Genetics*, 193, 327–45.

Dikeman, M. E. (1990), 'Genetic effects on the quality of meat from cattle', *Proceedings of the 4th World Congress on Genetics Applied to Livestock Production* XV, 521–30.

Dikeman, M. E., Pollak, E. J., Zhang, Z., Moser, D. W., Gill, C. A. and Dressler, E. A. (2005), 'Phenotypic ranges and relationships among carcass and meat palatability traits for fourteen cattle breeds, and heritabilities and expected progeny differences for Warner-Bratzler shear force in three beef cattle breeds', *Journal of Animal Science*, 83, 2461–7.

Erbe, M., Hayes, B. J., Matukumalli, L. K., Goswami, S., Bowman, P. J., Reich, M., Mason, B. A. and Goddard, M. E. (2012), 'Improving accuracy of genomic predictions within and between dairy cattle breeds with imputed high-density single nucleotide polymorphism panels', *Journal of Dairy Science*, 95, 4114–29.

Falconer, D. S. and Mackay, T. F. C. (1996), *Introduction to Quantitative Genetics*. Fourth edition. Longman, Harlow, UK.

FAO (2011), FAO in the 21st century: Ensuring food security in a changing world. www.fao.org/docrep/015/i2307e/i2307e.pdf (accessed 11 January 2016).

Garrick, D. J. (2011), 'The nature, scope and impact of genomic prediction in beef cattle in the United States', *Genetics Selection Evolution*, 43, 17.

Garrick, D. J., Taylor, J. F. and Fernando, R. L. (2009), 'Deregressing estimated breeding values and weighting information for genomic regression analyses', *Genetics Selection Evolution*, 41, 55.

Garrick, D. J. and Fernando, R. L. (2015), 'Genomic Prediction and Genome-wide Association Studies in Beef and Dairy Cattle', In: *The Genetics of Cattle*. 2nd edition, Eds D. J. Garrick and A. Ruvinsky, C. A. B. International, Oxon, 474–501.

Gianola, D. (2013), 'Priors in whole-genome regression: the Bayesian alphabet returns', *Genetics*, 194, 573–96.

Gianola, D., de los Campos, G., Hill, W. G., Manfredi, E. and Fernando, R. L. (2009), 'Additive genetic variability and the Bayesian alphabet', *Genetics*, 183, 347–63.

Goddard, M. (2009), 'Genomic selection: prediction of accuracy and maximisation of long term response', *Genetica*, 136, 245–57.

Habier, H., Fernando, R. L., Kizilkaya, K. and Garrick, D. J. (2011), 'Extension of the Bayesian Alphabet for Genomic Selection', *BMC Bioinformatics*, 12, 186.

Hamill, R. M., Begonya, M., Rai, D. K. and Mullen, A. M. (2012), 'Omics Approaches to Meat Quality Management', OMICs *Technologies Tools for Food Science*, CRC Press 2012, 249–82.

Hayes, B. J., Bowman, P. J., Chamberlain, A. J. and Goddard, M. E. (2009a), 'Genomic selection in dairy cattle: progress and challenges', *Journal of Dairy Science*, 92, 433–43.

Hayes, B. J., Bowman, P. J., Chamberlain, A. C., Verbyla, K., and Goddard, M. E. (2009b), 'Accuracy of genomic breeding values in multi-breed dairy cattle populations', *Genetics Selection Evolution*, 41, 51.

Hickey, J. M., Kinghorn, B. P., Tier, B., van der Werf, J. H. J. and Cleveland, M. A. (2012), 'A phasing and imputation method for pedigreed populations that results in a single-step genetic evaluation', *Genetics Selection Evolution*, 44, 9–19.

Hozé, C., Fritz, S., Phocas, F., Boichard, D., Ducrocq, V. and Croiseau, P. (2014), 'Efficiency of multi-breed genomic selection for dairy cattle breeds with different sizes of reference population', *Journal of Dairy Science*, 97, 3918–29.

Ibanez-Escriche, N. and Simianer, H. (2016), 'Animal breeding in the genomics era', *Animal Frontiers*, 6, 4–5.

Johnston, D., Tier, B. and Graser, H. -U. (2012), 'Beef cattle breeding in Australia: opportunities and needs', *Animal Production Science*, 52, 100–6.

Kachman, S. D., Spangler, M. L., Bennett, G. L., Hanford, K. J., Kuehn, L. A., Pollak, E. J., Snelling, W. M., Thallman, R. M., Saatchi, M., Garrick, D. J., Schnabel, R. D., Taylor, J. F. and Pollak, E. J. (2013), 'Comparison of molecular breeding values based on within- and across-breed training in beef cattle', *Genetics Selection Evolution*, 45, 30.

Kelly, M., Tume, R., Newman, S. and Thompson, J. (2013), 'Genetic variation in fatty acid composition of subcutaneous fat in cattle', *Animal Production Science*, 53,129–33.

Kinghorn, B. P., Banks, R. G. and Simm, G. (2015), 'Genetic Improvement of beef cattle', In: *The Genetics of Cattle*. 2nd edition, Eds D. J. Garrick and A. Ruvinsky, C. A. B. International, Oxon, 451–73.

Leheska, J. M., Thompson, L. D., Howe, J. C., Hentges, E., Boyce, J., Brooks, J. C., Shriver, B., Hoover, L. and Miller, M. F. (2008), 'Effects of conventional and grass feeding systems on the nutrient composition of beef', *Journal of Animal Science*, 86, 3575–85.

MacNeil, M. D., Nkrumah, J. D., Woodward, B. W. and Northcutt, S. L. (2010), 'Genetic evaluation of Angus cattle for carcass marbling using ultrasound and genomic indicators', *Journal of Animal Science*, 88, 517–22.

Marshall, D. M. (1999), 'Genetics of meat quality', In: *The Genetics of Cattle*, Eds R. Fries and A. Ruvinsky, C. A. B. International, Wallingford, 605–36.

Mateescu, R. G. (2015), 'Genetics of meat quality', In: *The Genetics of Cattle*. 2nd edition, Eds D. J. Garrick and A. Ruvinsky, C. A. B. International, Oxon, 474–501.

Matukumalli, L. K., Lawley, C. T., Schnabel, R. D., Taylor, J. F., Allan, M. F., Heaton, M. P., O'Connell, J., Moore, S. S., Smith, T. P., Sonstegard, T. S. and Van Tassell, C. P. (2009), 'Development and characterization of a high density SNP genotyping assay for cattle', *PLoS ONE 4*, e5350.

McClure, M. C., Ramey, H. R., Rolf, M. M., McKay, S. D., Decker, J. E., Chapple, R. H., Kim, J. W., Taxis, T. M., Weaber, R. L., Schnabel, R. D. and Taylor, J. F. (2012), 'Genome-wide association analysis for quantitative trait loci influencing Warner–Bratzler shear force in five taurine cattle breeds', *Animal Genetics*, 43,662–73.

Melfsen, A., Hartung, E. and Haeussermann, A. (2012), 'Accuracy of milk composition analysis with near infrared spectroscopy in diffuse reflection mode', *Biosystems Engineering*, 112, 210–17.

Meuwissen, T. H. E., Hayes, B. J. and Goddard, M. E. (2001), 'Prediction of total genetic value using genome-wide dense marker maps', *Genetics*, 157, 1819–29.

Meuwissen, T. H. E., Hayes, B. J. and Goddard, M. E. (2016), 'Genomic selection: A paradigm shift in animal breeding', *Animal Frontier*, 6, 6–14.

Mullen, A. M., Stapleton, P. C., Corcoran, D., Hamill, R. M. and White, A. (2006), 'Understanding meat quality through the application of genomic and proteomic approaches', *Meat Science*, 74, 3–16.

Nkrumah, J. D., Crews Jr., D. H.,. Basarab, J. A., Price, M. A., Okine, E. K., Wang, Z., Li, C. and Moore, S. S. (2007), 'Genetic and phenotypic relationships of feeding behaviour and temperament with performance, feed efficiency, ultrasound, and carcass merit of beef cattle', *Journal of Animal Science*, 85, 2382–90.

Onogi, A., Ogino, A., Komatsu, T., Shoji, N., Shimizu, K., Kurogi, K., Yasumori, T., Togashi, K. and Iwata, H. (2015), 'Whole-genome prediction of fatty acid composition in meat of Japanese Black cattle', *Animal Genetics*, 46, 557–9.

Page, B. T., Casas, E., Heaton, M. P., Cullen, N. G., Hyndman, D. L., Morris, C. A., Crawford, A. M., Wheeler, T. L., Koohmaraie, M., Keele, J. W. and Smith, T. P. (2002), 'Evaluation of single-nucleotide polymorphisms in CAPN1 for association with meat tenderness in cattle', *Journal of Animal Science*, 80, 3077–85.

Pan, A., Sun, Q., Bernstein, A. M., Schulze, M. B., Manson, J. E., Stampfer, M. J., Willett, W. C. and Hu, F. B. (2012), 'Red meat consumption and mortality results from 2 prospective cohort studies', *Archives of Internal Medicine*, 172, 555–63.

Pannier, L., Mullen, A. M., Hamill, R. M., Stapleton, P. C. and Sweeney, T. (2010), 'Association analysis of single nucleotide polymorphisms in DGAT1, TG and FABP4 genes and intramuscular fat in crossbred Bos taurus cattle', *Meat Science*, 85, 515–18.

Piccoli, M. L., Braccini, J., Cardoso, F. F., Sargolzaei, M., Larmer, S. G. and Schenkel, F. S. (2014), 'Accuracy of genome-wide imputation in Braford and Hereford beef cattle', *BMC Genetics*, 15, 157.

Prieto, N., Roehe, R., Lavín, P., Batten, G. and Andrés, S. (2009), 'Application of near infrared reflectance spectroscopy to predict meat and meat products quality: A review', *Meat Science*, 83, 175–86.

Pringle, T. D., Williams, S. E., Lamb, B. S., Johnson, D. D. and West, R. L. (1997), 'Carcass characteristics, the Calpain proteinase system, and aged tenderness of Angus and Brahman crossbred steers', *Journal of Animal Science*, 75, 2955–61.

Pszczola, M., Strabel,T., Mulder,H. A. and Calus, M. P. L. (2012), 'Reliability of direct genomic values for animals with different relationships within and to the reference population', *Journal of Dairy Science*, 95, 389–400.

Raven, L-A., Cocks, B. G. and Hayes, B. J. (2014), 'Multibreed genome wide association can improve precision of mapping causative variants underlying milk production in dairy cattle', *BMC Genomics*, 15, 62.

Reverter, A., Johnston, D. J., Ferguson, D. M., Perry, D., Goddard, M. E., Burrow, H. M., Oddy, V. H., Thompson, J. M. and Bindon, B. M. (2003), 'Genetic and phenotypic characterisation of animal, carcass, and meat quality traits from temperate and tropically adapted beef breeds. 4. Correlations among animal, carcass, and meat quality traits', *Australian Journal of Agricultural Research*, 54,149–58.

Rolf, M. M., McKay, S. D., McClure, M. C., Decker, J. E., Taxis, T. M., Chapple, R. H., Vasco, D. A., Gregg, S. J., Kim, J. W., Schnabel, R. D. and Taylor, J. F. (2010), 'How the next generation of genetic technologies will impact beef cattle selection'. *Proceedings of the Beef Improvement Federations 42nd Annual Research Symposium and Annual Meeting*, Columbia, MO, USA, 46–56.

Saatchi, M., Garrick, D. J., Tait, R. G., Mayes, M. S., Drewnoski, M., Schoonmaker, J., Diaz, C., Beitz, D. C. and Reecy, J. M.(2013a), 'Genome-wide association and prediction of direct genomic breeding values for composition of fatty acids in Angus beef cattle', *BMC Genomics*, 14, 730.

Saatchi, M., McClure, M. C., McKay, S. D., Rolf, M. M., Kim, J., Decker, J. E., Taxis, T. M., Chapple, R. H., Ramey, H. R., Northcutt, S. L., Bauck, S., Woodward, B., Dekkers, J. C. M., Fernando, R. L., Schnabel, R. D., Garrick, D. J. and Taylor, J. F. (2011), 'Accuracies of genomic breeding values in American Angus beef cattle using K-means clustering for cross-validation', *Genetics Selection Evolution*, 43, 40.

Saatchi, M., Schnabel, R. D., Rolf, M. M., Taylor, J. F. and Garrick, D. J. (2012), 'Accuracy of direct genomic breeding values for nationally evaluated traits in US Limousin and Simmental beef cattle', *Genetics Selection Evolution*, 44, 38.

Saatchi, M., Ward, J. and Garrick, D. J. (2013b), 'Accuracies of direct genomic breeding values in Hereford beef cattle using national or international training populations', *Journal of Animal Science*, 91, 1538–51.

Sargolzaei, M., Chesnais, J. P. and Schenkel, F. S. (2014), 'A new approach for efficient genotype imputation using information from relatives', *BMC Genomics*, 15, 478.

Scollan, N. D., Richardson, R. I., De Smet, S., Moloney, A. P., Doreau, M., Bauchart, D. and Nuernberg, K. (2005), 'Enhancing the content of beneficial fatty acids in beef and consequences for meat quality', In: *Indicators of Milk and Beef Quality*, Eds J F. Hocquette and S. Gigli, Wageningen, The Netherlands: Academic Publishers, 151–62.

Simm, G., Lambe,N., Bunger, L., Navajas, E. A. and Roehe, R. (2009), 'Use of meat quality information in breeding programmes', In: *Improving the Sensory and Nutritional Quality of Fresh Meat*, Eds J. Kerry and D. A. Ledward, Cambridge, UK: Woodhead Publishing Ltd, 264–91.

Swan, A. A., Johnston, D. J., Brown, D. J., Tier, B. and Graser, H. -U. (2012), 'Integration of genomic information into beef cattle and sheep genetic evaluations in Australia', *Animal Production Science*, 52, 126–32.

Vahmani, P., Mapiye, C., Prieto, N., Rolland, D. C., McAllister, T. A., Aalhus, J. L. and Dugan, M. E. (2015), 'The scope for manipulating the polyunsaturated fatty acid content of beef: A review', *Journal of Animal Science and Biotechnology*, 6, 1–13.

Van Eenennaam, A. L., Li, J., Thallman, R. M., Quaas, R. L., Dikeman, M. E., Gill, C. A., Franke, D. E. and Thomas, M. G. (2007), 'Validation of commercial DNA tests for quantitative beef quality traits', *Journal of Animal Science*, 85, 891–900.

Van Eenennaam, A. L., Weigel, K. A., Young, A. E., Cleveland, M. A. and Dekkers, J. C. (2014), 'Applied animal genomics: Results from the field', *Annual Review of Animal Biosciences*, 2, 105–39.

Van Tassell, C. P., Smith, T. P. L., Matukumalli, L. K., Taylor, J. F., Schnabel, R. D., Lawley, C. T., Haudenschild, C. D., Moore, S. S., Warren, W. C. and Sonstegard, T. S. (2008), 'Simultaneous SNP discovery and allele frequency estimation by high throughput sequencing of reduced representation genomic libraries', *Nature Methods*, 5, 247–52.

Ventura, R. V., Lu, D., Schenkel, F. S., Wang, Z., Li, C. and Miller, S. P. (2014), 'Impact of reference population on accuracy of imputation from 6K to 50K SNP chips in purebred and crossbreed beef cattle', *Journal of Animal Science*, 92, 1433–44.

Verbeke, W., Perez-Cueto, F. J. A, de Barcellos, M D., Krystallis, A. and Grunert, K. G. (2010), 'European citizen and consumer attitudes and preferences regarding beef and pork', *Meat Science*, 84, 284–92.

Weber, K. L., Thallman, R. M., Keele, J. W., Snelling, W. M., Bennett, G. L., Smith, T. P., McDaneld, T. G., Allan, M. F., Van Eenennaam, A. L. and. Kuehn, L. A. (2012), 'Accuracy of genomic breeding values in multibreed beef cattle populations derived from deregressed breeding values and phenotypes', *Journal of Animal Science*, 90, 4177–90.

Wheeler, T. L., Cundiff, L. V., Shackelford, S. D. and Koohmaraie, M. (2010), 'Characterization of biological types of cattle (Cycle VIII): Carcass, yield, and longissimus palatability traits', *Journal of Animal Science*, 88, 3070–83.

White, S. N., Casas, E., Wheeler, T. L., Shackelford, S. D., Koohmaraie, M., Riley, D. G., Chase, C. C. Jr, Johnson, D. D., Keele, J. W. and Smith, T. P. (2005), 'A new single nucleotide polymorphism in CAPN1 extends the current tenderness marker test to include cattle of Bos indicus, Bos taurus, and crossbred descent', *Journal of Animal Science*, 83, 2001–8.

Wiggans, G. R., VanRaden, P. M. and Cooper, T. A. (2011), 'The genomic evaluation system in the United States: Past, present, future', *Journal of Dairy Science*, 94, 3202–11.

Williams, A. R. (2002), 'Ultrasound applications in beef cattle carcass research and management', *Journal Animal Science*, 80, E183–8.

Chapter 18

# The sustainability and 'carbon footprints' of conventional and alternative beef production systems

*Jude L. Capper, Livestock Sustainability Consultancy, UK*



## 1 Introduction

The world's population currently stands at over seven billion people and is predicted to surpass nine billion by 2050 (United Nations, 2011). Given the increased population, a 60% increase in food production is estimated by the Food and Agriculture Organization (FAO) of the United Nations to be required to fulfil global nutrient demands (FAO, 2013c). As the global population increases, disproportionate growth will be seen in low- to middle-income countries in conjunction with an increase in average income. A positive correlation exists between per capita income and demand for milk, meat and eggs (Delgado, 2003); thus increases in the average income of inhabitants of China, India and Africa over the next 40 years will result in considerable increases in animal-source food consumption (FAO, 2009; Masuda and Goldsmith, 2010) and livestock production will have to substantially increase to fulfil global demand. Indeed, the FAO (2013c) estimates a 1.1% per annum (p.a.) increase in global milk and 1.2% p.a. increase in meat production (including a 1.3% p.a. increase in bovine meat production) to supply global food demand by 2050. The projected population gains will confer increased competition for resources (land, water, energy) between agriculture and other uses; therefore, the need to 'produce more,

http://dx.doi.org/10.19103/AS.2016.0009.18

using less' (one of the most simple definitions of sustainability) will become increasingly important. This chapter will review the environmental sustainability of beef production in the United States and discuss a range of measures to enhance the sustainability of current and future beef systems.

## 2 Defining sustainability

The word 'sustainability' originated in the eighteenth-century forestry industry, where it was used to describe a system that did not harvest more than the forest yielded in new growth (Kuhlman and Farrington, 2010). However, a range of definitions now exist, from the aforementioned 'produce more, using less' to the complex; yet the most widely adopted may be that of the Brundtland Commission, which suggested that sustainable development '. . . meets the needs of the present without compromising the ability of future generations to meet their own needs' (United Nations World Commission on Environment and Development, 1987). This definition is founded upon concern for health and nutrition of the current and future populations, and the planet, and is, therefore, appropriate in the current global situation where 870 million people are food-insecure (FAO, 2012b).

Sustainability is often partitioned into three components: environmental responsibility (including resource use and greenhouse gas (GHG) emissions), economic viability (primarily for the agricultural producer) and social acceptability (United Nations, 2005), originally stemming from the Triple Bottom Line concept ('people, planet, profit') developed by Elkington (1994). For a region, industry or system to achieve sustainability, all three components must balance (Pope et al., 2004). However, this is not always practical in the short term. For example, in a region where per capita income is low and animal-source foods are not readily available to all, economic metrics (consumer retail price) and social acceptability (food security) may take precedence over environmental metrics. Some forms of short-term environmental degradation may be acutely toxic but reversible, yet other, less obviously detrimental actions may have environmental consequences that impact long-term system viability (Kuhlman and Farrington, 2010). In the case of natural resource use, it can be argued that an optimal rate of exhaustion exists in which the social cost of losing the resource is outweighed by the social benefit yielded over the period of use (Hotelling, 1931).

Furthermore, consumers now have a range of demands relating to livestock production systems that perhaps cannot simply be encompassed by 'social acceptability'. For example, animal welfare has become a significant concern for all stakeholders within livestock production and is often partitioned out as a separate issue. Indeed, in some marketing campaigns it is given higher priority than other sustainability attributes, for example, poultry production systems that are perceived to have higher welfare, but have lower yields, thus increasing both resource use and economic cost to consumers. A proportion of consumers also express concerns over the nutritive value and healthfulness of animal products. It can, therefore, be argued that a fourth dimension relating to animal and human health and welfare needs to be added to the current sustainability axis, or that the social acceptability sector needs to be explored further. With regard to available scientific literature concerning beef system sustainability, the social acceptability dimension is currently the least strongly represented of the three.

Livestock provide high-quality animal-source foods in conjunction with a myriad of economic and social benefits to rural and urban communities worldwide (Council for Agricultural Science and Technology (CAST), 2013). Nonetheless, the question is often raised as to whether the consumption of milk and meat is inherently unsustainable due to the quantities of resources used and waste emitted from the production of animal-source foods. Livestock system sustainability is, therefore, one of the most highly debated issues within food production.

Specific sustainability concerns vary considerably according to region. In the developing world, food security is the focal point, with limited food availability because of low agricultural yields, lack of producer education, inadequacies in transport and sanitary infrastructure, political instability and military conflict (Godfray et al., 2010; Pinstrup-Andersen, 2000). Milk, meat and eggs play important roles in maintaining the health and nutritional status of inhabitants of developing countries (Murphy and Lindsay, 2003; Neumann et al., 2002; Randolph et al., 2007; Smith et al., 2012) and the social and economic importance of livestock ownership in these regions cannot be underestimated, as they provide significant benefits in terms of fertilizer, draught power, income/capital assets and cultural status (Godfray et al., 2010; McDermott et al., 2010; Randolph et al., 2007; Sansoucy, 1995; Smith et al., 2012). According to the FAO (2012a), almost one billion households worldwide rely on livestock for their livelihood.

Within the developed world, environmental responsibility is arguably the greatest concern. Climate change has become a social and political concern, with many governments mandating reductions in GHG emissions in an attempt to reduce the impact of global warming. Although global warming and climate change are often used interchangeably, the former refers to an increase in temperature of the earth's crust, usually (although not always) with reference to the effect of human activity on said temperature. By contrast, climate change is a combination of changes in weather and environmental conditions resulting from global warming and other side effects of warming (Kennedy and Lindsey, 2015). Climate change may be caused by human activity (e.g. increased release of GHG into the atmosphere) or by natural processes such as carbon dioxide being released from oceans during ice ages. However, it seems clear (Joos et al., 1999) that the ongoing cycle of global warming has sped up as a result of human-caused release of heat-trapping GHG (principally methane ($CH_4$), nitrous oxide ($N_2O$) and carbon dioxide).

Resource use, waste outputs and GHG emissions from livestock production have been scrutinized by a variety of groups ranging from global scientific associations (FAO, 2006) to activist groups (Environmental Working Group, 2011; Goodland and Anhang, 2009; Koneswaran and Nierenberg, 2008). The scientific consensus is that animal agriculture uses a considerable amount of resources (both renewable and non-renewable) and makes a significant contribution to GHG emissions (FAO, 2006; Pelletier and Tyedmers, 2010) that may increase global warming; however, the magnitude of livestock's global contribution is under debate. Leaver (2011) suggested that land, water and energy are the three major limiting resources to increasing future food production. This appears to damn ruminant systems, which are often cited (Pimental, 2003, 2007; Dye Gussow, 1994; Goodland, 1997) as threats to sustainable food production, given the quantities of resources employed and waste produced per unit of food. Yet, beef and dairy systems offer a number of positive environmental impacts, not least is the conversion of human-inedible by-product feeds into animal protein.

# 3 GHG emissions and carbon footprints

A myriad of environmental issues are associated with livestock production systems, including resource use (land, water, energy); air and water quality issues; impacts on biodiversity and native wildlife and indirect effects such as deforestation and land degradation from feed production. These issues are all valid concerns and should be considered as important issues to be raised in any discussion of beef system sustainability; yet GHG emissions are arguably the most widely discussed environmental sustainability issue with regard to livestock production. In the interests of brevity, this chapter will primarily focus on GHG emissions, yet the reader should not assume that other environmental issues are of lesser importance, despite the tendency on the part of media and scientific reporters to concentrate on carbon footprints as the sole arbiter of environmental impacts.

## 3.1 What is a carbon footprint?

It is important to recognize that almost every human activity is associated with the release of GHG, from breathing to using electricity or driving a car; food production and consumption is no exception. The phrase 'carbon footprint' simply refers to the quantity of GHG associated with a specific activity (e.g. driving for a certain distance or manufacturing a product) expressed in $CO_2$-equivalents ($CO_2$-eq). Carbon dioxide is used as a reference gas, with a global warming potential (GWP) of 1.0, whereas $CH_4$ has a GWP of 25 as it is 25 times more potent at causing global warming than $CO_2$, and $N_2O$ has a GWP of 298 (IPCC, 2006). Thus, an activity that emits any of these gases through, for example, fuel combustion, enteric fermentation, bacterial activity (e.g. $N_2O$ from soils), etc, can be associated with a calculable and verifiable carbon footprint. The majority of GHG emissions from livestock production originate from manure production and storage ($CH_4$ and $N_2O$), feedstuff production ($N_2O$ and $CO_2$), transportation/fossil fuel combustion ($CO_2$) and, in the case of ruminant animals, enteric fermentation ($CH_4$).

## 3.2 Global GHG emissions

The FAO report 'Livestock's Long Shadow' (2006) was an initial founder of the now widespread suggestion that livestock production is detrimental to the environment due to its contribution to GHG emissions. The report stated that 18% of global anthropogenic GHG emissions were derived from animal agriculture, and that this compromised a greater share of emissions than the entire transport sector. The methodology and results of the FAO report have been discussed at length, and although Pitesky et al. (2009) credited the FAO authors for their sound recommendations as to the importance of improving productivity and efficiency in order to reduce future GHG emissions, the documented discrepancies in methodology between analyses of animal agriculture and other sources of GHG emissions led to a marked overestimation of the global contribution made by animal agriculture. The FAO authors subsequently admitted that the report contained flawed methodology and that the 18% comparison was incorrect (Black, 2010), yet the statistic is invariably quoted in both scientific and lay media when examining the environmental impact of livestock production.

Although the proportion of global GHG emissions allocated to animal agriculture by the FAO (2006) report was inaccurate, it is within the ranges reported by other authors.

Bellarby et al. (2008) estimated that total agriculture accounts for 17–32% of global anthropogenic GHG emissions and Pelletier and Tydermers (2010) suggested that animal agriculture accounted for 14% of global GHG emissions. Global averages are somewhat meaningless except as a reference point as they do not represent any specific region or system and thus offer little opportunity for practical mitigation below the global level. By contrast, the World Watch Institute (Goodland and Anhang, 2009) propounded to correct perceived errors within the FAO (2006) report, concluding that animal agriculture accounts for 51% of global GHG emissions. Given the myriad of GHG sources arising from all human and animal activities worldwide, it is difficult to attach any significant degree of credence to the latter estimate; yet it has gained a certain amount of media attention.

## 3.3 GHG emissions from meat compared to other foods

The consensus view is that meat is associated with greater GHG emissions per unit of food compared to plant-based proteins, and the number of papers citing reduced meat consumption as the principal solution to reducing global GHG emissions are evidence of the spotlight shining upon the livestock industry. In a scientific review of food-product life-cycle analyses (LCA), Nijdam et al. (2012) noted that ruminant meats (9–150 kg $CO_2$-eq/kg product) had GHG emissions at least an order of magnitude higher than milk, vegetable-based meat substitutes or legumes (1–2 kg $CO_2$-eq/kg product), with pork (4–11 $CO_2$-eq/kg product), poultry (2–6 kg $CO_2$-eq/kg product) and eggs (2–6 kg $CO_2$-eq/kg product) falling between these ranges. Similar patterns were seen when GHG emissions were reported per kg of protein, with beef at 45–640 kg $CO_2$-eq/kg product, compared to pork at 20–55 kg $CO_2$-eq/kg product and legumes at 4–10 kg $CO_2$-eq/kg product (Nijdam et al., 2012). These wide ranges reflect the diversity of efficiencies and systems worldwide, from the intensive to extensive. Generally, an increase in system intensity would be associated with lower GHG emissions per kg of food. It should be noted that both animal and plant proteins are also associated with by-products that have an economic and social value. Thus, the edible and inedible offal from animal carcasses (e.g. liver, feet, hides) or the fibrous by-products from crop production (e.g. cottonseed meal, wheat straw, sugar beet pulp) are often not accounted for in these calculations as the main GHG burden is attributed to the major product.

Increased GHG emissions associated with animal versus plant-source foods are attributed to the relative inefficiency of converting vegetable matter into animal gain, versus direct human consumption of plant-based foods, although this is complicated by the proportion of GHG emissions resulting from the digestion and metabolism of human-inedible feedstuffs (e.g. pasture and forages) grown on land that could not support production of human food crops (Council for Agricultural Science and Technology (CAST), 2013). Relative GHG emissions associated with ruminant versus monogastric production are due to differences in feed efficiency between the species in combination with variation in feed digestion mechanisms. Monogastric animals rely upon a single stomach and, although individual species may have gastrointestinal adaptations that allow fibrous feeds to be digested, they do not produce cellulase enzymes that facilitate the digestion of cellulose (Hofmann, 1989). By contrast, the ruminant rumen (the largest of the four stomach compartments) acts as a fermentation vessel full of microorganisms that produce enzymes capable of digesting plant cellulose and hemicellulose (Hofmann, 1989). This biological advantage, which allows cattle to grow and thrive upon human-inedible forages and feedstuffs, confers a negative environmental trade-off in that methane production is an almost inevitable consequence of ruminal fermentation. Fibre fermentation by methanogenic bacteria and protozoa within the rumen leads to 86–255 g $CH_4$ being emitted per animal on a daily basis (DeRamus et al., 2003); thus, enteric $CH_4$ production has a considerable impact upon the magnitude of beef's GHG emissions compared to other animal-source foods.

## 3.4 GHG contributions to beef's carbon footprint

Scientific literature reveals considerable regional variation in GHG emissions per kg of beef, from 9.0 kg $CO_2$-eq to 24.2 kg $CO_2$-eq in intensive systems, compared to extensive systems at 12.0–726 kg $CO_2$-eq/kg beef (Table 1). A certain amount of the variation may be attributed to methodological differences (Bertrand and Barnett, 2011). Without a single global analysis of beef production's carbon footprint, it is difficult to draw firm conclusions as to the variation within and between regions. However, as noted in an analysis by Nijdam et al. (2012), the considerable range is mainly due to production system variation. For example, analyses comparing differing systems demonstrated reductions in GHG emissions per unit of beef conferred by increased harvest weight, growth rate or conversion from grass-fed to feedlot-finishing systems (Capper, 2011, 2012; Capper and Hayes, 2012; Pelletier et al., 2010; Stackhouse-Lawson et al., 2012a,b; White and Capper, 2014).

As shown in Fig. 1, enteric $CH_4$ contributes the greatest proportion of total GHG emissions per unit of beef, followed by $N_2O$ from manure management and fertilizer

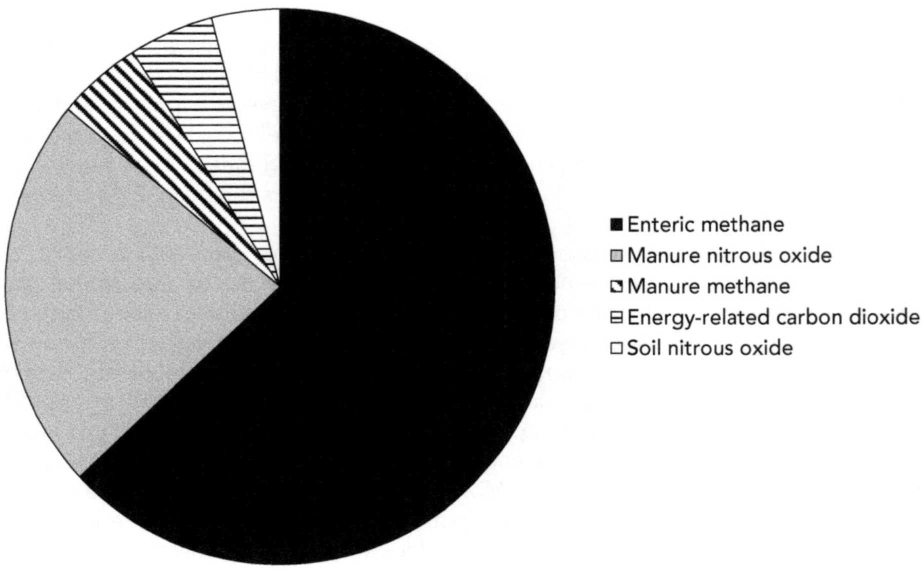


**Figure 1** Proportional contributions of different greenhouse gases to the carbon footprint per unit of beef.

Source: data from Beauchemin et al. (2011).

**Table 1** Variation in greenhouse gas emissions per unit of product from differing beef production systems and regions

| Beef production system | GHG emissions (kg $CO_2$-eq/kg HCW beef) | Region | Reference |
|---|---|---|---|
| Beef and dairy-bred, feedlot-finished | 17.9 | USA | Capper (2011) |
| Beef and dairy-bred, feedlot-finished, with technology | 16.0 | USA | Capper (2012) |
| Beef and dairy-bred, feedlot-finished, natural | 18.0 | USA | Capper (2012) |
| Beef-bred, grass-finished | 26.8 | USA | Capper (2012) |
| Beef and dairy-bred, feedlot-finished | 16.0 | USA | Capper and Hayes (2012) |
| Beef-bred, feedlot-finished | 23.5* | USA | Pelletier et al. (2010) |
| Beef-bred, grass-finished | 33.1** | USA | Pelletier et al. (2010) |
| Beef-bred, feedlot-finished, natural | 24.2 | USA | Stackhouse-Lawson et al. (2012) |
| Beef-bred, feedlot-finished, with implants | 22.6 | USA | Stackhouse-Lawson et al. (2012) |
| Beef-bred, calf-fed, feedlot-finished, with implants | 21.2 | USA | Stackhouse-Lawson et al. (2012) |
| Dairy-bred, feedlot-finished, with implants | 10.7 | USA | Stackhouse-Lawson et al. (2012) |
| Beef and dairy-bred, feedlot-finished | 20.1 | USA | White and Capper (2014) |
| Beef-bred, yearling-fed, feedlot-finished | 23.5* | Canada | Conestoga-Rovers & Associates (2010) |
| Beef-bred, calf-fed, feedlot-finished | 22.4* | Canada | Conestoga-Rovers & Associates (2010) |
| Beef-bred, feedlot-finished | 16.5* | Canada | Vergé et al. (2008) |
| Beef-bred, feedlot-finished | 22.0 | Canada | Beauchemin et al. (2010) |
| Suckler beef | 27.3 | Europe | Nguyen et al. (2010) |
| Dairy-bred, feedlot finished | 16.0–19.9 | Europe | Nguyen et al. (2010) |
| Suckler beef | 19.4† | Ireland | Casey and Holden (2006) |
| Beef-bred, feedlot-finished | 18.1** | Ireland | Casey and Holden (2006) |
| Dairy-bred, feedlot finished | 12.6** | Ireland | Casey and Holden (2006) |
| Beef-bred, grass-finished | 25.7† | France | Veysset (2010) |
| Beef-bred, feedlot-finished | 9.0 | Australia | Peters et al. (2010) |
| Beef-bred, grass-finished | 12.0 | Australia | Peters et al. (2010) |
| Various | 10.1–12.7 | Australia | Ridoutt et al. (2011) |
| Beef-bred, grass-finished | 28.0 | Brazil | Cederburg et al. (2009) |
| Beef-bred, grass-finished | 44.0 | | Cederburg et al. (2011) |
| Beef-bred, grass-finished, including deforestation (regional average) | 180 | Brazil | Cederburg et al. (2011) |
| Beef-bred, grass-finished, including on-farm deforestation | 726 | Brazil | Cederburg et al. (2011) |
| Beef-bred, feedlot finishing stage only (no cow-calf or stocker operations included) | 25.5§ | Japan | Ogino (2004) |

* Based on conversion from live weight with dressing percentage for feedlot-fed beef of 63%.
** Based on conversion from live weight with dressing percentage for Irish feedlot-fed beef of 60%.
† Based on conversion from live weight with dressing percentage for grass-fed beef of 58%.
§ Based on retail conversion from hot carcass weight (HCW) beef of 70%.

application (Beauchemin et al., 2010; Bonesmo et al., 2013; Cederberg et al., 2009a; Conestoga-Rovers & Associates, 2010; Nguyen et al., 2010; Pelletier et al., 2010). For example, Cederburg (2009b) cited that $CH_4$ contributed 76% and $N_2O$ supplied 22% of GHG emissions from Brazilian beef production; Nguyen (2010) reported that GHG emissions from European beef fattening systems comprised 73–88% enteric $CH_4$ and Beauchemin (2010) found that enteric $CH_4$ and $N_2O$ from soil and manure management contributed 63% and 27% of Canadian beef GHG emissions, respectively. By contrast, $CO_2$ emissions from fuel combustion comprise a minor proportion of the total emissions (Beauchemin et al., 2010; Bonesmo et al., 2013; Capper, 2011, 2012; Capper and Hayes, 2012; Cederberg et al., 2009a).

Within the beef production system, the cow-calf sector makes the greatest contribution to total GHG emissions per unit of beef, ranging from 63 to 80% of GHG emissions (Beauchemin et al., 2010; Pelletier et al., 2010; Stackhouse-Lawson et al., 2012a). This does not mean that cow-calf systems are less environmentally responsible than other sectors within the industry – it is simply a function of the high ratio of animal inputs to outputs within this sector. The cow-calf herd contains all the animals required to support the production of calves; thus, for every calf sold to a stocker or feedlot operation, another 1.6 animals are present in the cow-calf herd – not simply the mature cow required to produce the calf, but also a proportion of a mature bull, replacement heifer and replacement bull (Capper, 2012). Furthermore, the supporting herd of cows, bulls and replacement animals in the cow-calf operation are maintained year-round. This is in contrast to feedlot operations that contain only growing animals, and for a relatively short period of time (typically 4–7 months), with similar total gains (236 kg in the cow-calf operation from birth to weaning, compared to 231 kg from feedlot entry to harvest), thus having a far smaller animal input/output ratio. For example, a cow-calf enterprise weaning calves at 7 months could be considered to have an animal input/output ratio of 2.2 animals/236 kg live weight output compared to 0.3 animals/231 kg live weight output for a feedlot operation with an average of 4–5 months on feed (Capper, 2013f).

# 4 Trends in beef sustainability

The US beef industry was founded upon the conversion of pasture and forages into animal protein and continues in this fashion into the present day. During the first half of the twentieth century, cattle were harvested either at weaning (at 200–250 kg) or directly from pasture, with little or no supplementary feed inputs (Corah, 2008). Large-scale cattle feeding operations were first developed in the 1940s in the Great Plains, and the first scientific papers describing the effects of feeding corn to weaned cattle or yearlings were published in the 1950s. The modern cattle industry still utilizes pasture and forages as the basis for beef production, with the average finished beef steer spending 50–75% of its life grazing forages.

The total beef production in the United States climbed steadily from 2.8 billion kg in 1930 until reaching a fairly stable plateau at ~12 billion kg between 2006 and 2012, as shown in Fig. 2 (Capper, 2011a). Although the number of cattle harvested annually rose from 19.9 million in 1944 to 35.6 million in 1970, annual harvest numbers remained fairly constant at ~34 million head per year from the 1970 to 2010 (after which drought had a significant effect on the number of head harvested per year) – the growth in annual beef

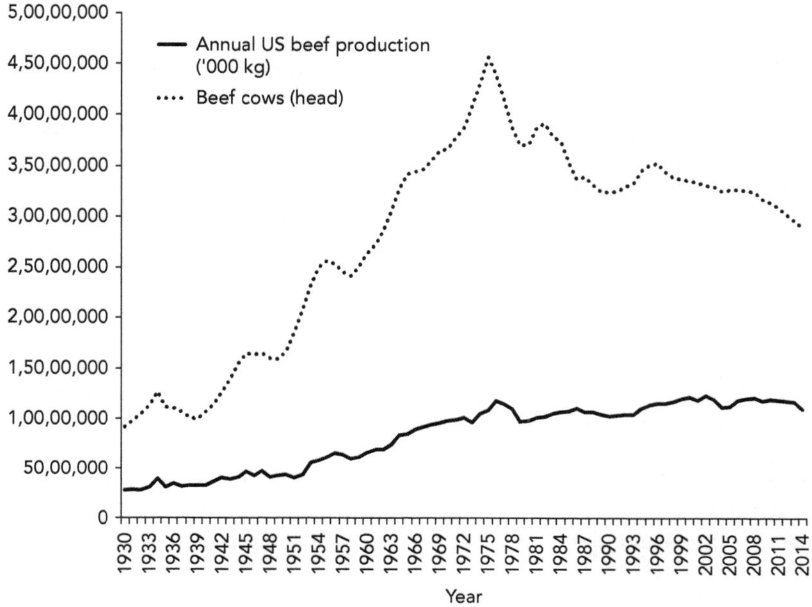


**Figure 2** Trends in annual US beef production over time. Source: data from USDA (2015).

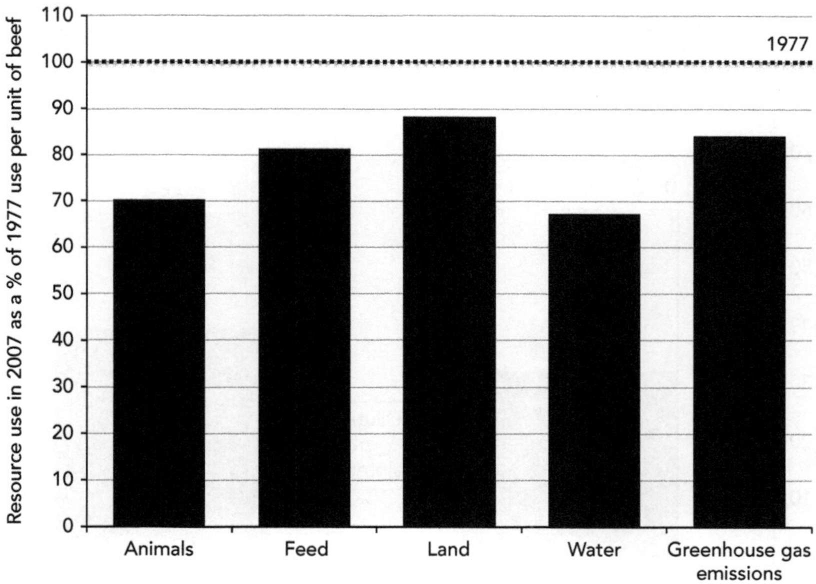


**Figure 3** Reductions in resource use and greenhouse gas emissions conferred by improved US beef industry productivity between 1977 and 2007.

Source: data from Capper (2011).

production resulting from an increase in beef yield per animal rather than total cattle numbers harvested (USDA/NASS, 2011).

Productivity gains over time have improved resource use and the environmental sustainability of beef production. Capper (2011) quantified the effects of productivity improvements between 1977 and 2007 upon the environmental impact of the US beef industry and concluded that improvements in nutrition, management and genetics reduced both resource inputs and GHG emissions per unit of beef. Increased growth rates (1.18 kg/day in 2007 vs. 0.72 kg/day in 1977) reduced the average days from birth to harvest from 609 days (1977) to 485 days (2007), despite the increase in average harvest weights (607 kg per animal in 2007 vs. 468 kg per animal in 1977). In combination (Fig. 3), these productivity gains resulted in 19% less feed, 33% less land, 12% less water and a 16% reduction in GHG emissions per unit of beef (Capper, 2011).

The US national beef and dairy herd declined to 89.3 million head in 2011 as a result of widespread drought and lack of forage (USDA, 2015); yet, as demonstrated by historical trends, it is possible to maintain total beef output from a smaller herd. The mechanism behind the previously discussed improvements in resource use and GHG emissions between 1977 and 2007 is known as the 'reduction and dilution of maintenance' concept (Capper, 2011). Every animal in the beef herd (cow, calf, bull, steer and heifer) has a daily maintenance nutrient requirement that can be considered as a proxy for resource use (e.g. feed, land, water) and GHG emissions. As meat yield per animal increases, fewer cattle need to be harvested to maintain beef production and thus the size of the total supporting population (cows, heifers and bulls within cow-calf operations) is reduced.

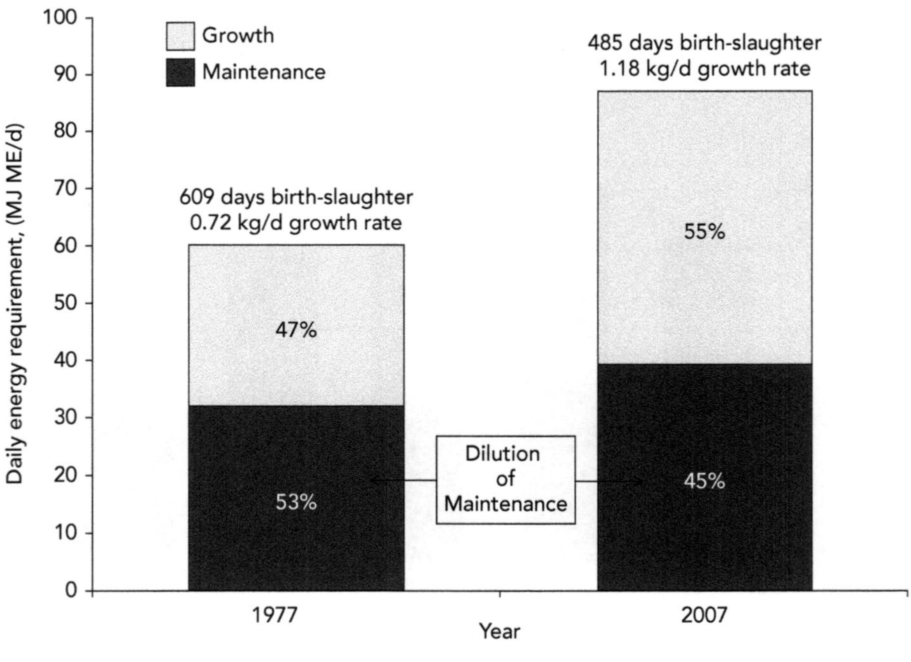


**Figure 4** The 'dilution of maintenance' concept.

Source: data from Capper (2011).

Therefore, the maintenance nutrient requirement of the population decreases ('reduction of maintenance') and fewer resources are required to produce a set quantity of beef.

The dilution of maintenance concept is illustrated in Fig. 4. As growth rate increases, the proportion of daily energy allocated to maintenance decreases. The resulting decrease in total days from birth to harvest reduces the total maintenance requirement of the growing-finishing population (Capper, 2011). It is important to note that dilution of maintenance only reduces the resource use of growing animals within the population (calves, stocker and feedlot animals), whereas reduction of maintenance affects the resource use within the entire population (all animals within the cow-calf, stocker and feedlot sector).

# 5 Sustainability of alternative (grass-fed, natural, organic) beef production systems

Consumers often claim to be interested in purchasing 'environmentally friendly' or 'green' foods including organic or natural products, although these interests are often not demonstrated by actual purchasing decisions (Vermeir and Verbeke, 2006). However, consumer perceptions of sustainable animal products are not always comparative with documented or calculated environmental impacts of meat production systems as defined by scientific literature. 'Intensive' (grain finishing) production systems are often perceived as being environmentally unfavourable based on energy, fossil fuel, fertilizer or antibiotic use; yet such claims often fail to account for the unfavourable environmental impacts of reduced productivity in many extensive (grass and forage finishing) systems.

In the majority of cases, forages and grasses are capable of fulfilling the nutrient requirements of mature cows, bulls and replacement heifers; yet growing cattle need proportionally more nutrients per day if they are to fulfil their genetic potential for carcass growth (NRC, 2000). Nutrient-dense diets fed within a feedlot finishing system will be able to satisfy these requirements more effectively than a pasture-based diet, just as children will grow faster on a relatively high energy and protein diet than one containing only low-energy vegetables. Therefore, when a grass-fed system (intensive) is compared to conventional beef production (extensive), cattle performance may not differ considerably within cow-calf and stocker sectors, but cattle have considerably lower productivity in the finishing sector (Capper, 2012), with direct and indirect consequences that filter back down the chain.

In an average US grass-fed system, cattle have an average growth rate of 0.67 kg/day and are slaughtered at an average of 486 kg live weight, compared to growing at 1.21 kg/day and being slaughtered at 569 kg live weight in a conventional or intensive feedlot-finishing system (Capper, 2012). The reduction in HCW from grass-fed cattle means that a greater number of finished cattle are required to maintain total beef production. This has direct effects on the total herd size required to produce a set quantity of beef, because for every finished animal, the supporting population contains 1.6 animals. Thus, if the number of finishing cattle increases, the size of the supporting population also increases. In addition, the reduced growth rate means that cattle take more days to attain harvest weight in the grass-fed system, at 679 days compared to 444 days in the feedlot system (Capper, 2012a). To produce the entire annual US beef supply of 11.8 billion kg from grass-fed systems would, therefore, require a 77% increase in the national herd (Fig. 5).

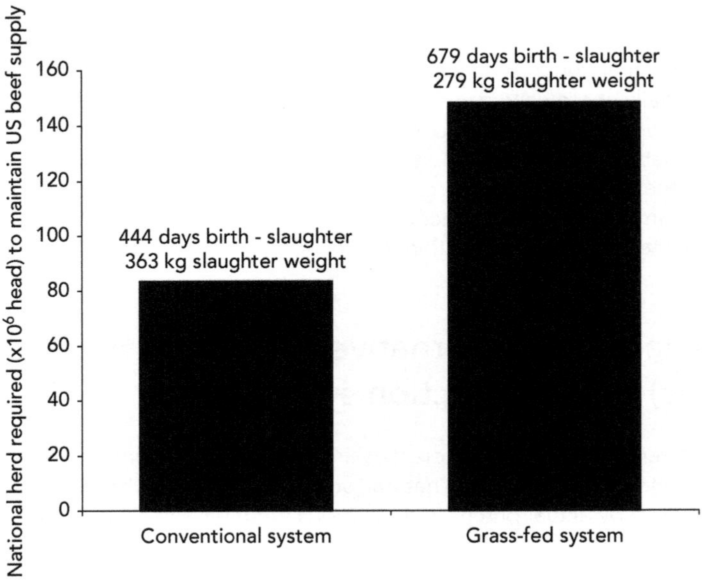


**Figure 5** National herd sizes required to produce the annual US beef supply from conventional or grass-fed production systems.

Source: data from Capper (2012).

Clear trends exist as to resource use in grass-fed versus feedlot systems. Both Capper (2012) and Pelletier (2010) concluded that land use was greater in grass-fed beef production systems, with an 81% increase per unit of HCW beef in the former study and a 5% increase in the latter study. In a comparison of six different Australian beef production systems, Ridoutt (2014) also reported that land use was doubled per unit of beef in systems that raised cattle to a greater age (24–36 months) on unimproved pasture than in systems that produced yearling cattle on improved pasture, although this result was somewhat skewed by the heavier finishing weights in the former system (340 vs. 205 kg). Capper (2012) reported a 302% increase in water use per kg of beef in grass-fed production systems compared to feedlot finishing and noted that this was highly sensitive to irrigation water within the grass-fed system. For example, if the original assumption that 50% of pasture used to finish cattle was irrigated at average US rates was changed to 15%, the corresponding increase in water use in the grass-fed versus feedlot-finished comparison was equal to a 39.8% increase.

With regard to GHG, feeding forage-based diets (compared to grain-based) also increases $CH_4$ emissions on a per animal, per day basis (Grainger and Beauchemin, 2011). Physiologically, this occurs because of the greater starch concentration within the grain-based diet, which promotes propionic acid production, inhibits growth of methanogenic microorganisms and reduces concentrations of rumen protozoa (Murphy et al., 1982; Van Kessel and Russell, 1996; Williams and Coleman, 1988). Various management strategies may be employed to reduce $CH_4$ emissions of grazing cattle; yet the daily increase in emissions multiplied by the increased number of days required to finish grass-fed cattle has a significant effect upon GHG emissions per unit of beef. Pelletier et al. (2010) cited GHG emissions of 33.1 kg $CO_2$-eq/kg HCW for grass-fed beef compared to 23.5 kg $CO_2$-eq/kg HCW for feedlot-finished beef, whereas Capper (2012) reported 26.8 kg $CO_2$-eq/kg HCW for grass-fed beef and 16.0 kg $CO_2$-eq/kg HCW for feedlot-finished beef.

Sequestration of carbon dioxide into grassland (i.e. $CO_2$ being removed from the air and trapped in the soil) is often cited as a major advantage of grass-fed systems, as well-managed grazing land can potentially sequester more carbon than cropping land used to produce grain or legume-based feeds for the feedlot (Bruce et al., 1999). However, a question arises as to whether grassland sequestration can outweigh greater quantities of enteric $CH_4$ emitted from grazing cattle, that is, what is the effect upon net GHG emissions? Including carbon sequestration in the analysis by Pelletier et al. (2010) reduced the GHG emissions of grass-fed beef to a level 15% lower than feedlot-finished beef. By contrast, Capper (2012) reported that the extent of sequestration required to offset GHG emissions from reduced productivity in a grass-fed system (0.5 metric tonnes/ha) was threefold higher than had been reported in literature to date.

Organic beef production comprises a small niche market within the US industry at present, and there is a paucity of data relating to organic environmental impacts compared to other systems. If comparisons within dairy production may be used as a guide, organic dairying was predicted by Capper et al. (2008) to increase animal numbers by 25%, land use by 30% and GHG emissions by 13% per unit of milk, primarily as a consequence of reduced milk yield per cow. However, the reader should note that, to date, there are no analyses examining the less tangible environmental impacts of organic or grass-fed versus feedlot-finishing systems, for example, biodiversity, soil health, land degradation and animal or human health. It is important to realize that, although resource use is a critical component of environmental impact, there are other metrics that may be as, or even more, important in some regions. Therefore, the increases in resource use observed with less intensive systems should not be considered as evidence that these systems are inherently less sustainable than their intensive cohorts, but that areas of opportunity for improvement exist in all systems.

# 6 Opportunities for improving and maintaining beef sustainability

Sustainable intensification, which may be defined as 'a move towards continuous improved productivity and efficiency, while fulfilling the environmental, economic and social balance required for a sustainable system', is crucial in order to reduce the environmental impact of beef production while improving productivity and global food security (FAO, 2006). Gains may be achieved through management practices that reduce specific impacts, for example, soil testing to assess fertilizer requirements, installation of anaerobic digesters to convert waste into energy, the use of feed additives (see Grainger and Beauchemin, 2011 for an excellent review) to reduce GHG emissions or by recycling water. However, the economic or practical feasibility of these practices may not be operation size or system-independent, and a greater mitigation effect may be conferred by whole-system approaches that allow cattle to perform to their genetic potential. If a beef system working at optimal efficiency would potentially produce a set quantity of meat in a given time, every productivity loss within the system will reduce meat yield and increase both economic cost and environmental impact per unit of food produced (Capper and Hayes, 2012). Furthermore, productivity-based approaches are irrespective of system (e.g. intensive vs. extensive, conventional vs. organic), operation size or region, and thus can be implemented globally with due consideration for the markets and tools available within that area.

## 6.1 Improving growth or finishing weight

Several comparative studies have shown beneficial environmental and economic effects of improving beef productivity (Capper, 2012; Capper and Hayes, 2012; Pelletier et al., 2010; White and Capper, 2012). For example, White and Capper (2012) reported that improving average daily gain (ADG) by 15% reduced land, water and feed use per unit of HCW beef by 2%, 13% and 4%, respectively, while GHG emissions were reduced by 7%. Similarly, improving carcass weight by 15% led to 13%, 16% and 9% reductions in land, water and feed use per unit of HCW beef, respectively, compared to controls, and GHG emissions were reduced by 10%. These reductions were accompanied by improvements in production sector (cow-calf to feedlot) profitability of 7% for the ADG increase and 14% for the carcass weight increase. Although to date no published literature exists on the benefits of improving the per cent meat yield versus simply increasing HCW, it is only logical that improving per cent meat yield would also decrease land, water and feed use per harvested animal.

## 6.2 Growth-enhancing technologies

Currently, 29 products with a label claim for growth or productivity enhancement (productivity-enhancing technologies (PET)) are approved for use and marketed in the United States: 27 steroidal implants (used at any point within the beef sector in growing animals) and 2 beta-adrenergic agonists (βAA; used within the feedlot only; Johnson et al., 2013). A number of studies have evaluated the productivity effects of implants or βAA alone, or in combination (Abney et al., 2007; Baxa et al., 2010; Beckett et al., 2009; Elam et al., 2009; Gruber et al., 2007; Laudert et al., 2005; Montgomery et al., 2009a,b; Parr et al., 2010; Scramlin et al., 2010; Selk et al., 2006; Thomson et al., 2009; Vasconcelos et al., 2008; Vogel et al., 2009); yet few have extended the analyses to examine the effects upon beef sustainability metrics. Capper (2013b) examined the individual and additive effects of implant and βAA use using a whole-system beef model and reported that the most significant environmental mitigation effects were conferred by the use of both technologies. The population size required to produce 363 kg beef was reduced from 3.08 animals (control) to 2.92 (βAA-only), 2.72 (implants-only) or 2.62 (both technologies). Compared to the control group, production systems using both implants and βAA used 13.4% less land and 13.2% less water; and GHG emissions were reduced by 10.7% (Fig. 6). The reductions in GHG emissions seem relatively minor; yet it should be noted that the majority of GHG emissions associated with a unit of beef are contributed by the cow-calf operation, which was unaffected by either technology. Changes in the economic cost of production followed the same trends as environmental impacts, with decreases of 3.8% (βAA-only), 7.5% (implants-only) and 9.9% (both technologies), compared to the control system. Similar environmental results were reported by Stackhouse-Lawson et al. (2012b), with 4–9% reductions in GHG emissions per unit of beef resulting from using implants or implants plus βAA in a whole-system production simulation.

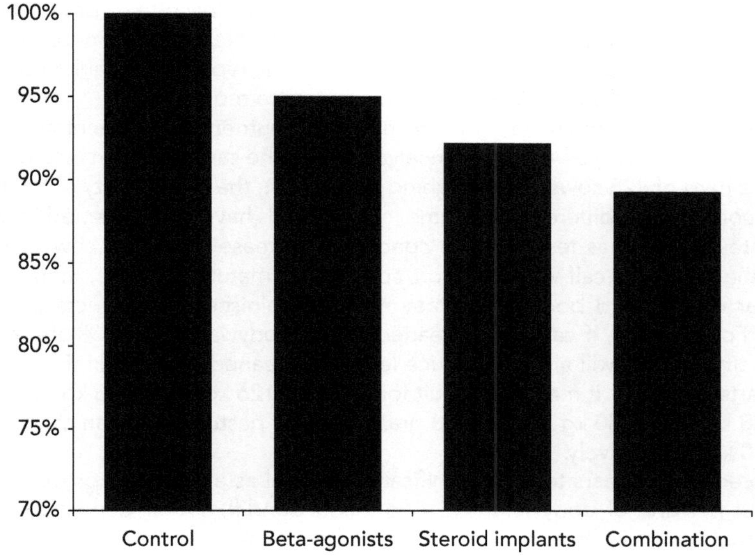


**Figure 6** The impact of using steroid implants, beta-adrenergic agonists or their combination on relative greenhouse gas emissions per unit of beef.

Source: data from Capper (2013b).

Implants and βAA demonstrably improve the environmental and economic sustainability of beef production, yet consumer perceptions regarding pharmaceutical use within livestock production might restrict their future usage (Capper and Hayes, 2012; Neumeier and Mitloehner, 2013). Cooprider et al. (2011) compared productivity of conventional cattle with 'never-ever three' cattle (raised without antibiotics, production-enhancing technologies or animal by-products in feed, although the latter would seldom be used in cattle enterprises) within the calf-fed feedlot system and demonstrated a 34% increase in ADG, 42-day reduction in finishing period length and 22% reduction in GHG emissions per unit of feedlot weight gain in the conventional group. Economically, this was equivalent to a 21% reduction in feed and technology cost per unit of conventional gain; thus, consumers would have to be prepared to pay a premium of at least \$142.52/animal (\$0.39/kg HCW beef) at the packer level to cover additional feed costs associated with productivity losses in never-ever three cattle.

## 6.3 Feed efficiency

Given that beef production is dependent on the establishment and maintenance of a supporting population of beef cows, heifers and bulls, output per unit of herd bodyweight may be employed as a metric to improve beef sustainability. This is dependent upon either increasing output from a constant herd bodyweight or improving the production efficiency of individual animals, such that efficiency is increased and output may be maintained through a herd of smaller animals. Although these may superficially appear to be the same concept, improved environmental and economic sustainability of the US livestock industries has historically concentrated on the former approach (e.g. producing heavier carcasses from fast-growing animals) with little focus on the productivity or maintenance requirements of the supporting population (cows, bulls, replacement animals) that make up a considerable portion of the total herd maintenance requirement.

Every unit of bodyweight mass within the herd has a nutrient requirement; thus, a herd of 500 cows each weighing 544 kg will conceivably have the same maintenance requirement as that of a herd of 428 cows each weighing 635 kg. Yet, the productivity of the two herds might be considerably different. Over time, many breeds have been selected for increased growth rate and this has resulted in a concurrent increase in mature cow size (Capper, 2011). If the weight of calf weaned is proportional to mature cow size, then the ratio of output per unit of herd bodyweight may not be diminished by an increase in mature cow size. For example, if calves are weaned with a bodyweight of 50% of the maternal weight, a smaller cow will always produce less total weaned calf output than her heavier counterparts. However, it may be difficult for a cow at 126 kg $W^{0.75}$ (635 kg live weight) or one at 133 kg $W^{0.75}$ (680 kg live weight) grazing native pastures to wean calves weighing 317 or 340 kg, respectively.

Cross-breeding appears to have significant potential as a tool to improve productivity and output per unit of bodyweight (Gregory and Cundiff, 1980). Cundiff et al. (1974a) reported that heterosis (hybrid vigour conferred by crossbreeding) significantly reduced the interval from parturition to oestrus (important as a fertility metric), increased the calf crop weaned by 6.4% through improved conception and pregnancy rate in crossbred cows and increased calf weaning weight by 14.5% (23 kg). Maternal heterosis (i.e. the effect of a cow being crossbred vs. straightbred) improved weight at birth, 100 days and weaning, and conformation – principally as a consequence of improved milk production (Cundiff et al., 1974b). When crossbred cows were used to produce crossbred calves, the weight of calves weaned per cow increased by 23% (Cundiff et al., 1974a). Around half (48.7%) of the US beef producers have crossbred cattle in their commercial cow-calf enterprise (USDA, 2009); thus, the advantages of crossbreeding have been recognized by commercial producers. However, to date, no scientific studies have quantified the potential sustainability impacts of these productivity gains.

Ward (1980) suggested that future beef production sustainability would depend, in part, upon reducing the feed requirements per unit of animal bodyweight. Feed efficiency is often used as a divisive argument for the sustainability of differing animal proteins as monogastrics require less feed to produce one unit of gain than ruminant animals (Wilkinson, 2011). The similar, yet not identical, feed/gain ratio has been used as a productivity metric within beef production (specifically the feedlot sector) for many years; yet selection for this metric often leads to an increase in mature cow size and feed intake due to the genetic correlation with growth rate (Archer et al., 1999). Therefore, the current focus on improving feed efficiency has shifted towards selecting cattle for low residual feed intake (RFI), that is, breeding high-efficiency cattle that eat less food than predicted by published feeding standards (Archer et al., 1999; Herd and Arthur, 2008).

## 6.4 Reproduction

Around 89% of the US beef cows bear a live calf every year (USDA, 2009), with this number declining to 50–60% in extensive systems characteristic of some regions of South America, South Africa and Northern Australia. Given that the cow-calf operation contributes up to 80% of GHG emissions per unit of beef (Beauchemin et al., 2010) and

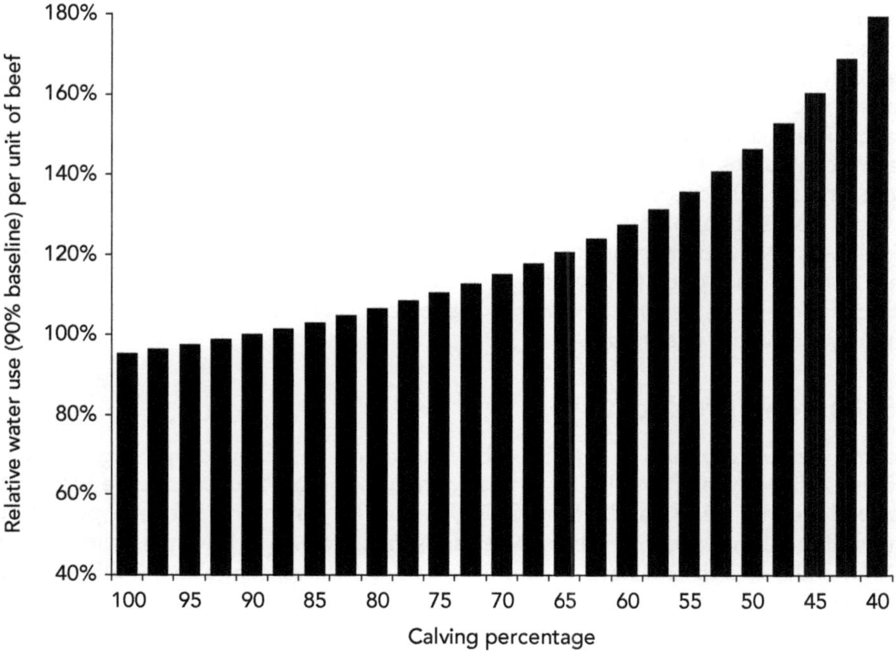


**Figure 7** The impact of changes in calving rate on relative water use per unit of beef. Source: data from Capper (2013a).

that productivity improvements post-calving may not compensate for the resource use and GHG emissions associated with maintaining a non-productive cow, management practices and technologies that increase calving rate offer significant opportunities to improve sustainability. Capper (2013a) modelled the impact of improving calving rate within the US beef production from 60% to 90%, to the ideal (100%). Compared to the ideal, at a 90% calving rate, land use was increased by 8.1% and water use by 5.2% per unit of beef, whereas at a 60% calving rate, land use was increased by 53.2% and water use by 34.1% per unit of beef (Fig. 7). As calving rate decreased, GHG emissions increased from 5459 kg $CO_2$-eq/363 kg beef (100%) to 5841 kg $CO_2$-eq/363 kg beef (90%) to 7945 kg $CO_2$-eq/363 kg beef (60%); and the decline in calving rate from 100% to 60% conferred a 36.3% increase in production costs per unit of beef. Beauchemin et al. (2011) also noted the effects of weaning a greater number of calves through a combination of improved calving rate and reduced pre-weaning mortality (90% weaning rate compared to 85%) and reported that GHG emissions could be reduced by 4% per unit of beef via this strategy.

Reproductive technologies, including artificial insemination, estrus synchronization and embryo transfer, allow genetically superior males to sire a greater number of offspring; cattle to calve within a tight time period (thus producing more uniform calves) and genetically superior females to produce more than one offspring per year (embryo transfer). Sexed semen offers an opportunity to produce heifers from genetically superior cows, and the recent advent of genetic marker techniques allows offspring performance to be assessed without waiting for cattle to reach maturity or to produce their own offspring. However, the sustainability impacts of these technologies are dependent upon their adoption by the industry. According to the USDA (2009), oestrus synchronization is used only in 7.9% of the US beef operations, artificial insemination in 7.6% of operations and embryo transfer in 1.6% of operations, and marker-assisted selection is still in its infancy. These data are not a representative indicator of the number of cattle exposed to these technologies, however, as the average cow-calf herd in the United States contains approximately 40 cows and the adoption rate of reproductive technologies is greater in larger herds than in small herds. The majority of beef producers cited labour and time inputs required by reproductive technologies to be the major factor dissuading them from their adoption (USDA, 2009). Thus, genetic improvement in the beef industry may continue to be second to other management practices in terms of improvement opportunities in the near future.

## 6.5 Lifetime productivity

Lifetime productivity may be improved by maximizing cow productivity (e.g. producing one live calf per year), but also by reducing the proportion of unproductive time within the total lifetime (Szabó and Dákay, 2009), that is, by optimizing the time period from birth to first calving, and intervals between subsequent calvings. Ideally, a heifer will calve at 24 months of age and have a 12-month calving interval (Engelken, 2008); however, these metrics are not applicable to all systems. For example, restricted grazing or forage availability may result in lower heifer growth rates; thus, achieving a 24-month age at first calving could result in significantly greater economic and nutritional input costs that might not be outweighed by improved lifetime productivity (Fordyce et al., 1994). The tradition that heifers should be bred at 60–65% of their target mature weight was recently challenged by Funston et al. (2012), who suggested that producers should instead focus on generating 'sound, functional, low-cost, and pregnant' heifers that are suited to the system – a philosophy that would certainly fit with the sustainability objectives of maximizing economic and resource efficiency.

Although not directly applicable to the US systems (where a 3-year age at first calving would be unusual), Nguyen et al. (2013) quantified the environmental gains resulting from reducing age at first calving from 3 to 2 years of age within French beef systems. Despite a reduction in the age at culling (6 years of age rather than 7), lifetime productivity was improved by reducing age at first calving (an average of 4.7 calves were produced per lifetime compared to 4.4 calves in the 3-year age at first calving scenario) and reduced environmental impacts per unit of carcass weight by 7.8–8.4%. It is interesting to note that both Nguyen et al. (2013) and Beauchemin et al. (2011) reported that improving cow longevity conferred negligible differences in GHG emissions per unit of carcass weight. Although the total number of calves born per cow lifetime increased, there was relatively little advantage of producing finished heifers at low live weights compared to culling a greater proportion of (heavier) mature cows. By contrast, Oishi et al. (2013) reported that both environmental and economic indices were improved by increasing the longevity of mature cows within cow-calf production systems in Japan.

## 6.6 Animal health

Animal health and welfare are significant concerns for both consumers and producers; however, there is often a dichotomy between the belief that animals should be entitled to veterinary care and concerns regarding pharmaceutical use. Appropriate veterinary treatment that allows animals to be free of pests and disease is not only morally appropriate, but also allows cattle to perform at a level appropriate to their genetic potential. Both Lawrence and Ibarburu (2007) and Capper (2013c) reported that judicious use of parasiticides in beef cattle improved performance, with associated positive environmental and economic impacts. Within the latter paper, withdrawing effective parasite control from the US cattle production systems increased the cattle population size required to produce a unit of beef by 18%, with concurrent increases in land use (16%), water use (15%), GHG emissions (13%) and economic costs of production (12%). Wileman et al. (2009) reported that morbidity was reduced from 55 to 29% and mortality from 3.8 to 1.8% in cattle receiving metaphylactic treatment upon feedlot entry. Increased rates of morbidity and mortality represent both productivity losses and economic losses to producers and thus are antagonistic to a culture of efficiency and continuous improvement. However, the underlying consumer concern regarding the use of pharmaceuticals within livestock production cannot and should not be underestimated.

# 7 The role of consumers

Globally, 30% of food is wasted through losses either pre- or postharvest, particularly at the consumer level (FAO, 2016). Although proportionally less meat is wasted compared to other foods (e.g. baked goods, vegetables), consumer behaviours should not be ignored in any discussion of beef sustainability. A number of researchers suggest that consumption of meat, milk and eggs should be curtailed in order to improve environmental sustainability. Berners-Lee et al. (2012) reported that converting from the average UK inhabitant's diet to a vegetarian or vegan diet containing equal daily quantities of protein would reduce diet-related GHG emissions by 22 or 26%, respectively. A global transition to diets containing less meat or a switch to a totally plant-based diet was also shown by Stehfest et al. (2009) to reduce both agricultural land use and emissions of methane and nitrous oxide. By contrast, although Vieux et al. (2012) noted that animal-source foods were associated with higher GHG emissions than plant-based foods, the perception that high nutritional-quality diets (high in plant-based foods) were associated with lower GHG emissions did not hold true, and that when expressed per calorie, fruit and vegetable production imparted GHG emissions similar in magnitude to animal-source foods.

Tobler et al. (2011) suggested that reducing meat consumption on a global scale would meet with considerable opposition, although Säll and Gren (2015) demonstrated that imposing a tax on meat and dairy products could potentially reduce the environmental impacts of Swedish livestock production by 12%, and would have a greater relative impact on meat consumption, because consumers' buying behaviours were more sensitive to shifts in meat price compared to dairy. The average US inhabitant consumes 75.9 kg of meat per year (FAO, 2013a). To take the simplistic view that a 1 day per week withdrawal from meat consumption would cut meat production by one-seventh, if every one of the United States' 316 million inhabitants adopted such a dietary change, the projected annual reduction in national GHG emissions would be less than one-third of one per cent (Capper, 2013d). Moreover, livestock production does not simply produce animal-source foods, but also multifarious by-products (both edible and inedible), including hides, fertilizer, tallow and pharmaceuticals. It is not yet clear what the global environmental impacts would result from replacing manure with inorganic fertilizers or pharmaceutical ingredients with synthetic alternatives; yet these consequences of reduced meat consumption should be accounted for in any whole-system analyses of dietary change. Garnett (2011) suggested that a multifactorial approach (including improving energy efficiency, land management and reducing food waste) is required to reduce the environmental impact of the food system, and that although reducing meat consumption is one potential avenue, significant gains may also be made by improving cattle productivity.

Ultimately, the only way to maintain beef sustainability is to ensure social acceptability – in the absence of social licence to operate and a market for both beef and associated by-products, the US beef industry will cease to exist. The positive correlation between resource efficiency and economic costs of production (Capper, 2013a,b; White and Capper, 2014) ensures that productivity gains may continue to provide an affordable product to consumers; yet there is a growing backlash against so-called 'cheap' food and the perception that large-scale operations are concerned more with the economic profits of cattle production than with environmental or animal welfare concerns. Nonetheless, survey data indicate that 95% of consumers buy food on the basis of three parameters: price, taste and nutrition, whereas a small proportion (4%) make lifestyle-based food choices (e.g. gourmet, organic, local) and only 1% believe that certain foods should be banned or restricted on the basis of specific technologies or production methods (Simmons, 2011). Whether these trends will continue in the future is debatable. The ability to choose from a wide variety of foods and production methods leads to an increase in selectivity on the part of consumers; thus, it is possible that the predicted increases in per capita income in currently developing countries will also lead to increasing demand for beef produced in alternative systems.

# 8 Conclusion

The sustainability of the beef industry will clearly continue to be an issue for all stakeholders into the future. However, as discussed within this chapter, there is no single answer and no one-size-fits-all system. Instead, improving productivity and efficiency throughout the industry, allowing producers access to the tools and technologies that confer yield and growth gains and communicating effectively with consumers on the environmental gains made by the beef industry are all important components of maintaining future beef sustainability (Capper and Yancey, 2015).

Industry-wide environmental responsibility has improved over time, as demonstrated by reductions in resource use and GHG emissions per unit of beef (although data are still lacking on other environmental metrics), and it appears clear that further opportunities are available in terms of improving output per unit of herd bodyweight. However, as noted by Smith et al. (2007), many positive and negative trade-offs may exist when attempting to maintain environmental sustainability in the future. For example, increasing animal and crop productivity may improve food security and spare land for other purposes; yet, intensifying water use may improve productivity but deplete reserves. Furthermore, enhancing carbon sequestration may increase soil productivity and reduce erosion, but restricting fertilizer use in an attempt to reduce $N_2O$ emissions and energy use might reduce crop yields. Nonetheless, the FAO (2013b) suggested that if producers within any given region adopted the practices of the 10% of producers with the lowest emissions while maintaining productive output, the livestock sector's GHG emissions could be cut by 30%. It is, therefore, in the industry's best interests to continue to develop new mechanisms to improve environmental responsibility.

Improving productivity tends to have positive effects on economic viability for producers (Capper, 2013a,b,c; White and Capper, 2014), and beef still appears to be a fundamental protein choice for many consumers, as recent increases in beef retail price indicate that demand is still high (Mintert et al., 2009). Maintaining and improving sustainability in the face of a consumer market that is increasingly sceptical about food production transparency and management practices is a considerable challenge, yet is not impossible. According to a recent report from Sullivan Higdon and Sink FoodThink (2013), only 22% of consumers believe that the agricultural industry is transparent about its practices, yet 53% of consumers trust farmers and ranchers (compared to academics at 51%, media at 21% and politicians at 8%). Therefore, the industry has a considerable opportunity to explain management systems and practices to consumers, thus increasing both transparency and the social acceptability of beef as a protein choice (Capper and Yancey, 2015). The range of improvement opportunities available to all types and sizes of production systems indicate that the future US beef industry has the potential to be sustainable, providing that a concerted effort to improve is adopted and executed by all producers and sectors within the chain.

# 9 Where to look for further information

## Reports and Journal articles

FAO. 2006. *Livestock's Long Shadow – Environmental Issues and Options*. FAO, Rome, Italy.

FAO. 2013b. *Tackling Climate Change Through Livestock – A Global Assessment of Emissions and Mitigation Opportunities*. Food and Agriculture Organization of the United Nations, Rome, Italy.

Kebreab, E. 2013. *Sustainable Animal Agriculture*. CABI Publishing, Wallingford, UK.

Pitesky, M. E., K. R. Stackhouse and F. M. Mitloehner. 2009. Clearing the air: Livestock's contribution to climate change. *Adv. Agron.* 103: 3–40.

SHS FoodThink. 2013. *Building Trust in What We Eat – Consumers Knowledge and Trust in Food Production and How Food Marketers Can Improve It*. Sullivan Higdon & Sink, Kansas City, MO.

Simmons, J. 2011. *Making Safe, Affordable and Abundant Food a Global Reality*. Elanco Animal Health, Greenfield, IN.

## Websites

International Livestock Research Institute: https://www.ilri.org/
Enough: https://www.enoughmovement.com/
CAST: http://www.cast-science.org/
Genetic Literacy Project: https://www.geneticliteracyproject.org/

# 10 References

Abney, C. S., J. T. Vasconcelos, J. P. McMeniman, S. A. Keyser, K. R. Wilson, G. J. Vogel and M. L. Galyean. 2007. Effects of ractopamine hydrochloride on performance, rate and variation in feed intake, and acid-base balance in feedlot cattle. *J. Anim. Sci.* 85: 3090–8.

Archer, J. A., E. C. Richardson, R. M. Herd and P. F. Arthur. 1999. Potential for selection to improve efficiency of feed use in beef cattle: A review. *Aust. J. Ag. Res.* 50: 247–61.

Baxa, T. J., J. P. Hutcheson, M. F. Miller, J. C. Brooks, W. T. Nichols, M. N. Streeter, D. A. Yates and B. J. Johnson. 2010. Additive effects of a steroidal implant and zilpaterol hydrochloride on feedlot performance, carcass characteristics, and skeletal muscle messenger ribonucleic acid abundance in finishing steers. *J. Anim. Sci.* 88: 330–7.

Beauchemin, K. A., H. Janzen, S. M. Little, T. A. McAllister and S. M. McGinn. 2010. Life cycle assessment of greenhouse gas emissions from beef production in western Canada: A case study. *Ag. Sys.* 103: 371–9.

Beauchemin, K. A., H. H. Janzen, S. M. Little, T. A. McAllister and S. M. McGinn. 2011. Mitigation of greenhouse gas emissions from beef production in western Canada – Evaluation using farm-based life cycle assessment. *An. Feed. Sci. Tech.* 166–7: 663–77.

Beckett, J. L., R. J. Delmore, G. C. Duff, D. A. Yates, D. M. Allen, T. E. Lawrence and N. A. Elam. 2009. Effects of zilpaterol hydrochloride on growth rates, feed conversion, and carcass traits in calf-fed Holstein steers. *J. Anim. Sci.* 87: 4092–100.

Bellarby, J., B. Foereid, A. Hastings and P. Smith. 2008. *Cool Farming: Climate Impacts of Agriculture and Mitigation Potential.* Greenpeace International, Amsterdam, The Netherlands.

Berners-Lee, M., C. Hoolohan, H. Cammack and C. N. Hewitt. 2012. The relative greenhouse gas impacts of realistic dietary choices. *Energy Policy* 43: 184–90.

Bertrand, S. and J. Barnett. 2011. Standard method for determining the carbon footprint of dairy products reduces confusion. *An. Front.* 1: 14–18.

Bonesmo, H., K. A. Beauchemin, O. M. Harstad, A. O. Skjelvag and A. D'Alton. 2013. Greenhouse gas emission intensities of grass silage based dairy and beef production: A systems analysis of Norwegian farms. *Livest. Prod. Sci.* 152: 239–52.

Bruce, J. P., M. Frome, E. Haites, H. Janzen, R. Lal and K. Pawtian. 1999. Carbon sequestration in soils. *J. Soil Water Conserv.* 54: 382–9.

Capper, J. L. 2011. The environmental impact of beef production in the United States: 1977 compared with 2007. *J. Anim. Sci.* 89: 4249–61.

Capper, J. L. 2012. Is the grass always greener? Comparing resource use and carbon footprints of conventional, natural and grass-fed beef production systems. *Animals.* 2: 127–43.

Capper, J. L. 2013a. The environmental and economic impact of calving rate within U.S. beef production. *Proceedings of the ADSA-ASAS Joint Annual Meeting.* Indianapolis, IN, USA.

Capper, J. L. 2013b. The environmental and economic impact of steroid implant and beta-adrenergic agonist use within U.S. beef production. *Proceedings of the ADSA-ASAS Joint Annual Meeting.* Indianapolis, IN, USA.

Capper, J. L. 2013c. The environmental and economic impact of withdrawing parasite control (Fenbendazole) from U.S. beef production. *Proceedings of the ADSA-ASAS Joint Annual Meeting.* Indianapolis, IN, USA.

Capper, J. L. 2013d. Should we reject animal source foods to save the planet? A review of the sustainability of global livestock production. *South African Journal of Animal Science* 43: 233–46.

Capper, J. L., E. Castañeda-Gutiérrez, R. A. Cady and D. E. Bauman. 2008. The environmental impact of recombinant bovine somatotropin (rbST) use in dairy production. *Proc. Natl. Acad. Sci. U.S.A.* 105: 9668–73.

Capper, J. L. and D. J. Hayes. 2012. The environmental and economic impact of removing growth-enhancing technologies from United States beef production. *J. Anim. Sci.* 90: 3527–37.

Capper, J. L. and J. W. Yancey. 2015. Communicating animal science to the general public. *An. Front.* 5: 28–35.

Cederberg, C., D. Meyer and A. Flysjo. 2009a. *Life Cycle Inventory of Greenhouse Gas Emissions and Use of Land and Energy in Brazilian Beef Production*. The Swedish Institute for Food and Biotechnology, Gothenburg, Sweden.

Cederberg, C., M. Persson, K. Neovius, S. Molander and R. Clift. 2011. Including carbon emissions from deforestation in the carbon footprint of Brazilian beef. *Environ. Sci. Technol.* 45: 1773–9.

Cederberg, C., U. Sonesson, M. Henriksson, V. Sund and J. Davis. 2009b. *Greenhouse Gas Emissions from Swedish Production of Meat, Milk and Eggs: 1990 and 2005*. Swedish Institute of Food & Biotechnology, Gothenburg, Sweden.

Conestoga-Rovers & Associates. 2010. *Evaluating Environmental and Economic Impact for Beef Production in Alberta Using Life Cycle Analysis*. Alberta Agriculture and Rural Development, Waterloo, Ontario.

Cooprider, K. L., F. M. Mitloehner, T. R. Famula, E. Kebreab, Y. Zhao and V. Eenennaam. 2011. Feedlot efficiency implications on GHG emissions and sustainability. *J. Anim. Sci.* 89: 2643–56.

Corah, L. R. 2008. Development of a corn-based beef industry. *J. Anim. Sci.* 86: 3635–9.

Council for Agricultural Science and Technology (CAST). 2013. Animal Feed vs. Human Food: Challenges and Opportunities in Sustaining Animal Agriculture Toward 2050. Issue Paper 53, CAST, Ames, IA.

Cundiff, L. V., K. E. Gregory and R. M. Koch. 1974a. Effects of heterosis on reproduction in Hereford, Angus and Shortforn cattle. *J. Anim. Sci.* 38: 711–27.

Cundiff, L. V., K. E. Gregory, F. J. Schwulst and R. M. Koch. 1974b. Effects of heterosis on maternal performance and milk production in Hereford, Angus and Shorthorn cattle. *J. Anim. Sci.* 38: 728–45.

Delgado, C. L. 2003. Rising consumption of meat and milk in developing countries has created a new food revolution. *J. Nutr.* 133: 3907S–10S.

DeRamus, H. A., T. C. Clement, D. D. Giampola and P. C. Dickison. 2003. Methane emissions of beef cattle on forages: Efficiency of grazing management systems. *J. Environ. Qual.* 32: 269–77.

Elam, N. A., J. T. Vasconcelos, G. G. Hilton, D. L. VanOverbeke, T. E. Lawrence, T. H. Montgomery, W. T. Nichols, M. N. Streeter, J. P. Hutcheson, D. A. Yates and M. L. Galyean. 2009. Effect of zilpaterol hydrochloride duration of feeding on performance and carcass characteristics of feedlot cattle. *J. Anim. Sci.* 87: 2133–41.

Elkington, J. 1994. Towards the sustainable corporation: Win-win-win business strategies for sustainable development. *Calif. Manage. Rev.* 36: 90–100.

Engelken, T. J. 2008. Developing replacement beef heifers. *Theriogenology* 70: 569–72.

Environmental Working Group. 2011. *Meat Eater's Guide to Climate Change and Health*. Environmental Working Group, Washington, DC.

FAO. 2006. *Livestock's Long Shadow – Environmental Issues and Options*. FAO, Rome, Italy.

FAO. 2009. *How to Feed the World in 2050*. FAO, Rome, Italy.

FAO. 2012a. *Livestock Sector Development for Poverty Reduction: An Economic and Policy Perspective*. FAO, Rome, Italy.

FAO. 2012b. *The State of Food Insecurity in the World*. FAO, Rome, Italy.

FAO. 2013a. FAOSTAT. http://faostat.fao.org (Accessed 1 June 2013).

FAO. 2013b. *Tackling Climate Change Through Livestock – A Global Assessment of Emissions and Mitigation Opportunities*. Food and Agriculture Organization of the United Nations, Rome, Italy.

FAO. 2013c. *World Agriculture Towards 2030/2050: The 2012 Revision*. United Nations Food and Agriculture Organisation, Rome, Italy.

FAO. 2016. SAVE FOOD: Global Initiative on Food Loss and Waste Reduction. http://www.fao.org/save-food/en/ (Accessed 8 January 2016).

Fordyce, G., K. W. Entwistle and L. A. Fitzpatrick. 1994. Developing cost-effective strategies for improved fertility in Bos indicus cross cattle. Final Report, Project NAP2:DAQ.062/UNQ.009, Meat Research Corporation, Sydney, Australia.

Funston, R. N., J. L. Martin, D. M. Larson and A. J. Roberts. 2012. Nutritional aspects of developing replacement heifers. *J. Anim. Sci.* 90: 1166–71.

Garnett, T. 2011. Where are the best opportunities for reducing greenhouse gas emissions in the food system (including the food chain)? *Food Policy* 36: S23–S32.

Godfray, H. C. J., J. R. Beddington, I. R. Crute, L. Haddad, D. Lawrence, J. F. Muir, J. Pretty, S. Robinson, S. M. Thomas and C. Toulmin. 2010. Food security: The challenge of feeding 9 billion people. *Science*. 327: 812–18.

Goodland, R. and J. Anhang. 2009. *Livestock and Climate Change. What if the Key Actors in Climate Change were Pigs, Chickens and Cows?* Worldwatch Institute, Washington, DC, USA.

Grainger, C. M. and K. A. Beauchemin. 2011. Can enteric methane emissions from ruminants be lowered without lowering their production? *An. Feed. Sci. Tech.* 166–7: 302–20.

Gregory, K. E. and L. V. Cundiff. 1980. Crossbreeding in beef cattle: Evaluation of systems. *J. Anim. Sci.* 51: 1224–42.

Gruber, S. L., J. D. Tatum, T. E. Engle, M. A. Mitchell, S. B. Laudert, A. L. Schroeder and W. J. Platter. 2007. Effects of ractopamine supplementation on growth performance and carcass characteristics of feedlot steers differing in biological type. *J. Anim. Sci.* 85: 1809–15.

Herd, R. M. and P. F. Arthur. 2008. Physiological basis for residual feed intake. *J. Anim. Sci.* 87: E64–E71.

Hofmann, R. R. 1989. Evolutionary steps of ecophysiological adaptation and diversification of ruminants: a comparative view of their digestive system. *Oecologica* 78: 443–57.

Hotelling, H. 1931. The economics of exhaustible resources. *J. Polit. Econ.* 39: 137–75.

IPCC. 2006. *IPCC Guidelines for National Greenhouse Gas Inventories.* Institute for Global Environmental Strategies (IGES) for the IPCC, Kanagawa, Japan.

Johnson, B. J., F. R. B. Ribeiro and J. L. Beckett. 2013. Application of growth technologies in enhancing food security and sustainability. *An. Front.* 3: 8–13.

Joos, F., G.-K. Plattner, T. F. Stocker, O. Marchal and A. Schmittner. 1999. Global warming and marine carbon cycle feedbacks on future atmospheric $CO_2$. *Science*. 284: 464–7.

Kennedy, C. and R. Lindsey. 2015. What's the difference between global warming and climate change? https://www.climate.gov/news-features/climate-qa/whats-difference-between-global-warming-and-climate-change (Accessed 10 June 2016).

Koneswaran, G. and D. Nierenberg. 2008. Global farm animal production and global warming: impacting and mitigating climate change. *Environ. Health Perspect.* 116: 578–82.

Kuhlman, T. and J. Farrington. 2010. What is Sustainability? *Sustainability* 2: 3436–48.

Laudert, S., G. Vogel, A. Schroeder, W. Platter and M. Van Koevering. 2005. Effects of ractopamine fed to finishing steers. I. Summary of six studies - Growth performance. *J. Anim. Sci.* 83 (suppl. 1): 112.

Lawrence, J. D. and M. Ibarburu. 2007. *Economic Analysis of Pharmaceutical Technologies in Modern Beef Production in a Bioeconomy Era*. Iowa State University, Ames, IA.

Leaver, D. 2011. Global food supply: A challenge for sustainable agriculture. *Nutr. Bull.* 36: 416–21.

Masuda, T. and P. D. Goldsmith. 2010. China's meat consumption: An income elasticity analysis and long-term projections Agricultural & Applied Economics Association 2010 AAEA, CAES, & WAEA Joint Annual Meeting. Agricultural & Applied Economics Association, Denver, CO.

McDermott, J. J., S. J. Staal, H. A. Freeman, M. Herrero and J. A. Van de Steeg. 2010. Sustaining intensification of smallholder livestock systems in the tropics. *Livest. Sci.* 130: 95–109.

Mintert, J. R., G. T. Tonsor and T. C. Schroeder. 2009. U.S. beef demand drivers and enhancement opportunities: A research summary. Paper 274. Range Beef Cow Symposium, University of Nebraska - Lincoln, NE, USA.

Montgomery, J. L., C. R. Krehbiel, J. J. Cranston, D. A. Yates, J. P. Hutcheson, W. T. Nichols, M. N. Streeter, D. T. Bechtol, E. Johnson, T. TerHune and T. H. Montgomery. 2009a. Dietary zilpaterol hydrochloride. I. Feedlot performance and carcass traits of steers and heifers. *J. Anim. Sci.* 87: 1374–83.

Montgomery, J. L., C. R. Krehbiel, J. J. Cranston, D. A. Yates, J. P. Hutchseon, W. T. Nichols, M. N. Streeter, R. S. Swingle and T. H. Montgomery. 2009b. Effects of dietary zilpaterol hydrochloride on feedlot performance and carcass characteristics of beef steers fed with and without monensin and tylosin. *J. Anim. Sci.* 87: 1013–23.

Murphy, M. R., R. L. Baldwin and L. J. Koong. 1982. Estimation of stoichiometric parameters for rumen fermentation of roughage and concentrate diets. *J. Anim. Sci.* 55: 411–21.
Murphy, S. P. and L. H. Lindsay. 2003. Nutritional importance of animal source foods. *J. Nutr.* 133: 3932S–5S.
Neumann, C., D. M. Harris and L. M. Rogers. 2002. Contribution of animal source foods in improving diet quality and function in children in the developing world. *Nutr. Res.* 22: 193–220.
Neumeier, C. J. and F. M. Mitloehner. 2013. Cattle biotechnologies reduce environmental impact and help feed a growing planet. *An. Front.* 3: 36–41.
Nguyen, T. L. T., J. E. Hermansen and L. Mogensen. 2010. Environmental consequences of different beef production systems in the EU. *J. Clean. Prod.* 18: 756–66.
Nguyen, T. T. H., M. Doreau, M. Eugene, M. S. Corson, F. Garcia-Launay, G. Chesneau and H. M. G. van der Werf. 2013. Effect of farming practices for greenhouse gas mitigation and subsequent alternative land use on environmental impacts of beef cattle production systems. *Animal.* 7: 860–9.
Nijdam, D., T. Rood and H. Westhoek. 2012. The price of protein: Review of land use and carbon footprints from life cycle assessments of animal food products and their substitutes. *Food Policy* 37: 760–70.
NRC. 2000. *Nutrient Requirements of Beef Cattle*. National Academy Press, Washington, DC.
Oishi, K., Y. Kato, A. Ogino and H. Hirooka. 2013. Economic and environmental impacts of changes in culling parity of cows and diet composition in Japanese beef cow–calf production systems. *Ag. Sys.* 115: 95–103.
Parr, S. L., K. Y. Chung, M. L. Galyean, J. P. Hutcheson, N. DiLorenzo, K. E. Hales, M. L. May, M. J. Quinn, D. R. Smith and B. J. Johnson. 2010. Performance of finishing beef steers in response to anabolic implant and zilpaterol hydrochloride supplementation. *J. Anim. Sci.* 89: 560–70.
Pelletier, N., R. Pirog and R. Rasmussen. 2010. Comparative life cycle environmental impacts of three beef production strategies in the Upper Midwestern United States. *Ag. Sys.* 103: 380–9.
Pelletier, N. and P. Tyedmers. 2010. Forecasting potential global environmental costs of livestock production 2000–50. *Proc. Natl. Acad. Sci. U.S.A.* 107: 18371–4.
Pinstrup-Andersen, P. 2000. Food policy research for developing countries: emerging issues and unfinished business. *Food Policy* 25: 125–41.
Pitesky, M. E., K. R. Stackhouse and F. M. Mitloehner. 2009. Clearing the air: Livestock's contribution to climate change. *Adv. Agron.* 103: 3–40.
Pope, J., D. Annandale and A. Morrison-Saunders. 2004. Conceptualising sustainability assessment. *Environ. Imp. Assess. Rev.* 24: 595–616.
Randolph, T. F., E. Schelling, D. Grace, C. F. Nicholson, J. L. Leroy, D. C. Cole, M. W. Demment, A. Omore, A. Zinsstag and M. Ruel. 2007. Invited review: Role of livestock in human nutrition and health for poverty reduction in developing countries. *Int. J. An. Sci.* 85: 2788–800.
Ridoutt, B. G., G. Page, K. Opie, J. Huang and W. Bellotti. 2014. Carbon, water and land use footprints of beef cattle production systems in southern Australia. *J. Clean. Prod.* 73: 24–30.
Säll, S. and I.-M. Gren. 2015. Effects of an environmental tax on meat and dairy consumption in Sweden. *Food Policy* 55: 41–53.
Sansoucy, R. 1995. Livestock - a driving force for food security and sustainable development. *World Anim. Rev.* 84/85: 5–17.
Scramlin, S. M., W. J. Platter, R. J. Gomez, W. T. Choat, F. T. McKeith and J. Killefer. 2010. Comparative effects of ractopamine hydrochloride and zilpaterol hydrochloride on growth performance, carcass traits, and longissimus tenderness of finishing steers. *J. Anim. Sci.* 88: 1823–9.
Selk, G. E., R. R. Reuter and G. L. Kuhl. 2006. Using growth-promoting implants in stocker cattle. *Vet. Clin. Food Anim. Pract.* 22: 435–49.
SHS FoodThink. 2013. *Building Trust in What We Eat – Consumers Knowledge and Trust in Food Production and How Food Marketers Can Improve It*. Sullivan Higdon & Sink, Kansas City, MO.
Simmons, J. 2011. *Making Safe, Affordable and Abundant Food a Global Reality*. Elanco Animal Health, Greenfield, IN.

Smith, J., K. Sones, D. Grace, S. MacMillan, S. Tarawali and M. Herrero. 2012. Beyond milk, meat, and eggs: Role of livestock in food and nutrition security. *An. Front.* 3: 6–13.

Smith, P., D. Martino, Z. Cai, D. Gwary, H. Janzen, P. Kumar, B. McCarl, S. Ogle, F. O'Mara, C. Rice, B. Scholes, O. Sirotenko, M. Howden, T. McAllister, G. Pan, V. Romanenkov, U. Schneider and S. Towprayoon. 2007. Policy and technological constraints to implementation of greenhouse gas mitigation options in agriculture. *Agric. Ecosyst. Environ.* 118: 6–28.

Stackhouse-Lawson, K. R., C. A. Rotz, J. W. Oltjen and F. M. Mitloehner. 2012a. Carbon footprint and ammonia emissions of California beef production systems. *J. Anim. Sci.* 90: 4641–55.

Stackhouse-Lawson, K. R., C. A. Rotz, J. W. Oltjen and F. M. Mitloehner. 2012b. Growth promoting technologies reduce the carbon footprint, ammonia emissions, and costs of California beef production system. *J. Anim. Sci.* 90: 4656–65.

Stehfest, E., L. Bouwman, D. P. van Vuuren, M. G. J. den Elzen, B. Eickhout and P. Kabat. 2009. Climate benefits of changing diet. *Climatic Change* 95: 83–102.

Szabó, F. and I. Dákay. 2009. Estimation of some productive and reproductive effects on longevity of beef cows using survival analysis. *Livest. Sci.* 122: 271–5.

Thomson, C., D. Rheinhardt and D. G. Renter. 2009. Analysis of modern technologies commonly used in beef cattle production: Conventional beef production versus nonconventional production using meta-analysis. *J. Anim. Sci.* 87: 3418–26.

Tobler, C., V. H. M. Visschers and M. Siegrist. 2011. Eating green. Consumers' willingness to adopt ecological food consumption behaviors. *Appetite* 57: 674–82.

United Nations. 2005. *2005 World Summit Outcome*, United Nations Publications, New York, NY, USA.

United Nations, Department of Economic and Social Affairs, Population Division. 2011. World Population Prospects: The 2010 Revision, Volume I: Comprehensive Tables. ST/ESA/SER.A/313., United Nations, New York, NY, USA.

USDA. 2009. Beef 2007–08 Part II: Reference of Beef Cow-calf Management Practices in the United States, 2007–08, USDA:APHIS:VS, CEAH, National Animal Health Monitoring System, Fort Collins, CO, USA.

USDA. 2015. Data and Statistics. http://www.nass.usda.gov/Data_and_Statistics/Quick_Stats/index.asp (Accessed 12 May 2015).

USDA/NASS. 2011. *Livestock Slaughter Annual Summary*. USDA, Washington, DC, USA.

Van Kessel, J. A. S. and J. B. Russell. 1996. The effect of pH on ruminal methanogenesis. *FEMS Microbiol. Immunol.* 20: 205–10.

Vasconcelos, J. T., R. J. Rathmann, R. R. Reuter, J. Leibovich, J. P. McMeniman, K. E. Hales, T. L. Covey, M. F. Miller, W. T. Nichols and M. L. Galyean. 2008. Effects of duration of zilpaterol hydrochloride feeding and days on the finishing diet on feedlot cattle performance and carcass traits. *J. Anim. Sci.* 86: 2005–15.

Vergé, X. P. C., J. A. Dyer, R. L. Desjardins and D. Worth. 2008. Greenhouse gas emissions from the Canadian beef industry. *Ag. Sys.* 98: 126–34.

Vermeir, I. and W. Verbeke. 2006. Sustainable food consumption: Exploring the consumer 'attitude – behavioral intention' gap. *J Agr. Environ. Eth.* 19: 169–94.

Vieux, F., N. Darmon, D. Touazi and L. G. Soler. 2012. Greenhouse gas emissions of self-selected individual diets in France: Changing the diet structure or consuming less? *Ecol. Econ.* 75: 91–101.

Vogel, G. J., G. C. Duff, J. Lehmkuhler, J. L. Beckett, J. S. Drouilliard, A. L. Schroeder, W. J. Platter, M. T. Van Koevering and S. B. Laudert. 2009. Effect of ractopamine hydrochloride on growth performance and carcass traits in calf-fed and yearling Holstein steers fed to slaughter. *The Professional Animal Scientist* 25: 26–32.

Ward, G. M. 1980. Energy, land and feed constraints on beef production in the 80's. *J. Anim. Sci.* 51: 1051–64.

White, R. R. and J. L. Capper. 2012. The environmental, economic and social implications of improving yield and average daily gain in beef production. *J. Anim. Sci.* 90 (Suppl. 3): 330.

White, R. R. and J. L. Capper. 2014. An environmental, economic and social assessment of improving cattle finishing weight or average daily gain within United States beef production. *J. Anim. Sci.* 91: 5801–12.

Wileman, B. W., D. U. Thomson, C. D. Reinhardt and D. G. Renter. 2009. Analysis of modern technologies commonly used in beef cattle production: Conventional beef production versus nonconventional production using meta-analysis. *J. Anim. Sci.* 87: 3418–26.

Wilkinson, J. M. 2011. Re-defining efficiency of feed use by livestock. *Animal.* 5: 1014–22.

Williams, A. G. and G. S. Coleman. 1988. The Rumen Protozoa. In P. N. Hobson and C. S. Stewart (eds), The Rumen Microbial Ecosystem, pp. 77–129. Springer, New York, NY, USA.

Chapter 19

# Controversies surrounding the impact of the fat content of beef on human health

*Jennifer Fleming and Penny Kris-Etherton, Penn State University, USA*



## 1 Introduction

Whether 'red' or 'processed meat' intake is related to adverse health outcomes in the US population and other developed countries is a topic under intense debate. Beef, lamb and pork all are typically labelled as 'red meats'; however, beef is the most prevalent source of red meat in the US diet (55%; beefUSA.org). Despite this, few studies have specifically investigated the relationship between beef consumption and cardiovascular disease (CVD), which will be the primary focus of this chapter. Some studies, of note, have evaluated the relationship between red meat (not distinguishing beef) and CVD. Like many other foods, guidelines for meat consumption in cardiovascular health were based historically on the research conducted on isolated nutrients (i.e. saturated fat and cholesterol). Epidemiologic studies have linked beef (as well as other red meat) consumption to increased CVD, cancer mortality, acute myocardial infarction and metabolic syndrome. In contrast, however, there also are numerous epidemiologic studies that have not reported an association between beef (and other red meat) consumption and CVD. It is important to note that epidemiologic studies of nutrition and chronic disease do not establish whether the relationships are real or simply random associations. Concerns about whether red or processed meat causes adverse outcomes cannot simply be justified because epidemiologic studies might not adequately account for fat content, meat processing treatments and cooking methods. These factors can modify the health effects of lean red meat (i.e. specifically, beef) in the diet. For example,

http://dx.doi.org/10.19103/AS.2016.0009.19

it is important to distinguish the health outcomes that result from processed meats versus unprocessed red meat, and for the purposes of this chapter, lean beef.

# 2 Beef and CVD

## 2.1 Observational trials

To date, a small number of individual studies suggest similar CVD risk for beef versus processed meats. Sinha et al. (2009), for example, reported an increase in CVD risk for men and women in the highest quintile of red (relative risk [RR], 1.27 and 1.50, respectively) and processed meat (RR, 1.09 and 1.38, respectively) intakes. Red meat included all types of beef and pork such as bacon, cold cuts, ham, hamburger, hot dogs, liver, sausage and steak. Likewise, in an analysis of 84 136 females in the Nurses' Health Study, Bernstein et al. (2010) reported a RR of 1.29 for the highest quintile of total red meat intake (hamburger, beef hot dog, processed meat, bacon, and beef/pork/lamb as a mixed and main dish) and a RR of 1.13 for red meat excluding processed meat.

In contrast, three recent meta-analyses of prospective studies all reported that processed meat consumption was significantly associated with a moderate increase in all-cause mortality, whereas the consumption of unprocessed red meat was not (Abete et al. 2014; Larsson and Orsini, 2014; Wang et al. 2015). In a meta-analysis of thirteen cohort studies, Abete et al. (2014) examined the association between consumption of meat (total, red, white and processed) and all-cause, CVD and ischaemic heart disease (IHD) mortality. Red meat was defined as fresh meat from beef, veal, lamb, or pork, hamburgers and meatballs. Processed meat was defined as any meat preserved by smoking, curing, salting or addition of chemical preservatives (i.e. bacon, sausage, salami, hot dogs and luncheon meats). In the analysis for all-cause mortality ($n = 10$), combining the risk estimates for the highest versus the lowest consumption category showed that the consumption of processed meat but not of total, red and white meats was positively associated with all-cause mortality (RR 1.22, 95% CI = 1.16–1.29). In an analysis of CVD mortality ($n = 9$), combined risk estimates for the comparison of the highest versus the lowest consumption category of processed meat (RR 1.18, 95% CI = 1.05–1.32) and red meat (RR 1.16, 95% CI = 1.03–1.32) showed positive associations with CVD mortality, yet neither was associated with IHD mortality. Given that IHD is the most prevalent form of CVD, and they share the same pathogenesis, the inconsistency in these results raises questions about the findings (Klurfeld, 2015). Abete et al. (2014) noted the possibility of residual confounding as well as the high heterogeneity among the studies and advised that their findings be interpreted with caution.

In a second meta-analyses of nine prospective cohort studies, Larsson and Orsini (2014) calculated the summary RR for all-cause mortality and processed meat or any meat preserved by smoking, curing, salting or addition of chemical preservatives (i.e. bacon, sausage, salami, hot dogs and luncheon meats) as 1.23 (95% CI = 1.17–1.28) from five studies; unprocessed red meat (unprocessed beef, pork, lamb or game, excluding fish and poultry) as 1.10 (95% CI = 0.98–1.22) from six studies; and total red meat (unprocessed red meat and processed meat combined) as 1.29 (95% CI = 1.24–1.35) from five studies. When a dose–response meta-analysis was performed, consumption of processed meat and total red meat, but not unprocessed red meat, was significantly associated with increased risk of all-cause mortality in a non-linear fashion (Larsson and Orsini, 2014). This suggests that even low intakes of processed meats increase all-cause mortality risk, whereas no association was observed with unprocessed red meat. The most recent meta-analysis by Wang et al. (2015) included nine studies with 17 prospective cohorts from populations around the world (the United States, $n = 3$; Australia, $n = 1$; Asia, $n = 2$; Europe, $n = 3$). For the total population, the highest category of unprocessed red meat (unprocessed beef, lamb or pork) consumption was not associated with an increase in the risk of all-cause mortality (RR 1.05, 95% CI = 0.93–1.19), CVD mortality (RR 1.06, 95% CI = 0.88–1.28) or cancer mortality (RR 1.03, 95% CI = 0.89–1.18), compared with the lowest category. In a dose–response analysis of the different populations, similar results were observed for the European and Asian populations. However, for the US population, each serving of unprocessed red meat consumption per day was significantly and positively associated with risk of all-cause mortality (RR 1.15, 95% CI = 1.12–1.19), CVD mortality (RR 1.19, 95% CI = 1.13–1.26) and cancer mortality (RR 1.12, 95% CI = 1.07–1.17). The difference among populations may be explained, in part, by the relatively higher consumption of beef, pork and lamb in the United States compared to other populations. Higher fish intake may have helped reduce disease progression among the Asian populations. Additionally, it has been suggested that food preparation techniques in the United States, such as the frying and grilling of meats, is associated with increased disease risk (Wang et al. 2015).

Although the results of several meta-analyses examining the consumption of red and/or processed meat intake consistently have shown that processed meat is related to increased risk of all-cause and CVD mortality, few studies have assessed total red meat consumption in relation to heart failure incidence (Ashaye et al. 2011; Nettleton et al. 2008; Kaluza et al. 2014). Of these, only one study examined the association with heart failure incidence separately for processed versus unprocessed meat consumption (Kaluza et al. 2014). In this study of 37 035 Swedish adult men, food intake was recorded at baseline only and subjects were followed for an average of 11.8 years. At follow-up, they found that the men who reported eating ≥75 g/d of processed meat compared with those who consumed <25 g/d had a 1.28 (95% CI = 1.10–1.48, $P$ trend = 0.01) higher risk of heart failure incidence and 2.43 (95% CI = 1.52–3.88, $P$ trend $< 0.001$) higher risk of heart failure mortality. In contrast, no association between consumption of unprocessed meat and incident or fatal heart failure was observed. A similar conclusion was reported in an update of the Micha et al. (2010) meta-analysis published in 2012 (Micha et al. 2012). In their original 2010 analysis (Micha et al. 2010), the authors separately evaluated the associations of red meat (defined as any unprocessed meat from beef, such as hamburgers, lamb, pork or game and excluding poultry and fish) and processed red meat (defined as any meat preserved by smoking, curing, or salting or addition of chemical preservatives, such as bacon, salami, sausages, hot dogs, or processed deli or luncheon meat) with the development of CVD events and diabetes. In their analyses of 17 prospective cohort studies and 3 case-control studies, they found that consumption of processed meats, but not unprocessed red meats, was associated with higher incidence of CHD and diabetes mellitus. Of the three studies that evaluated intake of red meat and stroke, no association was observed.

The 2012 update incorporated the findings from two large prospective cohort studies evaluating the associations between unprocessed red meat, which included beef, pork and lamb and risk of incident CHD (Bernstein et al. 2010) and CVD mortality (Pan et al. 2012). Their updated analyses showed that consumption of processed meat (50 g/d), which included bacon, hot dogs, sausage, salami and processed deli or luncheon meats, related to CHD with a RR of 1.42 (95% CI = 1.07–1.89), while consumption of unprocessed red meat (100 g/d) had a RR of 1.00 (95% CI = 0.81–1.23). In a similar analyses of nine studies, the risk of type 2 diabetes increased by 51% for each 50 g/day increase of processed meat consumption and 19% for each 100 g/day increase of unprocessed red meat consumption (Micha et al. 2012).

Based on the evidence to date, there remains uncertainty about the association between beef consumption and CVD mortality. Because a controlled trial of meat consumption (or any single food) and chronic disease endpoints would be nearly impossible to conduct, it is important to understand the inherent limitations of observational research. The influence of residual confounding (i.e. when a confounding factor has not been adequately adjusted for), for example, may explain some or all of the observed modest associations of red meat with CVD. The Sinha et al. (2009) study, for example, was criticized due to the observed association between eating habits and other health status markers. Those in the top quintile of red meat consumption were three times more likely to be smokers, half as likely to do regular vigorous exercise, much less likely to have a college degree, were substantially heavier and had higher caloric intakes than low red meat consumers (Schectman, 2009). Additionally, saturated fat intake was not included in the multivariate model, nor was any information regarding previous health status, other than smoking (de Abreu Silva and Marcadenti, 2009). In contrast, the observed association between processed meats and CVD is more plausible because much of it can be explained by the established effects of sodium on blood pressure (BP) (He et al., 2002) and the relationship between BP and other CVD risk factors (Singh et al., 2013). Compared to unprocessed meats, processed meats contain an average of 400% more sodium and 50% more nitrates per gram (Micha et al. 2010). Micha et al. (2012) estimated that the average sodium provided in one 50 g/day serving of processed meats would predict about 27% higher risk of CHD, or more than two-thirds of the observed 42% higher risk seen in cohort studies. This suggests that sodium content alone may be largely responsible for the observed CHD risk with processed meat consumption (Micha et al. 2012). It also must be recognized that cross-sectional studies, conducted at a single time point, do not establish cause and effect. Thus, the limitations and inconsistencies of observational research must be fully realized if these are the studies that are the basis for current dietary recommendations.

## 2.2 Clinical trials

Clinical trials examining the relationship between beef/lean beef consumption and CVD risk factors indicate that within the context of a heart-healthy diet, there is no adverse effect on major CVD risk factors and, in fact, a heart-healthy diet with lean beef has been shown to decrease major CVD risk factors. A meta-analysis of 124 RCTs on lipid and lipoprotein changes comparing beef against poultry or fish consumption found no significant differences; both decreased LDL-C similarly (Maki et al. 2012). Two clinical trials conducted in a free-living setting found comparable cholesterol-lowering effects of a National Cholesterol Education Program Step I diet (total fat $<30\%$, SFA $<10\%$ energy) containing lean red meat (LDL-C, –1.7 to –1.8%) or lean white meat (LDL-C, –2.0 to –2.9%) (Hunninghake et al. 2000; Davidson et al. 1999). Importantly, in both studies, subjects in the lean red meat group were more compliant to the cholesterol-lowering diet. Similarly, Nowson et al. (2009) examined the BP-lowering effects of a Dietary Approaches to Stop Hypertension (DASH)-like diet containing lean red meat in moderately hypertensive post-menopausal women. After 14 weeks of dietary intervention (lean red meat was provided), the investigators observed significant reductions in systolic BP (–5.6 mm Hg), with no changes in lipids or lipoproteins. The lack of an effect on lipids and lipoproteins was most likely due to the higher SFA intake (12–13% of total calories) in the DASH-like diet.

In addition to reductions in BP and LDL-C, controlled feeding intervention studies with lean beef have shown benefits in other important risk factors for heart disease, such as apolipoproteins A-I and C-III, and vascular health (Roussell et al. 2012, 2014).

In the Beef in an Optimal Lean Diet (BOLD) study, 36 moderately hypercholesterolemic individuals (LDL-C >110mg/dL) completed a randomized crossover controlled feeding study. Participants were fed 4 intervention diets; Healthy American Diet [33% total fat, 12% SFA, 49% carbohydrate (CHO), 19% protein (PRO), 0.7 oz beef/d]; DASH Diet (27% total fat, 6% SFA, 50% CHO, 19% PRO, 1.0 oz beef/d); BOLD (28% total fat, 6% SFA, 54% CHO, 19% PRO, 4.0 oz beef/d); and BOLD+ (28% total fat, 6% SFA, 46% CHO, 28% PRO, 5.4 oz beef/d) for five weeks with a 1 week washout between diets. The study was designed to evaluate a 'dose–response' effect of lean beef in the context of a blood cholesterol-lowering diet on lipids, lipoproteins, apolipoproteins and vascular health.

Significant reductions ($P < 0.05$) in total cholesterol (TC) and LDL-C were reported in response to the DASH (–19.0 ± 4.3 and –14.4 ± 3.7 mg/dL), BOLD (–18.6 ± 4.1 and –13.5 ± 3.6 mg/dL) and BOLD+ (–19.6 ± 4.1 and –13.5 ± 3.8 mg/dL) diets versus the HAD (–8.5 ± 4.1 and –5.5 ± 3.9 mg/dL). When compared with a DASH-like diet, the test diets containing lean beef demonstrated equivalent beneficial effects on several cardiovascular risk markers including lipids, lipoproteins and potential anti-inflammatory markers. Of note, compared with the HAD, the BOLD+ diet was the only treatment diet that significantly decreased apolipoprotein B, which is the predominant protein in non-HDL particles (i.e. the atherogenic lipoprotein particles). In addition to lipid parameters, the BOLD study was one of the first controlled feeding trials to examine the effect of varying macronutrient profiles (and protein sources) on vascular health. Measurements of vascular health included a BP measurement taken at baseline and at the end of each diet period as well as augmentation index (AI) – a measure of vascular reactivity. Interestingly, systolic BP was significantly reduced (–4.24 mm Hg) following the BOLD+ diet compared to HAD while no other significant reductions in systolic BP were observed. Similarly, the AI significantly decreased following the BOLD diet compared with the HAD, DASH and BOLD+ diets. While both lean beef diets produced favourable results, the inconsistency regarding the relationship between systolic BP and AI underscore the need for additional research examining the effects of diet and protein (both quantity and source) on vascular reactivity.

Collectively, the BOLD study demonstrated that lean beef can be included in a cholesterol-lowering diet that is low in SFA and meets contemporary food-based recommendations. This dietary pattern elicited reductions in TC and LDL-C that were equivalent to heart-healthy, plant protein-rich diets, such as the DASH diet. Of importance is that increasing lean beef while controlling SFA (<7% calories) did not attenuate the decrease in LDL-C. The study also showed that increasing total dietary protein by including lean beef and other animal proteins was an effective strategy for reducing BP. Thus, the decrease in two major risk factors for CVD would be expected to correspond to a lower overall CVD risk in response to a heart-healthy diet that includes lean beef.

## 3 Beef and diabetes

Understanding potential differences in the associations of processed and unprocessed meats with other important disease outcomes, such as diabetes, is relevant for informing dietary guidance. Research to date suggests that both processed and unprocessed red meat, regardless of fat, are linked to a higher incidence of diabetes; however, gram for gram, the risk is doubled for processed meats (Pan et al. 2011). Although several potential mechanisms have been identified, they require further investigation. For example, there is evidence to show the association may be mediated by the effects of haeme iron, which may increase the risk for diabetes (Rajpathak et al. 2009). Haeme iron acts as a pro-oxidant assisting in the reactive oxygen species production and thereby inducing oxidative stress. This cascade of events can cause damage to tissues, in particular the pancreatic beta cells, and interfere with insulin secretion (Rajpathak et al. 2009). In the Pan et al. (2011) analysis, red meat consumption was highly correlated with haeme iron intake (correlation coefficients ranged from 0.53 to 0.66), and as a result, the association between red meat and diabetes was further attenuated after adjustment for dietary haeme iron intake. However, there also is evidence to show that when lean beef is incorporated into a DASH-like diet, high in fruits and vegetables, significant improvements in cardiovascular risk factors are observed compared with an average American diet (Roussell et al. 2012). Thus, it is possible that incorporating lean beef into a diet rich in antioxidants may negate any pro-oxidant effects of haeme iron.

It is also possible that other constituents in processed meat, such as sodium and nitrites, might contribute to the higher risk associated with processed red meats. In addition to its effect on BP, sodium also may increase peripheral vascular resistance and impair arterial compliance (Sacks and Campos 2010), risk factors for both CVD and diabetes. A prospective study in Finland reported a 50% higher RR of type 2 diabetes for the highest versus lowest quintile of total meat consumption. Although the association remained after adjustment for confounding factors related to diabetes (RR 1.45, 95% CI = 1.16–1.81; $P$ trend $< 0.001$), the association was moderately attenuated by an additional adjustment for sodium (RR 1.28, 95% CI = 1.00–1.64). The authors concluded that the association between processed meat and diabetes may be largely explained by sodium (Männistö et al. 2010). Nitrites, used in the preservation of processed meat, can be converted into nitrosamines through the interaction with amino compounds in extremely acidic environments such as in the stomach (Pan et al. 2011). High temperatures, such as frying, also may contribute to the formation of nitrosamines, which have been shown to be toxic to pancreatic beta cells and to increase the risk of diabetes in animal studies (Tong et al. 2009). In human studies, blood nitrite concentrations have been used as a biomarker of endothelial dysfunction (Kleinbongard et al. 2006) and impaired insulin response (Pereira et al. 2008). Additional mechanisms being investigated for their role in the association between beef and diabetes include the formation of advanced glycation end products and trimethylamine *N*-oxide in red meat (Kim et al. 2015).

# 4 Fatty acid composition of beef

Like most foods containing fat, beef contains a variety of fatty acids. Despite the common reference to animal fats as 'saturated', lean beef contains less than 10 g total fat and no more than 4.5 g SFA per 3-oz serving. In general, 54% of the fatty acids in lean beef are monounsaturated (MUFA) or polyunsaturated (PUFA) fatty acids, with the majority being MUFA. Recently, researchers from Texas Tech University discovered differences in the subcutaneous (backfat) compared to intramuscular fat (marbling fat) of grain-fed cattle. Specifically, they found marbling adipocytes to be smaller in size and diameter than subcutaneous adipocytes. In terms of energy source, subcutaneous adipocytes use acetate, which is a volatile fatty acid produced in the rumen of cattle, whereas adipocytes in marbling utilize glucose. This new discovery may enable alterations in the fatty acid composition of beef by manipulating the ration and/or length of time cattle are fed. In addition, they report that marbling appears to be very high in the MUFA, oleic acid, which has been linked to health benefits. In general, longer feeding times are required for the development of marbling; however, extended feeding times might not be economically feasible for beef producers. Therefore, new techniques for increasing marbling and the MUFA (oleic acid) content of beef while achieving a lower total fat content , are now being developed. Genetic selection might be one option. Thus, their goal is to increase marbling via increases in MUFA while reducing the total fat content of beef. Although numerous studies have reported health benefits of plant-derived MUFA (i.e. olive oil and other vegetable oils), there is no evidence to date of health benefits of MUFA from beef, which is an area that warrants further investigation (Texas Tech University. 'Increasing marbling in beef without increasing overall fatness'. ScienceDaily. 5 May 2016. www.sciencedaily.com/releases/2016/05/160505223115.htm).

To our knowledge Gilmore et al. (2011) conducted the only trial designed to test the hypothesis that ground beef high in MUFA would reduce cardiovascular risk factors, similar to plant-based MUFA, an effect that would not be seen with consumption of ground beef low in MUFA. Ground beef from both pasture- and grain-fed cattle was used and total fat was matched at 24% of calories. Pasture-fed cattle contained higher concentrations of SFA, TFA and α-linolenic acid [ALA, 18:3(n-3)], whereas the grain-fed cattle contained higher proportions of oleic acid [18:1(n-9)]. Their hypothesis was confirmed as the high MUFA beef significantly increased HDL-C concentrations compared to the low MUFA beef and significantly decreased the LDL-C:HDL-C ratio compared to baseline. The authors note that the magnitude of the effect of high MUFA ground beef in increasing HDL-C concentration (~0.08 mmol/L) is comparable to that reported in a review by Kris-Etherton and Yu (1997) of the effects of high-oleic acid vegetable oils on HDL-C levels in human studies (Gilmore et al. 2011).

The main PUFA in beef are the essential fatty acids, linoleic acid (n–6) and α-linolenic acid (n–3). The main SFA in beef are palmitic acid (C16:0; approximately half) and stearic acid (C18:0; about one-third). Palmitic acid increases TC and LDL-C levels, and stearic acid has a neutral effect. There currently are 29 cuts of beef that meet USDA's guidelines to be labelled 'lean'. To select a lean cut, choose those with Loin, Round or Chuck in the name and labelled as Select or Choice. Some examples of healthy cuts are listed in Table 1. Portion control and cooking technique also are important to control calories, total fat and saturated fat.

Owing to increasing public awareness of the relationships between diet and health, beef producers are finding new ways to enrich beef with beneficial fatty acids through the manipulation of feeding practices and/or diet. For example, comparable beef cuts from grass-fed cattle are consistently lower in fat than those from grain-fed cattle, but the SFA content remains similar. The reduction in fat is due in large part to reductions in MUFA. However, the percentage of beneficial PUFA (i.e. n-3 PUFA, vaccenic acid [a precursor of CLA], conjugated linoleic acid [CLA]) in grass-fed beef is approximately 25% greater as compared to grain-fed beef. The subsequent reduction in the SFA to PUFA ratio and increase in the percentage of n-3 fatty acids is thought to exert beneficial effects on human health (Freitas et al. 2014). Likewise, despite little difference in the

**Table 1** Nutritional facts for lean cuts of beef – based on 3 oz cooked serving

| Cut | Calories (kcal) | Fat (g) | SFA (g) | MUFA (g) | PUFA (g) | Protein (g) |
|---|---|---|---|---|---|---|
| Sirloin, tip side or round tip<br>Steaks cut from the side of the sirloin tip. Very lean, but retains flavour. | 136 | 3.3 | 1.1 | 1.2 | 0.2 | 25 |
| Top round steak<br>Cut from the hip and considered more tender than other cuts from the round. Marinate before grilling. | 138 | 3.2 | 1.3 | 1.5 | 0.3 | 26 |
| Eye of round<br>Similar to cuts taken from the tenderloin, but can be slightly tough and dry. Prepare in the slow cooker or braised for optimum flavour. | 142 | 4.0 | 1.4 | 1.6 | 0.3 | 25 |
| Bottom round roast or steak<br>Taken from the outer part of the round that is more muscular and tough. Marinating or slow cooker method are recommended. | 150 | 5.5 | 1.9 | 2.3 | 0.2 | 24 |
| Top sirloin steak<br>Well-flavoured and moderately tender with no bones and little fat. Versatile, juicy and delicious. | 156 | 5.0 | 1.9 | 2.0 | 0.2 | 26 |

Source: National Nutrient Database for Standard Reference, Release 28 slightly revised May, 2016. Software v.2.5.4 The National Agricultural Library

amount of total saturated fat compared to conventionally fed cattle, grass-fed beef contains a higher proportion of the stearic acid, the SFA that has a neutral effect on cholesterol, and less of the cholesterol raising SFAs, myristic and palmitic acids (Daley et al. 2010). Additionally, studies examining the levels of antioxidant nutrients in response to grass versus grain feeding have found significantly higher levels of precursors for vitamins A (carotenes) and E in grass-fed cattle (Daley et al. 2010; Van Elswyk and McNeill 2014).

Although feedlot finishing on high-grain diets is standard practice, there is some evidence to show that 'finishing' on forage-based diets may provide some benefit and offers an alternative (to exclusive grass feeding) (Vahmani et al. 2015). Sensory qualities and stability also need to be considered. Changing the fatty acid composition of beef, and specifically PUFA content, can increase susceptibility to oxidation. Oxidation is known to reduce beef quality, colour, flavour and nutritional value (Scollan et al. 2014). Refer to chapter 14 on packaging systems for beef for further information.

Beyond grass and forage feeding, increasing the amount of PUFA in beef may be achieved through supplementing the diet with PUFA-rich oils or oilseeds. This strategy is not always effective, however, due to extensive biohydrogenation of PUFA in the rumen leading to significantly less absorption and availability of PUFA. For example, Gonzalez et al. (2014) found only minor increases in the PUFA concentration of beef from cattle fed PUFA supplemented feed. Trying to protect PUFA from biohydrogenation has led to the development of rumen protection technologies that involve either encapsulation of unsaturated fatty acids inside a microbial-resistant shell (such as formaldehyde-treated or lipid encapsulated) or alteration of fatty acid structure (such as calcium salts or fatty amides) to resist microbial enzymes' action (Jenkins and Bridges 2007). To date, the use of encapsulated oils has proven most successful. In most studies, the encapsulated oils have been rich in linolenic acid; however, this work recently has been extended to long chain n–3 fatty acids (Dunne et al. 2011). Encapsulation of n–3 oils to modify the fatty acid composition of beef warrants further investigation.

## 4.1 Saturated fat

The relationship between saturated fat and risk of CVD has been extensively studied in randomized controlled trials and these trials have demonstrated that SFA compared to MUFA, PUFA or carbohydrates increases total and LDL-C levels. Therefore, limiting SFA intake has been a widely accepted recommendation of the dietary guidelines. In particular, it is recommended that individuals consume no more than 10% of daily calories from SFA. Consequently, this has led to a public health emphasis on selecting lean meats and reducing red meat intake. Although years of recommendations to reduce SFA, and specifically, beef consumption, have achieved moderate reductions in the incidence of heart disease (Mozaffarian et al. 2016), a concomitant increase in consumption of refined carbohydrates along with decreasing intakes of fibre have led to an increase in the prevalence of type 2 diabetes. Moreover, the data illustrate that obesity also increased proportionately with the increased consumption of refined carbohydrates in the United States (Gross et al. 2004; ADA 2016). Consequently, when a reduction in saturated fat is achieved by substitution with refined carbohydrates, there may be an increase in CVD risk due to adverse effects on components of metabolic syndrome (Siri-Tarino et al. 2010). Therefore, a singular focus on reducing saturated fat may be counterproductive because dietary fat is typically replaced by refined carbohydrates (Hu et al. 2010). Prompted by this evidence, the science has shifted towards re-evaluating the relationship between saturated fat and its sources (i.e. beef as well as dairy products and coconut oil) and incidence of CVD.

According to the 2015–20 Dietary Guidelines for Americans, the top food sources contributing to SFA intake are mixed dishes including burgers and sandwiches, and snacks and sweets. Despite a decrease in SFA intake in recent decades, the current intake remains relatively high at a median of 11.1% of daily calories. Thus, SFA continues to be an area of public health concern and the relationship between SFA and chronic disease risk was re-evaluated as part of the 2015 Dietary Guidelines Advisory Committee Report (DGAC 2015). The report concludes that strong and consistent evidence from randomized controlled trials shows that replacing SFA with unsaturated fats, and PUFA specifically, reduces total and LDL-C. Replacing SFA with carbohydrates also reduces total and LDL-C; however, significant increases in triglycerides and reductions in HDL-C are observed. Furthermore, consistent evidence from prospective cohort studies shows that higher SFA intake as compared to total carbohydrates is not associated with CVD risk (DGAC 2015). Thus, the recommendation is to consume a low SFA diet that is moderate in total fat with emphasis on unsaturated fat, both PUFA and MUFA, rather than a low-fat, high-carbohydrate (refined) diet.

When evaluating the impact of dietary SFA on health, in addition to examining the replacement nutrient, it may be just as important to examine the food source. In the majority of studies linking red meat, and specifically the SFA in beef to CVD mortality, the reports do not distinguish between processed and unprocessed meats. In fact, far fewer studies have separately evaluated the effects of processed versus unprocessed meats.

In the large European Prospective Investigation into Cancer (EPIC) study, Rohrmann et al. (2013) examined the association of red meat (beef, pork, mutton/lamb, horse and goat), processed meat (all meat products, including ham, bacon and sausages) and white meat (poultry, including chicken, hen, turkey, duck, goose, unclassified poultry and rabbit) consumption with the risk of early death in 448 568 men and women. It should be noted that the investigators appropriately accounted for potential effects of residual confounding and random errors in diet assessment. After adjustment for diet and lifestyle factors, consumption of red meat was not significantly associated with CVD mortality (per 100 g/d, RR 1.09, 95% CI = 1.00–1.18), whereas consumption of processed meat was associated with a 30% higher risk per 50 g/d (RR 1.3, 95% CI 1.17–1.45). Although residual confounding in the form of dietary factors (fibre, whole grains, fish and legumes) is still possible, previous evidence suggests that meat intake is generally associated with low intakes of these foods. Therefore, not adjusting for these factors would likely overestimate the harmful associations of red and processed meats, as well as the protective associations with poultry (Micha et al. 2013). Interestingly, it also should be noted that the EPIC results demonstrate a slight J-shaped association with the lowest risk being among those subjects with low-to-moderate intakes of meat. This finding was observed for both red meat and poultry, which suggests that consumption of 0.7–2.8 oz/day (but not zero) of meat may be beneficial for health. This is comparable to the 2015–20 Dietary Guidelines for Americans recommendation of 26 oz/week of meat, poultry and eggs (based on a 2000 kcal diet).

Another method to evaluate potential confounding, as identified in the EPIC study, is the use of a 'negative control', or a health outcome on which the primary risk factor would likely have little effect. For example, when other causes of death were evaluated, intake of red meat was not associated with any other deaths, whereas intake of processed meat was associated with higher rates of cancer and other deaths (with a smaller magnitude than for CVD deaths) but not with respiratory or digestive deaths. The absence of associations of processed meat intake with biologically unrelated causes of death supports a low likelihood of confounding as an explanation for the observed higher risks of CVD and cancer deaths (Micha et al. 2013). These findings, taken together with previous studies, provide strong evidence that intake of processed meat increases CVD risk, whereas intake of red meat, primarily beef, has a relatively small effect, if any, regardless of SFA content.

## 4.2 Polyunsaturated fats (PUFA)

While the predominant fatty acid in beef is MUFA, beef can contribute to PUFA and, specifically, the essential fatty acid of linoleic acid (n–6) and CLA. Although beef is typically not considered a source of n–3 PUFA, it contains a higher amount of n-3 PUFA than white meat (chicken or pork). The effect of a diet rich in lean beef on the fatty acid composition of plasma phospholipids was investigated in a study of 33 healthy volunteers aged 22 to 66 years (Sinclair et al. 1994). Following consumption of 500 g/d of beef for two weeks as part of a low-fat diet, individuals were then divided into three groups and provided the same amount of lean beef with a 10% increase in dietary fat each week for the following two weeks. The additional fat came from beef fat, olive oil or safflower oil. Significant increases in the plasma phospholipid composition of eicosapentaenoic acid (EPA) and docosapentaenoic acid (DPA) were associated with both lean beef plus beef fat and lean beef plus olive oil, whereas lean beef plus safflower oil (>75% LA) led to a significant decrease in EPA and increase in DPA. EPA and its precursor, DPA, have been shown to reduce the inflammatory process that leads to many chronic conditions. This suggests that high levels of LA intake inhibit EPA incorporation into phospholipids. These results support previous findings that lean beef (as it is not high in linoleic acid) could affect the n-3 PUFA phospholipid status in humans, which may have health benefits. However, future research in this area is warranted.

Also present in beef is CLA, a collective term for a mixture of positional and geometric isomers of linoleic acid. CLA are derived from PUFA in animal feed that has been biohydrogenated in the rumen. The majority of CLA in the animal's tissues are produced from vaccenic acid. Evidence from *in vitro* and animal studies suggest that CLA has a wide range of potential health benefits including anti-cancer and anti-obesity effects, improvements in cardiovascular health, glucose tolerance, bone mass, immune system and function, and inflammation (Daley et al. 2010; Pariza et al. 2001; Dilzer and Park 2012). While ruminant meat is a major dietary source of CLA, more research is needed to determine whether the amount provided in a serving of meat can contribute to potential health benefits in humans.

# 5 Food-based dietary recommendations and nutrient density

The 2015–20 Dietary Guidelines for Americans recommends a variety of protein foods including lean meat, poultry, eggs and fish/seafood; beans, peas and nuts; and seeds and soy products. Three healthy eating patterns are included in the 2015–20 Dietary Guidelines for Americans, two of which recommend weekly consumption of 26.5 oz equivalents of lean red meat, poultry and eggs (Table 2). These values were derived from the 2015 Dietary Guidelines for Americans Committee Report, which proposed specific

**Table 2** Dietary guidelines for Americans: recommended healthy dietary patterns (at the 2000 kcal level)

| Food Group | Healthy US-Style Pattern | Healthy Vegetarian Pattern | Healthy Med-Style Pattern |
|---|---|---|---|
| Fruit | 2 c per day | 2 c per day | 2½ c per day |
| Vegetables | 2½ c per day | 2½ c per day | 2½ c per day |
| Grains | 6 oz eq per day | 6 ½ oz eq per day | 6 oz eq per day |
| Whole grains | 3 oz eq per day | 3 ½ oz eq per day | 3 oz eq per day |
| Dairy | 3 c per day | 3 c per day | 2 c per day |
| Protein foods | 5 ½ oz eq per day | 3 ½ oz eq per day | 6 ½ oz eq per day |
| Meat, poultry, eggs | 26 oz eq/wk | – | 26 oz eq/wk |
| Seafood | 8 oz eq/wk | – | 15 oz eq/wk |
| Nuts, seeds, soy | 5 oz eq/wk | 14 oz eq/wk | 5 oz eq/wk |
| Bean and peas | | 6 oz eq/wk | |
| Oils | 27 g per day | 27 g per day | 27 g per day |

Source: U.S. Department of Health and Human Services and U.S. Department of Agriculture. 2015–2020 Dietary Guidelines for Americans. 8th Edition. December 2015. Available at http://health.gov/dietaryguidelines/2015/guidelines/.

**Table 3** Dietary guidelines for Americans: committee report proposed healthy eating patterns with protein subgroups

| Food Group | Healthy US-Style Pattern | Healthy Vegetarian Pattern | Healthy Med-Style Pattern |
|---|---|---|---|
| Fruit | 2 cups per day | 2 cups per day | 2½ cups per day |
| Vegetables | 2½ c per day | 2½ c per day | 2½ c per day |
| Grains | 6 oz eq per day | 6 ½ oz eq per day | 6 oz eq per day |
| Whole grains | 3 oz eq per day | 3 ½ oz eq per day | 3 oz eq per day |
| Dairy | 3 c per day | 3 c per day | 2 c per day |
| Protein foods | 5½ oz eq per day | 3½ oz eq per day | 6½ oz eq per day |
| Meats (red and processed) | 12 ½ oz eq/wk | – | 12 ½ oz eq/wk |
| Poultry | 10 ½ oz eq/wk | – | 10 ½ oz eq/wk |
| Seafood | 8 oz eq/wk | – | 15 oz eq/wk |
| Eggs | 3 oz eq/wk | 3 oz eq/wk | 3 oz eq/wk |
| Nuts, seeds | 4 oz eq/wk | 7 oz eq/wk | 4 oz eq/wk |
| Processed soy (incl. tofu) | ½ oz eq/wk | 8 oz eq/wk | ½ oz eq/wk |
| Oils | 27 g per day | 27 g per day | 27 g per day |

Source: Dietary Guidelines Advisory Committee (DGAC). Scientific Report of the Dietary Guidelines Advisory Committee on the Dietary Guidelines for Americans. (Advisory Report) to the Secretaries of the U.S. Departments of Health and Human Services (HHS) and Agriculture (USDA). Washington, DC: USDA; 2015. Available at http://health.gov/dietaryguidelines/2015-scientific-report/.

subgroup amounts for the protein group (Table 3). Unlike other protein foods, inclusion of lean beef in a heart-healthy diet provides essential nutrients such as iron, zinc and B vitamins. Specifically, beef is the number one food source of zinc in the diet, providing 39% of the daily value (DV) per 3-oz portion. It is also an excellent source of vitamin B12 (37% of the DV per 3-oz serving) and selenium (24% of DV per 3-oz serving), and a good source of vitamin B6 (16% of the DV per 3-oz serving) and iron (14% DV). Compared to other protein sources, it would take 13.5 3-oz servings of salmon (2363 kcal), 7.5 3-oz servings of chicken breast (1050 kcal), 4.5 3-oz servings of tuna (491 kcal) or 6 tbsp peanut butter (570 kcal) to match the zinc, vitamin B12, riboflavin and protein in one 3-oz serving of lean beef, respectively. As evidenced by the EPIC study, meat is an important source of nutrients and it is possible that an inadequate intake of some of these nutrients may be associated with an increased risk for CVD morbidity and mortality.

# 6 Beef and other health research

In addition to the potential role of haeme iron in the development of type 2 diabetes, haeme is also thought to be a link between the risk of colon cancer and beef intake

(Bastide et al. 2011). Although the mechanisms involved in the development of colorectal cancer by haeme iron are poorly understood, some possible explanations do exist. In the gastrointestinal tract, nitrosation of peptide-derived amines form carcinogenic *N*-nitroso compounds. A study by Cross et al. (2003) showed that haeme iron may be responsible for the nitrosation observed in the gut. The formation of nitrosamines, nitrosamides and nitrosoguanidines can produce alkylating agents during metabolism leading to DNA damage of the intestinal mucosa. Another possible mechanism is the role of haeme iron in the formation of lipid oxidation end products. Haeme catalyses the peroxidation of dietary fats, leading to the formation of a variety of reactive compounds such as epoxides and aldehydes. The major aldehyde products, malondialdehyde and 4-hydroxynonenal, are risk factors for several human diseases (Kanner 2007). However, a healthy dietary pattern that meets all food-based dietary recommendations would be expected to be rich in antioxidants and protect against the formation of any lipid oxidation products from haeme iron.

# 7 Conclusion

In this chapter, we have discussed the controversies that exist in the scientific community, the general public and the media about saturated fat and, specifically beef, because of its contribution to saturated fat in the US diet. Adding to consumer confusion, a book titled *Big Fat Surprise* by journalist Nina Teicholz has created a lot of controversy by making a case for consumption of higher fat diets from animal sources high in saturated fat.

The evidence presented provides little support for conventional dietary recommendations to select meats solely on fat content; rather, the current recommendation advocates inclusion of modest amounts (~3 oz/d) of unprocessed lean beef to provide important bioavailable nutrients (i.e. iron and zinc), and reduce saturated fat while avoiding/limiting processed (e.g. sodium-preserved) meats such as bacon, sausage, salami and processed deli meats (Mozaffarian 2016). Dietary recommendations should continue to move away from total fat-based guidelines and instead focus on foods and dietary patterns that encourage consumption of whole foods that meet food-based recommendations, achieve nutrient adequacy and achieve recommendations for specific nutrients to limit (i.e. saturated fat, sodium and added sugars). In addition, reductions in processed meats and refined grains are important to meet current guidelines. Consumer health messages should prioritize avoidance of processed meats, regardless of whether it is red or white meat or low fat or high fat. An important message to convey is that lean beef can be included in a healthy dietary pattern in recommended amounts.

# 8 Where to look for further information

- National Cattlemen's Beef Association, www.beefusa.org.
- United States Department of Agriculture Food Safety and Inspection Service, https://www.fsis.usda.gov/wps/portal/fsis/topics/food-safety-education/get-answers/food-safety-fact-sheets/meat-preparation/beef-from-farm-to-table/ct_index.
- Center for Nutrition Policy and Promotion, https://www.cnpp.usda.gov/.

- United States Department of Agriculture Nutrient Data Laboratory, https://www.ars.usda.gov/northeast-area/beltsville-md/beltsville-human-nutrition-research-center/nutrient-data-laboratory/.

# 9 References

Abete, I., Romaguera, D., Vieira, A. R., Lopez de Munain, A. and Norat, T. 2014. Association between total, processed, red and white meat consumption and all-cause, CVD and IHD mortality: a meta-analysis of cohort studies. *Br. J. Nutr.* 112: 762–75.

American Diabetes Association (ADA). Statistics about diabetes. http://www.diabetes.org/diabetes-basics/statistics/ (accessed May 2016).

Ashaye, A., Gaziano, J. and Djoussé, L. 2011. Red meat consumption and risk of heart failure in male physicians. *Nutr. Metab. Cardiovasc Dis.* 21(12): 94194–6.

Bastide, N. M., Pierre, F. H. and Corpet DE. 2011. Heme iron from meat and risk of colorectal cancer: a meta-analysis and a review of the mechanisms involved. *Cancer Prev Res (Phila).* 4(2): 177–84.

Bernstein, A. M., Sun, Q., Hu, F. B., Stampfer, M. J., Manson, J. E. and Willett, W. C. 2010. Major dietary protein sources and risk of coronary heart disease in women. *Circulation.* 122: 876–83.

Cross, A. J., Pollock, J. R. A. and Bingham, S. A. 2003. Haem, not protein or inorganic iron, is responsible for endogenous intestinal N-nitrosation arising from red meat. *Cancer Res.* 63: 2358–60.

Daley, C. A., Abbott, A., Doyle, P. S., Nader, G. A. and Larson, S. 2010. A review of fatty acid profiles and antioxidant content in grass-fed and grain-fed beef. *Nutr. J.* 9: 10.

Davidson, M. H., Hunninghake, D., Maki, K. C., Kwiterovich Jr., P. O. and Kafonek, S. 1999. Comparison of the effects of lean red meat vs lean white meat on serum lipid levels among free-living persons with hypercholesterolemia: a long-term, randomized clinical trial. *Arch. Intern. Med.* 159(12): 1331–8.

de Abreu Silva, E. O. and Marcadenti, A. 2009. Higher red meat intake may be a marker of risk, not a risk factor itself. *Arch. Intern. Med.* 169(16): 1538–9.

Dietary Guidelines Advisory Committee (DGAC). 2015. Scientific Report of the Dietary Guidelines Advisory Committee on the Dietary Guidelines for Americans. (Advisory Report) to the Secretaries of the U.S. Departments of Health and Human Services (HHS) and Agriculture (USDA). Washington, DC: USDA. Available at http://health.gov/dietaryguidelines/2015-scientific-report/

Dilzer, A. and Park, Y. 2012. Implication of conjugated linoleic acid (CLA) in human health. *Crit. Rev. Food Sci. Nutr.* 52(6): 488–513.

Dunne, P. G., Rogalski, J., Childs, S., Monahan, F. J., Kenny, D. A. and Moloney, A. P. 2011. Long chain n-3 polyunsaturated fatty acid concentration and color and lipid stability of muscle from heifers offered a ruminally protected fish oil supplement. *J. Agric. Food Chem.* 59(9): 5015–25.

Freitas, A. K., Lobato, J. F., Cardoso, L. L., Tarouco, J. U., Vieira, R. M. and Dillenburg DR. 2014. Castro I. Nutritional composition of the meat of Hereford and Braford steers finished on pastures or in a feedlot in southern Brazil. *Meat Sci.* 96(1): 353–60.

Gilmore, L. A., Walzem, R. L., Crouse, S. F., Smith, D. R., Adams, T. H., Vaidyanathan, V., Cao, X. and Smith, S. B. 2011. Consumption of high-oleic acid ground beef increases HDL-cholesterol concentration but both high- and low-oleic acid ground beef decrease HDL particle diameter in normocholesterolemic men. *J. Nutr.* 141(6): 1188–94.

González, L., Moreno, T., Bispo, E., Dugan, M. E. and Franco, D. 2014. Effect of supplementing different oils: linseed, sunflower and soybean, on animal performance, carcass characteristics, meat quality and fatty acid profile of veal from 'Rubia Gallega' calves. *Meat Sci.* 96(2 Pt A): 829–36.

Gross, L. S., Li, L., Ford, E. S. and Liu, S. 2004. Increased consumption of refined carbohydrates and the epidemic of type 2 diabetes in the United States: an ecologic assessment. *Am. J. Clin. Nutr.* 79(5): 774–9.

He, F. J. and MacGregor, G. A. 2002. Effect of modest salt reduction on blood pressure: a meta-analysis of randomized trials. Implications for public health. *J. Hum. Hypertens.* 16(11): 761–70.

Hu, F. B. 2010. Are refined carbohydrates worse than saturated fat? *Am. J. Clin. Nutr.* 91: 1541–2.

Hunninghake, D. B., Maki, K. C., Kwiterovich Jr., P. O., Davidson, M. H., Dicklin, M. R. and Kafonek, S. D. 2000. Incorporation of lean red meat into a National Cholesterol Education Program Step I diet: a long-term, randomized clinical trial in free-living persons with hypercholesterolemia. *J. Am. Coll. Nutr.* 19(3): 351–60.

Jenkins, T. C. and Bridges, W. C. 2007. Protection of fatty acids against ruminal biohydrogenation in cattle. *Eur. J. Lipid Sci. Technol.* 109: 778–89.

Kaluza, J., Akesson, A. and Wolk, A. 2014. Processed and unprocessed red meat consumption and risk of heart failure: prospective study of men. *Circ. Heart Fail.* 7(4): 552–7.

Kaluza, J., Åkesson, A. and Wolk, A. 2015. Long-term processed and unprocessed red meat consumption and risk of heart failure: A prospective cohort study of women. *Int. J. Cardiol.* 193: 42–6.

Kanner, J. 2007. Dietary advanced lipid oxidation endproducts are risk factors to human health. *Mol. Nutr. Food Res.* 51: 1094–101.

Kim, Y., Keogh, J. and Clifton, P. 2015. A review of potential metabolic etiologies of the observed association between red meat consumption and development of type 2 diabetes mellitus. *Metabolism.* 64: 768–79.

Kleinbongard, P., Dejam, A., Lauer, T., Jax, T., Kerber, S., Gharini, P., Balzer, J., Zotz, R. B., Scharf, R. E., Willers, R., Schechter, A. N., Feelisch, M., and Kelm, M. 2006. Plasma nitrite concentrations reflect the degree of endothelial dysfunction in humans. *Free Radic. Biol. Med.* 40: 295–302.

Klurfeld, D. M. 2015. Research gaps in evaluating the relationship of meat and health. *Meat Sci.* 109: 86–95.

Larsson, S. C. and Orsini, N. 2014. Red meat and processed meat consumption and all-cause mortality: a meta-analysis. *Am. J. Epidemiol.* 179(3): 282–9.

Maki, K. C., Van Elswyk, M. E., Alexander, D. D., Rains, T. M., Sohn, E. L. and McNeill, S. 2012. A meta-analysis of randomized controlled trials that compare the lipid effects of beef versus poultry and/or fish consumption. *J. Clin. Lipidol.* 6(4): 352–61.

Männistö, S., Kontto, J., Kataja-Tuomola, M., Albanes, D. and Virtamo, J. 2010. High processed meat consumption is a risk factor of type 2 diabetes in the Alpha-Tocopherol, Beta-Carotene Cancer Prevention study. *Br. J. Nutr.* 103: 1817–22.

Micha, R., Wallace, S. K. and Mozaffarian, D. 2010. Red and processed meat consumption and risk of incident coronary heart disease, stroke, and diabetes mellitus: a systematic review and meta-analysis. *Circulation.* 121(21): 2271–83.

Micha, R., Michas, G. and Mozaffarian, D. 2012. Unprocessed red and processed meats and risk of coronary artery disease and type 2 diabetes – an updated review of the evidence. *Curr. Atheroscler Rep.* 14(6): 515–24.

Micha, R., Michas, G., Lajous, M. and Mozaffarian, D. 2013. Processing of meats and cardiovascular risk: time to focus on preservatives. *BMC Med.* 11: 136.

Mozaffarian, D. 2016. Dietary and policy priorities for cardiovascular disease, diabetes, and obesity: a comprehensive review. *Circulation.* 133(2): 187–225.

Mozaffarian, D., Benjamin, E. J., Go, A. S., Arnett, D. K., Blaha, M. J., Cushman, M., Das, S. R., de Ferranti, S., Després J-P, Fullerton, H. J., Howard, V. J., Huffman, M. D., Isasi, C. R., Jiménez, M. C., Judd, S. E., Kissela, B. M., Lichtman, J. H., Lisabeth, L. D., Liu, S., Mackey, R. H., Magid, D. J., McGuire, D. K., Mohler ER III, Moy, C. S., Muntner, P., Mussolino, M. E., Nasir, K., Neumar, R. W., Nichol, G., Palaniappan, L., Pandey, D. K., Reeves, M. J., Rodriguez, C. J., Rosamond, W., Sorlie, P. D., Stein, J., Towfighi, A., Turan, T. N., Virani, S. S., Woo, D., Yeh, R. W., Turner, M. B.; on behalf of the American Heart Association Statistics Committee and Stroke Statistics Subcommittee. 2016. Heart disease and stroke statistics – 2016 update: a report from the American Heart Association. *Circulation.* 133(4): e38–e360.

Nettleton, J. A., Steffen, L. M., Loehr, L. R., Rosamond, W. D. and Folsom, A. R. 2008. Incident heart failure is associated with lower whole-grain intake and greater high-fat dairy and egg intake in the Atherosclerosis Risk in Communities (ARIC) study. *J. Am. Diet. Assoc.* 108(11): 1881–7.

Nowson, C. A., Wattanapenpaiboon, N. and Pachett, A. 2009. Low-sodium Dietary Approaches to Stop Hypertension-type diet including lean red meat lowers blood pressure in postmenopausal women. *Nutr. Res.* 29(1): 8–18.

Pan, A., Sun, Q., Bernstein, A. M., Schulze, M. B., Manson, J. E., Willett, W. C. and Hu, F. B. 2011. Red meat consumption and risk of type 2 diabetes: 3 cohorts of US adults and an updated meta-analysis. *Am J Clin Nutr.* 94: 1088–96.

Pan, A., Sun, Q., Bernstein, A. M., Schulze, M. B., Manson, J. E., Stampfer, M. J., Willett, W. C. and Hu, F. B. 2012. Red meat consumption and mortality: results from 2 prospective cohort studies. *Arch Intern Med.* 172(7): 555–63.

Pariza, M. W., Park, Y. and Cook, M. E. 2001. The biologically active isomers of conjugated linoleic acid. *Prog Lipid Res.* 40(4): 283–98.

Pereira, E. C., Ferderbar, S., Bertolami, M. C., Faludi, A. A., Monte, O., Xavier, H. T., Pereira, T. V. and Abdalla, D. S. 2008. Biomarkers of oxidative stress and endothelial dysfunction in glucose intolerance and diabetes mellitus. *Clin. Biochem.* 41: 1454–60.

Rajpathak, S. N., Crandall, J. P., Wylie-Rosett, J., Kabat, G. C., Rohan, T. E. and Hu, F. B. 2009. The role of iron in type 2 diabetes in humans. *Biochim. Biophys. Acta* 1790: 671–81.

Rohrmann, S., Overvad, K., Bueno-de-Mesquita, H. B., Jakobsen, M. U., Egeberg, R., Tjønneland, A., Nailler, L., Boutron-Ruault, M. C., Clavel-Chapelon, F., Krogh, V., Palli, D., Panico, S., Tumino, R., Ricceri, F., Bergmann, M. M., Boeing, H., Li, K., Kaaks, R., Khaw, K. T., Wareham, N. J., Crowe, F. L., Key, T. J., Naska, A., Trichopoulou, A., Trichopoulos, D., Leenders, M., Peeters, P. H., Engeset, D., Parr, C. L., Skeie, G., Jakszyn, P., Sánchez, M. J., Huerta, J. M., Redondo, M. L., Barricarte, A., Amiano, P., Drake, I., Sonestedt, E., Hallmans, G., Johansson, I., Fedirko, V., Romieux, I., Ferrari, P., Norat, T., Vergnaud, A. C., Riboli, E., Linseisen, J. 2013. Meat consumption and mortality–results from the European Prospective Investigation into Cancer and Nutrition. *BMC Med.* 11: 63.

Roussell, M. A., Hill, A. M., Gaugler, T. L., West, S. G., Heuvel, J. P., Alaupovic, P., Gillies, P. J. and Kris-Etherton, P. M. 2012. Beef in an Optimal Lean Diet study: effects on lipids, lipoproteins, and apolipoproteins. *Am. J. Clin. Nutr.* 95(1): 9–16.

Roussell, M. A., Hill, A. M., Gaugler, T. L., West, S. G., Ulbrecht, J. S., Vanden Heuvel, J. P., Gillies, P. J. and Kris-Etherton, P. M. 2014. Effects of a DASH-like diet containing lean beef on vascular health. *J. Hum Hypertens.* 28(10): 600–5.

Sacks, F. M. and Campos, H. 2010. Dietary therapy in hypertension. *N. Engl. J. Med.* 362(22): 2102–12.

Schectman, J. 2009. ACP Journal Club. High consumption of red meat and processed meat was associated with increased risk for mortality. *Ann. Intern. Med.* 21;151(2): JC1–15.

Sinclair, A. J., Johnson, L., O'Dea, K. and Holman, R. T. 1994. Diets rich in lean beef increase arachidonic acid and long-chain omega 3 polyunsaturated fatty acid levels in plasma phospholipids. *Lipids.* 29(5): 337–43.

Scollan, N. D., Dannenberger, D., Nuernberg, K., Richardson, I., MacKintosh, S., Hocquette, J. F. and Moloney, A. P. 2014. Enhancing the nutritional and health value of beef lipids and their relationship with meat quality. *Meat Sci.* 97(3): 384–94.

Singh, G. M., Danaei, G., Farzadfar, F., Stevens, G. A., Woodward, M., Wormser, D., Kaptoge, S., Whitlock, G., Qiao, Q., Lewington, S., Di Angelantonio, E., Vander Hoorn, S., Lawes, C. M., Ali, M. K., Mozaffarian, D. and Ezzati, M. 2013. Global Burden of Metabolic Risk Factors of Chronic Diseases Collaborating Group; Asia-Pacific Cohort Studies Collaboration (APCSC); Diabetes Epidemiology: Collaborative analysis of Diagnostic criteria in Europe (DECODE); Emerging Risk Factor Collaboration (ERFC); Prospective Studies Collaboration (PSC). The age-specific quantitative effects of metabolic risk factors on cardiovascular diseases and diabetes: a pooled analysis. *PLoS ONE* 8(7): e65174.

Sinha, R., Cross, A. J., Graubard, B. I., Leitzmann, M. F. and Schatzkin, A. 2009. Meat intake and mortality: a prospective study of over half a million people. *Arch. Intern. Med.* 169: 562–71.

Siri-Tarino, P. W., Sun, Q., Hu, F. B. and Krauss, R. M. 2010. Saturated fatty acids and risk of coronary heart disease: modulation by replacement nutrients. *Curr. Atheroscler Rep.* 12(6): 384–90.

Teicholz, N. 2015. *The Big Fat Surprise: Why Butter, Meat and Cheese Belong in a Healthy Diet.* Simon & Schuster Paperbacks, New York, NY.

Tong, M., Neusner, A., Longato, L., Lawton, M., Wands, J. R. and de la Monte, S. M. 2009. Nitrosamine exposure causes insulin resistance diseases: relevance to type 2 diabetes mellitus, non-alcoholic steatohepatitis, and Alzheimer's disease. *J. Alzheimers Dis.* 17: 827–44.

U.S. Department of Health and Human Services and U.S. Department of Agriculture. 2015–2020 Dietary Guidelines for Americans. 8th Edition. December 2015. Available at http://health.gov/dietaryguidelines/2015/guidelines/.

Vahmani, P., Mapiye, C., Prieto, N., Rolland, D. C., McAllister, T. A., Aalhus, J. L. and Dugan, M. E. 2015. The scope for manipulating the polyunsaturated fatty acid content of beef: a review. *J. Anim. Sci. Biotechnol.* 6(1): 29.

Van Elswyk, M. E. and McNeill, S. H. 2014. Impact of grass/forage feeding versus grain finishing on beef nutrients and sensory quality: the U.S. experience. *Meat Sci.* 96(1): 535–40.

Wang, X., Lin, X., Ouyang, Y. Y., Liu, J., Zhao, G., Pan, A. and Hu, F. B. 2016. Red and processed meat consumption and mortality: dose-response meta-analysis of prospective cohort studies. *Public Health Nutr.* 19(5): 893–905.

# Index

Page references in *italics* refer to figures. Letter-by-Letter filing is followed

www.ingramcontent.com/pod-product-compliance
Ingram Content Group UK Ltd.
Pitfield, Milton Keynes, MK11 3LW, UK
UKHW021825100926
8556IPUK00003B/160

9 781786 76060